ALLE · ZEIT · WACH · 1842

G. N. Somero · C. B. Osmond
C. L. Bolis (Eds.)

# Water and Life

## Comparative Analysis of Water Relationships at the Organismic, Cellular, and Molecular Levels

With 108 Figures

Springer-Verlag
Berlin Heidelberg New York
London Paris Tokyo
Hong Kong Barcelona
Budapest

Prof. GEORGE NICHOLLS SOMERO
Department of Zoology
Oregon State University
Corvallis, OR 97331-2914
USA

Prof. CHARLES BARRY OSMOND
Research School of Biological Sciences
Australian National University
Box 475
Canberra 2601
Australia

Prof. CARLA LIANA BOLIS
Laboratory of Comparative Biology
University of Milan
Via Balzaretti 9
20133 Milan
Italy

ISBN-13:978-3-642-766842-8 e-ISBN-13:978-3-642-76682-4
DOI: 10.1007/978-3-642-76682-4

Library of Congress Cataloging-in-Publication Data. Water and life: comparative analysis of water relationships at the organismic, cellular, and molecular levels / G.N. Somero, C.B. Osmond, C.L. Bolis (eds.). p. cm. Papers presented at the 10th International Conference on Comparative Physiology, Sept. 15–17, 1990, Crans-sur-Sierre, Switzerland. Includes index. ISBN-13:978-3-642-766842-8
1. Water – Physiological effect – Congresses. I. Somero, George N. II. Osmond, C.B. (Charles Barry), 1939–. III. Bolis, C.L. IV. International Conference on Comparative Physiology (10th: 1990: Crans, Switzerland) QP535.H1W37 1992 574.19'212–dc20
91-27493


Softcover reprint of the hardcover 1st edition 1992



Typesetting: International Typesetters Inc., Makati, Philippines, and K+V Fotosatz GmbH, Beerfelden
31/3145-5 4 3 2 1 0 – Printed on acid-free paper

# Preface

*Ich kenne die Weise, ich kenne den Text,*
*Ich kenn' auch die Herren Verfasser;*
*Ich weiß, sie tranken heimlich Wein*
*Und predigten öffentlich Wasser.*
Heinrich Heine

The Tenth International Conference on Comparative Physiology was held on September 15–17, 1990 in Crans-sur-Sierre, Switzerland. Participants in this conference were charged with summarizing the achievements and the current state of the art for a wide variety of research areas in which water stress figures prominently. The objective of the conference organizers was to assemble in one place a group of scientists who share in common an interest in the water relationships of organisms, but whose research programs focus on widely diverse organisms – bacteria, fungi, plants, and animals – and levels of biological organization – anatomy, behavior, physiology, biophysics, biochemistry, and molecular biology. Through the formal presentations and the ensuing discussions, it was hoped that a great deal of creative cross-pollination could result. The chapters in this volume suggest that these objectives were well realized. The analyses of water relationships indeed include most of the major classes of organisms, and include analyses that range from behavioral escape from water stress to the fine-scale features of water-solute-protein interactions. A major theme of the conference was to show that common features of water structure and solute effects on water and macromolecular structure underlie many, perhaps most, of the water relationships of organisms. To the extent that this important general principle has been emphasized and further elucidated, the conference organizers will view their efforts as worthwhile.

This conference was made possible by very generous contributions from the Swiss Academy of Sciences, Swiss National Science Foundation, Swiss Physiology Society, International Commission for Comparative Physiology (IUPS, IUPAB, and IUBS), Fidia Research Laboratories (Italy), and Acqua (Italy). The editors, indeed, all participants, acknowledge too the hospitality and support shown by the Community of Chermignon (Valais). The editors also thank Dr. Jürgen Schmitt for drawing to our attention the above poem by Heine, which he – and the editors – believe summarizes especially well the ambience of this most enjoyable conference.

Summer 1991 N. Somero, C.B. Osmond, C.L. Bolis

# Contents

# List of Contributors

Authors marked with an asterisk did not attend the symposium.
You will find the addresses at the beginning of the respective contributions.

# I Osmotic Solutes: Evolution, Function, and Regulation

Chapter 1

# Adapting to Water Stress: Convergence on Common Solutions

G.N. Somero

## Introduction

Because all cells, at least those that are metabolically active, are approximately 85–95% water, it is a truism to state that any environmental factor that affects the activity, structure, or physical state of water poses a threat to life. The primary focus of this symposium is on the ways in which diverse organisms – archaebacteria, eubacteria, fungi, plants, and animals – cope with water stress that may arise from a wide variety of environmental phenomena, including the salinity of the medium bathing aquatic organisms, desiccation in terrestrial habitats due to elevated temperatures or low humidity, and the threat – or reality – of freezing due to extremely low air or water temperatures. Despite the diversity of organisms considered, and the varied and complex environmental stresses that affect water relationships, it will be seen that a small number of fundamental adaptive strategies are followed in virtually all cases. Thus, an important conclusion from the investigations reported in this volume is that the notion of "unity in diversity" that is a hallmark of the discipline of comparative biochemistry (Baldwin 1970) applies strikingly well to the adaptations used by diverse organisms to cope with water stress. The title of this opening chapter is, then, an attempt to emphasize that, in the preservation of a physiologically appropriate intracellular solution for metabolic activities and macromolecular structure, there appear to be but a few acceptable alternatives in terms of osmotic solute (osmolyte) composition. Convergent evolution in different groups of organism has consistently "discovered" the types of adaptations that are needed to ensure that the aqueous portion of the cell is a "fit" environment for life – a concept developed in 1913 by L.J. Henderson in his classic work, *The Fitness of the Environment.*

One goal of this symposium is to develop a detailed understanding of the basis for evolutionary selection of a limited suite of low molecular weight solutes for use as osmotic agents in the intracellular fluid. The selective basis for using only a few classes of osmolytes will be seen to involve some of the most fundamental properties of biological systems, properties that were established at the dawn of biological evolution when the development of metabolic systems based on aqueous solutions

Marine Biology Research Division, Scripps Institution of Oceanography, University of California, San Diego, 9500 Gilman Drive, La Jolla, California 92093–0202, USA

Somero et al. (Eds.)
Water and Life

of macromolecules commenced. To understand the selective basis for osmolyte evolution is to understand some of the most fundamental properties governing the thermodynamic relationships of aqueous phase macromolecular systems. We consider these fundamental properties after reviewing briefly the classes of osmolytes that have been selected in diverse organisms, and noting their effects – or noneffects – on cellular macromolecules.

## Organic Osmolytes: The Commonest Solutions to Osmotic Stress

Table 1 lists the commonly occurring organic osmolytes found in different procaryotic and eucaryotic cells (for detailed treatment of these distribution patterns, see Yancey et al. 1982; Borowitzka 1985). These families of organic solutes are the dominant contributors to the intracellular osmolyte pool in most osmotically concentrated cells. Inorganic ions, e.g., $K^+$, tend not to be the major contributors to

**Table 1.** Distributions of organic osmolytes

| Type of organism | Major Osmolytes |
|---|---|
| | Sugars and polyhydric alcohols |
| Cyanobacteria | Trehalose, sucrose, glycerol, mannitol, glucosylglycerol |
| Fungi | Glycerol, mannitol, arabitol |
| Algae | Glycerol, mannitol, glucose, sorbitol, sucrose |
| Vascular plants | Sorbitol, glucose, sucrose, mannitol |
| Animals | Trehalose and glycerol (dormant forms) sorbitol (mammalian kidney) |
| | Amino acids and amino acid derivatives |
| Eubacteria | Glutamate, proline, GABA |
| Vascular plants | Proline |
| Marine invertebrates | Various amino acids |
| Cyclostome fishes | Various amino acids |
| | Methylamines |
| Halophilic eubacteria | Glycine betaine |
| Vascular plants | Glycine betaine |
| Marine invertebrates | Glycine betaine, TMAO |
| Marine cartilaginous fishes | Glycine betaine, TMAO, sarcosine |
| Coelacanth | Glycine betaine, TMAO |
| Mammals (kidney) | Glycine betaine, glycerophosphorylcholine |
| | Urea |
| Marine cartilaginous fishes, coelacanth, crab-eating frogs, mammalian kidney, estivating amphibians, lungfishes and snails | |
| | β-Dimethylsulfoniopropionate |
| Unicellular marine algae and marine macroalgae | |

intracellular osmolarity in most water-stressed cells, the halophilic archaebacteria being a noted exception. When the concentration of $K^+$ is built up in response to osmotic stress, as in certain eubacteria (see Csonka 1989), this response may be only a transient adaptation that is superseded by the accumulation of organic solutes like proline or glycine betaine (N'N'N'-trimethylglycine; Fig. 1).

Perhaps the most noteworthy feature of the osmolyte distribution data in Table 1 is the occurrence throughout the different kingdoms of only a few classes of organic osmolytes: sugars, polyhydric alcohols, amino acids and their derivatives (e.g., taurine and β-alanine), β-dimethylsulfoniopropionate (DMSP), methylamines [glycine betaine and trimethylamine-N-oxide (TMAO)], and urea. There is no phylogenetic barrier to the accumulation of most types of organic osmolytes. For example, methylamines contribute importantly to osmotic balance in eubacteria, plants, fishes, invertebrates, and the mammalian kidney. Polyol osmolytes likewise exhibit a wide distribution among diverse organisms. These striking examples of convergent evolution reflect the common and fundamental types of interactions that occur among low molecular solutes, macromolecules and large molecular assemblages like membranes, and cellular water in all organisms.

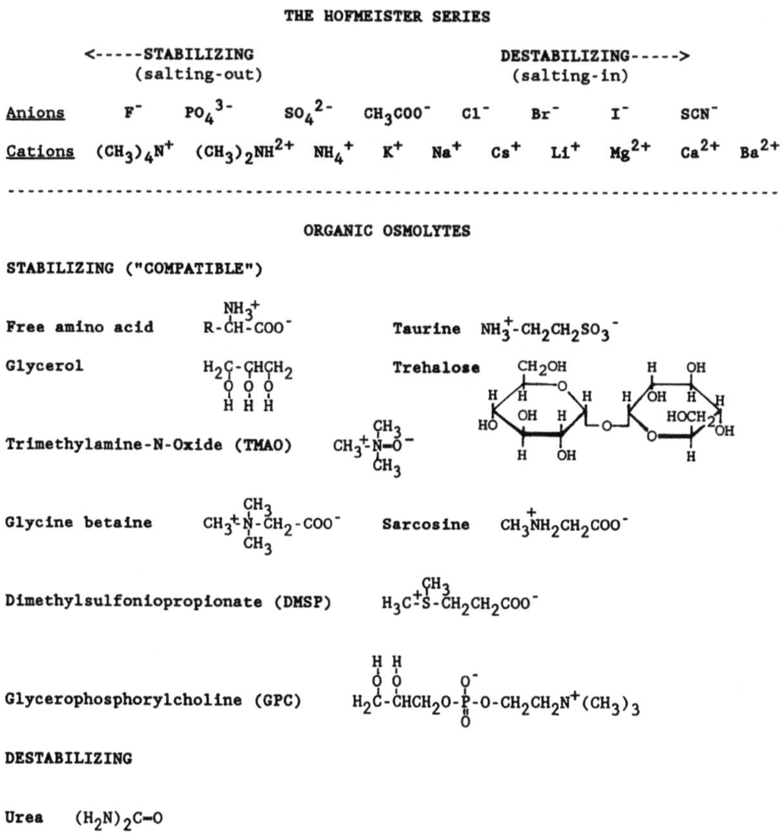


**Fig. 1.** The Hofmeister series and structures of commonly occurring organic osmolytes

## Solute-Sensitive Processes: What Biochemical Properties Must Be Conserved in the Face of Water Stress?

To understand the fundamental bases for the selection of these particular organic osmolytes, it is helpful to review some of the properties of cells that are highly sensitive to solute, e.g., salt, perturbation, yet which must be conserved if the cell is to remain metabolically active. Among the key properties of cells that are strongly disrupted by elevated salt concentrations are rates of enzymatic catalysis, binding of ligands by enzymes, protein subunit assembly, protein compartmentation, protein solubility, interactions between phospholipid bilayers and peripheral membrane proteins, and membrane fluidity and phase separation. Each of these key properties of macromolecular systems may be disrupted when intracellular salt concentrations reach high levels, e.g., concentrations greater than a few tenths molar (Yancey et al. 1982). A primary accomplishment of adaptation to water stress, then, is the avoidance of these multiple types of salt perturbation of critical biochemical systems.

A second major aspect of desiccation stress, one which will not be considered further in this chapter, but which will be the focus of the analyses of Drs. Crowe and Hand, is the accumulation of a water substitute in cells that undergo extremes of desiccation, e.g., cysts of the brine shrimp *Artemia*, which are almost fully desiccated. In these cysts, the integrity of proteins and membranes depends on the accumulation of trehalose (Fig. 1) which can provide a "water-like" physical environment which allows macromolecules and membranes to retain their native structures even in the near-absence of water. Thus, adaptation to desiccating conditions involves the selective accumulations of solutes that either allow the metabolic apparatus to function well in a concentrated intracellular milieu, or enable this apparatus to retain its native structure under conditions where most of the water is withdrawn from the cells, and the organism assumes a state of quiescence ("anhydrobiosis").

## Preserving Metabolic Function in a Concentrated Intracellular Milieu: the Halophilic Archaebacterial Solution

There appear to be only two basic types of adaptations that allow metabolic activity and macromolecular structure to be retained under conditions of elevated intracellular osmolarity. The less common of these two strategies of adaptation is manifested only in the halophilic archaebacteria (Lanyi 1974). In these salt-tolerant and salt-requiring cells, the intracellular concentration of inorganic ions (chiefly $K^+$ and $Na^+$) has been estimated to reach approximately 7 molal. At these salt concentrations the proteins of all other types of organisms would be grossly altered in structure, function, and solubility, and normal cellular structure and metabolic activity would be impossible. What adaptations allow the halophilic bacteria to thrive under – yet only under – these enormously high intracellular concentrations of inorganic ions?

Studies of a variety of different proteins from halophilic archaebacteria have revealed that these proteins typically require high salt concentrations for their normal structure and function. Figure 2 illustrates the extreme salt tolerance of malate dehydrogenase (MDH) from *Halobacterium halobium*, as contrasted with

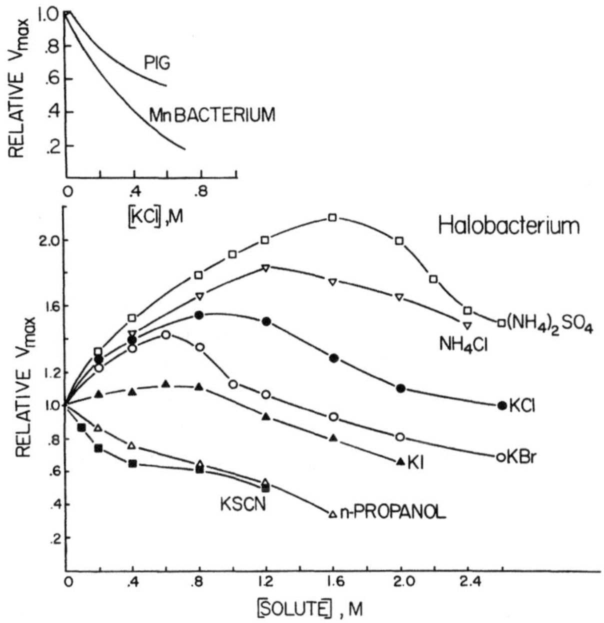


**Fig. 2.** The effects of different salts and n-propanol on the maximal velocity of the malate dehydrogenase (MDH) reaction of the halophilic archaebacterium *Halobacterium halobium*, and KCl effects on the activity of pig heart MDH (unpublished data of L.J. Borowitzka, S.L. French, and G.N. Somero)

the effects of KCl on MDHs of the pig and a marine manganese oxidizing eubacterium. The MDH of the halophilic bacterium not only tolerates vastly higher salt concentrations than the MDHs of the other two species, but it also exhibits peak activity at salt concentrations above 1 M in the case of certain salts, e.g., KCl, $NH_4Cl$, and $(NH_4)SO_4$. The halophile's MDH is just attaining peak function when the MDHs of the nonhalophiles are losing their activities.

The responses of MDH of this halophilic archaebacterium to salts reflect several important characteristics of the enzyme itself, and of the general interactions between proteins, small solutes, and water structure. First, the requirement of enzymes of halophilic archaebacteria for high salt concentrations in order to attain maximal activity is a reflection of the unusual amino acid composition of these proteins (Lanyi 1974). The proteins of these bacteria are exceedingly enriched in acidic amino acid residues (glutamyl and aspartyl residues) and depleted in basic residues (arginyl and lysyl residues). Thus, these proteins bear a high net negative charge, which prevents the proteins from folding into the native, compact structure needed for activity, unless a cationic counterion like $K^+$ is present at adequate concentrations. The activation of halophilic enzymes by increasing salt concentration is partly a reflection of the overcoming of charge repulsion among the acidic residues by salt titration, with the concomitant folding of the protein into a native, functionally active form.

This type of salt activation is completed by relatively low, i.e., a few tenths M, salt concentrations, a reflection of specific and strong ion binding to the carboxylate groups on the protein.

The further stimulation of activity by higher salt concentration is thought to be a reflection of the salting-out of the weakly hydrophobic groups that characterize the proteins of these halophilic organisms. In addition to bearing a large excess of negatively charged amino acid residues, proteins of halophilic archaebacteria contain, relative to proteins from other organisms, small percentages of strongly hydrophobic residues and a large percentage of weakly hydrophobic residues. The adaptive significance of this difference in hydrophobicity may involve the need to retain a satisfactory degree of protein structural flexibility in the face of salt conditions that tend to strongly stabilize protein structure. The high intracellular levels of $K^+$, for instance, will act to salt-out hydrophobic groups, making the proteins very rigid and insoluble. By reducing the inherent hydrophobicity of the proteins, these salting-out effects of $K^+$ are reduced, and the proteins retain the flexibility needed, for example, to undergo modulator-induced changes in conformation. The solubility of the proteins is enhanced by their high levels of net charge.

The activation of the proteins by different salts reflects the salts' abilities to salt-out hydrophobic groups. Ammonium sulfate, the strongest salting-out salt used in the study shown in Fig. 2, has the highest activating effect, with maximal activation being found at approximately 1.6 M salt. The effects of the other inorganic salts reflect their ranking in the Hofmeister series, the empirical ranking of ions based on their abilities to solubilize ("salt-in") or precipitate ("salt-out") proteins (see von Hippel and Schleich 1969 for review). Salts that are salting-in of hydrophobic groups, e.g., KBr, KI and KSCN, prevent the enzyme from folding into its native conformation, and are generally inhibitory of enzymatic activity. Only at low concentrations of KI and KBr do the charge-neutralizing effects of $K^+$ noted at low salt concentrations overcome the denaturing effects of the salting-in anions, $I^-$ and $Br^-$. For the strongly salting-in salt KSCN, the ability of the $SCN^-$ ion to solubilize hydrophobic groups is so strong that even the titration of the carboxylate groups by $K^+$ is unable to lead to activation of the enzyme.

The success with which halophilic archaebacteria have been able to exploit environments of extreme salinity attests to the effectiveness of this strategy of osmotic adaptation, which entails wholesale modification of proteins to allow life under high intracellular salt concentrations. However, the evolutionary success of these extreme halophiles can be viewed as having occurred at a considerable cost: these organisms are absolutely dependent for their survival on a high salt environment, and they are unable to thrive – or to survive – in the face of widely fluctuating salinities. This latter ability is a hallmark of the wide variety of halotolerant species, bacteria, plants, and animals, that have taken a different evolutionary route in developing tolerance of concentrated media. Osmotic adaptation in these halotolerant, and frequently very euryhaline, species has not entailed wholesale modification of the proteins of the cells, as in the halophilic archaebacteria, but rather has involved the selective accumulation of low molecular weight organic molecules that can be accumulated to extremely high concentrations without perturbing proteins. These organic osmolytes have appropriately been termed "compatible" solutes (Brown and Simpson 1972).

## Compatible Solutes: the Most Common Solution to Water Stress

Since the pioneering work by Brown and colleagues (e.g., Brown and Simpson 1972; Borowitzka and Brown 1974; Brown 1976), the widespread occurrence and physiological significance of compatible solutes has been documented in a wide array of osmotically concentrated organisms (Table 1). As illustrated in Fig. 3, solute compatibility with protein structure and function has been shown for polyhydric alcohols like glycerol, certain free amino acids and their derivatives (e.g., taurine), and methylamines like glycine betaine, osmolytes with widespread occurrence in the different kingdoms.

The effects (or noneffects) of compatible solutes on protein function are in contrast to the effects of even relatively benign inorganic salts like KCl and NaCl. Thus, for example, the activity of glucose-6-phosphate dehydrogenase of the halotolerant and euryhaline green alga *Dunaliella viridis* is strongly inhibited by KCl and NaCl, with virtually all activity lost by ~3 M combined salt concentration

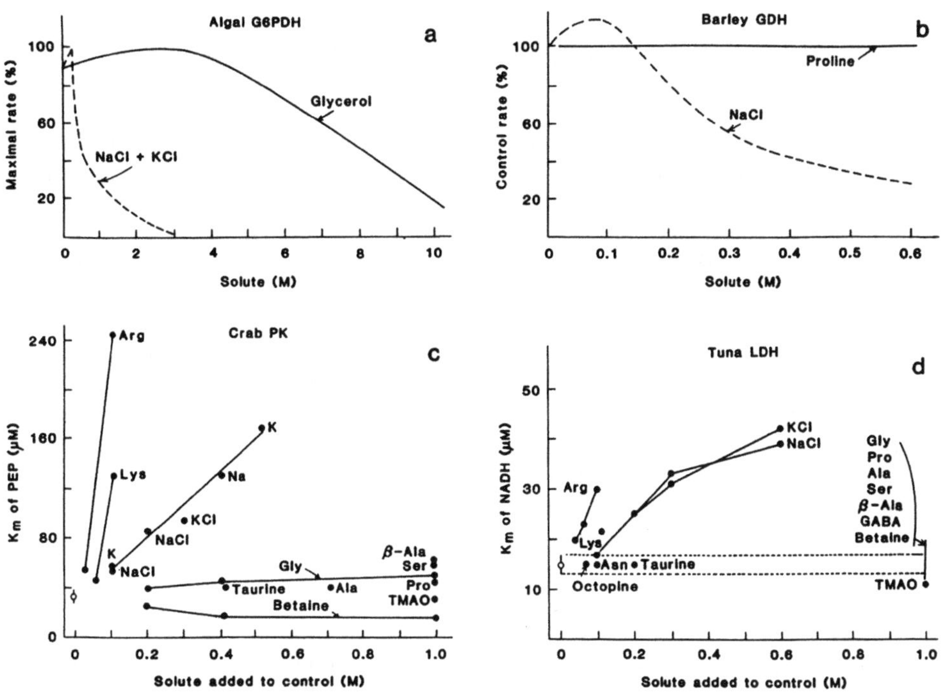


**Fig. 3a-d.** The effects of salts and organic osmolytes on enzymatic activities and Michaelis-Menten constants. **a** Salt and glycerol effects on the maximal velocity of the glucose-6-phosphate dehydrogenase reaction of the unicellular green alga *Dunaliella viridis* (Borowitzka and Brown 1974). **b** NaCl and proline effects on activity of glutamate dehydrogenase from barley (Stewart and Lee 1974). **c** Salt and organic solute effects on the $K_m$ of phosphoenolpyruvate (PEP) for pyruvate kinase (PK) of the marine crab *Pachygrapsus crassipes* (Bowlus and Somero 1979). **d** Salt and organic solute effects on the $K_m$ of NADH of muscle-type lactate dehydrogenase (LDH) from the bluefin tuna (Bowlus and Somero 1979). (Yancey et al. 1982)

(Borowitzka and Brown 1974; Fig. 3a). In contrast, glycerol has only minimal effects on the activity of the enzyme up to concentrations of approximately 4 molal. The compatible amino acid proline is without inhibitory effect on glutamate dehydrogenase of barley, whereas NaCl is slightly activating at low concentrations (~0.2 M and below) but strongly inhibitory at higher concentrations (Stewart and Lee 1974; Fig. 3b).

In a study of solute effects on the interactions between pyruvate kinase (PK) of the marine crustacean *Pachygrapsus crassipes* and the substrate phosphoenol-pyruvate (PEP), Bowlus and Somero (1979) found that KCl and NaCl were strongly perturbing of the apparent Michaelis-Menten constant ($K_m$) of PEP, as were the basic amino acids arginine and lysine, neither of which is, in fact, accumulated as an osmolyte (Fig. 3c). The amino acids and amino acid derivatives that are commonly used as osmolytes, e.g., proline, serine, and taurine, had no perturbing effect on the $K_m$ of PEP at concentrations up to 1 M, levels far higher than any single amino acid osmolyte is accumulated in marine invertebrates like *P. crassipes*. The two methylamines glycine betaine and TMAO decreased the $K_m$ of PEP slightly, albeit only at concentrations much above the physiological concentration range. This study demonstrated that free amino acids and their derivatives, and certain methylamines, can be varied widely in concentration without perturbing enzyme-substrate interactions, thus illustrating the fitness of these organic osmolytes for function in euryhaline species.

The study of PK from *P. crassipes*, like the studies of the enzymes of *Dunaliella* and barley, did not eliminate the possibility that the proteins of organisms that employ high concentrations of organic osmolytes are specifically adapted to be insensitive to these osmolytes, much as the proteins of halophilic bacteria are adapted for function in high concentrations of inorganic salts. To address the question of whether the lack of perturbation by organic osmolytes was a unique property of proteins that function in the presence of high concentrations of these solutes, parallel studies were done using enzymes from vertebrates whose cells do not contain high concentrations of organic osmolytes. As shown in Fig. 3d, lactate dehydrogenase (LDH) of the bluefin tuna displayed essentially the same responses to inorganic salts, free amino acids, and methylamines that were noted for PK of the marine crab. Therefore, the lack of effect of compatible solutes on protein function is not a reflection of special adaptations of the proteins of organisms that employ organic osmolytes, but, as discussed later, is due to fundamental aspects of protein-solute-water interactions that are universal (Arakawa and Timasheff 1985; Timasheff, this Vol.).

The use of compatible solutes would appear to be an evolutionarily simpler means for coping with high intracellular osmolarity than the restructuring of many, and perhaps all, of the intracellular proteins to give them tolerance of high salt concentration. By accumulating solutes that are essentially without effect on proteins, at least at physiological osmolyte concentrations, the required evolutionary changes in proteins to effect the compatible solute strategy are restricted to such adaptations as the acquisition of solute regulatory mechanisms that allow osmolyte concentrations to be modified in the face of changes in water relationships (Burg and Hanson, this Vol.). Relative to the pervasive modifications of proteins of halophilic archaebacteria, only relatively few proteins would need to be modified to effect regulation in organisms employing the compatible solute strategy.

## Urea + Methylamine Systems: Solutions to Urea Perturbation

Among the organic osmolytes that are found at high concentrations in several different types of animals (Table 1) is one, urea, that presents a difficult paradox in terms of rationalizing the selective value of the molecule in the context of protein structure and function. Urea is a strong perturbant of macromolecules, and is commonly used to denature proteins; it obviously does not merit categorization as a compatible solute. In cartilaginous fishes (sharks, skates, and rays) and the coelacanth (*Latimeria chalumnae*) urea concentrations range up to ~400 mM, a concentration known to strongly perturb protein function. Urea concentrations in the inner medulla of mammalian kidneys under antidiuresis may be even higher (Yancey 1988; this Vol.). Although some proteins of urea-rich fishes have been shown to be "urea-adapted" in that they require physiological concentrations of urea in order to attain proper function (LDH: Yancey and Somero 1978) or are more urea-resistant than the homologous proteins of teleost fishes or other species (Yancey 1985 and this Vol.; Ballantyne et al. 1987), most of the proteins of elasmobranchs that have been studied exhibit responses to urea that are similar to those shown by proteins from other types of organisms.

The ability of proteins of urea-rich species and tissues to sustain proper structure and function in the presence of potentially harmful concentrations of urea appears to be due mostly to the co-occurrence of a second type of organic osmolyte in the urea-rich intracellular fluids, the methylamines glycine betaine, sarcosine, TMAO, and glycerophosphorylcholine (GPC) (Table 1, Fig. 1). TMAO, sarcosine, and glycine betaine typically, although not invariably (Yancey and Somero 1978; Mashino and Fridovich 1987), counteract the perturbation of proteins by urea. GPC has not been so studied, but the structural similarities between GPC and the other methylamines, and the variation in GPC concentration with urea concentration in the kidney medulla, suggest that it too may be an effective urea counteractant (Yancey, this Vol.). Methylamine counteraction of urea effects is maximal at concentration ratios similar to those found physiologically, approximately 2:1–3:2 urea:methylamines (Yancey et al. 1982).

These counteracting solute effects are manifested in several ways, as illustrated in Fig. 4. The kinetic properties of enzymes, e.g., maximal velocity ($V_{max}$) and $K_m$ of substrate, may display complete, or nearly complete, counteracting effects. For example, the $K_m$ of adenosine diphosphate (ADP) for PK of the stingray (*Urolophis halleri*) was sharply increased by rising urea concentration (Yancey and Somero 1980; Fig. 4A). Methylamines reduced the $K_m$ in a concentration dependent manner. When TMAO and urea were present at the physiological ratio of 2:1, the $K_m$ of ADP was not different from the control (no added urea or methylamines) value. Note that it was the ratio of TMAO to urea that was critical, not the absolute concentrations. Counteraction was found at a 2:1 methylamine:urea ratio when the absolute concentrations of the solutes were varied over a fourfold range, e.g., [urea] from 200 to

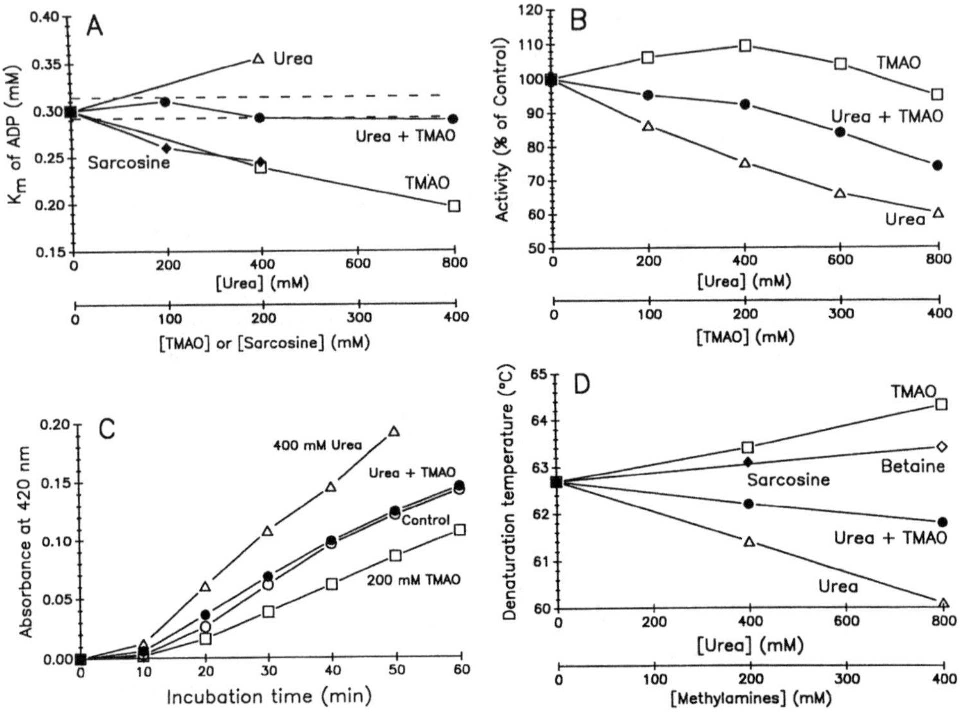


**Fig. 4A-D.** The counteracting effects of urea and methylamines (trimethylamine-N-oxide (TMAO), sarcosine, and glycine betaine) on protein structure and function. **A** Effects of urea and methylamines on the $K_m$ of ADP for pyruvate kinase of the stingray *Urolophis halleri*). The *horizontal dashed lines* represent the 95% confidence intervals around the control $K_m$ (no urea or methylamines). (Data from Yancey and Somero 1980). **B** The effects of urea and methylamines on the activity of actomyosin ATPase of the skate *Raja naevus* (Data from Yancey 1985). **C** The effects of urea and TMAO on the labeling of sulfhydryl groups of glutamate dehydrogenase by 4-chloro-7-nitrobenzofurazan (Data from Yancey and Somero 1979). **D** The effects of urea and methylamines on the thermal transition (denaturation) temperature of ribonuclease. (Data from Yancey and Somero 1979)

800 mM. Full counteraction of perturbation by urea was also found when a mixture of methylamines and urea simulating the intracellular fluid of this species was used (data not shown). This mixture consisted of 400 mM urea, 65 mM TMAO, 55 mM sarcosine, 50 mM β-alanine, and 30 mM glycine betaine (Yancey and Somero 1980).

Yancey (1985) found similar patterns when urea and methylamines were used in studies of the maximal velocity of the actomyosin ATPase of the skate (*Raja naevus*) (Fig. 4B); TMAO activated $V_{max}$, urea decreased $V_{max}$, and mixtures of TMAO and urea at physiological ratios exhibited counteraction. However, counteraction was not complete at physiological urea and TMAO concentrations, and the maximal stimulation of TMAO was at 200 mM, with higher TMAO concentrations leading to decreases in activity relative to the peak activity at 200 mM.

Protein structure, like protein function, typically is affected oppositely by urea and methylamines (Fig. 4C,D). The rate of labeling of sulfhydryl (-SH) groups of glutamate dehydrogenase (GDH) by the reagent 4-chloro-7-nitrobenzofurazan (Nbf) is reflected by the rate of increase in absorbance at 420 nm (Yancey and Somero 1979; Fig. 4C). In the presence of 400 mM urea, -SH labeling was rapid compared to control conditions (no urea or TMAO) or, especially, when 200 mM TMAO is present. When 400 mM urea and 200 mM TMAO both were present, the rate of -SH labeling by Nbf was the same as the control rate.

The thermal denaturation of ribonuclease (RNase) has proven to be a highly useful index of solute perturbation or stabilization of proteins (von Hippel and Schleich 1969). As shown in Fig. 4D, urea lowered the melting temperature ($T_m$) of RNase whereas TMAO, glycine betaine, and sarcosine increased $T_m$ (Yancey and Somero 1979). Partial counteraction between urea and TMAO was observed. These and other examples of the counteraction of urea perturbation by methylamines (see Anderson 1981; Yancey 1985) show that both protein structure and function are preserved in urea-rich animals as a result of the accumulation of solutes like TMAO which counteract urea's influences.

As discussed by Drs. Yancey and Burg in this symposium, the occurrence of urea-counteracting solutes is not restricted to lower organisms like marine elasmobranch fishes. Urea concentrations in the inner medulla of the mammalian kidney exceed those of elasmobranch fishes, and it is not surprising, therefore, that potential urea-counteracting solutes, e.g., glycine betaine and glycerophosphorylcholine, are present at high levels in the mammalian kidney. The physiological importance of counteracting solute effects on kidney cell function is suggested by the finding that urea inhibition of the colony forming ability of cultured kidney cells was counteracting by addition of glycine betaine to the cultures (Yancey and Burg 1990; Yancey, this Vol.).

## Similarities Between Osmolytes and Cryoprotectants

Many organisms face water stress due to the desiccating effects of extracellular ice formation. Some of these species are known to accumulate one or more types of low molecular weight organic solutes (Table 1). Freeze-tolerant insects may accumulate high concentrations of polyhydric alcohols, especially glycerol. Intertidal invertebrates from high latitudes also contain cryoprotectants, including some of the same organic molecules serving as osmolytes. Loomis et al. (1988) showed that taurine (Fig. 1) not only is a major osmolyte in the intertidal bivalve *Mytilus edulis*, but also is an effective cryoprotectant molecule during freeze-thaw cycles. Taurine was shown to prevent membrane damage and enzyme denaturation during freeze thaw cycles in vitro.

Based on a broad comparative study of the abilities of organic molecules to stabilize proteins during freeze-thaw cycles, Carpenter and Crowe (1988) found that the types of low molecular weight organic molecules most commonly found as osmolytes also possessed the greatest abilities to protect proteins against denaturation during freeze-thaw cycles. Certain other organic molecules which do not occur naturally, e.g., dimethylsulfoxide (DMSO) and polyethyleneglycols, also were cryoprotectants of proteins.

Carpenter and Crowe (1988) proposed that the mechanisms of solute stabilization of macromolecular structure that pertain for osmolytes in liquid phase systems also apply in the case of cryoprotectant effects, where ice-liquid mixtures occur. Of principal significance is the tendency of the stabilizing osmolytes/cryoprotectants to be excluded from the surface hydration zone surrounding proteins (cf. Arakawa et al. 1990; Timasheff, this Vol.).

## Are Compatible Solutes Invariably Nonperturbing?

The influences of solutes on proteins may depend on the concentration of the solute and on the temperature of the solution; solutes which are stabilizing of proteins at low temperature or at low concentrations may be perturbing at higher temperature or concentrations (Arakawa et al. 1990). An illustration of the constraints that high concentrations or high temperatures may impose on the accumulation of compatible organic solutes came from recent studies of β-dimethylsulfoniopropionate (DMSP) (Fig. 1) on protein structure and function. DMSP is a common organic osmolyte in many marine algae (Dickson et al. 1980, 1982; Edwards et al. 1987) and is thought to play a critical role in the global sulfur cycle. However, its effects on proteins had not been previously explored.

In our initial studies we obtained paradoxical results. DMSP, like the compatible solutes glycerol and TMAO, proved to be a strong stabilizer of protein structure under conditions of cold inactivation. When rabbit muscle phosphofructokinase (PFK) was incubated at 6 °C and pH 6.5, conditions favoring the dissociation of the active tetramer into inactive dimers, glycerol, TMAO and DMSP all reduced the loss of PFK activity (Hand and Somero 1982; Nishiguchi and Somero, in press; Fig. 5A). DMSP and TMAO were similarly effective in stabilizing the PFK tetramer.

Very different results were obtained when high temperatures were used to perturb protein structure. When muscle-type lactate dehydrogenase (LDH) was incubated at 50 °C, glycerol was found to be a strong structure stabilizer at all concentrations tested, but DMSP was effective only at relatively low concentrations, e.g., 0.2 M (Fig. 5B). At higher concentrations, e.g., 0.3 M, DMSP had only minimal stabilizing effect (Nishiguchi and Somero in press).

The types of effects noted with DMSP, while paradoxical for a so-called compatible solute, are not unprecedented. The effects of DMSP on proteins resemble those found for certain cryprotectants, e.g., the structurally similar molecule dimethylsulfoxide (DMSO), studied by Arakawa et al. (1990). These authors found that certain organic solutes that effectively protected proteins against damage during freeze-thaw cycles, notably DMSO, were destabilizing of protein structure at higher concentrations, and at higher temperatures. The model presented to account for these concentration- and temperature-dependent effects of DMSO and other cryoprotectants involves two aspects of protein-solute-water interactions that are of

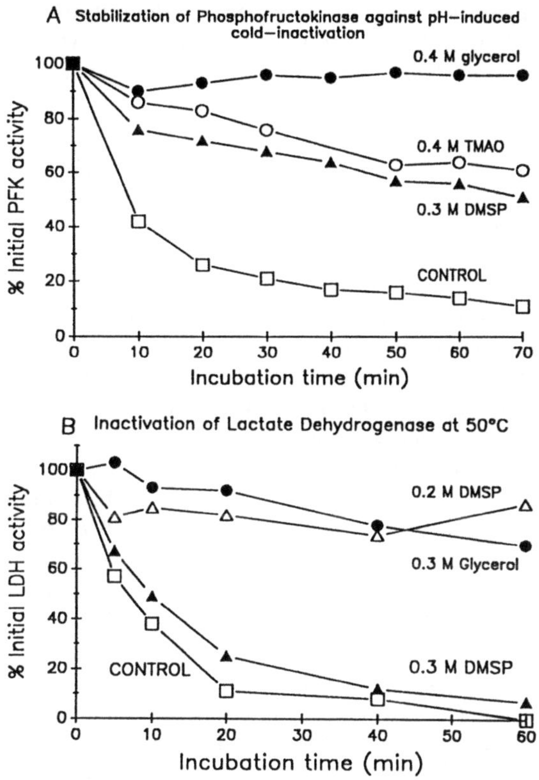


**Fig. 5.** **A** The stabilization of rabbit muscle phosphofructokinase against pH-induced cold inactivation by glycerol, TMAO, and β-dimethylsulfoniopropionate (DMSP). (TMAO data from Hand and Somero 1982; glycerol and DMSP data from Nishiguchi and Somero in press). **B** The effects of glycerol and β-dimethylsulfoniopropionate (DMSP) on the thermal stability of muscle-type lactate dehydrogenase (LDH) of the barracuda *Sphyraena argentea*. (Data from Nishiguchi and Somero, in press)

central importance in governing the compatibility of osmolytes (Arakawa et al. 1990).

As discussed in detail by Timasheff in this Volume, the stabilizing effects of compatible solutes derive largely from their inabilities to bind to proteins. Compatible solutes are excluded from the hydration shell surrounding a protein, which is an entropically unfavorable situation that favors minimization of protein surface area, i.e., favors retention of the compact, native protein structure and the aggregation and precipitation ("salting-out") of proteins. Strongly perturbing solutes like urea, guanidinium, and $SCN^-$ bind to proteins very effectively, e.g., by forming strong hydrogen bonds with peptide backbone linkages and hydrophobic interactions with non-polar side chains. These destabilizing ("salting-in") solutes favor maximization of protein surface urea, i.e., maximization of the number of solute binding sites.

For certain organic solutes, the relative tendencies to be excluded from the protein's hydration sphere, on the one hand, and to bind to the protein, on the other hand, vary with solution conditions, e.g., of temperature. Arakawa et al. (1990) proposed that DMSO and certain other cryoprotectants bind to proteins by forming hydrophobic interactions with nonpolar sidechains. Because hydrophobic interactions increase in strength as temperature is raised, solutes which have a nonpolar region like DMSO may become denaturants at high temperatures through their enhanced binding to nonpolar sidechains with concomitant unfolding of the native protein. High concentrations of solutes with the abilities to bind to hydrophobic groups likewise will denature proteins. Alcohols like propanol (Fig. 2) are known to have this effect.

Although to our knowledge the binding of DMSP to proteins has not been investigated, the structural similarities between DMSO and DMSP lead us to propose (Nishiguchi and Somero in press) that, at high concentrations or at high temperatures, DMSP can destabilize proteins, even though at low concentrations such as those found within algal cytoplasm, DMSP may be a compatible solute (Dickson et al. 1980, 1982; Edwards et al. 1987). There would seem, then, to be a blurring of the distinction between compatible and noncompatible solutes, in the sense that solute compatibility may not be invariably characteristic of a particular organic molecule. In fact, there must be upper limits to the concentration that any organic osmolyte can be accumulated before some type of biochemical perturbation occurs (e.g., Fig. 3a for the effects of very high glycerol concentrations). Osmolytes having the ability to form hydrophobic interactions with proteins at elevated temperature may be compatible only at relatively low concentrations compared to osmolytes lacking this property, e.g., glycerol and zwitterionic amino acids. The key point, however, is that under physiological conditions of temperature and solute concentration, compatibility is still the hallmark of osmolytes like DMSP that may, under nonphysiological in vitro conditions, disrupt protein structure.

The discovery that DMSP can be perturbing of proteins at high temperature raises the question of whether evolutionary selection of organic osmolytes is determined in part by the temperature regime of an organism. Glycerol, which appears to be stabilizing of proteins at both low (Fig. 5A) and high temperatures (Fig. 5B), is the compatible solute accumulated in heat-tolerant organisms, e.g., the unicellular green alga *Dunaliella* which occurs in desert salt ponds. Although the selection of compatible solutes is governed by a variety of factors, including ambient salinities and phylogenetic status (reviewed by Borowitzka 1985), the study of temperature's influence on the evolution of compatible solute systems might reveal an important further aspect of the evolutionary development of the internal milieu. Similarly, the finding that the selection of different polyol and carbohydrate osmolytes may be based on the level of salt stress facing the organism suggests that the total osmotic concentration built up in a cell may influence which organic osmolytes are most "fit." Solutes like DMSP, which may be compatible only at relatively low (few tenths M) concentrations, may be used only when salt stress is relatively low, or, in other cases, may be held at a stable concentration during conditions of hypersalinity that favor the accumulation of another type of solute which is compatible at high concentrations. The latter pattern of multiple osmolyte regulation has, in fact, been observed in the alga *Entermomorpha intestinalis* (Edwards et al. 1987). Under hypersaline stress, sucrose and proline concentrations increased, and DMSP, the major organic osmolyte under normal salinity conditions, did not rise in concentration. From data of this type, it appears clear that the selection and regulation of organic osmolytes is a complex phenomenon that is influenced by several factors, including the abilities of the solutes to interact with proteins under different conditions of temperature and salt stress. The regulatory patterns of methylamines and polyols in the medulla of the mammalian kidney (Burg, this Vol.) further emphasize the complexities of regulating multiosmolyte systems.

## Summary

Osmotically stressed cells generally accumulate one or more low molecular weight organic molecules as the dominant intracellular osmotic agents (osmolytes). Similar osmolytes are found in cells facing osmotic stress due to low environmental water availability, physiological activity (e.g., the inner medulla of the mammalian kidney), and as a consequence of ice formation in the extracellular fluids. The exceptions to this rule are the halophilic archaebacteria which accumulate high (up to ~7 molal) concentrations of inorganic ions, chiefly $K^+$ and $Na^+$. The proteins of the halophilic archaebacteria are extensively modified to function in the presence of high salt concentrations and, in fact, these proteins typically require high salt concentrations in order to attain native structure and high rates of function. In all other osmotically stressed species, the primary adaptive response is the accumulation of one or a few types or organic osmolytes – sugars, polyhydric alcohols (polyols), certain free amino acids or their derivatives, dimethylsulfoniopropionate, or methylamines (accumulated in conjunction with urea in some cases) – that lack strongly perturbing effects on macromolecules and, therefore, do not necessitate widespread modifications in proteins or membranes for the cell to cope with water stress. These "compatible" organic osmolytes can be used effectively in euryhaline species without perturbing macromolecular and membrane structure and function. Methylamines like trimethylamine-N-oxide, glycine betaine, and glycerophosphorylcholine have the additional property of offsetting the perturbing effects of urea on protein structure and function, a phenomenon termed the "counteracting solute effect." Although most organic osmolytes are protein stabilizers, some, e.g., β-dimethylsulfoniopropionate (DMSP), may be compatible solutes only at moderate concentrations and at relatively low temperatures. The striking convergence seen in the evolution of osmolyte systems reflects the unifying physicochemical principles that underlie the interactions between macromolecules, water, and small solutes.

*Acknowledgments.* Portions of this work were supported by National Science Foundation grant DCB88-12180 and by a John Simon Guggenheim Fellowship.

## References

Anderson PM (1981) Purification and properties of the glutamine- and N-acetyl-L-glutamate dependent carbamoyl phosphate synthetase from liver of *Squalus acanthias*. J Biol Chem 256:12228–12238

Arakawa T, Timasheff S (1985) The stabilization of proteins by osmolytes. Biophys J 47:411–414

Arakawa T, Carpenter JF, Kita YA, Crowe JH (1990) The basis for toxicity of certain cryoprotectants: a hypothesis. Cryobiology 27:401–415

Baldwin E (1970) An introduction to comparative biochemistry. Cambridge Univ Press, Cambridge

Ballantyne JS, Moyes CD, Moon TW (1987) Compatible and counteracting solutes and the evolution of ion and osmoregulation in fishes. Can J Zool 65:1883–1888

Borowitzka LJ (1985) Glycerol and other carbohydrate osmotic effectors. In: Gilles R, Gilles-Baillien M (eds) Transport processes, iono- and osmoregulation. Springer, Berlin Heidelberg New York, pp 437–453

Borowitzka LJ, Brown AD (1974) The salt relations of marine and halophilic species of the unicellular green alga, *Dunaliella*: the role of glycerol as a compatible solute. Arch Microbiol 96:37–52

Bowlus RD, Somero GN (1979) Solute compatibility with enzyme function and structure: Rationales for the selection of osmotic agents and end-products of anaerobic metabolism in marine invertebrates. J Exp Zool 208:137–152

Brown AD (1976) Microbial water stress. Bacteriol Rev 40:803–846

Brown AD, Simpson JR (1972) Water relations of sugar-tolerant yeasts: The role of intracellular polyols. J Gen Microbiol 72:589–591

Carpenter JF, Crowe JH (1988) The mechanism of cryoprotection of proteins by solutes. Cryobiology 25:244–255

Csonka LN (1989) Physiological and genetic responses of bacteria to osmotic stress. Microbiol Rev 53:121–147

Dickson RM, Wyn Jones RG, Davenport J (1980) Steady state osmotic adaptation in *Ulva lactuca*. Planta 150:158–165

Dickson DM, Wyn Jones RG, Davenport J (1982) Osmotic adaptation in *Ulva lactuca* under fluctuating salinity regimes. Planta 155:409–415

Edwards DM, Reed RH, Chudek JA, Foster R, Stewart WPD (1987) Organic solute accumulation in osmotically-stressed *Enteromorpha intestinalis*. Mar Biol 95:583–592

Hand SC, Somero GN (1982) Urea and methylamine effects on rabbit muscle phosphofructokinase. Catalytic stability and aggregation state as a function of pH and temperature. J Biol Chem 257:734–741

Henderson LJ (1913) The fitness of the environment. Macmillan, New York

Lanyi J (1974) Salt-dependent properties of proteins from extremely halophilic bacteria. Bacteriol Rev 38:272–290

Loomis SH, Carpenter JF, Crowe JH (1988) Identification of strombine and taurine as cryoprotectants in the intertidal bivalve *Mytilus edulis*. Biochim Biophys Acta 943:113–118

Mashino T, Fridovich I (1987) Effects of urea and trimethylamine-N-oxide on enzyme activity and stability. Arch Biochem Biophys 258:356–360

Nishiguchi M, Somero GN Temperature and concentration dependence of compatibility of the organic osmolyte β-dimethylsulfoniopropionate. Cryobiology (in press)

Stewart GR, Lee JA (1974) The role of proline accumulation in halophytes. Planta 120:279–289

von Hippel PH, Schleich T (1969) The effects of neutral salts on the structure and conformational stability of macromolecules in solution. In: Timasheff SN, Fasman GD (eds) Structure and stability of biological macromolecules. Marcel Dekker, New York, pp 417–574

Yancey PH (1985) Organic osmotic effectors in cartilaginous fishes. In: Gilles R, Gilles-Baillien M (eds) Transport processes, iono- and osmoregulation. Springer, Berlin Heidelberg New York, pp 424–436

Yancey PH (1988) Osmotic effectors in kidneys of xeric and mesic rodents: corticomedullary distributions and changes with water availability. J Comp Physiol 158:369–380

Yancey PH, Burg MB (1990) Counteracting effects of urea and betaine in mammalian cells in culture. Am J Physiol 258:R198–R204

Yancey PH, Somero GN (1978) Urea-requiring lactate dehydrogenases of marine elasmobranch fishes. J Comp Physiol 125:135–141

Yancey PH, Somero GN (1979) Counteraction of urea destabilization of protein structure by methylamine osmoregulatory compounds of elasmobranch fishes. Biochem J 183:317–323

Yancey PH, Somero GN (1980) Methylamine osmoregulatory solutes of elasmobranch fishes counteract urea inhibition of enzymes. J Exp Zool 212:205–213

Yancey PH, Clark ME, Hand SC, Bowlus RD, Somero GN (1982) Living with water stress: Evolution of osmolyte systems. Science 217:1214–1222

Chapter 2

# Compatible and Counteracting Aspects of Organic Osmolytes in Mammalian Kidney Cells in Vivo and in Vitro

P.H. YANCEY

## Introduction

Typical mammalian body fluids have osmotic pressures of about 300 mosm/kg, with inorganic ions as the major osmotic effectors (osmolytes) both extra- and intracellularly. Many non-mammalian species face much higher osmotic pressures, and comparative physiologists have long known that cells in such organisms generally use organic solutes to maintain osmotic balance. From eubacteria to lower vertebrates, these organic osmolytes appear to fall within a few categories of compounds, namely polyols, neutral amino acids and derivatives, dimethyl-sulfonioproprionate, and urea, usually in combination with methylamines (Yancey et al. 1982; Somero, this Vol.). Only recently has it been fully recognized that some mammalian tissues follow this widespread evolutionary pattern, in particular the kidney. As a consequence of the urine-concentrating mechanism, cells of the renal inner medulla may be exposed to extracellular urea and NaCl at well over 1000 mosm/kg. In the mid-1980s (Balaban and Knepper 1983; Finely 1984; Bagnasco et al. 1986), it was found that the inner kidney can contain high levels of the polyols sorbitol and (myo)-inositol, and the methylamines betaine and glycerophosphorylcholine (GPC; first detected by Ullrich 1959). These appear to be intracellular (since they are not found in urine or blood), and it was proposed that these compounds serve as osmolytes in renal cell volume maintenance (Fig. 1).

In the last few years, several laboratories have been examining hypotheses concerning these compounds in renal function. In addition to whole animals, isolated renal tubules and cells and cultured renal cell lines have facilitated this work (see below and Burg, this Vol.). Indeed, the existence of established renal cell lines has allowed for more detailed studies rarely available with other eukaryotes. Three related aspects concerning these organic solutes will be reviewed both here and in the following paper (Burg, this Vol.): their patterns and regulation as osmolytes, their properties as compatible solutes, and methylamines and urea as counteracting osmolytes.

Biology Department, Whitman College, Walla Walla, Washington 99362, USA

Somero et al. (Eds.)
Water and Life

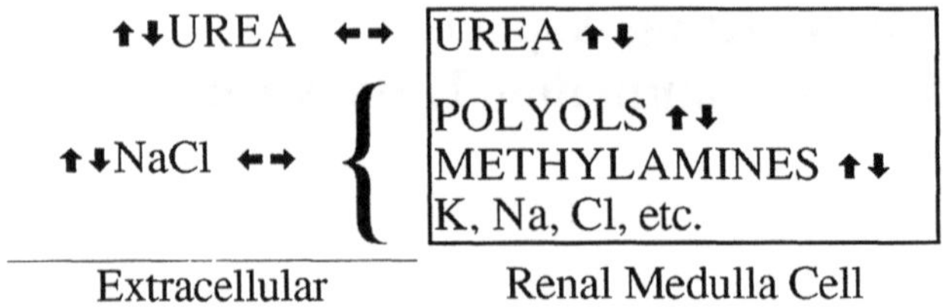


**Fig. 1.** Model of cell volume regulation in mammalian renal cells. Urea, which varies with animal diuretic state, is thought to equilibrate across membranes, thus creating no osmotic change. Similar variation in NaCl is thought to be largely extracellular, with intracellular ions remaining relatively constant. Thus, osmotic balance would be achieved by variations in intracellular polyols and methylamines matching the external NaCl variation

## Cell Volume Maintenance by Renal Polyols and Methylamines

Initial studies asked whether these organic solutes are indeed used to maintain renal osmotic balance in vivo. Several testable predictions can be made. Firstly, NaCl and urea are generally found in an increasing gradient from renal cortex to inner medulla in antidiuretic animals (Ullrich and Jarausch 1956). Since urea is thought to equilibrate across membranes, the main cause of osmotic imbalance for a medullary cell should be variations in NaCl, which is thought remain largely extracellular (Beck et al. 1984; Balaban and Burg 1987). To maintain cell volume, the sum total concentration of all non-urea (intracellular) osmolytes should exhibit a gradient similar to that for (extracellular) NaCl (Fig. 1). In general, this prediction is met by the renal contents of polyols and methylamines, as illustrated for antidiuretic rabbit in Fig. 2C (Yancey and Burg 1989): total sorbitol, inositol, betaine, and GPC follow a pattern quite similar to that for NaCl in kidneys sectioned into seven regions from cortex to papillary tip (in these studies, the data are for total contents of extracted kidneys, rather than intra- or extracellular concentrations). Several other studies, in which Na was not measured, also show similar gradients for organic osmolytes (Bagnasco et al. 1986; Oates and Goddu 1987; Wirthenson et al. 1987, 1989; Gullans et al. 1988).

As a better test of this prediction, we analyzed sodium as well as organic osmolyte contents in laboratory rabbit and rat, and in three wild rodent species ranging in water adaptation from the xeric pocket mouse (which never drinks water) to the mesic montane vole. In Fig. 3A, total contents of methylamines plus polyols are plotted against sodium contents for kidneys of antidiuretic animals (held 2 or more days without water) (Yancey 1988; Yancey and Burg 1989; Yancey et al. 1990b). The correlation is remarkably similar across all kidney regions (cortex to papilla) and species.

To function as osmolytes above the 150 mM Na found in "normal" mammalian plasma (Prosser 1973), concentrations of methylamines plus polyols should be held at about 1.7 times that of Na (0.85 for Na plus Cl; not 1.0, since inorganic ions have lower activity coefficients). To estimate this ratio, the slope of 0.66 found for contents (Fig. 3A) must be corrected for differing extra- and intracellular spaces, which are unknown for most species. However, in the rat, cells of cortex and outer medulla

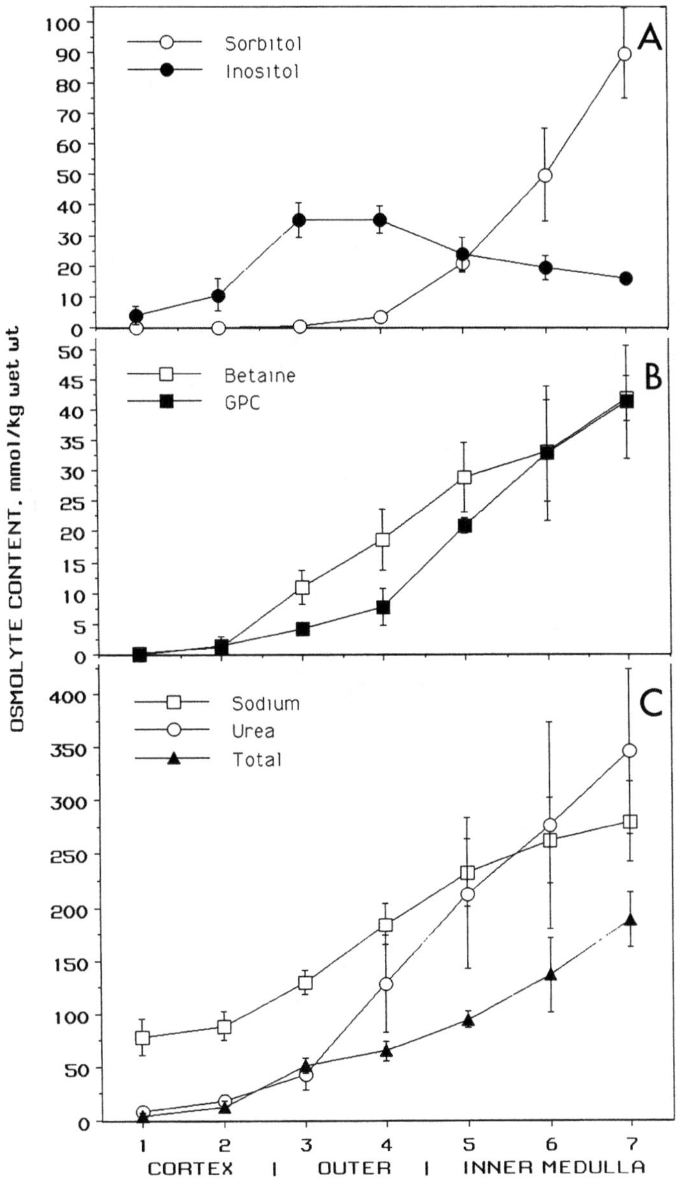


**Fig. 2A-C.** Distributions of osmolytes in antidiuretic rabbit kidneys cut in seven sections from outer cortex (1st section) to tip of inner medulla (7th section), as indicated on the *abscissa*. Animals (n = 5) had been without water for 2 days. *Error bars* indicate SD. *Total* in **C** indicates sum of methylamines and polyols, added for each individual animal before calculating means and SD. (After Yancey and Burg 1989)

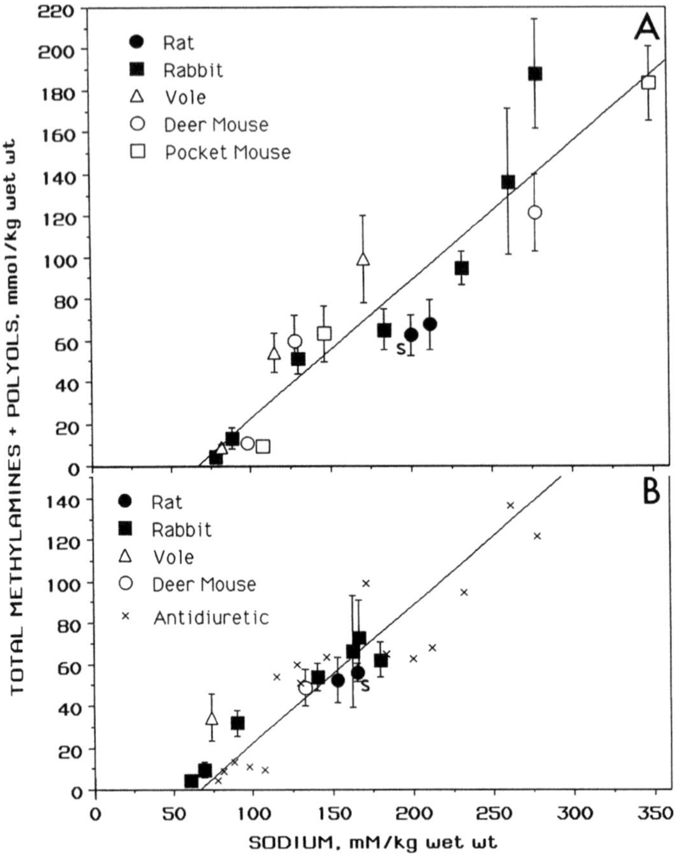


**Fig. 3A-B.** Correlation between renal contents of sodium and total methylamines and polyols, for rat (Yancey et al. 1990b), rabbit (Yancey and Burg 1989), and wild rodents (Yancey 1988), the mesic montane vole, *Microtus montanus*, the deer mouse *Peromyscus m. gambeli*, and the xeric desert pocket mouse *Perognathus parvus*. Each *point* represents four to seven animals, with *upper points* from inner medulla, *middle ones* from outer medulla, *lower* from cortex. *Error bars* are SD; those for Na are not shown, but were generally of the same size as those for total organic osmolytes (see Fig. 2 for rabbit Na errors). *S points* indicate rats on sorbinil, an aldose reductase inhibitor (see text). Data replotted from above references. **A** Antidiuretic animals, with the line fitted by linear regression: *Total* = 0.66 × Na – 43.5 ($r^2$ = 0.87). **B** Diuretic animals (on water 3 or more days), with the line and some points (X) replotted from A for comparison)

occupy about 60% of the space, but only about 33% in inner medulla (Pfaller 1982). If these values are used to estimate concentrations in Fig. 3A (e.g., for inner medulla by dividing total methylamines plus polyols by 0.33 and Na by .67), the slope becomes 1.5 ($r^2$ = 0.70) not far from the 1.7 predicted (see also Yancey 1988), with a Na intercept of 177, not far from "normal" blood at about 150 mM Na.

A second prediction can be made: as renal sodium levels vary within an animal due to changing water intake, polyols and methylamines should be up- or downregulated to maintain the same ratio as seen in the antidiuretic animals. Several studies have shown this prediction to be met: animals given access to water (sometimes sweetened) or in other ways made diuretic generally have reduced contents of renal polyols and methylamines compared to water-deprived animals (Cohen et al. 1982; Bagnasco et al. 1986; Oates and Goddu 1987; Wirthersohn et al. 1987, 1989; Gullans et al. 1988; Wolff et al. 1989; Yancey 1988; Yancey and Burg 1989; Yancey et al. 1990b). Again, as a better test, we analyzed sodium contents. Results for these studies are plotted in Fig. 3B for laboratory and wild animals, shown with the regression line of Fig. 3A for antidiuretic animals. Again, the correlation of Na is very similar across species and is nearly identical to that for antidiuretic animals.

In vitro preparations confirm these results: in freshly isolated tubules from rat inner medulla, cell contents of GPC, sorbitol, and inositol (betaine was not measured) varied directly with extracellular NaCl, but not urea (Wirthensohn et al. 1989). Similar results were found for sorbitol (no other osmolytes were measured) in freshly isolated cells from rat inner medullary ducts (Grunewald and Kinne 1989).

It is likely, therefore, that renal cell volume is held roughly constant by regulation of total polyols and methylamines. Two complications are found in this process. First, individual osmolytes may vary in different, poorly understood ways. In particular, sorbitol is almost absent from outer medulla, while myo-inositol is highest in outer medulla in rabbit (Fig. 2A; Yancey and Burg 1989) and all other species examined (Cohen et al. 1982; Oates and Goddu 1987; Yancey 1988; Wirthensohn et al. 1989). The reason for these patterns is not known; however, the sum of the two polyols follows a gradient (not shown) very similar to that for Na (Fig. 2C). Also, substantial amounts of free amino acids, common osmolytes in marine invertebrates, are found in the renal medulla, but are apparently not changed with water variation (Gullans et al. 1988). Despite these differences in individual osmolytes, total cellular osmolytes must be regulated by the kidney to track extracellular osmotic changes. How this is accomplished is unknown.

The regulatory mechanisms for individual osmolytes in vivo are also poorly understood. Sorbitol has been most thoroughly studied. It is produced from glucose by aldose reductase, with the enzyme (Corder et al. 1977) and its mRNA (Cowley et al. 1990) present in a renal gradient matching that of sorbitol content. Antidiuretic treatment (of diabetes insipidus rats) results in increased mRNA, enzyme and sorbitol in inner medulla, while reduction of sorbitol in normal rats made diuretic may result from cellular release (Cowley et al. 1990). Inositol may be taken in by a Na-dependent membrane transporter rather than by synthesis (Seifter et al. 1990). Mechanisms of GPC (Kanfer and McCartney 1989) and betaine (Grossman and Hebert 1989) regulation in vivo remain unclear, with hints of both synthesis and uptake mechanisms from the latter (Lohr and Acara 1990). Except for sorbitol, no processes have been shown to respond to osmotic signals.

A second complicating feature of renal organic osmolytes concerns the time course of changes. Many studies have shown that epithelial cells in general regulate cell volume during short-term osmotic disturbances by inorganic ion transport (see e.g., Spring 1985). Studies with renal cells in vitro suggest that, if extracellular osmolality rises rapidly, organic osmolyte production rises slowly. This suggests that inorganic ions may be used initially with later replacement by organic solutes (see Burg, this Vol.). However, in vivo and in vitro, organic osmolytes can be released quite rapidly (through membrane permeability changes) as extracellular osmolality falls, indicating that inorganic ion transport may not be needed in this response in renal cells (Wirthensohn et al. 1987, 1989; Wolff et al. 1989; Burg, this Vol.).

Osmotic regulation of osmolytes (contents and time courses) has been best elucidated with cultured cell lines. Two renal lines have revealed much: (1) Madin-Darby canine kidney (MDCK) cells are a long-established renal epithelial line, which when grown in hyperosmotic medium accumulate betaine, GPC and inositol (Nakanishi et al. 1988); and (2) PAP-HT25 cells are a renal epithelial line from rabbit inner medulla, which in standard hyperosmotic medium will accumulate sufficient amounts of sorbitol to account for cell volume maintenance (Bagnasco et al. 1987) (see Burg, this Vol. for details).

## Compatibility Properties of Renal Osmolytes

Why are polyols and methylamines used as osmolytes? One major explanation is the "compatible-osmolytes" hypothesis, which states that, unlike NaCl, KCl, or urea, certain polyols, free amino acids, and related compounds do not perturb protein functions, and are therefore safe to accumulate in cells over a wide concentration range. This property has been amply confirmed with isolated proteins from a variety of eukaryotic species (Brown and Simpson 1972; Clark 1985; Yancey et al. 1982; Somero, this Vol.), and for betaine in living bacteria (see below). We have examined two mammalian renal enzymes, argininosuccinase and uricase. We found sorbitol and inositol (and betaine for one) to be fully compatible with them, i.e., they do not alter kinetic properties (Fig. 4; Blykowski, Whitney and Yancey, unpublished work). This is in contrast to urea and NaCl, which are highly perturbing as expected (Fig. 4).

Testing the compatibility hypothesis on living cells or whole organisms requires defined manipulation of intracellular osmolyte contents. One can predict that cells will suffer if they are prevented from accumulating compatible osmolytes under osmotic stress. In bacteria, addition of extracellular betaine or its precursors greatly increases growth in high-osmolality medium (Chambers and Kunin 1985; Landfald and Strøm 1986). For mammalian renal cells, we devised a test with aldose reductase inhibitors, a group of drugs which prevent the conversion of glucose to sorbitol. In standard growth medium made hyperosmotic with NaCl, PAP-HT25 cells (see above) appear to rely almost exclusively on sorbitol for cell volume maintenance. As measured by ability to form clonal colonies in culture, these cells accumulate sorbitol and grow well in normal medium with about 250 mM NaCl (compared to normal nonrenal blood at 150 mM). However, if an aldose reductase inhibitor is added to this medium, cell growth drops dramatically in parallel with declining cell sorbitol content (Fig. 5) (as a control, the drug had no effect in medium at 150 mM NaCl, in which PAP cells have little or no sorbitol). In the presence of drug, the lack of sorbitol synthesis would result in higher concentrations of other solutes, either by cell shrinkage or by ions entering from the medium (no other detectable organic solutes

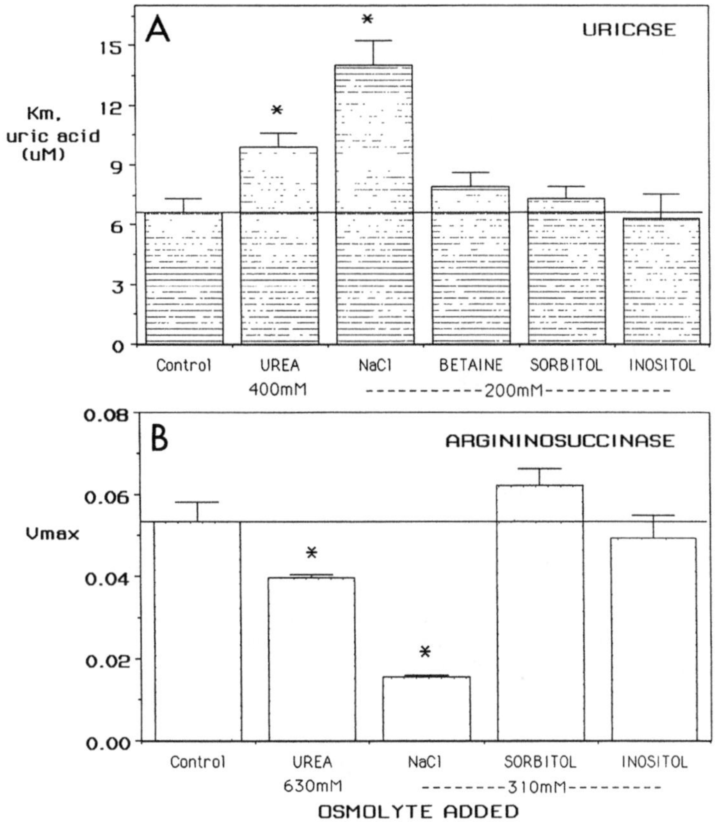


**Fig. 4A,B.** Effect of osmolytes on kinetic properties of two renal enzymes (Blykowski M, Whitney L and Yancey PH, unpublished). *Error bars* are SE; * indicates significant effect. **A** Bovine uricase, assayed at pH 8.5 in Tris buffer. Apparent Michaelis constant ($K_m$) for uric acid is plotted; maximal velocity was not affected by the solutes. **B** Porcine argininosuccinase, assayed according to Yancey and Somero (1980) with argininosuccinate as substrate. Maximal velocity is absorbance units $min^{-1}$; $K_m$ was not altered by the osmolytes. For betaine effects, see Fig. 8

increase in content), a situation clearly detrimental. These results support the concept of sorbitol as a compatible osmolyte.

More recently, we fed an aldose reductase inhibitor to rats, and analyzed their kidneys for osmolytes. Like the PAP cells (Fig. 5), sorbitol content of the inner medulla declined greatly with drug treatment (Fig. 6). Unlike the results with PAP cells, however, betaine contents rose (Fig. 6) in almost perfect compensation for the reduced sorbitol, such that there was no noticeable osmotic damage, and relationship to Na remained constant (Yancey et al. 1990b; "S" points in Fig. 3).

A recent study with PAP cells confirms these results in vivo for rats: if PAP cells are provided betaine (normally absent) in growth medium, they will accumulate it when sorbitol production is suppressed by an aldose reductase inhibitor. The large

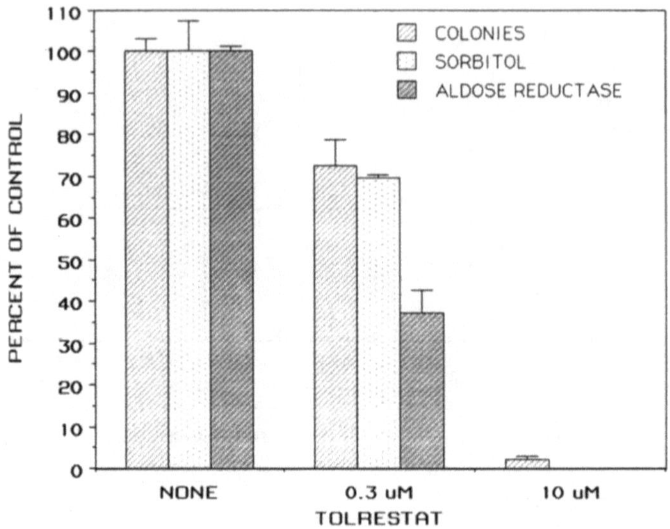


**Fig. 5.** Effects of tolrestat, an aldose reductase inhibitor (Ayerst), on PAP-HT25 rabbit papillary cells grown initially in medium with 150 mM NaCl then switched to 250 mM NaCl. Shown are colony-forming efficiency (described in Fig. 7), cell sorbitol content, and aldose reductase activity in cell extracts, normalized to 100% for no drug (*error bars* are SD). All effects of the drug were highly significant. (Yancey et al. 1990a)

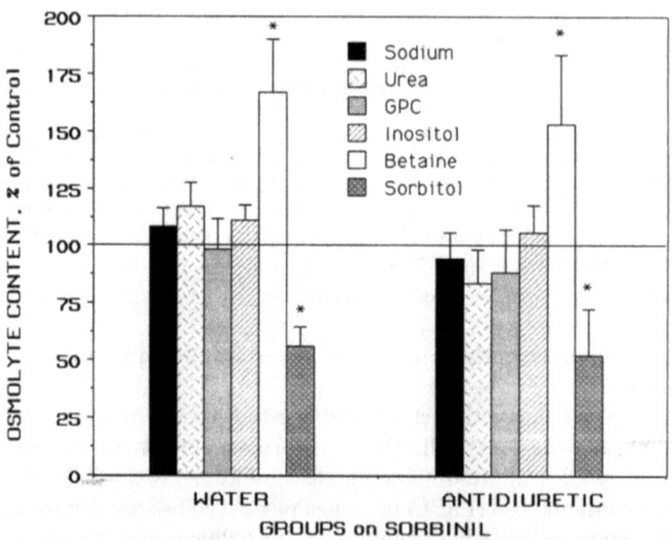


**Fig. 6.** Osmolyte contents of rat renal inner medullas, from animals given an aldose reductase inhibitor, sorbinil (Pfizer) in food for 10 days. *WATER* group had ad libitum water; *ANTIDIURETIC* group had no water for the final 3 days. Values, in mmol/kg wet wt, are normalized to values for the appropriate control groups (water or antidiuretic, no drug). *Error bars* are SD; * indicates significant change. (Yancey et al. 1990b)

decrease in cell viability previously seen (Fig. 5) is largely reversed by the betaine (see Burg, this Vol.). Thus, both betaine and sorbitol may serve as interchangeable compatible osmolytes in the kidney.

We have also compared the effects on clonal colony growth of MDCK cells of two osmolytes which rapidly cross cell membranes – urea and glycerol. Glycerol has been well studied as a compatible solute in protists (Brown and Simpson 1972), while urea is a well-known (noncompatible) protein destabilizer (Yancey et al. 1982). With MDCK cells grown with these osmolytes in the medium, glycerol is clearly far more compatible than urea at higher osmolalities (Fig. 7A). Other osmolytes such as sorbitol do not readily cross membranes, and require other approaches for testing (above).

## Counteracting Properties of Renal Osmolytes

Another major explanation for renal methylamine is the "counteracting-osmolytes" hypothesis, which states that methylamines act as protein stabilizers capable of offsetting the well-known destabilizing effects of urea. The hypothesis began with the observation that high urea concentrations in animals are generally associated with high methylamine concentrations. The counteracting property has been demonstrated with isolated proteins from a variety of vertebrate species and for demembranated shark muscle fibers, and is usually most effective at about a 2:1 urea:methylamine ratio (Yancey et al. 1982; Yancey 1985; Somero, this Vol.). This hypothesis was first developed with elasmobranchs, which contain about a 2:1 urea-methylamine ratio in their cells (Yancey and Somero 1979) (a similar hypothesis on methylamine counteraction of salt inhibition was independently developed by Pollard and Wyn Jones 1979; see also Clark 1985).

Many mammalian (including renal) enzymes are known to be inhibited by urea (Gutman and Katzper-Shamir 1971; Dousa 1972; Yancey and Somero 1980), and thus could benefit from the presence of the methylamines betaine and GPC. We have examined two mammalian renal enzymes, argininosuccinate and uricase. We found betaine to effectively counteract urea inhibition of catalytic rate for the former (Fig. 8; Arnell, Blykowski and Yancey, unpublished), although it did not counteract $K_m$ effects on the latter enzyme (not shown). Neither sorbitol nor inositol was capable of such counteraction (GPC was not tested; see below).

Tests of the hypothesis in vivo are more difficult. Two approaches have been made. First, the hypothesis predicts that total methylamines should be regulated to maintain about a 1:2 concentration ratio to urea inside cells. Analysis of kidney contents fits this prediction well for the methylamine GPC: in both antidiuretic (Fig. 9A) and diuretic (Fig. 9B) mammals from several species, we found that a nearly constant ratio is maintained. This strong linear correlation between GPC and urea (confirmed by Cowley et al. 1990) is not found for any other osmolyte, nor is it found between GPC and NaCl. In vitro results with MDCK cells (see Burg, this Vol.) support the conclusion that renal cells specifically regulate GPC to match urea.

Betaine, a methylamine directly shown to counteract urea effects on proteins, is also found in a linear correlation with urea in antidiuretic and diuretic animals, but

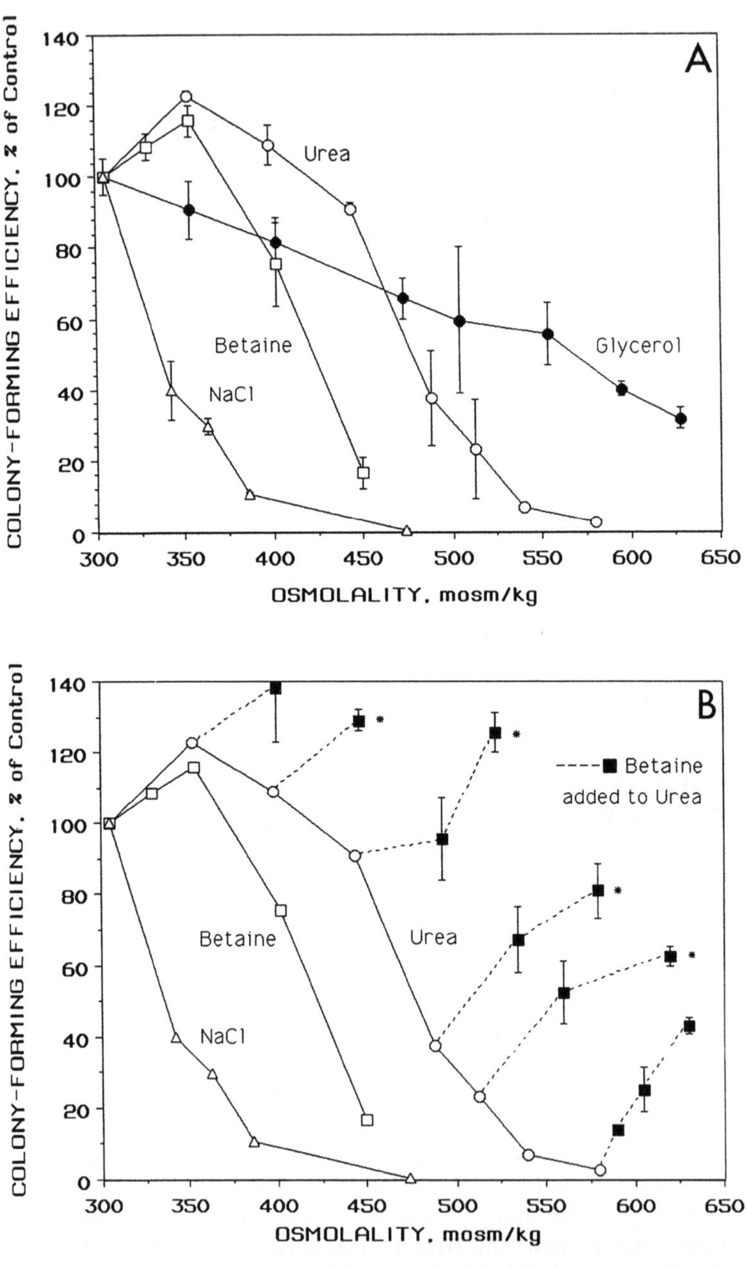


**Fig. 7A,B.** Effects of solutes added to growth medium on colony-forming efficiency of MDCK cells. Values, measured as number of colonies formed divided by number of single cells seeded on plates, are normalized to control (no test solutes added to normal growth medium at 305 mosm/kg). *Error bars* are SD. **A** Effects of urea, glycerol, betaine, or NaCl alone; all but NaCl equilibrated across cell membranes. **B** Effects of adding betaine and urea together; * indicates 2:1 urea-betaine ratio. (Yancey and Burg 1990)

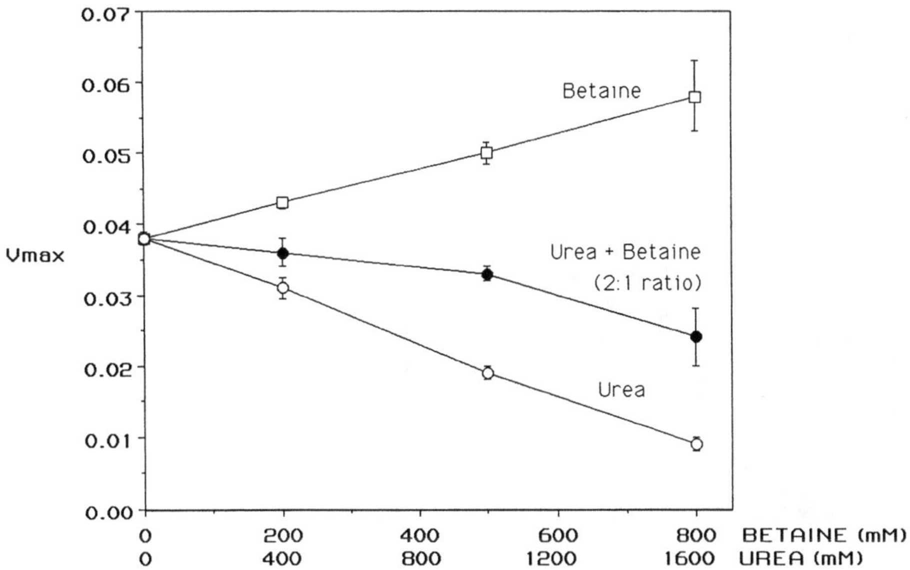


**Fig. 8.** Effects of betaine, urea, or a 2:1 mixture on maximal velocity of porcine argininocuccinase (Arnell T, Blykowski M and Yancey PH, unpublished), assayed according to Yancey and Somero (1980). See also Fig. 4B

the ratio differs in the two states: betaine is not always downregulated in diuresis in proportion to urea decrease (Yancey 1988; Yancey and Burg 1989). Nor does urea stimulate betaine accumulation as it does GPC accumulation in MDCK cells (see Burg, this Vol.). Summing the two methylamines and using estimation methods discussed above for total osmolytes, varying intracellular ratios of about 2–3 are estimated (Balaban and Knepper 1983; Yancey 1988). While this should provide strong urea counteraction, the lack of a constant ratio is as yet unexplained. Thus, the role of betaine (another than as an osmotic effector) remains uncertain, complicated by the observations (above) that betaine may functionally substitute for sorbitol as a compatible osmolyte.

The betaine question is further complicated by its urea-counteracting abilities in vivo. A second testable prediction of the hypothesis is that, like proteins, cells will function better with an intracellular mixture of urea and methylamine than with urea or methylamine alone (Yancey et al. 1982). We found that urea and betaine will penetrate MDCK cells, such that we could control intracellular contents (Yancey and Burg 1990). As shown in Fig. 7A, addition of urea or betaine alone greatly reduced cell colony growth, except for some stimulation at lower concentrations. However, addition of betaine to urea partly or fully restored normal growth (Fig. 7B), clearly supporting the counteracting-osmolytes hypothesis [GPC has not been tested with either proteins or cells because of difficulty in making high concentrations of pure product, but given its apparent regulation with urea in the kidney (above), such studies are clearly needed].

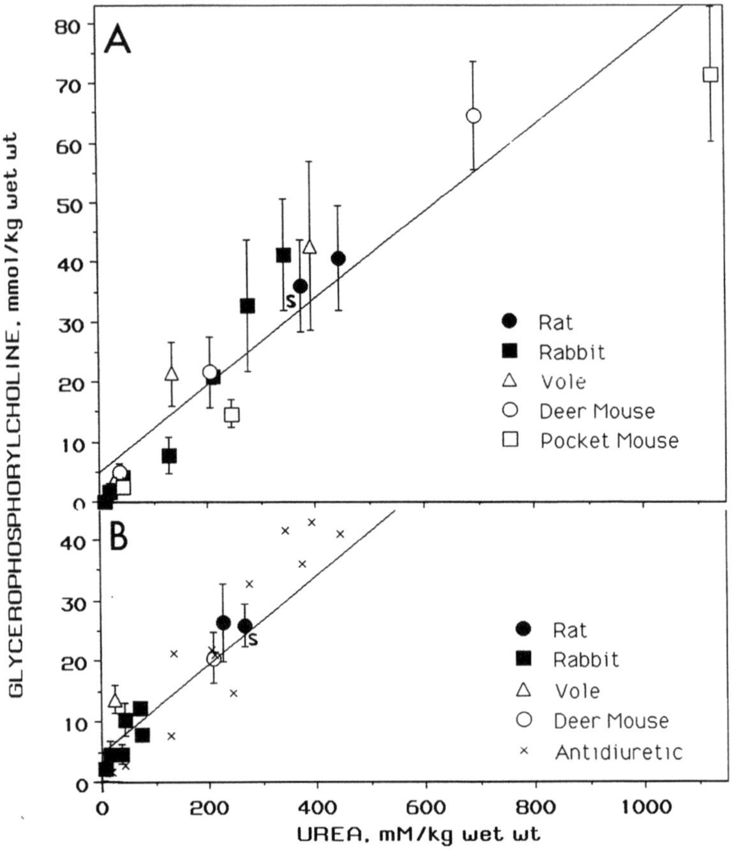


**Fig. 9A,B.** Correlation between renal contents of urea and glycerophosphorylcholine (GPC), for species as in Fig. 3 (Yancey 1988; Yancey and Burg 1989; Yancey et al. 1990b). No other osmolyte yields similar consistent correlation with urea. *Upper points* from inner medulla, *middle ones* from outer medulla, *lower* from cortex. *Error bars* are SD; those for urea are not shown, but were generally of the same size as those for total organic osmolytes. *S points* indicate rats on sorbinil, an aldose reductase inhibitor (see text). Data replotted from above references. **A** Antidiuretic animals, with the line fitted by linear regression giving GPC = 0.072 × urea + 4.9 ($r^2 = 0.89$). **B** Diuretic animals (on water 3 or more days), with the line and some points (X) replotted from **A** for comparison

Another possibility for urea counteraction lies in the polyols. Although we have found no counteraction on kinetic properties of two renal enzymes (above), sorbitol and inositol (but not glycerol) can offset urea denaturation of some proteins (Gerlsma 1968; Shifrin and Parrott 1975). We found that MDCK cells grew better in the presence of 200 mM urea and 50 mM inositol than with either solute alone, similar to though not as strong as the betaine-urea effect in Fig. 7B (Yancey and Burg, unpublished). More study is obviously needed.

## Conclusions

It is clear that the mammalian kidney, in its use of polyols and methylamines with urea, has evolved or retained a widespread, ancient mechanism for cell volume regulation. Many of the discoveries with kidneys merely confirm observations found with other organisms, but the use of cells lines has provided new insights. In addition to those just discussed, detailed in vitro cell studies on osmolyte regulation will be examined in the following paper (Burg, this Vol.).

## References

Bagnasco S, Balaban R, Fales HM, Yang Y-M, Burg M (1986) Predominant osmotically active organic solutes in rat and rabbit renal medullas. J Biol Chem 261:5872–5877

Bagnasco S, Uchida S, Balaban R, Kador P, Burg M (1987) Induction of aldose reductase and sorbitol in renal inner medullary cells by elevated extracellular NaCl. Proc Natl Acad Sci USA 84:1718–1720

Balaban R, Burg MB (1987) Osmotically active organic solutes in the renal inner medulla. Kidney Int 31:562–564

Balaban R, Knepper MA (1983) Nitrogen-14 nuclear magnetic resonance spectroscopy of mammalian tissues. Am J Physiol 245:C439–C444

Beck R, Dorge A, Rick R, Thurau K (1984) Intra- and extracellular element concentrations of rat renal papilla in antidiuresis. Kidney Int 25:397–403

Brown A, Simpson J (1972) Water relations of sugar-tolerant yeasts: the role of intracellular polyols. J Gen Microbiol 72:589–591

Chambers ST, Kunin CM (1985) The osmoprotective properties of urine for bacteria: the protective effect of betaine and human urine against low pH and high concentrations of electrolytes, sugars and urea. J Infect Dis 152:1308–1315

Clark ME (1985) The osmotic role of amino acids: discovery and function. In: Gilles R, Gilles-Ballien M (eds) Transport processes, iono- and osmoregulation. Springer, Berlin Heidelberg New York

Cohen MAH, Hruska KA, Daughaday WA (1982) Free myo-inositol in canine kidneys: selective concentration in the renal medulla. Proc Soc Exp Biol Med 169:380–385

Corder CN, Collins JG, Brannan TS, Sharma J (1977) Aldose reductase and sorbitol dehydrogenase distribution in rat kidney. J Histochem Cytochem 25:1–8

Cowley BD Jr, Ferraris JD, Carper D, Burg MB (1990) In vivo osmotic regulation of aldose reductase mRNA, protein, and sorbitol content in rat renal medulla. Am J Physiol 258:F154–F161

Dousa TP (1972) Effect of renal medullary solutes on vasopressin-sensitive adenyl cyclase. Am J Physiol 222:657–661

Finley KD (1984) The role of methylamines and free amino acids in protection of mammalian kidney proteins. Masters Thesis, San Diego State Univ, Calif

Gerlsma SY (1968) Reversible denaturation of ribonuclease in aqueous solutions as influenced by polyhydric alcohols and some other additives. J Biol Chem 243:957–961

Grossman EB, Hebert SC (1989) Renal inner medullary choline dehydrogenase activity: characterization and modulation. Am J Physiol 256:F107-F112

Grunewald RW, Kinne RKH (1989) Intracellular sorbitol content in isolated rat renal inner medullary collecting duct cells: regulation by extracellular osmolality. Pfluegers Arch 414:178–184

Gullans SR, Blumenfeld JD, Balschi JA, Kaleta M, Brenner RM, Heilig CW, Hebert SC (1988) Accumulation of major organic osmolytes in rat renal medulla in dehydration. Am J Physiol 255:F626–F634

Gutman Y, Katzper-Shamir Y (1971) The effect of urea, sodium and calcium on microsomal ATPase activity in different parts of the kidney. Biochim Biophys Acta 233:133–136

Kanfer JN, McCartney DG (1989) GPC phosphodiesterase and phosphomonoesterase activities of renal cortex and medulla of control, antidiuresis and diuresis rats. FEBS Lett 257:348–350

Landfald B, Strøm AR (1986) Choline-glycine betaine pathway confers a high level of osmotic tolerance in *Escherichia coli*. J Bacteriol 165:849–855
Lohr J, Acara M (1990) Effect of dimethylaminoethanol, an inhibitor of betaine production, on the disposition of choline in the rat kidney. J Pharmacol Exp Ther 252:154–158
Nakanishi T, Balaban RS, Burg MB (1988) A survey of osmolytes in renal cell lines. Am J Physiol 255:C181–C191
Oates PJ, Goddu KJ (1987) A sorbitol gradient in rat renal medulla. Kidney Int 31:448
Pfaller W (1982) Structure function correlation on rat kidney. In: Hild W, van Limborgh J, Ortmann R, Pauly JE, Schiebler TH (eds) Advances in Anatomy, Embryology and Cell Biology, vol 70. Springer, Berlin Heidelberg New York
Pollard A, Wyn Jones RG (1979) Enzyme activities in concentrated solutions of glycinebetaine and other solutes. Planta 144:291–298
Prosser CL (1973) Inorganic ions. In: Prosser CL (ed) Comparative Animal Physiology. Saunders, Philadelphia London Toronto
Seifter JL, Rivera J, Hughes R, Heilig C, Gullans S (1990) Properties of a $Na^+$-dependent myoinositol transporter in vesicles from rat renal outer medulla. Kidney Int 37:588
Shifrin S, Parrott CL (1975) Influence of glycerol and other polyhydric alcohols on the quaternary structure of an oligomeric protein. Arch Biochem Biophys 166:426–432
Spring (1985) Determinants of epithelial cell volume. Fed Proc 44:2526–2529
Ullrich KJ (1959) Glycerophosphorylcholinumsatz und Glycerophosphorylcholindiesterase in der Säugetier-Niere. Biochem Z 331:98–102
Ullrich KJ, Jarausch KH (1956) Untersuchungen zum Problem der Harnkonzentrierung und-verdünnung. Über die Verteilung der Elektrolyten (Na, K, Ca, Mg, Cl, anorganischem Phosphat), Harnstoff, Aminosäuren und exogenem Kreatinin in Rinde und Mark der Hundeniere bei verschiedenen Diuresezuständen. Pflügers Arch Ges Physiol 262:537
Wirthensohn G, Beck F, Guder WG (1987) Role and regulation of glycerophosphorylcholine in rat renal papilla. Pfluegers Arch 409:411–415
Wirthensohn G, Lefrank S, Schmolke M, Guder W (1989) Regulation of organic osmolyte concentrations in tubules from rat renal inner medulla. Am J Physiol 256:F128–F135
Wolff SD, Stanton TS, James SL, Balaban RS (1989) Acute regulation of the predominant organic osmolytes of the rabbit renal inner medulla. Am J Physiol 257:F676–681
Yancey PH (1985) Organic osmotic effectors in cartilaginous fishes. In: Gilles R, Gilles-Ballien M (eds) Transport processes, iono- and osmoregulation. Springer, Berlin Heidelberg New York
Yancey PH (1988) Osmotic effectors in kidneys of xeric and mesic rodents: corticomedullary distributions and changes with water availability. J Comp Physiol 158B:369–380
Yancey PH, Burg MB (1989) Distributions of major organic osmolytes in rabbit kidneys in diuresis and antidiuresis. Am J Physiol 257:F602–607
Yancey PH, Burg MB (1990) Counteracting effects of urea and betaine on colony-forming efficiency of mammalian cells in culture. Am J Physiol 258:R198–204
Yancey PH, Somero GN (1979) Counteraction of urea destabilization of protein structure by methylamine osmoregulatory compounds of elasmobranch fishes. Biochem J 182:317–323
Yancey PH, Somero GN (1980) Methylamine osmoregulatory compounds in elasmobranch fishes reverse urea inhibition of enzymes. J Exp Zool 212:205–213
Yancey PH, Clark ME, Hand SC, Bowlus RD, Somero GN (1982) Living with water stress: the evolution of osmolyte systems. Science 217:1212–1222
Yancey PH, Burg MB, Bagnasco SM (1990a) Effects of NaCl, glucose and aldose reductase inhibitors on cloning efficiency of renal cells. Am J Physiol 258:C156–163
Yancey PH, Haner RG, Freudenberger T (1990b) Effects of an aldose reductase inhibitor on osmotic effectors in rat renal medulla. Am J Physiol 259:F733–738

Chapter 3

# Molecular Basis for Accumulation of Compatible Osmolytes in Mammalian Cells

M.B. Burg

## Introduction

In the kidney medulla the blood and interstitial fluid NaCl and urea concentrations are much higher than in the rest of the body and vary considerably with the operation of the renal concentrating mechanism (Jamison and Kriz 1982). Systematic screening by mass spectrometry and nuclear magnetic resonance (NMR) spectrometry identified four predominant compatible organic osmolytes in the renal medullas of rabbits and rats, namely sorbitol, glycerophosphorylcholine (GPC), betaine, and myo-inositol (inositol) (Bagnasco et al. 1986). The same solutes are present in the renal medullas of other mammals.

This brief review will consider the mechanisms by which renal cells accumulate these four osmolytes in response to variations in extracellular NaCl and urea.

## Tissue Culture Systems for Studying Renal Organic Osmolytes

Study of the effects of hyperosmolality on renal tissue cultures provides insight into the kidney osmolytes that is difficult to attain in vivo. Tissue culture has the advantage that it is easier to make measurements and control specific variables. The initial tissue culture studies of renal organic osmolytes utilized GRB-PAP1 (rabbit renal inner medullary epithelial) cells, and in particular one of the strains (PAP-HT25) of that line that grows, albeit slowly, in medium made hyperosmotic to 600 mosmol by adding NaCl (Bagnasco et al. 1987). These cells were originally screened for organic osmolytes using proton NMR, which is a relatively insensitive method, and only sorbitol was identified as a regulated organic osmolyte (Bagnasco et al. 1987). A survey of other cell lines revealed that MDCK (dog kidney epithelial) cells accumulate betaine, GPC, and inositol, but not sorbitol (Nakanishi et al. 1988). More recently, under conditions that were modified, based on the MDCK cell studies, and using HPLC for the measurements, substantial amounts of betaine, GPC, and inositol were found in PAP-HT25 cells (Moriyama et al. 1990a,b). MDCK and PAP-HT25 cells have been used to investigate the mechanisms of accumulation and control of the renal organic osmolytes.

National Heart, Lung and Blood Institute, Bethesda, MD 20892, USA

Somero et al. (Eds.)
Water and Life

## Sorbitol

### *Hypertonicity Induces Sorbitol Accumulation by Increasing the Activity of Aldose Reductase, an Enzyme that Catalyzes Synthesis of Sorbitol*

Sorbitol is synthesized from glucose in a reaction catalyzed by aldose reductase. When PAP-HT25 cells grown at a normal osmolality are switched to high NaCl medium, there is a 6 h lag, following which aldose reductase activity increases gradually for 2 to 3 days and sorbitol accumulation follows (Uchida et al. 1989).

The rise in aldose reductase activity is due to an increase in the amount of the enzyme protein (Bedford et al. 1987), and the protein rises because its synthesis rate is increased (Moriyama et al. 1989). Aldose reductase synthesis and degradation were measured by $^{35}$S-methionine pulse-chase, followed by immunoprecipitation. Synthesis rate begins to increase 3 h after the osmotic pressure is elevated. By 24 h it peaks at approximately 15 times the baseline value, then it decreases to a steady level approximately six times greater in hypertonic medium than in medium of normal osmolality. The decrease in synthesis rate after 24 h coincides with the accumulation of sorbitol. If the rise in sorbitol is prevented by using the aldose reductase inhibitor, Tolrestat, the rate of synthesis continues to increase beyond 24 h. Conversely, when the experiment is designed so that the cells accumulate sorbitol from an exogenous source, not requiring intracellular synthesis, translation of aldose reductase declines (Moriyama et al. 1989). Thus, intracellular sorbitol somehow exerts a negative feedback on its own continued production by affecting aldose reductase protein synthesis rate. The degradation rate of aldose reductase protein is slow (half-time approximately 6 days) and does not differ between normal and hypertonic medium. Therefore, the rise of aldose reductase protein in hypertonic medium is entirely due to increased synthesis.

In order to investigate how hypertonicity increases aldose reductase protein synthesis, aldose reductase cDNA was cloned from PAP-HT25 cells, and the cloned cDNA was used as a probe to measure mRNA abundance (Garcia-Perez et al. 1989). When the medium is made hypertonic by adding NaCl, aldose reductase mRNA increases. The time course and magnitude of the increase closely match the increase in aldose reductase protein synthesis rate. Therefore, the increased translation of enzyme protein is due to greater abundance of its mRNA. Nuclear run-on assays show that the increase in aldose reductase mRNA abundance follows a rise in its transcription initiation rate (Smardo et al. 1990). It is not yet clear whether changes in aldose reductase mRNA stability also play a role.

When osmolality is decreased from a high to normal level, both aldose reductase activity and protein fall slowly (half-time of approximately 6 days) back to their baseline level (Bagnasco et al. 1988). In contrast, the abundance of aldose reductase mRNA decreases rapidly, returning close to the baseline value within one day (Garcia-Perez et al. 1989)

### *Decreased Extracellular Tonicity Induces Rapid Efflux of Sorbitol*

The level of sorbitol in PAP-HT25 cells decreases very rapidly by efflux from the cells when medium NaCl is lowered and the cells swell (Bagnasco et al. 1988). PAP-HT25 cells grown in medium made hyperosmotic by adding NaCl have a very low permeability to sorbitol. However, when the medium osmolality is decreased abruptly from 600 to 300 mosmol/kg by lowering the NaCl concentration, sorbitol efflux immediately increases more than 150-fold. Approximately half of the cell sorbitol appears in the medium within 15 min. By this time the sorbitol efflux rate constant has decreased again to only twice the value in the original hyperosmotic medium. The cell sorbitol continues to fall after 15 min, but more slowly. It is less than 25% of the initial value after 24 h and is close to baseline after 3 days.

The rapid efflux of sorbitol from PAP-HT25 cells results from increased sorbitol permeability (Siebens and Spring 1989). While sorbitol is exiting the cells at a fast rate, there is also a large increase in the influx of $^{14}$C-sorbitol. When PAP-HT25 cells grown in 500 mosmol/kg medium are switched to a 300 mosmol/kg solution, $^{14}$C-sorbitol uptake increases 71-fold. The initial increase in permeability occurs within 30 s, and, if the osmolality is then restored to the original value again, sorbitol permeability falls within 30 s. The permeability pathway is selective. Permeability to some other polyols is enhanced, although to a smaller extent (sorbitol > mannitol > inositol), but there is little change in permeability to sucrose, raffinose, or L-glucose. The pathway does not saturate with concentrations of sorbitol up to 315 mM, nor is there competitive inhibition by high concentrations of other polyols. Therefore, the $K_m$, if any, is very high. A large number of putative inhibitors fail to prevent the increase in sorbitol permeability. The most effective inhibitor is quinidine. One mM quinidine inhibits the osmoregulated increase in sorbitol permeability by 82%. The increase in permeability to sorbitol is also blocked by reducing the temperature to 0 °C. Presumably, sorbitol efflux is mediated by a carrier or a selective channel, but the nature of this novel transporter and its control remain to be elucidated.

Whatever its basis, the rapid efflux of sorbitol ensures that the cells do not remain swollen when the osmolality decreases. In this sense, the response is analogous to the rapid efflux of KCl from many different cell types during volume regulatory decrease (Eveloff and Warnock 1987) and the efflux of taurine from Erlich cells that also follows a fall in osmolality (Hoffman 1985).

Sorbitol can be catabolized to fructose in a reaction catalyzed by sorbitol dehydrogenase (Hers 1960). This enzyme is present in the kidney, but it is much lower in the medulla than in the cortex (Heinz et al. 1975; Clampitt and Hart 1978; Chauncey et al. 1988). Its activity is low in PAP-HT25 cells and is not measurably affected by medium osmolality. Also, no fructose is detectable in the cells under any of the conditions studied, consistent with the absence of significant sorbitol degradation. Thus, there is no evidence for significant sorbitol degradation in the cell cultures. On the other hand, sorbitol dehydrogenase activity increases in rabbit inner medullary collecting ducts after a day of diuresis, suggesting that sorbitol catabolism may play a role in lowering inner medullary sorbitol in vivo (Sands and Schrader 1990).

### *High Intracellular Ionic Strength Induces Aldose Reductase*

The signal for induction of aldose reductase activity and subsequent sorbitol accumulation has been examined by comparing the effect of raising medium osmolality with different solutes (Uchida et al. 1989). As already indicated, high extracellular NaCl increases aldose reductase activity. Raffinose, a trisaccharide that does not enter cells readily, is as effective NaCl, indicating that neither extracellular Na nor Cl is necessary. In contrast, raising the osmolality with neither urea or glycerol, which penetrate these cells readily, fails to increase aldose reductase activity. Therefore, increased osmolality per se is not a sufficient signal.

When osmolality is increased by adding raffinose, cell water decreases, and cell sodium and potassium concentrations increase. The changes persist for at least 24 h until sorbitol content begins to rise. Addition of urea, on the other hand, does not cause any significant change in cell water or electrolyte concentration, just as it fails to induce increased aldose reductase activity. Based on this correlation, the trigger for induction of aldose reductase appears to be prolonged extracellular hypertonicity.

The mechanism by which prolonged extracellular hypertonicity increases renal medullary cell aldose reductase activity might well involve either the reduction in cell volume or the increase in intracellular Na or K, both of which precede the rise in enzyme activity. To distinguish between these possibilities, ouabain was added. Ouabain generally decreases cell potassium and increases cell sodium and volume. Medium osmolality was increased by adding NaCl either without ouabain or with different concentrations of the drug. Aldose reductase activity did not correlate significantly with cell volume or sodium. It did correlate, however, with cell potassium concentration ($r = 0.72, P < 0.05$) and even better with the sum of cell potassium plus sodium concentration ($r = 0.99, P < 0.001$ (Uchida et al. 1989). Thus, the signal that induces increased aldose reductase activity apparently is a rise in the total concentration of sodium plus potassium salts (or ionic strength) in the cells.

### *Cellular Adaptation to Hypertonicity Affects Sorbitol Accumulation*

Based on a model originally proposed for yeast (Brown and Edgley 1980), the different phases of cellular adaptation involved in osmoregulation by mammalian renal cells have been identified (Burg 1988). The different states of adaptation are illustrated for sorbitol in Fig. 1.

Once aldose reductase has been induced, it is very stable. Its half-life is approximately 1 week regardless of the osmolality. This is the adapted state. Although adapted cells do not quickly alter their aldose reductase activity, they can vary sorbitol rapidly by changing permeability to it (Fig. 1, steps 1 to 4). Extrapolating to renal medullary cells in vivo from the results in cell culture, the following sequence can be visualized. When a diuresis ensues and the renal medullary extracellular NaCl concentration falls, the cells swell (Fig. 1, step 1), and sorbitol permeability increases (step 2). The resultant efflux of sorbitol and water restores cell volume, but without any change in aldose reductase (step 3). Although its production continues, sorbitol

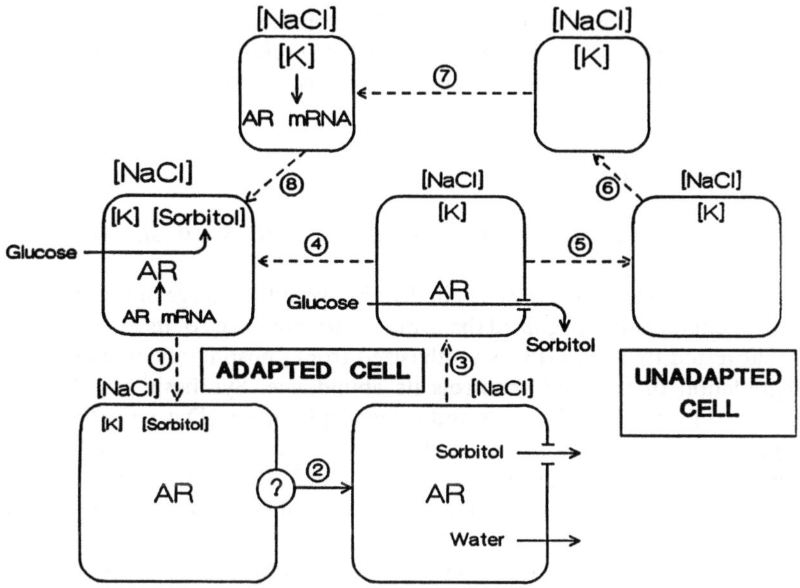


**Fig. 1.** Osmoregulation of sorbitol in renal medullary cells. In the state of chronic antidiuresis, the cells are adapted, meaning that they contain high levels of aldose reductase. During a transient diuresis, sorbitol concentration falls rapidly by efflux (steps *1* to *3*). Then, when antidiuresis is reestablished, sorbitol is accumulated again by synthesis, catalyzed by the aldose reductase, which has persisted during the diuresis (step *4*). If the diuresis is prolonged, however, as in diabetes insipidus, aldose reductase falls, and the cells become unadapted (step *5*). Then, for sorbitol to reaccumulate, antidiuresis must induce synthesis for new aldose reductase (steps *6* to *8*), a process requiring more than 1 day. See the text for additional details. (Garcia-Perez and Burg 1990)

does not accumulate again in the cells because permeability to it remains high enough so that it leaves the cells as fast as it is formed. If, within a few days, the diuresis ends and extracellular NaCl increases, the permeability falls and the sorbitol that is being synthesized is retained in the cells, increasing its level over a few hours to compensate for the higher NaCl (step 4). Sorbitol is able to reaccumulate rapidly because the machinery for synthesizing it has been maintained during the transient diuretic state.

On the other hand, when the diuretic state is prolonged and extracellular NaCl remains low, aldose reductase activity falls slowly over several days, as the enzyme degrades, and the cells eventually become unadapted (step 5). This occurs in pathological states like diabetes insipidus. If the condition causing the diuresis is corrected, the renal medullary extracellular NaCl increases, as the kidney makes more concentrated urine (step 6). This triggers the transcription of aldose reductase mRNA and translation of enzyme protein (step 7), returning the cells to the adapted state after several days (step 8).

The model in Fig. 1, refers specifically to sorbitol. However, the concept of adapted and unadapted states applies to all four renal organic osmolytes. Although the mechanisms for accumulating GPC, betaine, and inositol differ from that for sorbitol, all four systems are induced and decay relatively slowly. Yet, cellular levels of all four solutes can decrease rapidly by efflux, when extracellular tonicity falls.

## Glycerophosphorylcholine

### *High NaCl or Urea Induces Accumulation of GPC, Synthesized from Choline*

When Ullrich (Ullrich 1959) originally discovered the large amount of GPC in dog renal medulla and recognized the osmotic consequences of its intracellular location, he suggested that the GPC is synthesized from phosphatidylcholine (PC) and that degradation of the GPC controls its abundance. Synthesis of GPC from PC is catalyzed by phospholipase A and lysophospholipase (Dawson 1955), and GPC is degraded in kidneys to choline and α-glycerolphosphate in a reaction catalyzed by GPC: choline phosphodiesterase (Baldwin and Cornatzer 1968). Ullrich reasoned that GPC diesterase activity controls the level of GPC, based on two observations. (1) GPC diesterase activity is higher in the renal cortex than in the medulla, which could explain the higher level of GPC in the medulla, and (2) addition of NaCl and urea to homogenates of cortical tissue inhibits GPC diesterase activity. Therefore, the high levels of urea and NaCl in the renal medulla during antidiuresis could be inhibiting GPC diesterase, and thus elevating GPC. On the other hand, recent measurements failed to show differences in renal medullary GPC diesterase between diuretic and antidiuretic rats (Kanfer and McCartney 1989), which does not support this theory.

GPC is higher in MDCK cells when they are grown through many generations in medium made hyperosmotic by adding NaCl or NaCl plus urea (Nakanishi et al. 1988). When MDCK cells grown at normal osmolality are switched to the hyperosmotic media, GPC rises gradually over a week or more (Nakanishi and Burg 1989a). This increase in GPC was studied in detail to determine the mechanism involved.

Tissue culture media routinely contain choline, and it was present in the experiments described above. When cells grown in choline-containing medium of normal osmolality are switched to hyperosmotic medium containing no choline, GPC rises during the first 2 days, but then it falls back to the baseline level (Nakanishi et al. 1989a). In contrast, if choline is present in the hyperosmotic medium, the level of GPC continues to rise. Thus, GPC apparently is synthesized from choline that has been taken up from the medium, and, even when there is no exogenous choline, some metabolic intermediate already formed in the cells (presumably PC) can provide for a transient increase in GPC. Choline is taken up into MDCK cells by sodium-independent transport, which is not affected by medium osmolality. Therefore, transport of choline is not rate-limiting for osmoregulation of GPC accumulation.

Also, synthesis of GPC from PC also does not seem to be the rate-controlling step. High urea and NaCl decrease the rate of synthesis of GPC from $^{14}$C-choline by MDCK cells in hyperosmotic medium, rather than increasing it, as would be expected from increased phospholipase activity (Zablocki et al. 1990). Also, phospholipase A activity is not elevated by hyperosmolality (Zablocki et al. 1990). These observations point to degradation of GPC as the controlling step, as predicted by Ullrich. Consistent with this theory, GPC diesterase activity is decreased in MDCK cells adapted to high NaCl plus urea (Nakanishi and Burg 1989a; Zablocki et al. 1990).

When the elevated osmolality is decreased rapidly, GPC exits from the cells. The efflux of GPC from MDCK cells after lowering the osmolality is relatively small (only 10 to 15% during the first 3 h) compared to the much more rapid fall in sorbitol in PAP-HT25 cells, discussed earlier. Presumably, the loss of GPC occurs because of an increase in permeability, as is the case for sorbitol, but the GPC permeability has not been directly measured.

Like sorbitol, GPC increases when NaCl is added to the medium. However, GPC accumulation is also triggered by elevating urea alone (Nakanishi et al. 1989a; Nakanishi and Burg 1989a,b; Moriyama et al. 1990b), whereas sorbitol accumulation is not. Elevated urea concentration does not increase the intracellular ionic strength. Thus, high intracellular ionic strength is not necessary for GPC accumulation, as it is for sorbitol accumulation (see above). The fact that urea causes GPC to rise in particularly meaningful when we consider that GPC is a methylamine and that methylamines are "counteracting osmolytes" that can oppose the harmful effects of high urea.

## Betaine

### *Metabolism and Transport of Betaine*

Betaine is absorbed from the diet. It is also produced by oxidation of choline in various organs, including the kidney (Mann and Quastel 1937; Haubrich et al. 1975; Acara 1979; Lohr and Acara 1990). Within the kidney, betaine may be synthesized from choline by proximal tubule (Wirthensohn and Guder 1982) and renal medullary cells (Grossman and Hebert 1989; Lohr and Acara 1990), both of which contain choline dehydrogenase activity. The newly synthesized betaine initially accumulates in the cells that produce it, but within an hour it also appears in the blood and, to a much lesser extent, in the urine.

Synthesis of betaine by renal medullary cells is a potential pathway for its osmotically regulated accumulation. However, following salt loading, rats accumulate large amounts of betaine in their renal medullas, without significant change in the renal medullary mitochondrial choline dehydrogenase activity. Also, the activity does not differ significantly between the Brattleboro strain of rats that have congenital diabetes insipidus, and normal rats (Grossman and Hebert 1989). Thus, there is little evidence that betaine synthesis in renal medullary cells has an important role in its osmotic regulation

On the other hand, betaine can also be transported into kidney cells. It is concentrated in slices of rat renal cortex (Sung and Johnstone 1969; Evered and Nunn 1970; Lohr and Acara 1990) by a sodium-dependent process (Sung and Johnstone 1969), and there is sodium-dependent betaine transport in brush border vesicles prepared from rabbit renal cortex (Wright and Wunz 1989). Betaine is also concentrated by slices from rat renal inner and outer medulla (Lohr and Acara 1990).

### *Osmotic Regulation of Betaine Transport in Renal Cells*

Betaine accumulation is osmotically regulated in tissue cultures of MDCK, LLC-$PK_1$ and LLC-$PK_3$ cells adapted to long-term growth in medium made hyperosmotic by addition of NaCl to 615 mosmol/kg. Under those conditions the cells contain more betaine than in medium of normal osmolality (315 mosmol) (Nakanishi et al. 1988). The origin of the betaine is apparent in studies using defined medium (Nakanishi et al. 1990). When MDCK cells are grown at normal osmolality in a defined medium that contains no betaine, none is detectable in the cells (Nakanishi et al. 1990). Following a switch to medium made hypertonic to 500 mosmol/kg by adding NaCl and containing 50 μm betaine, cell betaine rises over a week or more (Nakanishi et al. 1990). If betaine is left out of the medium, the cells do not accumulate it, even at the higher osmolality. The results are essentially the same with PAP-HT25 cells (Moriyama et al. 1990a,b). In retrospect (Nakanishi et al. 1990), it is apparent that the medium routinely used in the earlier chronic experiments, although not formulated to contain betaine, contains approximately 18 μM betaine, derived from the 10% of fetal bovine serum which supplements it. The intracellular betaine concentration in the high NaCl medium is approximately 50 mM (Nakanishi et al. 1988). That is more than 1000 times higher than the 0.018 mM betaine in the medium. Since the accumulation of betaine by renal cells in tissue culture is supported by low concentrations of betaine in the medium, active betaine transport is implied.

MDCK cells take up betaine by active, sodium-dependent transport from the medium (Nakanishi et al. 1990). The kinetics are complex. Under isotonic conditions two transport sites are observed with $K_m$'s for betaine of 0.12 and 5.6 mM. When the cells are grown chronically for a year in medium made hypertonic to 615 mosmol/kg by addition of NaCl, the $V_{max}$'s of both sites increase almost tenfold without a significant change in the $K_m$'s. The higher medium NaCl directly accounts for only a small part of the rise in $V_{max}$. If the medium NaCl is reduced to a normal level during the transport measurement (hypertonicity maintained by mannitol), the $V_{max}$ still increases more than sevenfold. Cells acutely exposed (1–7 days) to hypertonic medium show increased sodium-dependent betaine uptake, which is maximal after 1 day, then decreases as betaine and the other osmolytes accumulate in the cells. In this respect the acute time course of increased betaine transport by MDCK cells resembles the time course of increased aldose reductase in PAP-HT25 cells following hypertonicity.

The increase in $V_{max}$ without change in $K_m$ is consistent with an increased number of functioning transporters. In support of this hypothesis, when mRNA, taken from MDCK cells exposed to hypertonic medium, is injected into frog oocytes, sodium-dependent $^{14}$C-betaine transport into the oocytes rises. However, mRNA from cells maintained in isotonic medium does not have this effect. Most likely, hypertonicity increases the abundance of mRNA coding for the betaine transporter in MDCK cells, and more betaine transporters are synthesized (Robey et al. 1990).

Betaine rises only slowly in MDCK and PAP-HT25 cells when the medium is made hypertonic, but it decreases much more rapidly when the medium is made isotonic again. Two mechanisms are involved in the decrease. (1) Betaine exits rapidly from the cells. When MDCK cells, that are grown in medium made hypertonic by adding NaCl, are switched to an isotonic medium, they lose 30% of their betaine into the medium within 15 min and over 40% in 1 h (Nakanishi and Burg 1989b). The betaine efflux from PAP-TH25 cells is even greater, approximately 70% in 10 min (Moriyama et al. 1990a,b). (2) Active betaine transport into the cells decreases when the osmolality falls. When the osmolality is returned to normal, sodium-dependent betaine flux into MDCK cells falls approximately 25% within 3 h, 50% by 1 day, and back close to baseline by the end of 2nd day (Nakanishi and Burg 1989b). Thus, lower medium osmolality reduces cell betaine both because of rapid efflux from the cells and a slower decrease in active betaine influx.

Whether cell betaine increases in hyperosmotic medium or not, depends on which solute is used to raise the osmolality. In MDCK cells, raffinose, added for 7 days, increases betaine approximately as much as does NaCl, but neither glycerol nor urea has any effect (Nakanishi et al. 1990). Similarly, 4 days of high NaCl increases betaine in PAP-HT25 cells, but urea does not (Moriyama et al. 1990a,b), and chronically high NaCl increases betaine in both MDCK and LLC-$PK_1$ cells, but chronically high urea does not (Nakanishi et al. 1988). The pattern is similar to that for induction of aldose reductase in PAP-HT25 cells. NaCl and raffinose, which induce accumulation of the osmolytes, are osmotically effective since they are excluded from the cells. Glycerol and urea, on the other hand, enter the cells rapidly, and are not osmotically effective. Thus, the important factor apparently is the tonicity of the medium, not the absolute osmolality.

In MDCK and PAP-HT25 cells, high urea not only fails to increase cell betaine, but actually decreases it. Also, in PAP-HT25 cells the rise in cell betaine when urea plus NaCl is added, is smaller than when NaCl is added alone, and $V_{max}$ for betaine transport rises less when MDCK cells are grown in chronically high NaCl plus urea, than in high NaCl (Nakanishi et al. 1990). Thus, in cell culture urea apparently inhibits betaine accumulation. This contrasts to the increase in GPC accumulation which is caused by urea.

## Inositol

### *Metabolism and Transport of Inositol*

Inositol is present in the diet and is produced in the gut by digestion of phytic acid and phospholipids. It is also synthesized from glucose in various organs, including the kidney (Eisenberg 1967). The kidney is the only organ that disposes of important amounts of inositol. Although little inositol normally is excreted in the urine, the kidney cortex and outer medulla (but not inner medulla) catabolize it to $CO_2$ (Howard and Anderson 1967). In the fasting state, a normal human kidney synthesizes approximately 2 g of inositol per day and disposes of approximately 1 g. Urinary excretion accounts for only 6% of the renal disposal of inositol. Oxidation of inositol accounts for the rest.

In addition to synthesizing inositol, many different kinds of cells can amass it by active transport. The normal plasma level of inositol is 25 to 60 μM in different species. Cells in many organs including the kidney, accumulate inositol from the extracellular fluid (Lewin et al. 1976) by active, sodium-dependent transport. The inositol transporters in different tissues and species have common characteristics. Their $K_m$'s are between 63 μM and 140 μM, and they are inhibited by phlorizin (Caspary and Crane 1970). They are also inhibited competitively by high concentrations of glucose. In contrast, inositol has no effect on D-glucose transport (Caspary and Crane 1970; Hammerman et al. 1980). Apparently, there are separate glucose and inositol transport systems, and glucose competes for the inositol transporter, but not vice versa.

Between transport and synthesis, the intracellular free inositol is high in many organs, including the kidney (Dawson and Freinkel 1961; Holub 1986). Inositol concentration is approximately 1000 times higher in the cells of the renal medulla than in the peripheral blood. Inositol is a major osmotic effector in the kidney medulla, but probably not in other tissues. High inositol may be important for synthesis of inositol phospholipids (Troyer et al. 1986), which have many functions, including endogenous activation of Na,K-ATPase in rabbit renal microsomes (Mandersloot et al. 1978). Thus, the elevated level of inositol in the renal outer medulla could serve both for osmotic regulation and to produce enough inositol phospholipids to support the high Na,K-ATPase activity in that region of the kidney. However, although increased inositol raises the amount of phosphatidyl inositol in yeast (Kelley et al. 1988), it apparently does not in MDCK cells (Shayman and Wu 1990).

There is no direct evidence distinguishing whether the osmotic regulation of inositol in the renal medulla in vivo occurs by net synthesis or by transport. However, there apparently is active transport of inositol in kidney medullas. Rat renal medullary slices take up inositol to high levels from the medium, and there is sodium-dependent inositol transport in membrane vesicles from rat renal outer medulla (Seifter et al. 1990). The characteristics of this transport are similar to those of the inositol transporter in renal cortex and other tissues, already described.

### *Osmotic Regulation of Inositol Transport in Renal Cells*

Many types of human and some mouse cells require exogenous inositol to grow in tissue culture (Eagle et al. 1957), and media generally are formulated to contain inositol. However, there are exceptions, such as mouse fibroblasts (Eagle et al. 1957) and L1210 leukemia cells (Moyer et al. 1988), that can synthesize enough inositol de novo to grow in inositol-free medium. Regardless of their capacity to synthesize inositol, many kinds of cells actively transport inositol in tissue culture. These include MDCK cells (Nakanishi et al. 1989b). In tissue culture, as in vivo, active inositol transport is sodium-dependent and is inhibited by phlorizin, or high glucose. All of the inositol transport systems studied in tissue culture have at least one high affinity site with $K_m$ between 12 and 41 μM. Some cells, including MDCK (Nakanishi et al. 1989b), also have a second, lower affinity transport site. Hypertonicity increases the level of inositol in renal cells in tissue culture, including MDCK cells (Nakanishi et al. 1988; Nakanishi et al. 1989b; Shayman and Wu 1990), LLC-$PK_1$ cells (Nakanishi et al. 1988), LLC-$PK_3$ cells (Nakanishi et al. 1988), GRB-MAL1 cells (Nakanishi et al. 1988), PAP-HT25 cells (Moriyama et al. 1990a,b), and rat inner medullary collecting tubule cells (Berl et al. 1990).

Inositol increases in MDCK cells when NaCl, alone, or both NaCl and urea are added to the medium. The additional inositol is taken up from the medium. If inositol is absent from the hypertonic medium, cell inositol falls, rather than increases (Nakanishi et al. 1989b). More rapid transport accounts for the cellular accumulation. When the MDCK cells are grown through many passages in medium made hyperosmotic to 915 mosmol by addition of NaCl plus urea, the inositol transport $V_{max}$ increases almost threefold without any significant change in $K_m$. The increased sodium gradient provided by higher medium NaCl accounts for only part of the rise in $V_{max}$. If the medium NaCl is equalized during the transport measurement (hypertonicity maintained by mannitol), the $V_{max}$ still rises almost twofold. In the remainder of the studies summarized here, medium NaCl was equalized during the transport measurements, so that differences in inositol transport are attributable to changes in the transporters themselves and not to altered sodium gradients.

MDCK cells acutely exposed to medium made hyperosmotic with NaCl plus urea to 700 mosmol show increased sodium-dependent inositol uptake (Nakanishi et al. 1989b; Yamauchi et al. 1990). $V_{max}$ increases without change in $K_m$. The transport is maximal after 1 day, then decreases as inositol accumulates in the cells (Nakanishi et al. 1989b). Apparently, there is a feedback by which increased cell inositol reduces its own transport. If the culture medium contains no inositol, so that inositol cannot accumulate following the increase in osmolality, the transport rate, measured by addition of radioactive inositol for a few minutes, continues to rise after the 1st day. Also, if the osmolality is increased with raffinose, the transport rate increases to higher levels than when the osmolality is increased with NaCl. Presumably, the greater sodium gradient provided by elevated medium NaCl allows the cells to accumulate the needed amount of inositol with less increase in intrinsic transport capacity than when hypertonicity is achieved with raffinose.

As already discussed for betaine, the increase in $V_{max}$ for inositol transport without any change in $K_m$ probably results from an increase in the number of functioning transporters. Cycloheximide prevents the hypertonicity-induced increase in transport in rat inner medullary collecting tubule cells (Berl et al. 1990), indicating that increased protein synthesis is necessary. When mRNA from MDCK cells grown in hypertonic medium is injected into frog oocytes, $^3$H-inositol uptake by the oocytes increases (Kwon et al. 1990b). In contrast, mRNA from cells grown in isosmotic medium does not have this effect (Kwon et al. 1990a). Apparently, mRNA for the inositol transporter increases following hypertonicity, resulting in synthesis of additional transporters.

Whether or not MDCK cell inositol rises depends on the nature of the solute added to make the medium hyperosmotic. Either chronic (through many passages)

(Nakanishi et al. 1988) or acute (4-day) (Nakanishi et al. 1989b; Nakanishi et al. 1990) addition of NaCl or NaCl plus urea increases MDCK cell inositol. Acute addition of raffinose, sorbitol, or dextrose has the same effect (Nakanishi et al. 1989b; Nakanishi et al. 1990). However, acute addition of urea (Nakanishi et al. 1989b; Nakanishi et al. 1990) or glycerol (Nakanishi et al. 1990) does not increase MDCK cell inositol. The results are essentially the same in PAP-HT25 cells (Moriyama et al. 1990a,b). Thus, the characteristics of inositol accumulation induced by hypertonicity are similar to those of sorbitol and betaine accumulation, discussed earlier. The stimulus apparently is increased osmotic pressure, which occurs when poorly permeating solutes, like NaCl and raffinose, are added to the medium, and not when rapidly permeating solutes like urea and glycerol are added.

Like the other organic osmolytes, inositol rises slowly over a week or more in MDCK cells when they are switched to medium made hyperosmotic with NaCl plus urea. In contrast, when MDCK cells grown in medium made hypertonic by adding NaCl, are switched to an isotonic medium, there is a rapid efflux of inositol (Nakanishi and Burg 1989b). The inositol efflux is less than that of sorbitol and betaine. The cells lose approximately 15% of their inositol into the medium within 15 min and over 20% in 1 h. The result is essentially the same in PAP-HT25 cells (Moriyama et al. 1990a,b). The greater efflux presumably is due to increased inositol permeability. $^{3}$H-inositol influx into PAP-HT25 cells increases at the same time that the efflux of inositol is occurring (Siebens and Spring 1989). After the osmolality has been reduced for a while, the active inositol transport into the cells also begins to decrease. Compared to the initial hypertonic condition, sodium-dependent inositol influx falls approximately 30% by 3 h after the osmolality decreases and back close to baseline by 12 h (Nakanishi and Burg 1989b). Thus, reduced medium osmolality causes cell inositol to fall both because of a rapid efflux from the cells and a slower decrease in active inositol influx.

## Determinants of the Relative Amounts of Different Organic Osmolytes in Renal Medullary Cells

The organic osmolytes in renal medullary cells appear to be independently regulated (Gullans et al. 1988). It is hard to analyze the basis for this selectivity in living animals because of the difficulty of selectively controlling the pertinent variables. Cell culture experiments are somewhat easier in this regard. The results of the cell culture experiments that are presented in the following sections help to explain at least some of the variations seen in vivo.

### *High Urea Increases Cell GPC, but Reduces Betaine*

Urea stimulates accumulation of GPC in cell culture, but inhibits accumulation of betaine, whereas NaCl stimulates both. After 4 days of exposure to hyperosmotic medium, MDCK cell GPC increases approximately equally whether the high osmolality is caused by adding urea or NaCl (Nakanishi and Burg 1989a; Nakanishi et al. 1990). The combination of NaCl plus urea does not clearly increase GPC more than either alone (Nakanishi and Burg 1989a; Nakanishi et al. 1990). In PAP-HT25 cells GPC accumulation is stimulated more by urea than by NaCl, and most by a combination of the two (Moriyama et al. 1990a). In contrast, betaine accumulation is stimulated by high NaCl, but is inhibited by high urea (Nakanishi et al. 1990; Moriyama et al. 1990a).

There is evidence for similar interactions in vivo. Thus, betaine accumulation is exaggerated when renal inner medullary NaCl is selectively elevated, while GPC increases more with high urea, as follows: (1) The amount of GPC in renal medullary cells is directly related to urea, whereas betaine is not (Yancey 1988). (2) Administration of hypertonic NaCl to rats increases renal inner medullary betaine, but not GPC (Heilig et al. 1989), associated with a larger increase in medullary $Na^+$ than in urea (Wald et al. 1989). (3) Administration of vasopressin to Brattleboro rats or dehydration of normal rats or rabbits increases medullary urea more than $Na^+$. Then, GPC increases more than betaine.

### *The Organic Osmolytes Are Affected Differently by the Magnitude of Hypertonicity*

This is illustrated by experiments in which the medium bathing PAP-HT25 cells is made hypertonic by adding NaCl in 50 mosmol/kg increments between 300 and 500 mosmol/kg (Moriyama et al. 1990a). The relative increments in cell sorbitol and inositol differ greatly over this range. Inositol increases most between 300 and 350 mosmol/kg and is predominant at 350 mosmol/kg. Above that, however, sorbitol increases more and it greatly predominates at 500 mosmol/kg. In the higher range of osmolality small increases in NaCl do not affect cell inositol, but do raise betaine and GPC, as well as sorbitol, significantly.

In inner medullas, the percent increase of sorbitol content exceeds that of inositol during antidiuresis, corresponding to the greater change in sorbitol found in cell culture in the upper range of osmolality. In contrast, during diuresis caused by furosemide, medullary osmolality decreases to a low level, and the fractional decrease of inositol equals or exceeds that of sorbitol, corresponding to the changes in the lower range of NaCl in tissue culture.

### *Decreased Tonicity Causes Efflux of All the Individual Osmolytes, but at Different Rates*

When medium NaCl is decreased, the efflux of betaine from MDCK cells is more rapid than that of inositol (Nakanishi and Burg 1989b) and efflux of GPC is relatively slow (Nakanishi and Burg 1989a). Similarly, in PAP-HT25 cells, efflux of betaine also is greatest and that of GPC is least (Moriyama et al. 1990a).

When animals are given the diuretic furosemide, renal inner medullary NaCl and urea fall, accompanied by decreases in cell betaine, sorbitol, inositol, and GPC. Presumably, the organic osmolytes exit from the cells in vivo, secondary to lower extracellular NaCl, like in tissue cultures (Bagnasco et al. 1988), but there is no direct evidence for this. In dehydrated rabbits the pattern of the decrease of the organic osmolytes relative to each other 40 min after furosemide (Wolff et al. 1989) is the same as in the rabbit PAP-HT25 cells after lowering NaCl in tissue culture (Moriyama et al. 1990a). In rats the relative decreases after furosemide are somewhat different: inositol ≈ sorbitol > betaine > GPC after 2 h (Cowley et al. 1990) and GPC ≈ inositol > sorbitol after 24 h (Wirthensohn et al. 1989). The differences between rabbits and rats may be due to species variation. In any case, the fall does differ between individual osmolytes, changing the relative amounts of them that remain.

### *A Change in One Osmolyte May Cause Reciprocal Changes in the Others*

The compatible osmolytes hypothesis predicts that the sum of the various compatible organic osmolytes will balance extracellular hypertonicity. For this relation to be maintained, a change in the concentration of one osmolyte has to be compensated by a reciprocal change in one or more of the others. Evidence that this occurs in cell culture appeared in the course of testing whether addition of betaine to hypertonic medium might result in its accumulation by PAP-HT25 cells. The resulting increase in cell betaine was reciprocated by a substantial decrease in cell sorbitol and smaller decreases in GPC and inositol (Moriyama et al. 1990a). Similarly, when intracellular sorbitol is elevated by adding glucose (the substrate from which sorbitol is synthesized), betaine and inositol fall. Also, when sorbitol is reduced by inhibition of aldose reductase, then betaine, inositol, and GPC rise (Moriyama et al. 1990a,b).

Thus, with constant medium osmolality, a change in one organic osmolyte leads to reciprocal changes in at least some of the others. Situations that might alter the concentration of a particular organic osmolyte independent of osmolality include (1) altered variability of the compound itself for transport, (2) altered availability of a substrate or cofactor required for its synthesis, and (3) changes in enzyme activity. Mechanisms for the ensuing alteration of other organic osmolytes include osmotic constraints, metabolic interactions, or changes in a common signal.

Although there are some interactions between virtually all of the organic osmolytes, the strongest and most consistent relationship is between sorbitol and betaine (Moriyama et al. 1990a,b). A possible connection, explaining their reciprocally relation is that increased intracellular ionic strength may signal the accumulation of both. As discussed earlier, high intracellular ionic strength induces sorbitol accumulation by increasing synthesis of aldose reductase (Moriyama et al. 1989), which then catalyzes synthesis of more sorbitol from glucose (Uchida et al. 1989). As sorbitol accumulates, ionic strength falls again. Ionic strength may also control the rate of synthesis of the betaine transporter, and the resultant accumulation of betaine may decrease ionic strength. Thus, betaine accumulation could reduce sorbitol by inhibiting synthesis of aldose reductase. In support of this hypothesis, addition of betaine to the medium raises its level in the cells and reduces the amount of aldose reductase (Moriyama et al. 1990a,b). Similarly, reducing sorbitol could elevate the intracellular ionic strength and signal an increase in cell betaine. This could help explain the increase in betaine that occurs when sorbitol is reduced by inhibition of aldose reductase (Moriyama et al. 1990a,b). In addition, betaine also inhibits aldose reductase directly, but to a minor extent (Moriyama et al. 1990a,b).

A reciprocal relation between sorbitol and betaine has also been observed in vivo. When dehydrated rats are treated with aldose reductase inhibitors, the decrease in renal medullary sorbitol is accompanied by an increase in betaine (Yancey et al. 1990).

In contrast, when medium (and cell) inositol concentration are varied, none of the other organic osmolytes change significantly (Moriyama et al. 1990a,b). Further, when the other organic osmolytes are primarily altered, either there is no significant change in cell inositol or the change is relatively small (Moriyama et al. 1990a,b). Cell inositol level is not much affected by the levels of the other organic osmolytes and vice versa.

Cell GPC falls when choline is removed from the medium (Nakanishi and Burg 1989a). Associated with the decrease in GPC there are increases in betaine and sorbitol, but the changes are small (Moriyama et al. 1990a,b). Conversely, there is only a small change in GPC when betaine or sorbitol is altered, and no change in GPC when inositol is changed. Thus, the level of GPC may affect other organic osmolytes, and vice versa, but the changes are small compared to the nearly reciprocal relation between sorbitol and betaine.

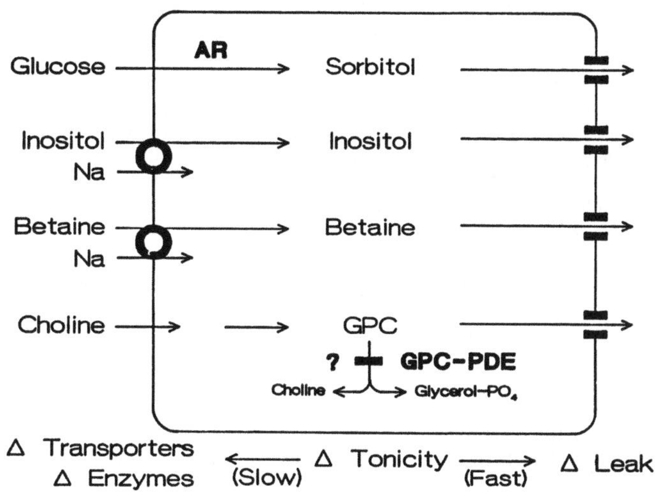


**Fig. 2.** Mechanisms of organic osmolyte accumulation and loss in renal medullary cells. When the extracellular osmolality is altered, transporters and enzymes change (Δ) slowly, (at *left*) but permeability changes rapidly (at *right*). (Garcia-Perez and Burg 1990)

## Summary and Conclusions

Sorbitol, inositol, GPC and betaine are the predominant organic osmolytes in renal medullary cells. They protect the cells from harmful effects of the high interstitial NaCl and urea concentrations that occur in the renal medulla during operation of the urinary concentrating mechanism. Their levels correlate with extracellular NaCl concentration, and in the case of GPC, also with urea. Sorbitol is synthesized from glucose, catalyzed by aldose reductase (Fig. 2). Inositol and betaine are transported into the cell. GPC is synthesized from choline. The transcription of aldose reductase and the transport of betaine and inositol are regulated, dependent on the degree of hypertonicity. Maintenance of appropriate levels of the organic osmolytes contributes to survival and growth of the renal medullary cells in their hyperosmolal environment.

## References


Acara M (1979) Effect of ethanol on the renal excretion and metabolism of choline in the isolated perfused rat kidney. Drug Metab Dispos 7:113–117

Bagnasco S, Balaban R, Fales H, Yang Y-M, Burg M (1986) Predominant osmotically active organic solutes in rat and rabbit renal medullas. J Biol Chem 261:5872–5877

Bagnasco S, Uchida S, Balaban R, Kador P, Burg M (1987) Induction of aldose reductase and sorbitol in renal inner medullary cells by elevated extracellular NaCl. Proc Natl Acad Sci USA 84:1718–1720

Bagnasco SM, Murphy HR, Bedford JJ, Burg MB (1988) Osmoregulation by slow changes in aldose reductase and rapid changes in sorbitol flux. Am J Physiol 254:C788–C792

Baldwin JJ, Cortnatzer WE (1968) Rad kidney glycerylphosphorylcholine diesterase. Biochim Biophys Acta 164:195–204

Bedford JJ, Bagnasco SM, Kador PF, Harris HW, Burg MB (1987) Characterization and purification of a mammalian osmoregulatory protein, aldose reductase, induced in renal medullary cells by high extracellular NaCl. J Biol Chem 262:14255–14259

Berl T, Veis J, Molitoris B, Teitelbaum I, Mansour I (1990) Hyperosmolality (H)-induced protein enhances myoinositol (MI) uptake in rat inner medullary collecting tubule (RIMCT) cells. Kidney Int 37:576

Brown AD, Edgley M (1980) Osmoregulation in yeast. In: Rains DW, Valentine RC, Hollaender A (eds) Genetic engineering of osmoregulation. Plenum Press, New York Lond, p 75

Burg MB (1988) Role of aldose reductase and sorbitol in maintaining the medullary intracellular milieu. Kidney Int 33:635–641

Caspary WF, Crane RK (1970) Active transport of myo-inositol and its relation to the sugar transport system in hamster small intestine. Biochim Biophys Acta 203:308-316

Chauncey B, Leite MV, Goldstein L (1988) Renal sorbitol accumulation and associated enzyme activities in diabetes. Enzyme 39:231–234

Clampitt RB, Hart RJ (1978) The tissue activities of some diagnostic enzymes in ten mammalian species. J Comp Pathol 88:607–621

Cowley BD Jr, Ferraris JD, Carper D, Burg MB (1990) In vivo osmoregulation of aldose reductase mRNA, protein and sorbitol in renal medulla. Am J Physiol 258:F154–F161

Dawson RMC (1955) The role of glycerylphosphorylcholine and glyceryl-phosphorylethanolamine in liver phospholipid metabolism. Biochem J 59:5–8

Dawson RMC, Freinkel N (1961) The distribution of free mesoinositol in mammalian tissues, including some observations on the lactating rat. Biochem J 78:606–610

Eagle H, Oyama VI, Levy M, Freeman AE (1957) myo-Inositol as an essential growth factor for normal and malignant human cells in tissue culture. J Biol Chem 226:191–205

Eisenberg F (1967) D-myo-inositol 1-phosphate as product by cyclization of glucose 6-phosphate and substrate for a specific phosphatase in rat testis. J Biol Chem 242:1375–1382

Eveloff JL, Warnock DG (1987) Activation of ion transport systems during cell volume regulation. Am J Physiol 252:F1–F10

Evered DF, Nunn PB (1970) Transport pathway for glycine betaine in rat kidney cortex in vitro. Biochem J 118:40P–41P

Garcia-Perez A, Burg MB (1990) Importance of organic osmolytes for osmoregulation by renal medullary cells. Hypertension 16:595–602

Garcia-Perez A, Martin B, Murphy HR, Uchida S, Murer H, Cowley BD, Handler JS, Burg MB (1989) Molecular cloning of cDNA coding for kidney aldose reductase: Regulation of specific mRNA accumulation by NaCl-mediated osmotic stress. J Biol Chem 264:16815–16821

Grossman EB, Hebert SC (1989) Renal inner medullary choline dehydrogenase activity: characterization and modulation. Am J Physiol 256:F107–F112

Gullans SR, Blumenfield JD, Balschi JA, Kaleta M, Brenner RM, Heilig CW, Hebert SC (1988) Accumulation of major organic osmolytes in rat renal inner medulla in dehydration. Am J Physiol 255:F626–F634

Hammerman MR, Sacktor B, Daughaday WH (1980) Myo-inositol transport in renal brush border vesicles and its inhibition by D-glucose. Am J Physiol 239:F113–F120

Haubrich DR, Wang PFL, Wedeking PW (1975) Distribution and metabolism of intravenously administered choline[methyl-3H] and synthesis in vivo of acetylcholine in various tissues of guinea pigs. J Pharmacol Exp Ther 193:246–255

Heilig C, Stromski M, Gullans S (1989) Methylamine and polyol responses to salt loading in renal inner medulla. Am J Physiol 257:F1117–F1123

Heinz F, Schlegel F, Krause PH (1975) Enzymes of fructose metabolism in human kidney. Enzyme 19:85–92

Hers HG (1960) L'aldose-reductase. Biochim Biophys Acta 37:120–126

Hoffmann EK (1985) Regulatory volume decrease in Erlich ascites tumor cells: role of inorganic ions and amino compounds. Mol Physiol 8:167–184

Holub BJ (1986) Metabolism and function of myo-inositol and inositol phospholipids. Annu Rev Nutr 6:563–597

Howard CF, Anderson L (1967) Metabolism of myo-inositol in animals. II. Complete catabolism of myo-inositol-14C by rat kidney slices. Arch Biochem Biophys 118:332–339

Jamison RL, Kriz W (1982) Urinary concentrating mechanism: structure and function. Oxford Univ Press, Oxford

Kanfer JN, McCartney DG (1989) GPC phosphodiesterase and phosphomonoesterase activities of renal cortex and medulla of control, antidiuresis and diuresis rats. FEBS Lett 257: 348–350

Kelley MJ, Bailis AM, Henry SA, Carman GM (1988) Regulation of phospholipid biosynthesis in *Saccharomyces cerevisiae by inositol.* J Biol Chem 263:18078–18085

Kwon HM, Yamauchi A, Garcia-Perez A, Burg MB, Handler JS (1990a) Evidence that hypertonicity increases MDCK cell content of mRNA for the Na/myo-inositol cotransporter. Clin Res 38:442A

Kwon HM, Yamauchi A, Preston AS, Garcia-Perez A, Burg MB, Handler JS (1990b) Expression cloning of cDNA for the renal sodium/myo-inositol cotransporter. Kidney Int 37:584

Lewin LM, Yannai Y, Sulimovici S, Kraicer PF (1976) Studies on the metabolic role of myo-inositol. Distribution of radioactive myo-inositol in the male rat. Biochem J 156:375–380

Lohr J, Acara M (1990) Effect of dimethylaminoethanol, an inhibitor of betaine production, on the disposition of choline in the rat kidney. J Pharmacol Exp Ther 252:154–158

Mandersloot JG, Roelofsen B, Grier DJ (1978) Phosphatidylinositol as the endogenous activator of the (Na + K)-ATPase in microsomes of rabbit kidney. Biochim Biophys Acta 508: 478–485

Mann PJG, Quastel JH (1937) CXVI. The oxidation of choline by rat liver. Biochem J 31:869–878

Moriyama T, Garcia-Perez A, Burg M (1989) High extracellular NaCl stimulates synthesis of aldose reductase, an osmoregulatory protein, in renal medullary cells. Kidney Int 35:499
Moriyama T, Garcia-Perez A, Burg MB (1990a) High urea induces accumulation of different organic osmolytes than does high NaCl in renal medullary cells. Kidney Int 37:586
Moriyama T, Garcia-Perez A, Burg MB (1990b) Factors affecting the ratio of the different organic osmolytes in renal medullary cells. Am J Physiol 259:F847–F858
Moyer JD, Malinoswki N, Napier EA, Strong J (1988) Uptake and metabolism of myo-inositol by L1210 leukemia cells. Biochem J 254:95–100
Nakanishi T, Burg MB, (1989a) Osmoregulation of glycerophosphorylcholine content of mammalian renal cells. Am J Physiol 257:C795–C801
Nakanishi T, Burg MB (1989b) Osmoregulatory fluxes of myo-inositol and betaine in renal cells. Am J Physiol 257:C964–C970
Nakanishi T, Balaban RS, Burg MB (1988) Survey of osmolytes in renal cell lines. Am J Physiol 255:C181–C191
Nakanishi T, Turner RJ, Burg MB (1989a) Osmoregulatory accumulation of betaine by MDCK cells in hyperosmotic medium involves increased betaine transport. Kidney Int 35:499
Nakanishi T, Turner RJ, Burg MB (1989b) Osmoregulatory changes in myo-inositol transport by renal cells. Proc Natl Acad Sci USA 86:6002–6006
Nakanishi T, Turner RJ, Burg MB (1990) Osmoregulation of betaine transport in mammalian renal medullary cells. Am J Physiol 258:F1061–F1067
Robey RB, Ellis DE, Kwon HM, Handler JS, Garcia-Perez A, Burg MB (1990) Expression of $Na^+$-dependent betaine uptake in *Xenopus* oocytes injected with mRNA from osmotically stressed MDCK cells. J Am Soc Nephrol 1:705
Sands JM, Schrader DC (1990) Regulation of intracellular sorbitol in microdissected rat inner medullary collecting duct segments by state of hydration. Clin Res 38:22A
Seifter JL, Rivera J, Hughes R, Heilig C, Gullans S (1990) Properties of a $Na^+$-dependent myo-inositol (MI) transporter in vesicles from rat renal outer medulla. Kidney Int 37:588
Shayman JA, Wu D (1990) Myo-inositol does not modulate PI turnover in MDCK cells under hyperosmolar conditions. Am J Physiol 258:F1282–F1287
Siebens A, Spring K (1989) A novel sorbitol transport mechanism in cultured renal papillary epithelial cells. Am J Physiol 257:F937–F946
Smardo FM Jr, Burg MB, Garcia-Perez A (1990) Osmotic regulation of kidney aldose reductase (AR) gene transcription. FASEB J 4:A2096
Sung CP, Johnstone RM (1969) Evidence for the existence of separate transport mechanisms for choline and betaine in rat kidney. Biochim Biophys Acta 173:548–553
Troyer DA, Schwertz DW, Kreisberg JI, Ventkatachalam MA (1986) Inositol phospholipid metabolism in the kidney. Ann Rev Physiol 48:51–71
Uchida S, Garcia-Perez A, Murphy H, Burg MB (1989) Signal for induction of aldose reductase in renal medullary cells by high external NaCl. Am J Physiol 256:C614–C620
Ullrich KJ (1959) Glycerylphosphorylcholinumsatz und glycerylphosphorylcholindiesterase in der Saugetier-Niere. Biochem Z 331:98–102
Wald H, Scherzer P, Popovtzer MM (1989) Inhibition of thick ascending limb $Na^+$-$K^+$-ATPase activity in salt loaded rats by furosemide. Am J Physiol 256:F549–F555
Wirthensohn G, Guder WG (1982) Studies on renal choline metabolism and phosphaditylcholine synthesis. In: Morel F (ed) Biochemistry of Kidney Functions. INSERM Symp 21, Elsevier Biomed Press, Amsterdam p 119
Wirthensohn G, Lefrank S, Schmolke M, Guder W (1989) Regulation of organic osmolyte concentrations in tubules from rat renal inner medulla. Am J Physiol 256:F128–F135
Wolff SD, Stanton TS, James SL, Balaban RS (1989) Acute regulation of the predominant organic solutes of the rabbit renal inner medulla. Am J Physiol 257:F676–F681
Wright SH, Wunz TM (1989) Na-dependent betaine transport in renal brush border membrane vesicles (BBMV). FASEB J 3:A858
Yamauchi A, Kwon HM, Uchida S, Preston AP, Handler JS (1990) The myo-inositol transporters regulated by tonicity are basolateral in cultured kidney (MDCK) cells. Clin Res 38:274A

Yancey PH (1988) Osmotic effectors in kidneys of xeric and mesic rodents: corticomedullary distributions and changes with water availability. J Comp Physiol 158:369–380

Yancey PH, Haner RG, Freudenberger TH (1990) Effects of an aldose reductase inhibitor on organic osmotic effectors in rat renal medulla. Am J Physiol 259:F733–F738

Zablocki K, Miller SPF, Garcia-Perez A, Burg MB (1990) Role of synthesis and degradation of glycerophosphorylcholine (GPC) in its osmotic regulation by renal medullary cells. FASEB J 4:A2059

Chapter 4

# Compatible Solute Synthesis and Compartmentation in Higher Plants

A.D. Hanson

## Introduction

This chapter presents a brief overview of the compartmentation, biosynthesis and degradation of compatible solutes in higher plants, using selected examples to illustrate the present state of knowledge and to highlight some unanswered questions. Reviews by Wyn Jones (1984), Rhodes (1987), and McCue and Hanson (1990b) also cover these topics, and are sources of additional literature citations (see also Somero, this Vol.).

Figure 1 presents a schematic comparison of what is known about the compatibility and compartmentation of various organic solutes accumulated by higher plants, about the metabolism of these solutes, and about the regulation of accumulation. The first point to note is the lack of experimental data on whether, and to what extent, some of the solutes are compatible. This applies to polyols and tertiary sulfonium compounds, and also to quaternary ammonium compounds other than glycine betaine. Figure 1 also indicates that compartmentation and metabolism have been thoroughly investigated only for glycine betaine and proline, and that control of accumulation is not well understood for any type of compatible solute. However, a little information on control is available for proline, and somewhat more for glycine betaine.

Accordingly, this chapter first uses glycine betaine to discuss compartmentation, metabolism and regulation. Then, some additional general questions about the function and evolution of compatible solutes in plants are raised, using as illustrations some quaternary compounds accumulated by species of the genus *Limonium*.

## Glycine Betaine Compartmentation and Metabolism

### *Compartmentation*

This has been investigated in members of the Chenopodiaceae using several methods: (1) cytochemical localization in freeze-substituted shoot sections (Hall et al. 1978); (2) analysis of shoot tissues differing in degree of vacuolation (Gorham and

MSU-DOE Plant Research Laboratory, Michigan State University, East Lansing, MI 48824, USA

Somero et al. (Eds.)
Water and Life

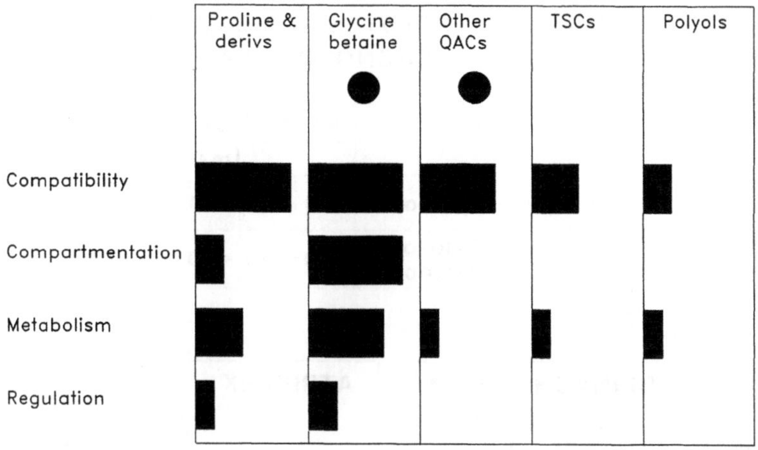


**Fig. 1.** Comparison of the information available in four areas for some putative compatible solutes in higher plants. The *sizes of the bars* are based on the relative amounts of literature in each area. The *filled circles* denote solutes discussed in this chapter. *QACs*, quaternary ammonium compounds; *TSCs*, tertiary sulfonium compounds. Proline derivatives include proline betaine and hydroxyproline betaines

Wyn Jones 1983; Stumpf and O'Leary 1985); (3) analysis of isolated chloroplasts (Robinson and Jones 1986; Schröppel-Meier and Kaiser 1988) and vacuoles (Matoh et al. 1987). The composite picture from such studies is that glycine betaine accumulates mainly in extravacuolar compartments, that its concentration in these compartments in moderately salinized plants (200–300 mM NaCl) reaches at least 200–300 mM, and that the chloroplasts alone can contain 20–40% of the total leaf glycine betaine (Fig. 2).

Many questions remain about glycine betaine compartmentation. It is not known whether the above picture for salinized Chenopodiaceae can be extrapolated to glycine betaine accumulators from other families (e.g., Poaceae, Asteraceae, Solanaceae), or to drought-stressed plants. The levels of glycine betaine accumulated by mitochondria, nongreen plastids and other organelles have not been determined. Most importantly, nothing is known about the transport systems responsible for the cytoplasmic localization of glycine betaine, and for its export from the site of synthesis in the chloroplast (see below).

### *Biosynthesis and Degradation*

In vivo isotope labeling studies with species from several families (Chenopodiaceae, Amaranthaceae, Convolvulaceae, Solanaceae, Poaceae) confirm that glycine betaine is synthesized in higher plants by a two step oxidation of choline, via betaine aldehyde (Rhodes 1987; Weretilnyk et al. 1989). In vivo radiolabeling experiments (Ladyman et al. 1980; Hanson and Wyse 1982) and chemical analyses (Ahmad and

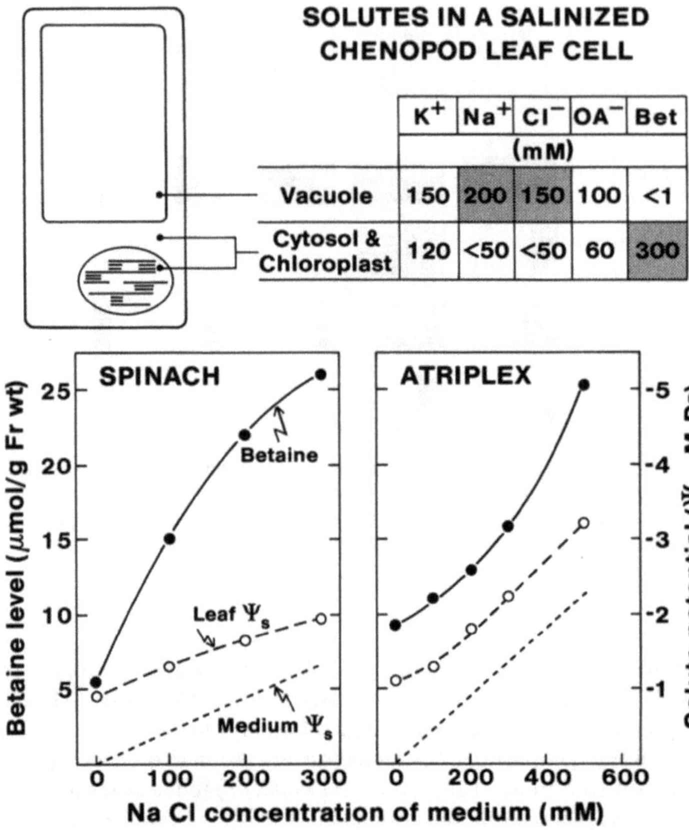


**Fig. 2.** Glycine betaine as a compatible cytoplasmic solute in plant cells. The lower frames illustrate typical patterns of osmotic adjustment and betaine accumulation in leaves of salt stressed plants of the family Chenopodiaceae. Spinach (*Spinacia oleracea*) is moderately salt tolerant, and *Atriplex spongiosa* is a halophyte. In both species, as salt is added to the growth medium, the leaves accumulate solutes, maintaining their solute potential below that of the medium by a fairly constant 0.6 to 1 MPa. This differential provides the driving force for growth and water uptake. The major solutes which accumulate in salinized leaves are $Na^+$, $Cl^-$, organic acids ($OA^-$), and glycine betaine (Bet). Salinization causes relatively little change in vacuolar and cytoplasmic $K^+$ levels. Analyses of the levels of these solutes in various cell compartments are shown schematically in the upper frame, for a leaf cell when salinized of these Chenopodiaceae with about 200 mM NaCl. As indicated by the shaded boxes, $Na^+$ and $Cl^-$ accumulate mainly in the vacuole, and betaine mainly in the cytoplasm (chloroplasts + cytosol). The vacuole occupies about 90% of the cell volume, and the chloroplast and cytosol about 5% each. Hence osmotic adjustment of the bulk of the cell water is achieved with $Na^+$ and $Cl^-$, which are readily available from the salinized growing medium, but are toxic to metabolism. Glycine betaine, which is non-toxic but energetically expensive, is used for osmotic adjustment only in the crucial metabolic compartments. [Sources of data are given in the text and in McCue and Hanson (1990b).]

Wyn Jones 1979; Ladyman et al. 1980; McDonnell and Wyn Jones 1988) also show that glycine betaine is not appreciably degraded by species from the Poaceae and Chenopodiaceae, but is readily translocated in the phloem.

The enzymes of biosynthesis have been studied in the Chenopodiaceae spinach and sugar beet, where they are chloroplastic (Hanson et al. 1985). As shown in Fig. 3, the first is a ferredoxin-dependent choline monooxygenase (CMO), which converts choline to betaine aldehyde hydrate (Brouquisse et al. 1989; Lerma et al. 1988). CMO is localized in the chloroplast stroma (Brouquisse et al. 1989). The dependence on reduced ferredoxin leads to a marked stimulation of choline oxidation by light in both intact leaves (Lerma et al. 1988) and isolated chloroplasts (Weigel et al. 1988). In maize, Rhodes and Rich (1988) have identified a naturally occurring recessive mutation which produces betaine deficiency. Although CMO activities in the wild-type and mutant have not yet been tested, in vivo radiolabeling studies locate the biochemical lesion at the choline → betaine aldehyde step (Lerma et al. 1990).

The second enzyme of betaine synthesis is a substrate-specific betaine aldehyde dehydrogenase (BADH) with a marked preference for $NAD^+$; it has been purified to homogeneity from spinach and antibodies prepared (Arakawa et al. 1987; Pan 1988; Weretilnyk and Hanson 1989). Spinach BADH exists as a dimer of identical ~60 kDa subunits which are encoded by a nuclear gene (Weretilnyk and Hanson 1988; Weretilnyk and Hanson 1989). Glycine betaine accumulators from several diverse families contain BADHs immunologically related to the spinach enzyme (Weretilnyk et al. 1989). Most of the BADH activity in spinach resides in a stromal isozyme, although there may also be a minor cytosolic form (Weigel et al. 1986; Weretilnyk and Hanson 1988). BADH cDNA clones containing the entire coding region have been isolated from spinach (Weretilnyk and Hanson 1990) and sugar beet (McCue and Hanson 1990a).

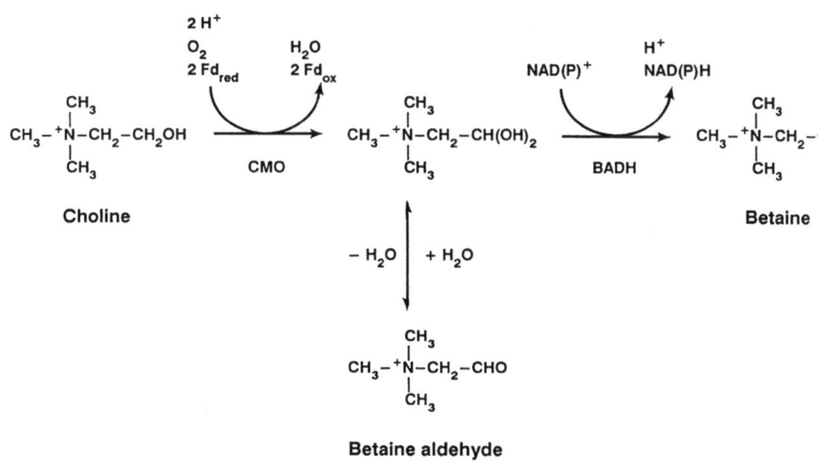


**Fig. 3.** Pathway for glycine betaine synthesis. The product of the CMO reaction is the hydrate (*gem* diol) form of betaine aldehyde, which is in spontaneous equilibrium with the carbonyl form

Knowledge of synthesis and degradation of glycine betaine is lacking in several areas. CMO has not yet been purified and thoroughly characterized. It is not known whether CMO occurs in plants other than Chenopodiaceae. These are important points because CMO is an unusual enzyme, quite unlike the dehydrogenases or oxidases which catalyze choline oxidation in other organisms. The apparent lack of metabolism of glycine betaine in plants is also intriguing because many other organisms can use glycine betaine as a methyl donor or as a nitrogen and carbon source.

### *Regulation of Glycine Betaine Accumulation*

The activities of both CMO and BADH are increased several-fold by gradual salinization or by salt shock in spinach leaves (Weigel et al. 1986; Brouquisse et al. 1989). In the case of BADH, use of antibody and cDNA probes has shown this rise in activity to reflect an increase in BADH protein level (Weretilnyk and Hanson 1989), and to be associated with the accumulation of BADH mRNA (Weretilnyk and Hanson 1990). These observations indicate that regulation of glycine betaine accumulation involves control of gene expression, and raise several questions about the mechanism of this control. The first questions concern the nature of the signal perceived by the stressed leaf. Three possibilities are: (1) a decline in water status (turgor) transmitted from the root via the xylem water potential; (2) a biochemical signal synthesized or released by the roots and transported upwards; (3) an increase in extra- or intracellular NaCl concentration or ionic strength due to a higher flux of NaCl arriving in the xylem stream. Experimental evidence does not yet rule any of these in or out. However, it is worth noting that gradual salinization protocols (50-mM increments at 3-day intervals) elicit maximal responses, yet would be expected to have only small effects on turgor. Also, experiments with sugar beet show that salinization induces BADH mRNA accumulation in roots as well as leaves (McCue and Hanson, unpublished), which suggests that an inter-organ biochemical messenger may not be needed. These considerations focus attention on possibility (3) – response to $Na^+$, $Cl^-$ or ionic strength. Note that although ideas about perception of osmotic stress in plants are currently dominated by possibilities (1) and (2), there is evidence that specific ions, or ionic strength, are involved in osmoregulation of gene expression in bacteria and animal cells (Csonka 1989; Uchida et al. 1989).

## Quaternary Ammonium Compounds in *Limonium* spp.

### *Occurrence and Compatibility*

*Limonium* spp. (Plumbaginaceae) are among the most salt-tolerant of higher plants; they have salt glands on the leaves which are capable of massive NaCl excretion (Lüttge 1975). Figure 4 shows the major quaternary ammonium compounds found in various *Limonium* spp. Some species (*L. perezii, L. sinuatum*) accumulate glycine betaine, whereas others accumulate β-alanine betaine; these include *L. vulgare* (Larher and Hamelin 1975), *L. nashii*, and *L. latifolium*. Thus, in some species the

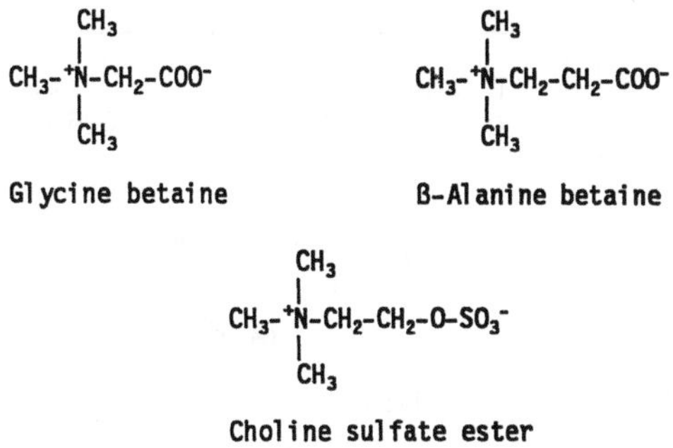


**Fig. 4.** Quaternary ammonium compounds accumulated by members of the genus *Limonium* (Plumbaginaceae)

prevalent angiosperm betaine is replaced by its higher homolog. All species tested also accumulate the sulfate ester of choline. Choline sulfate has previously been found in salt-excreting mangrove species, but not in those unable to excrete salt (Benson and Atkinson 1967). Both β-alanine betaine and choline sulfate behave as compatible solutes comparable in effectiveness to glycine betaine when bioassayed in *Escherichia coli* (Fig. 5). Taken together, these results raise the three following questions.

### *Are All Betaines Functionally Equivalent?*

Because accumulation of large amounts of glycine betaine occurs widely among diverse higher plant families, and because the capacity to synthesize small amounts of betaine is even more widespread (Wyn Jones 1984; Weretilnyk et al. 1989), glycine betaine could be considered the archetypal angiosperm betaine. It might therefore be argued that glycine betaine accumulation in the genus *Limonium* is a primitive condition, and that acquisition of the ability to accumulate β-alanine betaine is an evolutionary advance. It is reasonable (although not necessary) to invoke some functional advantage of β-alanine betaine over glycine betaine to account for this. Such an advantage could be physicochemical or metabolic; respective examples would be superior exclusion from protein surfaces (Low 1985), or readier reclamation of the nitrogen for use in growth following stress relief.

### *Do S-Containing Compatible Solutes Have Dual Roles?*

It may not be fortuitous that choline sulfate accumulation has so far been found only in plants with salt glands. Although the salt glands of *Limonium* spp. and mangroves secrete large amounts of $Cl^-$ salts they are unable to secrete $SO_4^{2-}$ (Lüttge 1975; Popp 1984). Because seawater contains approximately 30 mM $SO_4^{2-}$ in addition to 500 mM $Cl^-$, and only the $Cl^-$ entering the shoot can be excreted by the salt glands, $SO_4^{2-}$ accumulation in leaves can result (Popp 1984). As $SO_4^{2-}$ is toxic to photosynthesis (Kaiser et al. 1986), its esterification to choline may achieve two ends: detoxification and synthesis of a beneficial osmolyte. Excess sulfur in salt marsh plants may also

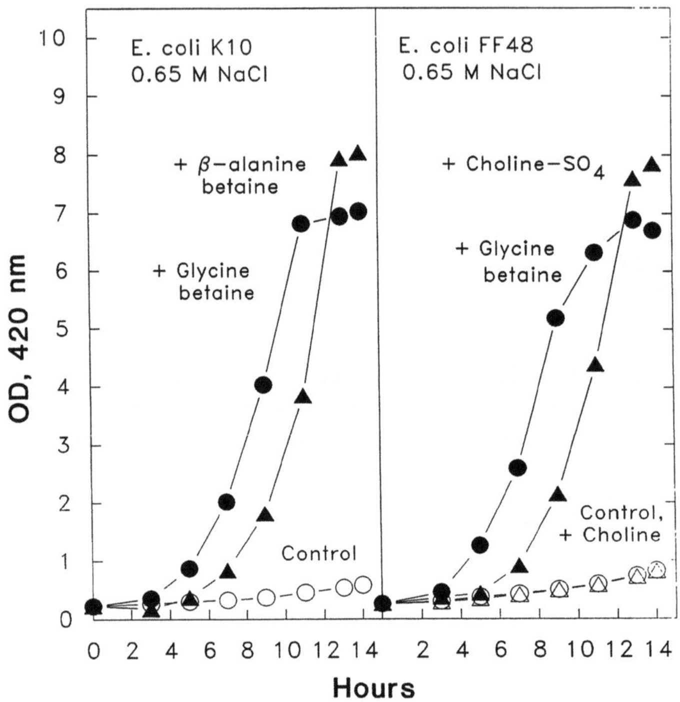


**Fig. 5.** Evidence that β-alanine betaine and choline sulfate are compatible solutes: stimulation of the growth of *E. coli* in minimal medium containing 0.65 M NaCl. Both compounds decreased the doubling time to approximately 1.6 h, as did the benchmark compatible solute, glycine betaine. Choline sulfate was tested in strain FF48, which lacks choline dehydrogenase (Styrvold et al. 1986), to exclude the possibility of growth stimulation due to hydrolysis of choline sulfate followed by oxidation of choline to betaine

come from uptake of sulfide, which can be oxidized to $SO_4^{2-}$ within the plant (Carlson and Forrest 1982; van Diggelen et al. 1986), and so contribute to $SO_4^{2-}$ accumulation. Note that similar considerations about possible dual roles apply to dimethylsulfoniopropionate (van Diggelen et al. 1986) and other sulfonium compounds.

### *Has Cross-Species Gene Transfer Occurred in Osmolyte Evolution?*

The accumulation of β-alanine betaine in some species of *Limonium*, and the absence of this unusual compound from others suggests the acquisition of novel genetic information during evolution of this genus. β-Alanine betaine is probably synthesized via methylation of β-alanine (Larher 1976), which is ubiquitous in plants (as a moiety of coenzyme A and acyl carrier protein), so that the novel genetic information required is presumably for a methyltransferase or methyltransferases. Because β-alanine betaine has not been reported from higher plants other than

*Limonium* and the related genus *Armeria*, but is known in several marine algae (Blunden et al. 1981) it is tempting to speculate that the methyltransferase(s) were acquired by lateral gene transfer from an alga. If β-alanine betaine indeed has some functional advantage over glycine betaine in saline environments (see above), such an acquisition might have an immediate selective advantage that would favor its spread through a population. A parallel argument can be made for dimethylsulfoniopropionate, which also appears to have a very limited distribution in angiosperms, and to occur quite widely in marine algae (Wyn Jones 1984).

## References


Ahmad N, Wyn Jones RG (1979) Glycinebetaine, proline and inorganic ion levels in barley seedlings following transient stress. Plant Sci Lett 15:231–237

Arakawa K, Takabe T, Sugiyama T, Akazawa T (1987) Purification of betaine-aldehyde dehydrogenase from spinach and preparation of its antibody. J Biochem 101:1455–1488

Benson AA, Atkinson MR (1967) Choline sulfate and phosphate in salt excreting plants. Fed Proc 26:394

Blunden G, El Barouni MM, Gordon SM, McLean WFH, Rogers DG (1981) Extraction, purification and characterization of Dragendorff-positive compounds from some British marine algae. Bot Mar 24:451–456

Brouquisse R, Weigel P, Rhodes D, Yocum CF, Hanson AD (1989) Evidence for a ferredoxin-dependent choline monooxygenase from spinach chloroplast stroma. Plant Physiol 90:322–329

Carlson PR, Forrest J (1982) Uptake of dissolved sulfide by *Spartina alterniflora*: evidence from natural sulfur isotope abundance ratios. Science 216:633–635

Csonka LN (1989) Physiological and genetic responses of bacteria to osmotic stress. Microbiol Rev 53:121–147

Gorham J, Wyn Jones RG (1983) Solute distribution in *Suaeda maritima*. Planta 157:344–349

Hall JL, Harvey DMR, Flowers TJ (1978) Evidence for the cytoplasmic location of betaine in leaf cells of *Suaeda maritima*. Planta 140:59–62

Hanson AD, Wyse R (1982) Biosynthesis, translocation, and accumulation of betaine in sugar beet and its progenitors in relation to salinity. Plant Physiol 70:1191–1198

Hanson AD, May AM, Grumet R, Bode J, Jamieson GC, Rhodes D (1985) Betaine synthesis in chenopods: localization in chloroplasts. Proc Natl Acad Sci USA 82:3678–3682

Kaiser WM, Schröppel-Meier G, Wirth E (1986) Enzyme activities in artificial stroma medium. Planta 167:292–299

Ladyman JAR, Hitz WD, Hanson AD (1980) Translocation and metabolism of glycine betaine by barley plants in relation to water stress. Planta 150:191–196

Larher F (1976) Sur quelques particularités du métabolisme azoté d'une halophyte: *Limonium vulgare* Mill. PhD Thesis, Univ Rennes, France

Larher F, Hamelin J (1975) L'acide β-triméthylaminopropionique des rameaux de *Limonium vulgare* Mill. Phytochemistry 14:205–207

Lerma C, Hanson AD, Rhodes D (1988) Oxygen-18 and deuterium labeling studies of choline oxidation by spinach and sugar beet. Plant Physiol 88:695–702

Lerma C, Bolaños J, Rhodes D, Hanson AD (1990) Betaine deficiency in maize: metabolic basis and relation to osmotic adjustment. Plant Physiol 93S:108

Low PS (1985) Molecular basis of the biological compatibility of nature's osmolytes. In: Gilles R, Gilles-Baillien M (eds) Transport processes, iono- and osmoregulation. Springer, Berlin Heidelberg New York, p 469

Lüttge U (1975) Salt glands. In: Baker DA, Hall JL (eds) Ion transport in plant cells and tissues. North-Holland, Amsterdam Oxford, p 335

Matoh T, Watanabe J, Takahashi E (1987) Sodium, potassium, chloride and betaine concentrations in isolated vacuoles from salt-grown *Atriplex gmelinii* leaves. Plant Physiol 84:173–177

McCue KF, Hanson AD (1990a) cDNA cloning and analysis of betaine aldehyde dehydrogenase, a salt-inducible enzyme in sugar beet. Plant Physiol 93S:98

McCue KF, Hanson AD (1990b) Drought and salt tolerance: towards understanding and application. Trends Biotechnol 8:358–362

McDonnell E, Wyn Jones RG (1988) Glycinebetaine biosynthesis and accumulation in unstressed and salt-stressed wheat. J Exp Bot 39:421–430

Pan S-M (1988) Betaine aldehyde dehydrogenase in spinach. Bot Bull Acad Sin 29:255–263

Popp M (1984) Chemical composition of Australian mangroves. I. Inorganic ions and organic acids. Z Pflanzenphysiol 113:395–409

Rhodes D (1987) Metabolic responses to stress. In: Davies DD (ed) The biochemistry of plants, vol 12. Academic Press, Lond New York, p 201

Rhodes D, Rich PJ (1988) Preliminary genetic studies of the phenotype of betaine deficiency in *Zea mays* L. Plant Physiol 88:102–108

Robison SP, Jones GP (1986) Accumulation of glycinebetaine in chloroplasts provides osmotic adjustment in salt stress. Aust J Plant Physiol 13:659–668

Schröppel-Meier G, Kaiser WM (1988) Ion homeostasis in chloroplasts under salinity and mineral deficiency. Plant Physiol 87:822–827

Stumpf DK, O'Leary JW (1985) The distribution of $Na^+$, $K^+$ and glycinebetaine in *Salicornia bigelovii*. J Exp Bot 36:550–555

Styrvold OB, Falkenberg P, Landfald B, Eshoo MW, Bjørnsen T, Strøm AR (1986) Selection, mapping and characterization of osmoregulatory mutants of *Escherichia coli* blocked in the choline-glycine betaine pathway. J Bacteriol 165:856–863

Uchida S, Garcia-Perez A, Murphy H, Burg M (1989) Signal for induction of aldose reductase in renal medullary cells by high external NaCl. Am J Physiol 256:C614–C620

van Diggelen J, Rozema J, Dickson DMJ, Broekman R (1986) β-Dimethylsulfoniopropionate, proline and quaternary ammonium compounds in *Spartina anglica* in relation to sodium chloride, nitrogen and sulphur. New Phytol 103:573–586

Weigel P, Weretilnyk EA, Hanson AD (1986) Betaine aldehyde oxidation by spinach chloroplasts. Plant Physiol 82:753–759

Weigel P, Lerma C, Hanson AD (1988) Choline oxidation by intact spinach chloroplasts. Plant Physiol 86:54–60

Weretilnyk EA, Hanson AD (1988) Betaine aldehyde dehydrogenase polymorphism in spinach: genetic and biochemical characterization. Biochem Genet 26:143–151

Weretilnyk EA, Hanson AD (1989) Betaine aldehyde dehydrogenase from spinach leaves: purification, in vitro translation of the mRNA, and regulation by salinity. Arch Biochem Biophys 271:56–63

Weretilnyk EA, Hanson AD (1990) Molecular cloning of a plant betaine-aldehyde dehydrogenase, an enzyme implicated in adaptation to salinity and drought. Proc Natl Acad Sci USA 87:2745–2749

Weretilnyk EA, Bednarek S, McCue KF, Rhodes D, Hanson AD (1989) Comparative biochemical and immunological studies of the glycine betaine pathway in diverse families of dicotyledons. Planta 178:342–352

Wyn Jones RG (1984) Phytochemical aspects of osmotic adaptation. Recent Adv Phytochem 18:55–78

Chapter 5

# Osmotic Control of Transcription of the *proU* Operon of *Salmonella typhimurium*

D.G. Overdier[1,2], S. Fletcher[1], and L.N. Csonka[1]

## Introduction

Organisms generally respond to increases or decreases in the external osmolarity by parallel increases or decreases in the intracellular concentrations of solutes. As a result of this adaptive process, cells can maintain their volumes within relatively narrow limits despite large variations in the external osmolarity (Cram 1976; Yancey et al. 1982). Because the solutes used to regulate the intracellular osmolarity can be present at high concentrations, they need to be relatively inert with respect to their effects on macromolecular functions, and for this reason they have been termed compatible solutes (Brown and Simpson 1972).

Like organisms from all other kingdoms, bacteria can carry out processes of osmotic adaptation. Because of their structural simplicity as unicellular organisms and their ready amenability to genetic manipulation, bacteria have been very useful for the elucidation of general osmoregulatory processes (Csonka 1989). Only a limited number of compatible solutes are found in bacteria: $K^+$, glutamate, proline, glycinebetaine and other related zwitterionic quaternary amines, and trehalose (see references in Csonka 1989).

The enteric bacteria *Salmonella typhimurium* and *Escherichia coli* accumulate proline and glycinebetaine to high concentrations in media of high osmolarity by transport from the medium (Britten and McClure 1962; Csonka 1983; LeRudulier and Bouillard 1983). Although these organisms can synthesize proline to meet the needs for protein synthesis, they are incapable of producing it at a sufficient level to function as a compatible solute (Csonka 1981). *E. coli* can synthesize glycinebetaine only from exogenous choline, and *S. typhimurium* cannot synthesize this quaternary amine at all (LeRudulier and Bouillard 1983; Andersen et al. 1988). Therefore, these organisms are dependent on transport for the accumulation of these two compounds, and these two compounds, when present in the growth medium, exert a profound stimulatory effect on the growth rate of enteric bacteria in media of high osmolarity (Christian 1955; LeRudulier and Bouillard 1983). Proline and glycinebetaine are taken up by *S. typhimurium* and *E. coli* via two permeases: the ProP and the ProU

[1]Department of Biological Sciences, Purdue University, West Lafayette, IN 47907, USA
[2]Present address: Department of Biochemistry, A-312, College of Medicine West, The University of Illinois at Chicago, Chicago, IL 60612, USA

Somero et al. (Eds.)
Water and Life

systems. These were first discovered as transport systems for proline (Csonka 1983; Ratzkin et al. 1978; Stalmach et al. 1984), but subsequently they were shown to have a higher affinity for glycinebetaine as a substrate than for proline (Cairney et al. 1985a,b; May et al. 1986). The activities of both of these transport systems are regulated by osmotic stress. The former system, which is synthesized at a nearly invariant level, is stimulated by high osmolarity by some post-transcriptional mechanism (Cairney et al. 1985a; Dunlap and Csonka 1985; Milner et al. 1988), while the latter one is activated by high osmolarity as a result of a several-hundred-fold induction of transcription of the *proU* operon which encodes its components (Cairney et al. 1985b; Dunlap and Csonka 1985; Gowrishankar 1985).

Besides the *proU* operon, there are only a few genes in enteric bacteria whose transcription is under osmotic control. The greatest range of osmotic control (>1000-fold) is exhibited by the *kdpABC* operon, which specifies the components of a high affinity $K^+$ transport system (Laimins et al. 1981). The *ompF* and *ompC* genes, which encode two porins or channel proteins for the outer membrane, are subjecct to a reciprocal osmotic control, with the transcription of the *ompC* gene being enhanced and that of the *ompF* gene diminished in media of low osmolarity, (Hall and Silhavy 1981). The *bet* operon, found in *E. coli* but not in *S. typhimurium*, which encodes proteins for the transport of choline and its conversion to glycinebetaine, is induced by high osmolarity (Eshoo 1988). In addition to these operons, there are about a dozen other genetic loci in *E. coli* which exhibit osmotic control (Gutierrez et al. 1987).

The regulatory proteins which control the transcription of the above genes are known in only two cases. The transcriptional control of the *ompC* and *ompF* genes is mediated by the products of the *envZ* and *ompR* genes (Hall and Silhavy 1981), and that of the *kdpABC* operon by the products of the *kdpD* and *kdpE* genes (Walderhaug et al. 1987). These proteins belong to a family of transcriptional regulatory proteins (Ronson et al. 1987; Epstein et al. 1990) having extensive amino acid sequence similarities. The *envZ* and *kdpD* gene products are located in the membrane, and they were proposed to sense the osmolarity of the medium (Igo et al. 1990) and the turgor pressure exerted by the membrane (Walderhaug et al. 1987, respectively. The *envZ* gene product has been shown to be a kinase which can phosphorylate the *ompR* gene product, the transcriptional activator of the *ompC* and *ompF* genes (Igo et al. 1990). By analogy, a similar mechanism was proposed to accomplish the transcriptional control of the *kdpABC* operon by the *kdpD* and *kdpE* proteins (Epstein et al. 1990), but this has not yet been demonstrated experimentally.

Although there is a great deal of biochemical information on the transcriptional activation of the *kdpABC* and especially the *ompC* and *ompF* genes, the fundamental issue of how cells "sense" the external osmolarity is an open question. Our laboratory is pursuing this issue via our investigations of the mechanism that carries out the transcriptional regulation of the *proU* operon of *S. typhimurium*. This research is conducted with a bacterium, but because of the underlying similarities in the cellular responses of all organisms to osmotic stress, it is likely to provide insights that will be applicable to higher forms of life.

## Osmotic Control of Transcription of the *proU* Operon

The induction of the *proU* operon by hyperosmotic shock is very rapid; it can be detected within eight minutes after increases in the osmolarity of the medium (Dunlap and Csonka 1985). Increased transcription of the operon can be elicited by high concentrations of any solute which cannot diffuse across the cytoplasmic membrane, but not by solutes which can traverse the membrane without the aid of transport systems (such as glycerol, ethanol, or methanol) (Csonka 1983; Dunlap and Csonka 1985). As can be seen in Fig. 1, the steady state level of expression of the *proU* operon in exponentially growing *S. typhimurium* is dependent only on the osmotic strength (or water potential) of the growth medium but not on the nature of the inducing solute.

Like all Gram-negative bacteria, *S. typhimurium* possesses two cellular compartments: the cytoplasm, covered by the inner (or cytoplasmic) membrane, and the periplasmic space, surrounded by a second membrane, the outer membrane. The outer membrane has channels (made up of the OmpC and OmpF porin proteins) which enable solutes of molecular weight <600 to diffuse into the periplasm (Nikaido and Vaara 1987). Sucrose and NaCl can pass through the porins into the periplasmic space but they cannot traverse the cytoplasmic membrane, and polyethylene glycol-8000 is too large to pass into the periplasmic space. It is noteworthy that despite these differences in their permeability into the periplasmic space, these two types of solutes were equally effective, on an osmolar basis, in inducing the *proU* operon (Fig. 1).

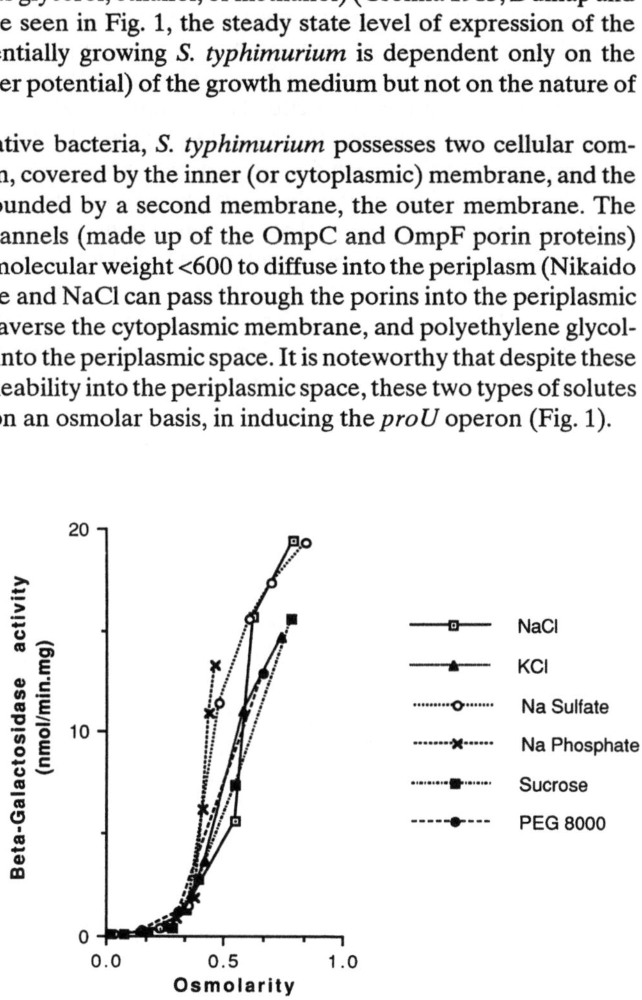


**Fig. 1.** The expression of the *proU* operon as a function of external osmolarity. These data were obtained with *S. typhimurium* strain TL671 (Φ*proU1844–lacZ*) grown in K medium containing solutes at the indicated concentrations for at least ten generations, and the β-galactosidase activity assayed in exponentially growing cells. For experimental details, see Overdier et al. (1989)

## The Structure of the Transcriptional Control Region of the *proU* Operon

The nucleotide sequence of the transcriptional control region of the *proU* operon of *S. typhimurium* was determined by Stirling et al. (1989) and by our group (Overdier et al. 1989), and that of *E. coli* by Gowrishankar (1989). The corresponding regions of the chromosome of the two organisms show >80% nucleotide identity, which suggests that the mechanism for the regulation of the *proU* operon is probably similar if not identical in the two species. With an in vitro deletion analysis of the promoter of the *proU* operon, we were able to narrow down the region required for normal osmotic control of expression of the *proU* operon to a 260 base pair DNA fragment, which carries residues from 60 bases upstream to 200 bases downstream of the transcription initiation site. The minimum size of the transcription control region may be even shorter, but we did not obtain deletion endpoints closer to the transcription start site which did not alter the expression of the *proU* promoter.

We made an interesting discovery concerning the control mechanism of the *proU* operon in that we found that deletion of sequences between nucleotides 33 and 200 downstream from the promoter resulted in high level of expresion of the *proU* promoter even in the absence of high osmolarity (Overdier et al. 1989; D. Overdier and L. Csonka, in preparation). This observation suggests that there is an important site with negative regulatory function which is located downstream of the promoter. The structure of the transcriptional control region of the *proU* operon is presented in a schematic form in Fig. 2.

A common form of transcriptional control at sites located downstream of the promoter in bacteria entails a transcriptional termination or pausing mechanism. In order to test whether these types of mechanism function in the *proU* operon, we placed the putative regulatory site of the operon downstream of the *tac* promoter, and it proved to be ineffective as a regulator of transcription originating at this promoter (D. Overdier and L. Csonka, in preparation). Therefore, the transcriptional control of the *proU* operon does not entail a transcription termination or pausing mechanism. We do not yet know how the regulatory site is involved in the osmotic control of expression of the *proU* operon, but we are considering the possibility that it may be the binding site for some transcriptional repressor protein which can interact with RNA polymerase to inhibit initiation of transcription at the *proU* promoter in media of low osmolarity but allow it in media of high osmolarity. Because this regulatory site is approximately 150 to 200 bases from the promoter, the latter type of mechanism probably constitutes an example of transcriptional control

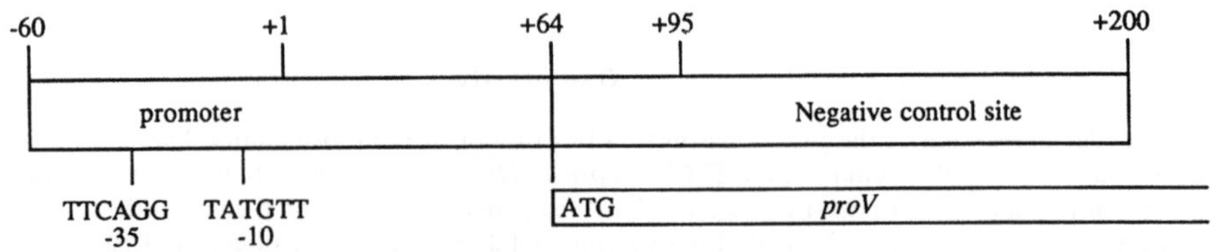


**Fig. 2.** The structure of the transcriptional control region of the *proU* operon of *S. typhimurium*

at a distance, which was observed in eukaryotes (Levine and Manley 1989). We are conducting experiments to test this suggestion.

## What Is the Osmoregulatory Signal for the *proU* Operon?

One of the interesting issues concerning the transcriptional regulation of the *proU* operon is the nature of the signal that the cells perceive for the osmotic control of transcription of the *proU* operon. The results shown in Fig. 1 were obtained with cells that were grown for at least ten generations in the media of indicated osmolarities, with the shift from low to high osmolarity having been carried out 12 to 36 h prior to the measurement of the expression of the *proU* operon. When bacteria are exposed to hyperosmotic shock, they lose water during the course of plasmolysis (Koch 1984), which has been suggested to be accompanied by a loss of turgor (Epstein 1986). The plasmolysis phase lasts only approximately 30 minutes, after which the cells are restored to close to their initial volume as a result of the accumulation of compatible solutes by transport and synthesis (Dinnbier et al. 1988). Because cell growth resumes after this period of osmotic adjustment, the return to normal cytoplasmic volume must be accompanied by a restoration of turgor to a value that is sufficient to maintain cell growth. The fact that the *proU* operon remains induced at a high level in cells that have completed osmotic adaptation raises the intriguing question of what is the biochemical signal for the continued high level expression of the *proU* operon which is present in cells that have adapted to media of high osmolarity but absent from cells grown in media of low osmolarity.

In exponentially growing *E. coli*, the intracellular concentrations of $K^+$ and glutamate are proportional to the osmolarity of the medium (Epstein and Schultz 1965; Richey et al. 1987). It has been proposed that the intracellular concentration of $K^+$ (perhaps together with that of glutamate) is the regulatory signal for the transcriptional regulation of the *proU* operon (Epstein 1986; Higgins et al. 1988; Sutherland et al. 1986), such that the recognition of the *proU* transcriptional control region by some regulatory protein or perhaps RNA polymerase is sensitive to the K-glutamate concentration. There is some evidence for this model, because the expression of the *proU* promoter in cell-free transcription translation systems was found to be stimulated by 0.3 M K-glutamate (Ramirez et al. 1989; Javanovich et al. 1989).

Although these in vitro studies of Ramirez et al. (1989) and Jovanovich et al. (1989) are consistent with K-glutamate being the regulatory signal for the expression of the *proU* promoter, we obtained some results which cast some doubt on the specificity of this interaction. In our tests of the effect of K-glutamate on the expression of the *proU* promoter in a cell-free transcription-translation system, we too found that ~0.3 M K-glutamate stimulated the expression of the intact *proU* promoter. However, the dependence of the expression of the *proU* promoter carried on a DNA fragment which also contained the downstream regulatory site on the $K^+$-glutamate concentration was identical to that of the *proU* promoter carried on a DNA fragment which did not include the downstream regulatory site (L. Csonka and S. Kustu, in preparation). Consequently, the stimulation of expression of the *proU* promoter by K-glutamate observed in vitro is not necessarily relevant to the in vivo control of the operon. K-glutamate has been shown to stimulate a number of biochemical processes, including the recognition of a number of promoters by RNA polymerase (Leirmo et al. 1987; Richey et al. 1987), and therefore the stimulation of the expression of the *proU* promoter by 0.3 M K-glutamate may be a manifestation of this generalized stimulatory effect of this solute.

## Is Turgor the Signal for the Regulation of the Intracellular $K^+$ Concentration?

While the model that the concentration of $K^+$ is the regulatory signal for the expression of the *proU* operon can explain the persistent induction of this operon in cells growing in media of high osmolarity, it raises a second question: how is the intracellular $K^+$ concentration regulated by the osmolarity of the growth medium? Epstein (1986) suggested that the intracellular concentration of this ion is governed by turgor pressure, such that a reduction of turgor upon plasmolysis resulting from an increase in external osmolarity causes an increase in the rate of $K^+$ uptake. As a consequence of this homeostatic regulation, the turgor pressure can be maintained within a critical range which is required for cell growth.

Turgor is the pressure exerted by the cytoplasmic membrane on the surrounding cell wall, and it is generated by the diffusion of water into the cells because of the fact that the osmolarity of the cytoplasm is greater than that of the medium. The model proposed by Epstein (1986) is based on the assumption that in Gram-negative bacteria, the cytoplasmic membrane is in contact with the cell wall (Walderhaug et al. 1987), as in Gram-positive species or in plant cells with rigid cell walls. However, for the sake of argument, we would like to present the possibility that this may not be the case in Gram-negative bacteria, because of the periplasmic space, which is situated between the cell wall and the cytoplasmic membrane. The periplasmic space constitutes a rather large fraction (20–40%) of the total cell volume, regardless of the osmolarity of the medium (Richey et al. 1987). As it was argued by Stock et al. (1977), the fact that the periplasm is separated from the cytoplasm by an elastic lipid bilayer implies that there cannot be a very large pressure difference between the two compartments. Because phospholipid bilayers cannot withstand a stress above 0.1 M Pa, we can estimate, according to the arguments made by Nobel (1983), that the difference in the pressure of the cytoplasm and the periplasm of a bacterium (radius = 0.5 µm) can be at most $3 \times 10^{-3}$ M Pa, which is insignificant compared to the 0.3–0.5 M Pa that is thought to be the intracellular pressure in Gram-negative bacteria (Csonka 1989). If our view of the structure of the periplasmic space is valid, then the cytoplasmic membrane itself cannot exert a turgor pressure on the cell wall, and therefore any proteins which are embedded into the cytoplasmic membrane, such as the osmoregulatory EnvZ and KdpD proteins, cannot detect fluctuations in turgor pressure. (In fact, if the above conclusion is correct, then the whole issue of what is meant by turgor pressure in a Gram-negative bacterium may have to be reexamined.)

Thus, the cytoplasmic compartment may behave as a bubble within a bubble, like the vacuoles within plant cells or the wall-less animal cells, whose surfaces cannot generate a turgor pressure. However, both vacuoles in plant cells and the cells of animals are capable of regulating their volumes in face of fluctuations in the external osmolarity (Kregenow 1981; Uchida et al. 1990; Gutknecht 1968). The ability of wall-less compartments or cells to carry out osmotic adaptation implies that changes in turgor pressure are not necessarily prerequisites for the triggering of the osmoregulatory responses, and therefore it is not inconceivable that some other signal besides turgor loss is the primary signal for osmoregulation in Gram-negative bacteria.

The osmoregulatory processes in animal cells appear to be directed at maintaining the cellular volume within a narrow range (Kregenow 1981). Since volume is related inversely to solute concentration, a possible mechanism for volume regulation is that the osmoregulatory processes are governed by protein(s) which sense the concentrations of some key signal molecules. A second possible mechanism could be that changes in volume are detected as changes in cell surface area. In this second possibility, some protein which is sensitive to the surface tension of the membrane [e.g., a stretch-activated ion channel (Martinac et al. 1987)] might be the primary osmoregulator which sets off the necessary chain of responses to maintain membrane surface area to a constant value. Conceivably, either of these two mechanisms could operate in Gram-negative bacteria in the regulation of the responses to osmotic stress.

*Acknowledgments.* We thank A. Hanson for stimulating discussions. The work presented here was funded by the U.S. Public Health Service under Grant #R01-GM3194401.

## References

Andersen PA, Kaasen I, Styrvold OB, Boulnois G, Strøm AR (1988) Molecular cloning, physical mapping and expression of *bet* genes governing the osmo-regulatory choline-glycinebetaine pathway of *Escherichia coli*. J Gen Microbiol 134:1737–1746

Britten RJ, McClure FT (1962) The amino acid pool in *Escherichia coli*. Bacteriol Rev 26:292–335

Brown AD, Simpson JR (1972) Water relations of sugar-tolerant yeast: the role of intracellular polyols. J Gen Microbiol 72:589–591

Cairney J, Booth IR, Higgins CF (1985a) *Salmonella typhimurium proP* gene encodes a transport system for the osmoprotectant betaine. J Bacteriol 164:1218–1223

Cairney J, Booth IR, Higgins CF (1985b) Osmoregulation of gene expression in *Salmonella typhimurium*: *proU* encodes an osmotically induced betaine transport system. J Bacteriol 164:1224–1232

Christian JHB (1955) The influence of nutrition on the water relations of *Salmonella oranienburg*. Aust J Biol Sci 8:75–82

Cram WJ (1976) Negative feedback of transport in cells: The maintenance of turgor, volume and nutrient supply. In: Lüttge U, Pitman MG (eds) Encyclopedia of plant physiology, New Series 2A. Springer, Berlin Heidelberg New York, pp 284–316

Csonka LN (1981) Proline over-production results in enhanced osmotolerance in *Salmonella typhimurium*. Mol Gen Genet 182:82–86

Csonka LN (1983) A third L-proline permease in *Salmonella typhimurium* which functions in media of elevated osmotic strength. J Bacteriol 151:1433–1443

Csonka LN (1989) Physiological and genetic responses of bacteria to osmotic stress. Microbiol Rev 53:121–147

Dinnbier U, Limpinsel E, Schmid R, Bakker EP (1988) Transient accumulation of potassium glutamate and its replacement by trehalose during adaptation of growing cells of *Escherichia coli* K-12 to elevated sodium chloride concentrations. Arch Microbiol 150:348–357

Dunlap VJ, Csonka LN (1985) Osmotic regulation of L-proline transport in *Salmonella typhimurium*. J Bacteriol 163:296–304

Epstein W (1986) Osmoregulation by potassium transport in *Escherichia coli*. FEMS Microbiol Rev 39:73–78

Epstein W, Schultz SG (1965) Cation transport in *Escherichia coli*. V. Regulation of cation content. J Gen Physiol 49:221–234

Epstein W, Walderhaug MO, Polarek JW, Hesse JH, Dorus E, Daniel JM (1990) The bacterial Kdp potassium ATPase and its relation to other transport ATPases, such as the sodium potassium ATPase and calcium ATPase in higher organisms. Philos Trans R Soc Lond B 326:479–487

Eschoo MW (1988) *lac* fusion analysis of the *bet* genes of *Escherichia coli*: regulation by osmolarity, temperature, oxygen, choline, and glycinebetaine. J Bacteriol 170:5208–5215

Gowrishankar J (1985) Identification of osmoresponsive genes in *Escherichia coli*: evidence for participation of potassium and proline transport systems in osmoregulation. J Bacteriol 164:434–445

Gowrishankar J (1989) Nucleotide sequence of the osmoregulatory *proU* operon of *Escherichia coli*. J Bacteriol 171:1923–1931

Gutierrez C, Barondess J, Manoil C, Beckwith J (1987) The use of transposon Tn *phoA* to detect genes for cell envelope proteins subject to a common regulatory stimulus. J Mol Biol 195:289–297

Gutknecht J (1968) Salt transport in *Valonia*: inhibition of potassium uptake by small hydrostatic pressures. Science 160:68–70

Hall MN, Silhavy TJ (1981) Genetic analysis of the *ompB* locus in *Escherichia coli* K-12. J Mol Biol 151:1–15

Higgins CF, Dorman CJ, Stirling DA, Wadell L, Booth IR, May G, Bremer E (1988) A physiological role for DNA supercoiling in the osmotic regulation of gene expression in *S. typhimurium* and *E. coli*. Cell 52:569–584

Igo MM, Slauch JM, Silhavy TJ (1990) Signal transduction in bacteria: kinases and control of gene expression. New Biol 2:5–9

Jovanovich S, Record MT Jr, Burgess RR (1989) In an *Escherichia coli*-coupled transcription-translation system, expression of the osmoregulated gene *proU* is stimulated at elevated potassium concentrations and by an extract from cells grown at high osmolality. J Biol Chem 264:7821–7825

Koch AL (1984) Shrinkage of growing *Escherichia coli* cells by osmotic stress. J Bacteriol 159:919–924

Kregenow FM (1981) Osmoregulatory salt transporting mechanisms: control of cell volume in anisotonic media. Annu Rev Physiol 43:493–505

Laimins LA, Rhoads DB, Epstein W (1981) Osmotic control of *kdp* operon expression in *Escherichia coli*. Proc Natl Acad Sci USA 78:464–468

Leirmo S, Harrison C, Cayley DS, Burgess RR, Record MT Jr (1987) Replacement of KCl by K glutamate dramatically enhances protein-DNA interactions in vitro. Biochemistry 26:7157–7164

LeRudulier D, Bouillard L (1983) Glycine betaine, an osmotic effector in *Klebsiella pneumoniae* and other members of the *Enterobacteriaceae*. Appl Environ Microbiol 46:152–159

Levine M, Manly JL (1989) Transcriptional repression of eukaryotic promoters. Cell 59:405–408

Martinac B, Beuchner M, Delcour A, Adler K, Kung C (1987) Pressure-sensitive ion channel in *Escherichia coli*. Proc Natl Acad Sci USA 84:2297–2301

May G, Faatz E, Villarejo M, Bremer E (1986) Binding protein dependent transport of glycine betaine and its osmotic regulation in *Escherichia coli* K12. Mol Gen Genet 205:225–223

Milner JL, Grothe S, Wood JM (1988) Proline porter II is activated by a hyperosmotic shift in both whole cells and membrane vesicles of *Escherichia coli* K-12. J Biol Chem 263:14900–14905

Nikaido H, Vaara M (1987) Outer membrane. In: Neidhardt FC, Ingraham JL, Low KB, Magasanik B, Schaechter M, Umbarger HE (eds) *Escherichia coli* and *Salmonella typhimurium*: cellular and molecular biology. Am Soc Microbiol Washington DC, pp 7–22

Nobel PS (1983) Biophysical plant physiology and ecology. Freeman, San Francisco

Overdier DG, Olson ER, Erickson BD, Ederer MM, Csonka LN (1989) Nucleotide sequence of the transcriptional control region of the osmotically regulated *proU* operon of *Salmonella typhimurium* and identification of the 5′ endpoint of the *proU* mRNA. J Bacteriol 171:4694–4706

Ramirez RM, Prince WS, Bremer E, Villarejo M (1989) In vitro reconstitution of osmoregulated expression of *proU Escherichia coli*. Proc Natl Acad Sci USA 86:1151–1157

Ratzkin B, Grabnar M, Roth J (1978) Regulation of the major proline permease in *Salmonella typhimurium*. J Bacteriol 133:737–743

Richey B, Cayley DS, Mossing MC, Kolka C, Anderson CF, Farrar TC, Recort MT Jr (1987) Variability in the intracellular ionic environment of *Escherichia coli*: differences between in vitro and in vivo effects of ion concentrations on protein-DNA interactions and gene expression. J Biol Chem 262:7157–7164

Ronson CW, Nixon BT, Ausubel FM (1987) Conserved domains in bacterial regulatory proteins that respond to environmental stimuli. Cell 49:579–581

Stalmach ME, Grothe S, Wood JM (1984) Two proline porters in *Escherichia coli* K-12. J Bacteriol 156:481–486

Stirling DA, Hulton CSJ, Waddell L, Park SF, Stewart GSAB, Booth IR, Higgins CF (1989) Molecular characterization of the *proU* loci of *Salmonella typhimurium* and *Escherichia coli* encoding osmoregulated glycine betaine transport systems. Mol Microbiol 3:1025–1038

Stock JB, Rauch B, Roseman S (1977) Periplasmic space in *Salmonella typhimurium* and *Escherichia coli*. J Biol Chem 252:7850–7861

Sutherland L, Cairney J, Elmore MJ, Booth IR, Higgins CF (1986) Osmotic regulation of transcription: induction of the *proU* betaine transport gene is dependent on accumulation of intracellular potassium. J Bacteriol 168:805–814

Uchida S, Garcia-Perez A, Murphy H, Burg M (1990) Signal for induction of aldose reductase in renal medullary cells by high external NaCl. Am J Physiol 256:C614–C620

Walderhaug MO, Dosch DC, Epstein W (1987) Potassium transport in bacteria. In: Rosen BP, Silver S (eds) Ion transport in prokaryotes. Academic Press, San Diego, Lond New York, pp 85–130

Yancey PH, Clark ME, Hand SC, Bowlus RD, Somero GN (1982) Living with water stress: evolution of osmolyte systems. Science 217:1214–1227

Chapter 6

# A Physicochemical Basis for the Selection of Osmolytes by Nature

S.N. TIMASHEFF

## Introduction

It is a striking fact of nature that compounds used as cellular osmolyte systems by a variety of plant and animal vertebrate and invertebrate systems are confined to a small number of chemical structures, all, or most of which are distributed over the entire gamut of organisms (Yancey et al. 1982; Somero 1986; Somero, this Vol.). These comprise sugars and other polyols, amino acids and amino acid derivatives, methylamines, and in some cases urea, frequently in combination with methylamines. All are electrically neutral molecules. With the exception of urea all of these organic osmolytes are "compatible solutes" (Brown and Simpson 1972; Clark 1985), i.e., they do not disturb cellular structure and function. Among amino acids, arginine and lysine are not used as osmolytes (Yancey et al. 1982). They are known to be "incompatible" (Somero 1986), in that they interfere with some biochemical processes. Nor are amino acids that contain large hydrophobic side chains used. It is noteworthy that in a case where arginine is released it is immediately converted to the "compatible" solute octopine (Hochachka et al. 1977).

Urea, which is an efficient protein denaturant and, therefore, an "incompatible" solute, when used as osmolyte is frequently accompanied by a "compatible" methylamine, such as trimethylamine-N-oxide (TMAO), sarcosine, and betaine (Yancey and Somero 1979). In fact, the proportions of these compounds found in the organisms appear to be such that the deleterious destabilizing effect of urea is compensated exactly by the stabilizing action of TMAO, in such a manner that the net effect is one of preserving cellular activity, such as enzyme activity, at the normal level with rates neither slowed down nor hyperactivated (Yancey and Somero 1979). A similar compensation of deleterious salt effects by betaine has been reported (Pollard and Wyn-Jones 1979).

The general use by a broad spectrum of biological species of a small number of compounds to maintain osmotic equilibrium raises the question: what is the common principle which restricts selection to these molecules? Since they act at high concentration and, therefore, are nonspecific, this question may be reworded: what is the physicochemical advantage of these molecules over others which are not used? An early insight was provided by the observation that the chemical structures of the "compatible" methylamines are similar to those of protein stabilizing methylam-

Graduate Department of Biochemistry, Brandeis University, Waltham, MA 02254, USA

Somero et al. (Eds.)
Water and Life

monium salts (Clark 1985; Wyn-Jones et al. 1977). Hence the proposal that these compounds are "compatible" because they favor the native structure of biological macromolecules (Yancey and Somero 1979; Pollard and Wyn-Jones 1979; Clark and Zounes 1977; Bowlus and Somero 1979; Borowitzka and Brown 1974; Yancey and Somero 1980). The finding that a number of compounds which are used as osmolytes are also known to stabilize native protein structure when present at high concentration by being preferentially excluded from the immediate domain of the protein (Arakawa and Timasheff 1985b) has identified the physicochemical principle which cells use in selecting these substances that they store at high concentration for the purpose of maintaining osmotic pressure.

## Osmolytes, Preferential Exclusion, and Structure Stabilization

Nature selects molecules to act as osmolytes on the basis of the criteria: (1) that they should not affect specifically any enzyme or other cellular processes; (2) that they should not be electrically charged so as not to upset the electrostatic balance of cellular components; (3) that they should act as native protein structure stabilizers and be preferentially excluded from contact with cellular components so as not to perturb their structures. Molecules which satisfy universally the last two criteria are sugars (Lee and Timasheff 1981; Arakawa and Timasheff 1982), glycerol (Gekko and Timasheff 1981; Na and Timasheff 1981), many polyols (Gekko and Morikawa 1981; Gekko and Koga 1984), as well as many amino acids (Arakawa and Timasheff 1983; Arakawa and Timasheff 1984a) and their derivatives (Arakawa and Timasheff 1985b) (Reviewed in Low 1985).

What is preferential exclusion and why is it linked with stabilization of native protein structure? When a protein molecule is immersed into a solvent consisting of water and another chemical species (a co-solvent), the interactions between the protein and the solvent components may lead to three possible situations: (1) the co-solvent is present at the protein surface in excess over its concentration in the bulk (this is what constitutes binding); (2) water is present in excess at the protein surface; this means that the protein has a higher affinity for water than for the co-solvent (this situation is referred to as preferential hydration, or preferential exclusion of the co-solvent); (3) the protein is indifferent to the nature of molecules (water or co-solvent) with which it comes in contact, so that no solvent concentration perturbation occurs at the protein surface. The first two cases are illustrated in Fig. 1, which depicts schematically a dialysis equilibrium experiment. It is evident that the final equilibrium state reflects the relative affinities of the protein for water and the co-solvent. As a consequence, the parameter $\nu_3$, which is referred to as "binding" in the standard Scatchard notation, as in a Scatchard plot (moles of ligand bound per mole of protein), is properly called *preferential interaction*. When $\nu_3$ is positive, we have *preferential binding* of ligand (co-solvent in the present case); when it is negative, we have *preferential hydration*, alternately known as *preferential exclusion* of ligand.

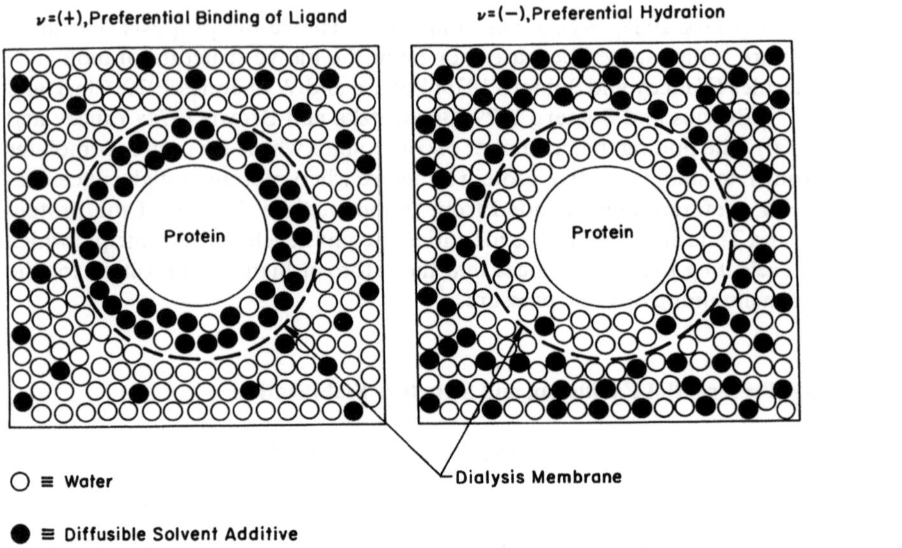


**Fig. 1.** Schematic representation of solvent component distribution at dialysis equilibrium. *Left* preferential binding; *right* preferential hydration. (Na and Timasheff 1981)

If we use the notation: component 1 = water, component 2 = protein, component 3 = co-solvent, the "binding" of component 3 to the protein measured by dialysis equilibrium is thermodynamically and operationally defined as:

$$\nu_3 = \left(\frac{\partial m_3}{\partial m_2}\right)_{T,\mu_1,\mu_3} = \frac{(\text{conc of comp 3 inside bag}) - (\text{conc in bulk})}{\text{conc of protein}}, \tag{1}$$

where $m_i$ = is the molal concentration of component i, T is Kelvin temperature and $\mu_i$ is the chemical potential of component i,

$$\mu_i = \mu_i^o + RT\ln a_i = \mu_i^o + RT\ln m_i\gamma_i, \tag{2}$$

where R is the universal gas constant, $a_i$ is the activity of component i, and $\gamma_i$ is its activity coefficient. The subscripts $\mu_1$, $\mu_3$ indicate that the system is at chemical (dialysis) equilibrium, i.e., the activities of water and the co-solvent are identical inside and outside the bag. Depending on whether the interaction conforms to case 1 or case 2, the numerical value of the measured binding, $\nu_3$, will be positive or negative. If the concentration is expressed in terms of grams i per gram of water, $g_i = m_iM_i/1000$, where $M_i$ is the molecular weight of component i, the preferential hydration is given by:

$$\left(\frac{\partial g_1}{\partial g_2}\right)_{T,\mu_1,\mu_3} = -\frac{1}{g_3}\left(\frac{\partial g_3}{\partial g_2}\right)_{T,\mu_1,\mu_3}. \tag{3}$$

Table 1 gives a list of binding values measured with stabilizing compounds which are also utilized as osmolytes. All are negative, i.e., in their presence the protein has a higher affinity for water than for the osmolyte. How does negative binding, i.e., preferential exclusion of component 3, lead to structure stabilization? Within a close approximation the thermodynamic definition of binding is:

$$\left(\frac{\partial m_3}{\partial m_2}\right)_{T,\mu_1,\mu_3} = -\frac{(\partial \mu_2/\partial m_3)_{T,P,m_2}}{(\partial \mu_3/\partial m_3)_{T,P,m_2}} \tag{4}$$

$$\left(\frac{\partial \mu_3}{\partial m_3}\right)_{T,P,m_2} = RT\left[\frac{1}{m_3} + \left(\frac{\partial \ln \gamma_3}{\partial m_3}\right)_{T,P,m_2}\right].$$

The numerator of the term on the right hand side is the perturbation of the chemical potential of the protein by the cosolvent. If the interaction is favorable, the chemical potential (Gibbs free energy) is lowered, the derivative is negative and, as a conse-

**Table 1.** Interactions of osmolytes with lysozyme[a]

| | Transition temperature $T_m$, °C | $\Delta T_m$/M of additive | $(\partial g_1/\partial g_2)_{T,\mu_1,\mu_3}$ g/g |
|---|---|---|---|
| None (Control) | 68–80 | — | — |
| Trimethylamine oxide (0.5 M) | 70–81 | 3 | — |
| L-Pro (pH 7.9) | 72–84 | 4 | 0.32 |
| L-Ser | 74–86 | 7 | 0.44 |
| Taurine (pH 6.8) | 72–84 | 9 | 0.38 |
| γ-amino butyric acid | 73–86 | 6 | 0.63 |
| Sarcosine | 73–86 | 6 | 0.48 |
| α-Alanine | 71–84 | 4 | 0.56 |
| β-Alanine | 73–85 | 6 | 0.60 |
| Glycine | 75–85 | 7 | 0.65 |
| Betaine | 72–83 | 4 | 0.60 |
| Valine (0.5 M) | 67–80 | –2 | 0.79 |
| Glucose | — | — | 0.35 |
| Arginine HCl (1.5 M)[b] | — | — | 0.23 |
| Lysine HCl | 72–84[c] | 4 | 0.40 |
| Lysine HCl (BSA) | — | — | 0.22 |
| Sucrose (1 M)[d] | 53.0 (48.0) | 5 | 0.44 |
| Glycerol (1.4 M)[d] | 43.5 (41.0) | 2 | 0.14 |

[a]The pH of the solutions was 6.0 and the additive concentrations 0.7 M, except where indicated. (Data taken from Arakawa and Timasheff 1985b; Arakawa and Timasheff 1983; Lee and Timasheff 1981; Gekko and Timasheff 1981; Arakawa and Timasheff 1984a; Arakawa and Timasheff 1982).
[b]Arakawa and Timasheff, unpublished.
[c]Value extrapolated from data in 1.7 M guanidine hydrochloride.
[d]The protein was ribonuclease A; the transition temperature reported is $T_m$; the value in parenthesis is the value with no co-solvent. The difference in the control values of $T_m$ reflect differences in the conditions of the experiments.

quence, the binding is positive. If the interaction is unfavorable, the chemical potential is increased, the derivative is positive, and the measured binding is negative. The denominator expresses the nonideality of the co-solvent. Let us now consider the denaturation reaction:

$$N \underset{}{\overset{K}{\rightleftharpoons}} D. \tag{5}$$

If the equilibrium constant, K, varies with the concentration (activity) of the co-solvent, then the Wyman linkage relation (Wyman 1964) states:

$$\left(\frac{\partial \ln K}{\partial \ln a_3}\right)_{T,P} = \nu_3^D - \nu_3^N = \Delta\nu_3, \tag{6}$$

where $\Delta\nu_3$ is the difference between the preferential binding of component 3 to the protein in the denatured and native states. If the preferential exclusion is greater in the denatured than the native state, $\Delta\nu_3$ is negative, and the equilibrium is shifted to the left, i.e., we observe protein stabilization. This situation is depicted schematically in Fig. 2.

The total free energy of stabilization by a co-solvent, i.e., the increase in the standard free energy of denaturation, $\Delta G^{o,N\text{-}D}$, due to the addition of the stabilizing co-solvent, is given by the difference between the transfer free energies, $\Delta\mu_{2,tr}$, of the protein from water to the co-solvent in the denatured and native states (Na and Timasheff 1981):

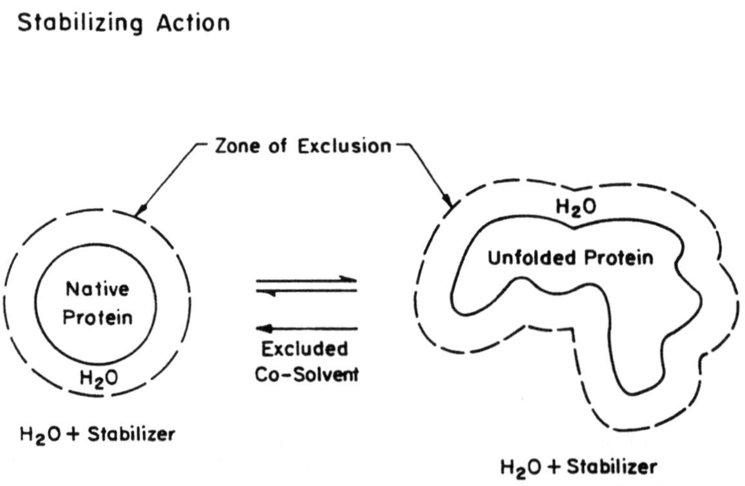


**Fig. 2.** Pattern of protein-solvent interactions in the native ⇌ denatured equilibrium when the interaction is that of nonspecific exclusion of the co-solvent by, for example, the increase of water surface tension by the co-solvent. Since in the asymmetric denatured state the surface of protein solvent contact is greater per protein molecule than in the compact native globular state, co-solvent exclusion is greater and the equilibrium is displaced toward native. (Arakawa et al. 1990)

$$\Delta G_3^{o,N\text{-}D} - \Delta G_w^{o,N\text{-}D} = \Delta\mu_{2,tr}^D - \Delta\mu_{2,tr}^N$$

$$\Delta\mu_{2,tr} = \int_0^{m_3} \left(\frac{\partial\mu_2}{\partial m_3}\right)_{T,P} dm_3 \quad , \tag{7}$$

where $\Delta G^{o,N\text{-}D}$ is the standard free energy of denaturation, the subscripts w and 3 signify water and mixed solvent, respectively. Therefore, if $(\partial\mu_2/\partial m_3)_{T,P}$ is more positive in the denatured state than in the native state, the transfer free energy will also be more positive in the denatured state, and the standard free energy of denaturation will be more positive in the presence of the co-solvent than in aqueous medium, i.e., the native structure of the protein will be stabilized.

## Preferential Exclusion, Compatibility, and Compensation

An examination of Table 1 reveals that all "compatible" osmolytes for which preferential interaction measurements have been made are preferentially excluded from the domain of native globular proteins. They can, therefore, be expected to be stabilizers of native protein structure. This, in fact, is the case. Are noncompatible molecules, then, necessarily preferentially bound to the proteins? As seen in Table 1, Arg • HCl and Lys • HCl are both preferentially excluded from globular proteins in the native state. Yet, they are found to be "incompatible." The reasons for this are easily gleaned from Eq. (6). The effect of an additive, co-solvent, on protein denaturation is given by the difference between the preferential binding (exclusion) of co-solvent to (from) the protein in the native and denatured states. Both arginine and lysine can bind to proteins by their positive charges (Arakawa and Timasheff 1983; Arakawa and Timasheff unpublished data) even though both are preferentially excluded by the general nonspecific mechanism of increase in water surface tension, which will be discussed below. The interaction observed on dialysis equilibrium is the net balance between all interactions between a protein and the co-solvent, both binding and exclusion:

$$\nu_3 = \nu_3^{\text{Binding}} + \nu_3^{\text{Exclusion}} \tag{8}$$

Thus, if binding to the denatured state, $\nu_3^{\text{Binding}}$, by, e.g., formation of hydrogen bonds between the guanidinium group of arginine and newly exposed peptide groups, becomes great on denaturation, then $\Delta\nu_3$ for the denaturation reaction may become positive, which would lead to protein structure destabilization. Nature then designates such compounds as incompatible substances. The same is true for valine, which slightly destabilizes lysozyme, even though it is preferentially excluded from the native protein (Arakawa and Timasheff 1985b). In this case, the destabilization may be ascribed to the affinity of the strongly hydrophobic side chain of valine for newly exposed nonpolar residues in the unfolded state of the protein.

The nature of interaction with proteins of the guanidinium ion, which is classified as "incompatible," is similar to that of arginine, but it is much stronger. Guanidinium ion binds to peptide groups and aromatic side chains (Lee and

Timasheff 1974). This is what renders it a denaturant. Its denaturing activity, however, can be compensated by the stabilizing action of the $SO_4^{2-}$ ion. Figure 3 shows the effect of guanidinium salts on the denaturation of ribonuclease A (von Hippel and Wong 1965). It is evident that the denaturing ability of guanidine salts depends on the nature of the anion. Thus, Gua • HCl is a good denaturant, the acetate is a weak denaturant, while $Gua_2 \cdot SO_4$ is, in fact, a stabilizer. The data of Table 2 give an explanation for these observations (Arakawa and Timasheff 1984b). In this table, various cations and anions are classified according to their preferential interactions with proteins. A pattern of additivity is evident. In the case of anions, which follow the Hofmeister series, sulfate generates the strongest preferential hydration, for cations this place is occupied by sodium. As a result, $Na_2SO_4$ is excluded the most strongly. It is also the best stabilizer and salting-out salt. Guanidinium ion binds to proteins. In the presence of chloride, its binding is predominant and the salt is a

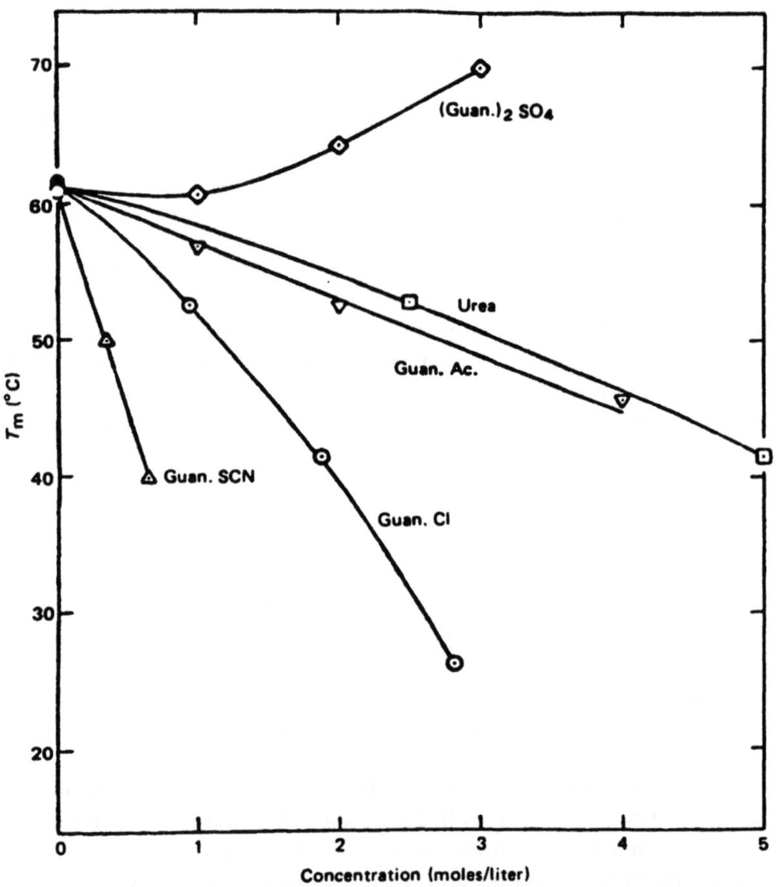


**Fig. 3.** Effect of guanidinium salts and urea on the transition temperature of ribonuclease (pH 7.0, 0.013 M sodium cacodylate, 0.15 M NaCl). (von Hippel and Wong 1965)

**Table 2.** Values of $(\partial m_3/\partial m_2)_{T,\mu_1,\mu_3}$ for the interactions of 1 M salts with bovine serum albumin. (Arakawa and Timasheff 1984b; Arakawa et al. 1990)

| Cation | Anion | | | |
|---|---|---|---|---|
| | $SCN^-$ | $Cl^-$ | $OAc^-$ | $SO_4^{2-}$ |
| $Na^+$ | 3[a] | –17 | –22 | –35 |
| $Mg^{2+}$, $Ca^{2+}$, $Ba^{2+}$ | — | –3 to 3 | –8 to –13 | –27[b] |
| Guanidinium | — | 18 | –6 | –16 |

[a]The cation was $K^+$.
[b]Value for $MgSO_4$.

denaturant. Its combination with sulfate results in weak exclusion and the consequent protein structure stabilization. There is, therefore, a clear pattern of additivity, or compensation, between the binding and exclusion of cations and anions. Sulfate, being the strongest excluded anion, overwhelms the binding of guanidinium and renders the salt a protein structure stabilizer. Thiocyanate, being a strongly bound anion, overwhelms the exclusion of sodium or potassium (Arakawa et al. 1990) and renders the salt into a protein denaturant (von Hippel and Wong 1965). Guanidine thiocyanate, which consists of two binding ions, is a very strong denaturant (von Hippel and Wong 1965). The same principle of additivity may be invoked to explain the compensation of urea by TMAO both in organisms and in pure enzyme systems (Yancey and Somero 1979). Urea is a denaturant; it is preferentially bound to the peptide groups of a protein. TMAO is a stabilizer; its preferential interactions with proteins have not been measured yet because of technical difficulties, but it can be expected to be strongly excluded by analogy with betaine and sarcosine (Arakawa and Timasheff 1985b).

## Preferential Exclusion of Osmolytes

The causes of preferential exclusion may be varied. The mechanisms which seem germane to osmolytes are those that are independent of the chemical nature of the protein surface, i.e., the interactions are nonspecific and the protein behaves as an inert solution component. Its role is only to present a surface. These mechanisms comprise steric exclusion, the increase of the surface tension of water by the co-solvent and the solvophobic effect. The first two are illustrated in Fig. 4A and B.

The steric exclusion mechanism is simply the consequence of the difference in size of water and co-solvent molecules, the co-solvent molecules being bulkier than water. If the co-solvent molecules are chemically inert toward the protein surface, they can approach the protein only until physical contact is made. At this closest approach, the centers of these molecules are removed from the protein surface by a distance equal to their effective radius (Re in Fig. 4A) (Arakawa and Timasheff 1985a). This creates around the protein surface a shell impenetrable to the co-solvent. Water, being a smaller molecule, does penetrate within this shell. As a conse-

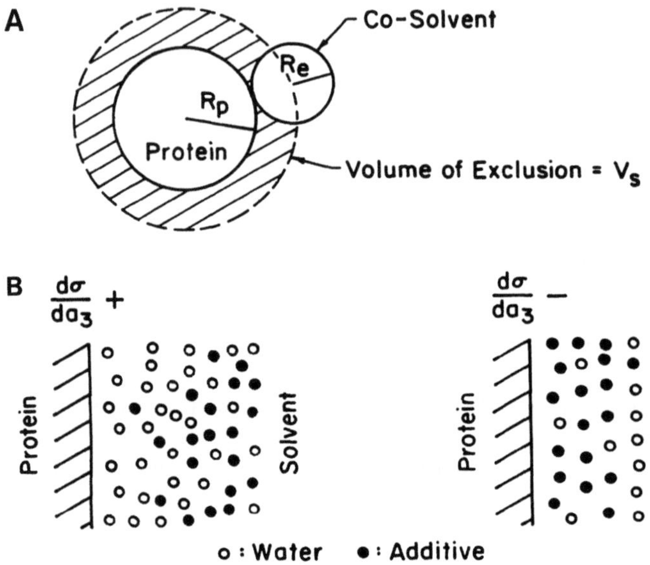


**Fig. 4A,B.** Schematic representation of non-specific mechanisms of preferential exclusion of co-solvents from proteins. **A** Steric exclusion: $R_p$ is the radius of the protein; $R_e$ is the radius of exclusion (effective radius of the co-solvent); $V_s$ is the volume in which there is an excess of water. (Arakawa and Timasheff 1985a). **B** Solvent component distribution at the protein-solvent interace; $\sigma$ is surface tension; $a_3$ is co-solvent activity. When the co-solvent raises the surface tension of water ($d\sigma/da_3$) is positive, the concentration of the co-solvent is lowered in the surface layer and the protein is stabilized. (Timasheff 1991)

quence, the solvent becomes enriched in water in the immediate vicinity of the protein. In a dialysis equilibrium experiment this is detected as negative binding, i.e., preferential hydration, which, by Eqs. (4) and (6), has the consequence that the native protein structure is stabilized (if unfolding of the protein does not generate any *new specific* interactions between the co-solvent and the newly exposed groups on the protein).

What appears to be the by far predominant mechanism of preferential exclusion is the increase of the surface tension of water by the co-solvents. This is dictated by the rule derived by J. Willard Gibbs in 1878, known as the Gibbs absorption isotherm (Gibbs 1878). In the present notation, it states that:

$$\left(\frac{\partial m_3}{\partial m_2}\right)_{T,P}^{\text{surface}} = -\frac{S_2 a_3}{RT}\left(\frac{\partial \sigma}{\partial a_3}\right)_{T,P}, \tag{9}$$

where $\sigma$ is surface tension and $S_2$ is the molar surface area of the protein; the superscript "surface" means preferential binding/exclusion due to the surface tension perturbation. Sugars (Landt 1931), nonhydrophobic amino acids (Pappenheimer et al. 1936; Bull and Breese 1974) and most salts (Melander and Horvath 1977) increase the surface tension of water. By the Gibbs adsorption isotherm, therefore, their concentration in the surface layer must be reduced relative to that of the bulk solvent. As a consequence, [see Eq. (4)], they must raise the chemical potential of the protein. Since in this, as well as in the previous mechanism, the role of the protein is simply to present an inert surface which increases on unfolding (Fig. 2), the net result must be stabilization of the native structure of the protein.

The third nonspecific mechanism of co-solvent exclusion is that of solvophobicity. Solvophobicity is a consequence of the reinforcement of the hydrophobic effect by a solvent additive due to water structure enhancement by the additive. Contact between nonpolar residues on the protein surface and water molecules, which are constantly fluctuating in and out of structured clusters, is thermodynamically unfavorable as defined by the hydrophobic effect (Tanford 1973). Such a contact becomes even more thermodynamically unfavorable in a solvophobic solvent system. In order to relieve this situation, a redistribution of solvent molecules takes place at the protein surface. This is the mechanism by which glycerol (Gekko and Timasheff 1981) and probably other polyhydroxy alcohols (Gekko and Morikawa 1981; Gekko and Koga 1984) are excluded. Polyols fit well the water lattice (Warner 1962) and their structure permits the formation of proper hydrogen bonds which reinforce the water structure. As a consequence, polyol molecules migrate away from the protein surface, which results in a deficiency of polyol in the immediate domain of the protein. Since in an unfolded protein the number of nonpolar residues exposed to solvent increases, this process should become even greater and, by Eq. (6), the native protein structure is stabilized.

All three mechanisms are based on a microphase separation of the solvent at the protein surface, with water molecules accumulating at that surface. This, however, does not signify that there is an impenetrable hydration layer consisting of water molecules ordered on the protein surface (the iceberg hypothesis) nor an increase in the water-protein affinity. These protein-solvent interactions are a function strictly of solvent properties. In the first mechanism, the key parameter is bulkiness of the co-solvent molecules coupled with their chemical inertness toward the protein. In the second one, the key factor is the effect of the co-solvent on the forces of cohesion between water molecules and the work required to form a surface. In the third case, it is the increase in solvent ordering which renders contact with protein nonpolar regions even more unfavorable than when the solvent is pure water. None of these effects is specific, in that none is dictated by direct interactions with sites on the protein surface. Nature seems to have chosen as osmolytes those molecules which by being neutral and inert toward the protein surface, perturb neither protein electrostatics, nor protein chemistry.

### Osmotic Pressure, Stabilization, and Solvent Exclusion

The quite limited number of chemical compounds chosen by organisms, both plant and animal, unicellular to vertebrates, for maintaining osmotic equilibrium, testifies to an extreme degree of selectivity. As has been noted, all are "compatible" substances and, with the exception of the halobacteria, which use inorganic ions, all are electrically neutral. All are known to be native protein structure protectants and, where measured, all are preferentially excluded from contact with protein surface. The sources of this exclusion cover a gamut of nonspecific thermodynamic phenomena. Sugars, and nonhydrophobic amino acids raise the surface tension of water and, therefore, are excluded from interfaces: glycerol and other polyols are repelled from nonpolar regions on the protein surface due to their enhancement of solvent order. The mechanism of exclusion of the methylamines is not known at present. It is noteworthy that the hydrophobic amino acids are not used. They depress the surface tension of water (Bull and Breese 1974) which should facilitate interaction with nonpolar regions newly exposed on protein unfolding. A facility to interact with peptide groups via hydrogen bonding and, thus, to overcome exclusion due to the surface tension effect seems to have eliminated arginine from the list. The urge by nature to achieve thermodynamic stabilization while maintaining osmotic equilibrium is brought out most elegantly in the balance between urea, a good denaturant, and methylamines which are stabilizers, as well as by the striking example of the chemical conversion of arginine to the innocuous compound octopine. What is the relation between osmotic pressure, structure stabilization, and preferential exclusion?

In a two-component system, the relation between osmotic pressure and solution composition is defined as:

$$\pi = \frac{nm_3}{m_1}\frac{RT}{\bar{V}_1}\phi , \tag{10}$$

where $\pi$ is osmotic pressure, n is the number of particles into which the solute (component 3) dissociates in solution (n = 1 for nondissociable solutes; n = 2 for KCl, etc.), $\bar{V}_1$ is the partial molal volume of the principal solvent (in our case water) and $\Phi$ is the osmotic coefficient, defined by:

$$\phi = -\frac{m_1}{nm_3}\ln a_1. \tag{11}$$

In the ideal case, for nondissociable solutes,

$$\pi = -\frac{m_3}{m_1}\frac{RT}{\bar{V}_1} \tag{12}$$

Combination of Eqs. (10) and (11) gives

$$\pi\bar{V}_1 = -RT\ln a_1 = -(\mu_1 - \mu_1^\circ). \tag{13}$$

Differentiation and application of the Gibbs-Duhem equation results in

$$\left(\frac{d\pi}{dm_3}\right)_{T,P} = \frac{m_3}{1000}\left(\frac{\partial\mu_3}{\partial m_3}\right)_{T,P}. \tag{14}$$

If Eq. (14) is combined with Eq. (4), we obtain:

$$\left(\frac{\partial m_3}{\partial m_2}\right)_{T,\mu_1,\mu_3} = -\frac{m_3}{1000}\frac{(\partial\mu_2/\partial m_3)_{T,P,m_2}}{(\partial\pi/\partial m_3)_{T,P,m_2}} \tag{15}$$

Equation (15) states that the degree of co-solvent exclusion is inversely proportional to the departure of osmotic pressure from the ideal Vant'Hoff case.

If the chemical nature of interactions does not change on protein unfolding, the parameter $(\partial\mu_2/\partial m_3)_{T,P,m_2}$ is a measure of protein stabilization [see Eq. (7)]. Therefore, the difference in osmotic pressure gradients with concentration of different co-solvents leads to different extents of co-solvent exclusion from the protein surface for a given degree of stabilization by co-solvents. This is illustrated by the calculations of Table 3. The concentration dependence of osmotic pressure for typical osmolytes is given in Fig. 5, where the data for glycine are calculated from known activity coefficients (Scatchard and Prentiss 1934; Smith and Smith 1937; Cohn and Edsall 1943). The "superosmolyte" data are totally simulated. They are similar to what should be observed for betaine, as deduced from the osmotic coefficients of that compound (Smith and Smith 1940). From these data, the preferential exclusions of the osmolytes at a 1 molal concentration ($m_3 = 1$) were calculated for a constant value of stabilization $(\partial\mu_2/\partial m_3)_{T,P,m_2} = 4000$ cal/mol$^2$, which is typical of values measured for small proteins, such as lysozyme and ribonuclease, in sucrose or amino acids. These are seen to vary by a factor of 1.65 between the two extremes. This means that the quantity of excess water around the protein also varies by the same factor. It is evident that the solvent composition perturbation at a solvent-cellular surface interface decreases as the deviation from ideality becomes more positive. Therefore, for any desired degree of structure stabilization a positive deviation of the osmotic pressure from ideal reduces the redistribution of solvent components in a cellular compartment. This is illustrated schematically in Fig. 6. Such a situation might seem to be advantageous to the integrity of the organism. The values of the preferential exclusion of osmolytes given in Table 3 lead to the preferential hydrations listed in column 6 of that table for a small protein, lysozyme, which has a radius of 15.8 Å. If

**Table 3.** Relation between osmotic pressure increment, protein stabilization and solvent exclusion[a]

| Osmolyte | $\left(\frac{\partial\pi}{\partial m_3}\right)$ atm/mol | $\left(\frac{\partial\pi}{\partial m_3}\right)_{T,P,m_2}$ cal/mol$^2$ | $\left(\frac{d\ln\gamma_3}{dm_3}\right)$ | $\left(\frac{\partial m_3}{\partial m_2}\right)_{T,\mu_1,\mu_3}$ mol/mol | $\left(\frac{\partial g_1}{\partial g_2}\right)^{b}_{T,\mu_1,\mu_3}$ g/g | L[b,c] Å |
|---|---|---|---|---|---|---|
| Ideal | 24.1 | 583 | 0 | –6.86 | 0.48 | 3.0 |
| Glycine | 21.1 | 514 | –0.118 | –7.78 | 0.54 | 3.4 |
| Sucrose | 29.6 | 716 | 0.229 | –5.59 | 0.39 | 2.5 |
| Betaine | 32.1 | 777 | 0.333 | –5.15 | 0.36 | 2.4 |
| "Superosmolyte" | 34.8 | 843 | 0.447 | –4.74 | 0.33 | 2.2 |

[a]All calculations were done for 20 °C and $m_3 = 1.0$.
[b]Calculated for lysozyme, mol. wt. = 14,300; $\bar{v}_2 = 0.70$; radius = 15.8 Å.
[c]L is the thickness of the hypothetical layer of water around a lysozyme molecule that corresponds to the preferential hydration value.

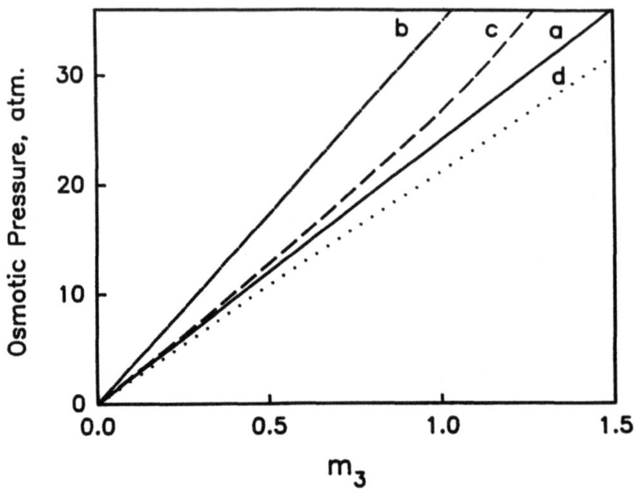


**Fig. 5.** Dependence of osmotic pressure on the molal concentration of co-solvent. The *solid line* (*a*) is the ideal case, $\gamma_3$ = 1.0 at all concentrations. The *dot-dashed line* (*b*) is calculated for a "superosmolyte"; the deviation from ideality is strongly positive. The *dashed line* (*c*) is the experimental data for sucrose (Berkeley and Hartley 1906; Morse et al. 1912; Frazer and Myrick 1916). The *dotted line* (*d*) is for glycine calculated from activity coefficient data; the deviation from ideality is negative. All values are for 20 °C

**Fig. 6.** Schematic representation of the effect of osmotic pressure on the thickness of the hypothetical solvent layer enriched in water along the surface of a cellular compartment for identical values of $(\partial\mu_2/\partial m_3)_{T,P,m_3}$, i.e., identical degrees of protein stabilization. It is shown that as $d\pi/dc_3$ decreases or becomes negative, the layer of preferential exclusion of the osmolyte becomes thicker

the preferential hydration is represented by a layer of water with a sharp boundary between it and bulk solvent, these values of preferential hydration correspond to hydration layers that vary from 3.4 Å in the presence of 1 molal glycine to 2.1 Å in the presence of the 1 molal "superosmolyte". Keeping in mind that, in the real situation, the solvent redistribution occurs over a gradient rather than at a sharp boundary, it becomes evident that in the presence of "superosmolytes" a much smaller volume of cellular fluid must be perturbed for the attainment of given desired values of osmotic pressure and structural stabilization. This may be a reason why evolutionarily "superosmolytes" have been selected by a number of systems.

*Acknowledgment.* This work was supported in part by NIH Grants CA-16707 and GM-14603. Communication No. 1721 from the Graduate Department of Biochemistry, Brandeis University.

## References

Arakawa T, Timasheff SN (1982) Stabilization of protein structure by sugars. Biochemistry 21:6536–6544

Arakawa T, Timasheff SN (1983) Preferential interactions of proteins with solvent components in aqueous amino acid solutions. Arch Biochem Biophys 224:169–177

Arakawa T, Timasheff SN (1984a) The mechanism of action of Na glutamate, glysine HCl, and PIPES in the stabilization of tubulin and microtubule formation. J Biol Chem 259:4979–4986

Arakawa T, Timasheff SN (1984b) Protein stabilization and destabilization by guanidinium salts. Biochemistry 23:5924–5929

Arakawa T, Timasheff SN (1985a) Mechanism of poly(ethylene glycol) interaction with proteins. Biochemistry 24:6756–6762

Arakawa T, Timasheff SN (1985b) The stabilization of proteins by osmolytes. Biophys J 47:411–414

Arakawa T, Bhat R, Timasheff SN (1990) Preferential interactions determine protein solubility in three-component solutions: The $MgCl_2$ system. Biochemistry 29:1914–1923

Berkeley the Earl of, Hartley EGJ (1906) On the osmotic pressure of some concentrated aqueous solutions. Trans R Soc Lond A 206:481–507

Borowitzka LJ, Brown AD (1974) The salt relations of marine and halophilic species of the unicellular green alga, Duneliella. The role of glycerol as compatible solute. Arch Microbiol 96:37–52

Bowlus RD, Somero GN (1979) Solute compatibility with enzyme function and structure: rationales for the selection of osmotic agents and end-products of anaerobic metabolism in marine invertebrates. J Exp Zool 208:137–152

Brown AD, Simpson JR (1972) Water relations of sugar-tolerant yeasts: the role of intracellular polyols. J Gen Microbiol 72:589–591

Bull HB, Breese K (1974) Surface tension of amino acid solutions: a hydrophobicity scale of the amino acid residues. Arch Biochem Biophys 161:665–670

Clark ME (1985) The osmotic role of amino acids: Discovery and function. In: Gilles R, Gilles-Baillien M (eds) Transport processes, iono- and osmoregulation. Springer, Berlin Heidelberg New York, pp 412–423

Clark ME, Zounes M (1977) The effects of selected cell osmolytes on the activity of lactate dehydrogenase from the euryhaline polychaete, *Nereis succinea*. Biol Bull, Woods Hole, Mass, 153:468–484

Cohn EJ, Edsall JT (1943) Proteins, amino acids and peptides. Reinhold, New York, p 218

Frazer JCW, Myrick RT (1916) Osmotic pressure of sucrose solutions at 30 °C. J Am Chem Soc 38:1907–1922

Gekko K, Koga S (1984) The stability of protein structure in aqueous propylene glycol amino acid solubility and preferential solvation of protein. Biochim Biophys Acta 786:151–160

Gekko K, Morikawa T (1981) Preferential hydration of bovine serum albumin in polyhydric alcohol-water mixtures. J Biochem Jpn 90:39–50

Gekko K, Timasheff SN (1981) Mechanism of protein stabilization by glycerol: preferential hydration in glycerol-water mixtures. Biochemistry 20:4667–4676

Gibbs JW (1878) On the equilibrium of heterogeneous substances. Trans Conn Acad 3:343–524

Hochachka PW, Hartline PH, Fields JHA (1977) Octopine as an end product of anaerobic glycolysis in the chambered nautilus. Science 195:72–74

Landt E (1931) The surface tensions of solutions of various sugars. Z Ver Dtsch Zuckerind 81:119–124

Lee JC, Timasheff SN (1974) Partial specific volumes and interactions with solvent components of proteins in guanidine hydrochloride. Biochemistry 13:257–265

Lee JC, Timasheff SN (1981) The stabilization of proteins by sucrose. J Biol Chem 256:7193–7201

Low PS (1985) Molecular basis of the biological compatibility of Nature's osmolytes. In: Gilles R, Gilles-Baillien M (eds) Transport processes, iono- and osmoregulation. Springer, Berlin Heidelberg New York, pp 469–477

Melander W, Horvath C (1977) Salt effects on hydrophobic interactions in precipitation and chromatography of proteins: An interpretation of the lyotropic series. Arch Biochem Biophys 183:200–215

Morse HN, Holland WW, Myers CN, Cash G, Zinn JB (1912) The omostic pressure of cane sugar solutions at high temperature. In: Washburn EW (ed) International Critical Tables (1928) 4:429. Am Chem J 48: 29–94

Na GC, Timasheff SN (1981) Interaction of calf brain tubulin with glycerol. J Mol Biol 151:165–178

Pappenheimer JR, Lepie MP, Wyman J Jr (1936) The surface tension of aqueous solutions of dipolar ions. J Am Chem Soc 58:1851–1855

Pollard A, Wyn-Jones RG (1979) Enzyme activities in concentrated solutions of glycine betaine and other solutes. Planta 144:291–298

Scatchard G, Prentiss SS (1934) Freezing points of aqueous solutions. VIII. J Am Chem Soc 56:2314–2319

Smith PK, Smith ERB (1937) The activity of aliphatic amino acids in aqueous solution at 25 °C. J Biol Chem 121:607–613

Smith PK, Smith ERB (1940) The activities of some hydroxy- and N-methylamino acids and proline in aqueous solution at 25 °C. J Biol Chem 132:57–64

Somero GN (1986) Protons, osmolytes, and fitness of internal milieu for protein function. Am J Physiol 251:R197–R213

Tanford C (1973) The hydrophobic effect: formation of micelles and biological membranes. Wiley, New York London

Timasheff SN (1991) Stabilization of protein structure by solvent additives. In: Ahern TJ, Manning M (eds) Stability of protein pharmaceuticals: in vivo pathways of degradation and strategies for protein stabilization, Vol 3

von Hippel PH, Wong K-Y (1965) On the conformational stability of globular proteins. J Biol Chem 240:3909–3923

von Hippel PH, Wong K-Y (1965) On the conformational stability of globular proteins. The effects of various electrolytes and nonelectrolytes on the thermal ribonuclease transition. J Biol Chem 240:3909–3923

Warner DT (1962) Some possible relationships of carbohydrates and other biological components with the water structure at 37 °. Nature (Lond) 196:1055–1058

Wyman J Jr (1964) Linked functions and reciprocal effects in hemoglobin: a second look. Adv Protein Chem 19:223–286

Wyn-Jones RG, Storey R, Leigh RA, Ahmad N, Pollard A (1977) A hypothesis on cytoplasmic osmoregulation. In: Marre E, Cifferi O (eds) Regulation of cell membrane activities in plants. North Holland, Amsterdam, p 121

Yancey PH, Somero GN (1979) Counteraction of urea destabilization of protein structure by methylamine osmoregulatory compounds of elasmobranch fishes. Biochem J 183:317–323

Yancey PH, Somero GN (1980) Methylamine osmoregulatory solutes of elasmobranch fishes counteract urea inhibition of enzymes. J Exp Zool 212:205–213

Yancey PH, Clark ME, Hand SC, Bowlus RD, Somero GN (1982) Living with water stress: evolution of osmolyte systems. Science 217:1214–1222

# II Dessication Stress

Chapter 7

# Membrane Integrity in Anhydrobiotic Organisms: Toward a Mechanism for Stabilizing Dry Cells

J.H. CROWE and L.M. CROWE

## Introduction

Water is normally thought to be required for maintenance of structure and function in biomolecules (reviewed in Tanford 1980; J.H. Crowe et al. 1987a; L.M. Crowe and J.H. Crowe 1988b). Nevertheless, numerous organisms are capable of surviving essentially complete dehydration, including some that are familiar in daily life, such as seeds of many plants, yeast cells, fungal spores, and the like (see Leopold 1986 for references), but also including some microscopic animals, such as certain nematodes, rotifers, tardigrades, and cysts of some crustacean embryos (for example, those of the brine shrimp, *Artemia*). The dry organisms may remain in this unique living state, which is known as "anhydrobiosis", for decades or perhaps even centuries under favorable conditions. When water again becomes available they may rapidly swell and resume active life.

Anhydrobiotic organisms commonly contain high concentrations of disaccharides, particularly trehalose (see J.H. Crowe et al. 1987a, 1988 for reviews). For example, when the nematode *Aphelenchus avenae* was dehydrated slowly it converted as much as 20% of its dry weight into this molecule (reviewed in J.H. Crowe and L.M. Crowe 1986). Survival of these and other animals in the absence of water was correlated with synthesis of the sugar. Similarly, baker's yeast (*Saccharomyces cerevisiae*) accumulates trehalose, and the presence of this molecule appears to be required for survival of these cells in the absence of water. Since yeasts are proving to have special uses in this field, we have devoted particular attention to them in this review, and will consider these cells at more length in the concluding sections. The same kinds of statements can be made about a large array of phylogenetically dissimilar organisms that exhibit anhydrobiosis, ranging from cysts of the brine shrimp, *Artemia* (reviewed in Clegg 1986) to the desert resurrection plant (Harding 1923; Leopold 1986). Surprisingly little is known about biochemical adaptations in higher plants, but according to emerging evidence, the analog of trehalose here appears to be sucrose (Leopold 1986; Caffrey et al. 1988; Hoekstra and van Roekel 1988; Koster and Leopold 1988; Hoekstra et al. 1989; Leprince et al. 1990).

---

Department of Zoology, University of California, Davis 95616, USA

Somero et al. (Eds.)
Water and Life

## Trehalose Stabilizes Dry Membranes and Proteins

### *Stabilization of Dry Membranes by Trehalose*

Sarcoplasmic reticulum isolated from muscle possess a characteristic morphology (as seen in freeze fracture) and easily measured biological activity (transport of Ca). When these vesicles were dried without trehalose they fused, the morphology was altered, and upon rehydration it was apparent that all Ca-transport capacity had been lost as a result of these damaging events (J.H. Crowe et al. 1983). However, when they were dried with trehalose at concentrations near those found in anhydrobiotic organisms, morphological damage was completely inhibited during drying, and upon rehydration vesicles were obtained that transported Ca at rates close to those seen in freshly prepared vesicles (J.H. Crowe et al. 1983). The dry vesicles may be stored for at least 6 months without loss of stability, so long as they are protected from oxygen (Mouradian et al. 1984). Subsequently, this stabilization procedure has been extended to other membranes, with similar results (J.H. Crowe et al. 1984; Mouradian et al. 1984; Reshkin et al. 1990).

More recently, when liposomes were dried with trehalose and then rehydrated, vesicles were obtained that had retained 100% of their original contents. Those dried without trehalose leaked all their contents during drying and rehydration (J.H. Crowe et al. 1984, 1987a, 1989a; L.M. Crowe et al. 1985; L.M. Crowe and J.H. Crowe 1988a). This result has had immediate applications in the pharmaceutical industry, where liposomes are being used for drug delivery, and in vitro and in vivo diagnostics. Subsequently, several other groups have confirmed and extended these results (Madden et al. 1985; Strauss et al. 1986; Ozer et al. 1988).

### *The Mechanism of Stabilization of Dry Bilayers*

Two primary stress vectors are responsible for destabilization of bilayers during drying: fusion and lipid phase transitions (reviewed in J.H. Crowe et al. 1987a, 1988). (a) Fusion. Trehalose and other sugars will inhibit fusion between the vesicles during drying, but inhibition of fusion alone is not sufficient to preserve the dry vesicles (L.M. Crowe et al. 1985, 1986). (b) Phase transitions. When the water that is hydrogen bonded to the polar head groups is removed the packing of the head groups increases, leading to increased van der Waals' interactions among the hydrocarbon chains (reviewed in J.H. Crowe and L.M. Crowe 1988; L.M. Crowe and J.H. Crowe 1988a,b). As result, the phase transition temperature increases enormously. In the case of the lipids used in the liposome dehydration experiments discussed above, the gel to liquid crystalline transition temperature ($T_m$) increases from about –10 °C to about 60 °C. Thus, the dry lipids would be in gel phase at room temperature – a temperature permissive of liquid crystalline phase when the lipids are fully hydrated. As a result, when the dry lipids are placed in water they would be expected to undergo a phase transition, from gel to liquid crystalline phase, as the polar headgroups become rehydrated. as hydrated bilayers undergo such phase transitions they become transiently leaky (reviewed in J.H. Crowe et al. 1989a). It follows that dry bilayers that are in gel phase might be expected to leak during rehydration.

Trehalose prevents this leakage by depressing the phase transition temperature of the dry lipids, maintaining them in liquid crystalline phase in the absence of water. Depression of $T_m$ by trehalose is strongly correlated with retention of trapped solute (J.H. Crowe et al. 1988, 1989a). As a result of this effect, the dry lipids do not undergo a transition from gel to liquid crystalline phase during rehydration, and they do not leak (J.H. Crowe et al. 1988, 1989a). This mechanism is represented pictorially in Fig. 1.

### *Mechanism of Interaction with Phospholipids*

The stabilizing effect of trehalose on phospholipid bilayers is due to direct interaction between -OH groups on the trehalose and the phosphate of membrane phospholipids (J.H. Crowe et al. 1984, 1987b; Lee et al. 1986, 1989; Strauss et al. 1986; Chandrasekhar and Gaber 1988; L.M. Crowe and J.H. Crowe 1988a; Tsvetkova et al. 1988; Quinn 1989; Tsvetkov et al. 1989). For instance, the elegant models from Gaber's laboratory at the Naval Research Laboratory in Washington, D.C. (Chandrasekhar and Gaber 1988) suggest that trehalose fits remarkably well between the polar head groups, with multiple sites of interaction, which suggests that the strong stabilizing properties of trehalose may be related to its stereochemistry. Alternatively, the reluctance of trehalose to crystallize may play an important role in its stabilizing effects (Caffrey et al. 1988).

### *Extension of the Phase Transition Hypothesis to Native Membranes*

More recently, we have shown that the model for stabilization of dry vesicles presented in Fig. 1 also applies to native membranes and intact cells, using an unusual method to measure the transitions. The most common method for measuring phase

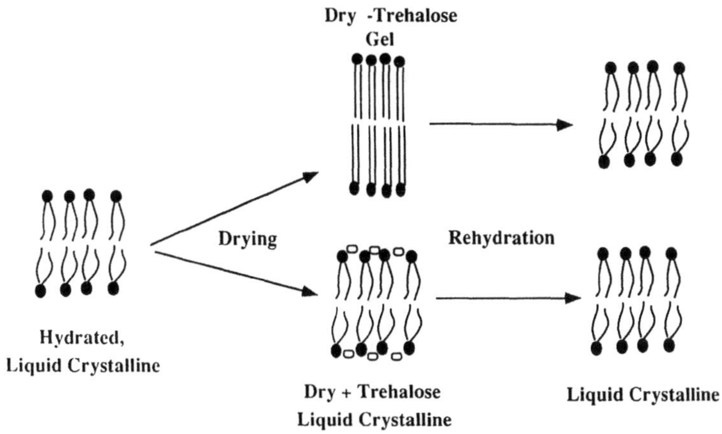


**Fig. 1.** Mechanism by which trehalose stabilizes phospholipid bilayers in the absence of water. (After J.H. Crowe et al. 1988)

transitions, differential scanning calorimetry, is often not useful for detecting transitions in the complex mixtures of lipids found in native membranes or in intact cells, since cooperativity of the transitions in such mixtures is low, with resulting diminished enthalpy. Consequently, it has been necessary to turn to other methods. We have adapted Fourier transform infrared spectroscopy for this purpose and have shown that the elevation in vibrational frequency in $CH_2$ groups in membrane lipids can be used to detect the phase transition in native membranes and in many eukaryotic cells (Fig. 2). The results show that when the sarcoplasmic reticulum vesicles described above were dried the average $T_m$ increased from about 0 °C to about 47 °C. But when the vesicles were dried in the presence of trehalose $T_m$ decreased steadily, reaching a stable value of about 21 °C (Fig. 3). The concentration of trehalose required to depress optimally $T_m$ in these intact, dry membranes corresponds closely with that required to produce maximal preservation (Fig. 4).

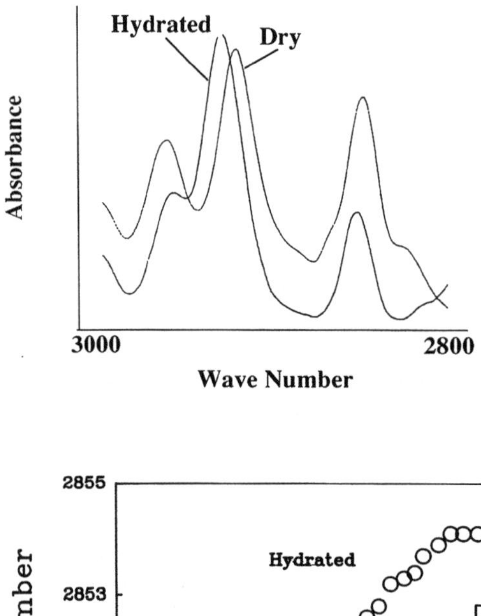


**Fig. 2.** Portions of infrared spectra for sarcoplasmic reticulum membranes in the hydrocarbon chain stretching region. Spectra are shown for dry and hydrated membranes, illustrating the shifts in vibrational frequency. (After Crowe and Crowe 1990)

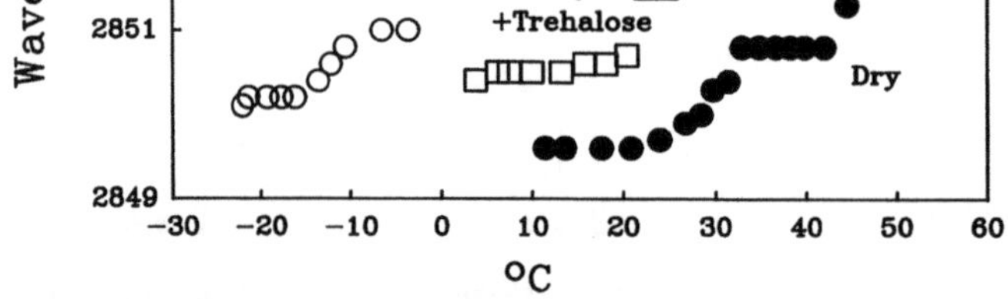


**Fig. 3.** Vibrational frequency as a function of temperature for the $CH_2$ asymmetric stretch in sarcoplasmic reticulum membranes hydrated or dried in the presence of the indicated amounts of trehalose. (After Crowe and Crowe 1990)

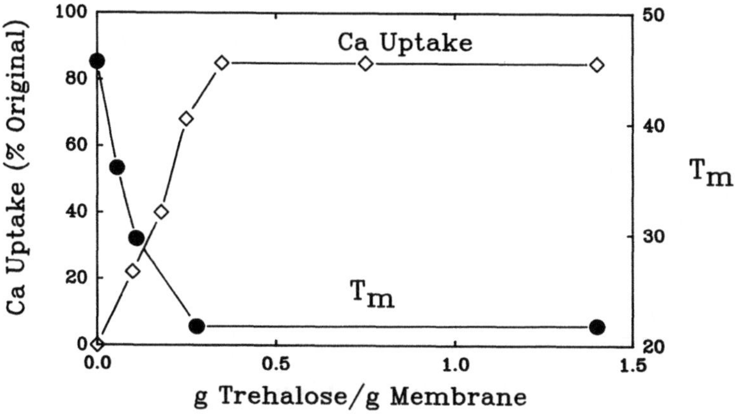


**Fig. 4.** Relationship between $T_m$ of dry sarcoplasmic reticulum membranes (prepared in the presence of various amounts of trehalose) and Ca transport by the rehydrated membranes. (After Crowe and Crowe 1990)

### *Stabilization of Dry Proteins*

Trehalose also preserves labile proteins during drying. For example, phosphofructokinase (PFK) from rabbit is a tetramer that irreversibly dissociates into inactive dimers during drying (Carpenter et al. 1987a,b; Carpenter and Crowe 1988a,b, 1989). Many molecules that stabilize proteins in solution (according to the mechanism described by Timasheff and his colleagues; see Arakawa et al. 1990 for a recent review) will prevent this dissociation when excess water is still present, but when the hydration shell of the protein is removed, the specificity for the stabilization becomes very high. Only disaccharides are effective at stabilizing the protein during extreme dehydration, and trehalose is the most effective. For instance, when PFK is dried partially, it is inactivated (Fig. 5). If proline, which stabilizes PFK in solution is added, the enzyme is stabilized during this partial dehydration, but as the dehydration proceeds this stabilization is lost (Fig. 5). If trehalose is added, by contrast, the enzyme is completely stabilized, even after extreme dehydration (Fig. 5). We have provided evidence that, like the situation with membranes, trehalose interacts directly with the dry protein, probably by hydrogen binding of -OH groups to polar residues in the protein (Carpenter and Crowe 1989), but it is not yet clear how this interaction leads to stabilization.

## Summary of Adaptations to Dehydration: Is Trehalose Sufficient?

One might expect the ability of cellular life to survive dehydration to involve a symphony of adaptations so complex that discovering the fundamental mechanisms by which cells normally survive drying to be daunting. Instead, results from our laboratory have shown that several labile subcellular components can be stabilized

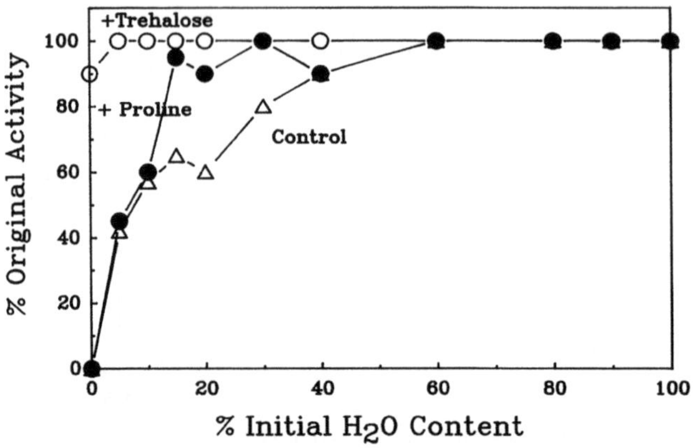


**Fig. 5.** Enzyme activity by phosphofructokinase previously dried in the presence of the indicated compounds and to the indicated water contents before the enzyme was rehydrated. Data from Carpenter and Crowe 1988a)

when they are dried in the presence of trehalose. The subcellular components with which this has been accomplished were derived from cells that in nature rarely experience (and never survive) dehydration – muscle from lobsters and rabbits, for example. Nevertheless, this single factor – addition of trehalose – preserved these parts of cells during drying. Based on these findings, we suggest that if it were possible to introduce trehalose into cells that do not normally synthesize this molecule, it might then be possible to remove all water from such cells without killing them. This intriguing possibility will occupy much of the remaining portion of this review, but before turning to it we wish to clarify a point of confusion about the relationship between the extreme dehydration that we are discussing here and freezing.

## Are Dehydration and Freezing Similar Stresses?

A large number of groups have recently reported that trehalose is an excellent cryoprotectant for isolated membranes and inact cells (Beker et al. 1984; Bhandal et al. 1985; Krag et al. 1985; Strauss et al. 1986; Anchordoguy et al. 1987; Borelli et al. 1987; Honadel and Killian 1988; Greiner et al. 1989; Hino et al. 1990), including, for example, plant cells in culture; human lens, islets of Langehans and arteries; and mouse embryos. These are interesting findings, but the fact that trehalose is a good cryoprotectant does not mean that freezing and the extensive dehydration that anhydrobiotic organisms experience are similar phenomena. In fact, we are convinced that they are distinctly different stress vectors, a suggestion that was the primary subject of a recent review (Crowe et al. 1990). Briefly, we pointed out the following: (1) glycerol and proline are excellent cryoprotectants, but are not effective at all at stabilizing dry membranes or proteins (reviewed in J.H. Crowe et al. 1990).

In fact, many molecules have cryoprotective properties, but only a few will protect against dehydration damage. That observation alone indicates strongly that freezing and dehydration cannot be one and the same phenomenon. (2) The extreme dehydration we are talking about here leads to removal of the minimal water of hydration for phospholipids and proteins (about 0.25 g $H_2O$/g dry mass – about 10 m $H_2O$/mol phospholipid). But there is good evidence that this same fraction of the water does not freeze (reviewed in J.H. Crowe et al. 1990). (3) The mechanism by which solutes stabilize a protein during freezing depends on the solution properties, as defined by Timasheff and his colleagues (see Timasheff, this Vol.) They have shown that if a solute is excluded from the domain of a protein in unfrozen solution it stabilizes the protein. They present convincing thermodynamic arguments as to why this is the case (reviewed in Arakawa et al. 1990). Carpenter and Crowe (1988a,b) have shown that this explanation for the stabilization of proteins in unfrozen solution applies also to stabilization during freezing, but that it clearly cannot explain stabilization during drying. When the bulk water is removed from around the protein, there is clearly no water left into which the solute can be excluded, and at this point the solute must come into direct contact with the protein surface. In fact, Carpenter and Crowe (1989) subsequently showed that stabilization during drying of the protein requires direct, specific interactions between the solute and polar residues in the protein.

## Extension of the Phase Transition Hypothesis to Intact Cells

### *Imbibitional Leakage and Phase Transitions*

When dry cells such as yeasts, seeds, pollen, etc. are placed in water they often leak their contents into the medium, a phenomenon known as "imbibitional leakage," as a result of which they are often killed. We have recently shown that this phenomenon, which has practical effects on survival of dehydration and rehydration by these economically important organisms, is conceptually identical to the process illustrated in Fig. 1. It has been possible to obtain FTIR spectra of hydrocarbons in intact pollen that survive dehydration (similar to those shown in Fig. 2) and to assign phase transitions in these hydrocarbons to membrane phospholipids (J.H. Crowe et al. 1989b,c). Using this method, we showed that in the dry pollen gel phase domains exist in membrane lipids at low temperatures. When the pollen is placed in water at such temperatures they leak their contents and are killed (Fig. 6; Hoekstra et al. 1989); but if the pollen is heated to higher temperatures, at which all membrane lipids are in liquid crystalline phase, the transition is clearly avoided. It has been possible to produce a hydration-dependent phase diagram for the intact pollen which predicts remarkably well the optimal imbibitional temperatures at which leakage is minimized and survival is optimized (Fig. 7). This result has clear and immediate applications in any situation where optimization of survival of dehydration by intact cells is needed.

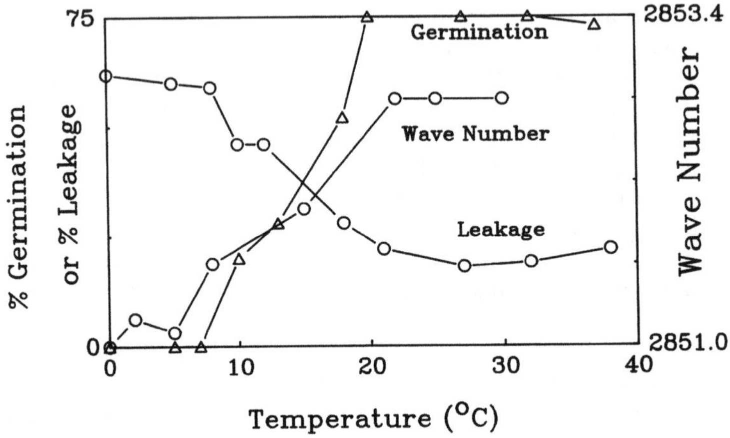


**Fig. 6.** Leakage of $NAD^+$ from pollen grains (*Typha latifolia*) rehydrated at the indicated temperatures, vibrational frequency in $CH_2$ asymmetric stretch as a function of temperature, and germination following rehydration at the same temperatures. (Data from J.H. Crowe et al. 1989b). From such plots $T_m$ and values for the half maximal germination ($G_m$) can be extracted

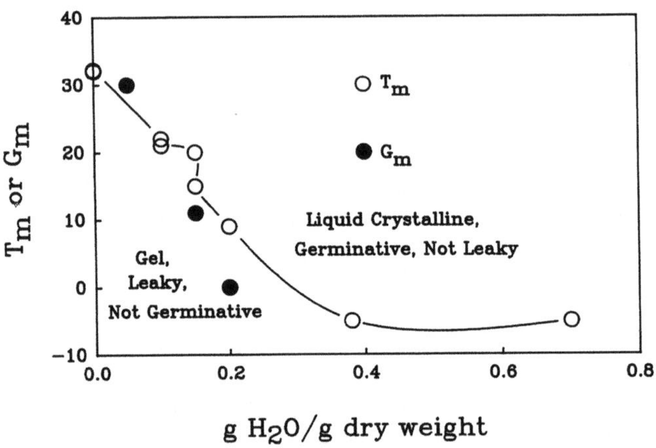


**Fig. 7.** Hydration-dependent phase diagram for pollen grains, showing effects of hydration on $T_m$ and $G_m$. (Data from J.H. Crowe et al. 1988). From such a phase diagram it is possible to predict whether the dry cells will survive rehydration

## *Sugars and Phase Transitions in Intact Cells*

There is some circumstantial evidence that sucrose, which constitutes about 25% of the dry weight in the pollen discussed above, depress $T_m$ in the dry cells. As shown in Fig. 7, $T_m$ rises from –6 °C to +32 °C when the pollen are dried – an increase of 38 °C. By contrast, when a pure phospholipid is dehydrated $T_m$ usually rises by a much larger amount – as much as 75 °C in the case of DPPC. When $T_m$ for polar lipids isolated from the pollen was measured, we found it to increase from –6 °C in excess water to 60 °C when dry, in other words an increase similar to that seen in pure phospholipids. It seems likely that sucrose in the intact pollen could be responsible for the apparent depression of $T_m$ from 60 to 32 °C. We have most recently set out to test this hypothesis, using yeast cells.

## A Role for Trehalose in Dry Yeasts?

### *The Role of Trehalose in Yeast Cells*

There is a large body of evidence that indicates a correlation between the accumulation of trehalose in yeast cells and their ability to germinate (e.g., Rapoport and Beker 1983; Streeter 1985; Brana et al. 1986; Martin et al. 1986; Oda et al. 1986; Attfield 1987; Gadd et al. 1987; Hottiger et al. 1987, 1989; Katohda et al. 1987, 1988; Mcbride and Ensign 1987; Van Laere et al. 1987; Donnini et al. 1988). It was originally proposed that the sugar served simply as a reserve carbohydrate that is mobilized when the organisms are in a catabolic state (reviewed in Panek and Bernardes 1983; Panek 1985). However, since there is some evidence that germination may occur without mobilization of trehalose (Donnini et al. 1988), there is some doubt that it is absolutely required for this process, and that it may be required for the reasons already suggested above.

The cells begin to accumulate trehalose in the stationary phase of growth in culture, and rapidly acquire the ability to survive dehydration as trehalose contents increase. Since the trehalose contents of these cells can be controlled experimentally by removing them from culture at the appropriate time and rapidly drying them, they provide an ideal experimental system for studying the role of trehalose in drying of intact cells. The correlation of trehalose content with survival in these cells appears clear, but its physiological role is far from clear, except by inference from the in vitro studies described above. We are currently exploring such a role in the studies described below.

### *Lipid Phase Transitions and Imbibitional Leakage in Dry Yeast*

There is excellent indirect evidence that the lipid phase transition model for pollen applies equally well to yeasts. Van Steveninck and Ledeboer (1974) showed that when dry yeast was placed directly in water the cells leaked $K^+$ extensively and showed poor survival. They also reported that if the cells were hydrated over water vapor to water contents ≥20% before they were placed in water leakage was avoided, a finding very similar to that shown for pollen in Fig. 7. Alternatively, if the dry cells were heated to 42 °C before they were placed in water leakage was also avoided. Based on these observations, they suggested that the leakage is probably due to lipid phase transitions – an hypothesis similar to that shown in Fig. 7 for pollen. Van Steveninck and Ledeboer (1974) had no way of testing their lipid phase transition hypothesis, and their suggestion was not pursued further until recently. The necessary experiments are still in progress, but the results in hand thus far are consistent with a lipid phase transition hypothesis: $T_m$ in hydrated yeast cells, as measured with FTIR, is about 10 °C, while that in the dehydrated cells appears to be about 30 °C. Thus, if the cells are placed in water at room temperature they would be expected to experience lipid phase transition during rehydration, but if they are rehydrated at temperatures above 30 °C, that phase transition will be obviated.

### *Potential Routes for the Introduction of Trehalose into Cells*

The data suggest that introduction of trehalose into the cytoplasm of cells that do not normally synthesize this molecule might be sufficient to confer resistance to damage from dehydration. The available routes are summarized in Fig. 8. (1) chemical permeabilization. Reversibly permeabilizing the plasma membrane of cells has long been done for purposes of, for example, introducing DNA into cells to transform them. We have attempted to use the same means to introduce trehalose into cells, with good success, but the cells do not survive dehydration afterwards. However, we strongly suspect that the harsh treatment involved in the permeabilization itself makes the cells more fragile and thus more susceptible to damage, so we have turned to other methods. (2) Genetic engineering. This approach involves obtaining the genes coding for enzymes responsible either for trehalose synthesis or for transport of the sugar across the plasma membrane. Work involving these approaches is summarized below.

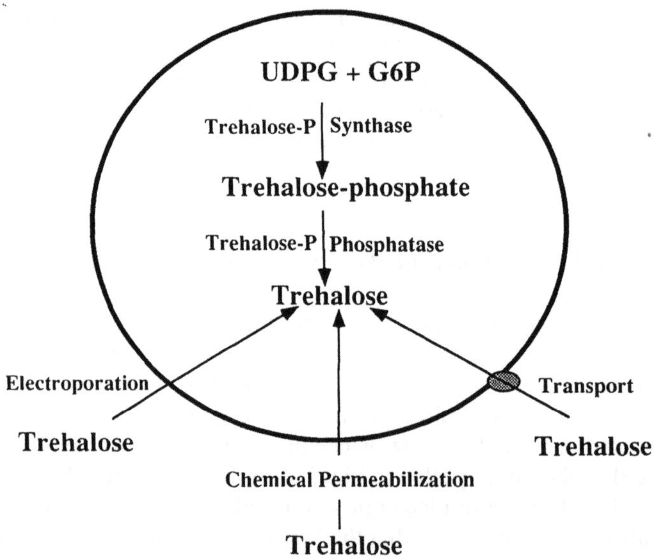


**Fig. 8.** Mechanisms for introducing trehalose into cells

## Yeasts: a Potent System for Genetic Manipulation of Trehalose Synthesis

### *Studies on Genetics of Trehalose Synthesis*

The simplicity of the classical pathway for trehalose synthesis (illustrated in Fig. 8) immediately suggests that the key enzyme should be amenable to genetic manipulation. In fact, important progress has already been made in this regard with yeast cells. Panek and her colleagues (Charlab et al. 1985; Panek et al. 1987; Coutinho et al. 1988) have shown that a mutant possessing a lesion in trehalose-phosphate synthase activity is unable to synthesize trehalose, and cannot survive dehydration (Coutinho et al. 1988). Similarly, a mutant lacking trehalose-phosphate phosphatase activity has also been constructed, but the physiological consequences of this deletion have not been investigated (Piper and Lockheart 1988). With these mutants in hand, it should now be possible to obtain the genes for the two key enzymes by complementation of the mutant phenotypes, using a genomic library. However, this proposal may be more complex than it sounds. Trehalosephosphate synthase has not yet been purified, but preliminary results from Anita Panek's laboratory suggest that the protein may be multimeric and may involve multiple loci and thus may not be amenable to cloning.

As an alternative to obtaining trehalose via trehalose phosphate synthase, evidence for the existence of an alternative pathway involving enzymatic conversion of maltose to trehalose exists (Murao et al. 1985; De Oliveira et al. 1986; Paschoalin et al. 1986). However, this pathway is not well understood as yet.

### *Survival of Drying in Trehalose-Phosphate Synthase Mutants*

Panek and her colleagues (Coutinho et al. 1988) have recently shown that addition of trehalose externally to the mutant yeast cells lacking the ability to synthesize trehalose restored their ability to survive dehydration. This elegant result, in which the only perturbation made to the system was addition of trehalose, again suggests a central role for this sugar. On first examination, it would seem unlikely that addition of trehalose just to the outside of the cells would stabilize their internal components, particularly since trehalose (like other sugars) crosses most membranes slowly, if at all, in the absence of a carrier. Such a carrier is known to exist in yeasts, as described below.

### *A Transport System for Trehalose in Yeasts*

Trehalose accumulation by yeast cells from external sources was documented more than a decade ago (Kotyk and Michaljanicova 1979; Kotyk et al. 1985), but until recently (Araujo et al. 1991; J.H. Crowe et al. 1991) there has been no further investigation of this transporter in yeasts. There is good evidence for a transport system for trehalose in bacteria (Marechal 1984; Postma et al. 1986; Poy and Jacobson 1990). It appears to have some distinct differences from that seen in yeasts (J.H. Crowe et al. 1991).

The trehalose transporter in yeasts does not follow Michaelis-Menten-type kinetics; instead of a hyperbolic relationship between velocity and substrate concentration, the curve is clearly sigmoidal, indicating considerable cooperativity (Fig. 9). Furthermore, there appear to be two forms of the transporter, a high affinity form that is saturable in the 3–5 mM range, and a lower affinity one that is saturable in the 25–30 mM range. The existence of high and low affinity transport systems for monosaccharides has been known for some time. There are still insufficient data to generalize about the nature of these two processes, but it appears that in the case of monosaccharides, the high affinity process depends on the activity of specific kinases, the roles of which are still unclear (see Poy and Jacobson 1990 for references). The low affinity process seems to be due to facilitated diffusion. Information about the mechanisms underlying similar low and high affinity transport process for disaccharides is almost completely lacking in yeasts. However, according to Kotyk and Michaljanicova (1979) both the low and high affinity systems depend on external pH, with an optimal extracellular pH of 5.5. This finding suggests that both systems may involve a proton symport-mediated active transport.

### *Conditions for Expression of the Trehalose Transporter*

The trehalose transporter showed low activity in log phase cells, but appeared in early stationary phase, coincident with the decline in glucose concentration in the growth medium. Trehalose transport activity increased fourfold by about 2 h after the complete depletion of glucose (Fig. 10). Thus, appearance of transport activity coincides with the period when these cells commence synthesis of trehalose (Panek 1985). The important point here is that the transporter appears to be inducible.

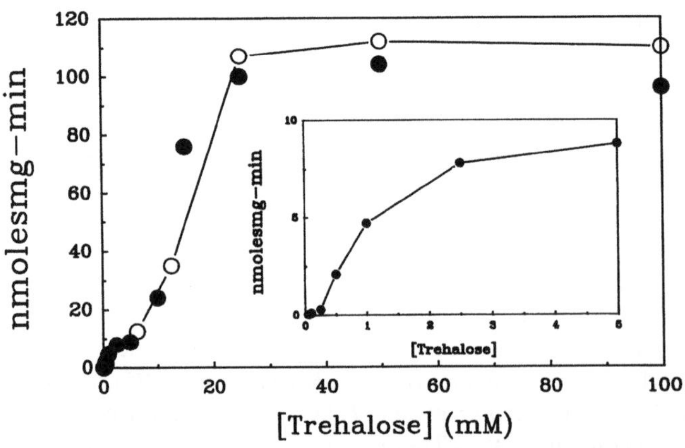


**Fig. 9.** Kinetics of trehalose transport by intact cells of baker's yeast, *Saccharomyces cerevisiae*. Open and closed circles represent two independent measurements. (Data from J.H. Crowe et al. 1990)

Furthermore, if glucose is added to medium containing cells in which the transporter is being expressed, within 30 min transport activity delines to preinduction levels (Fig. 10). When glucose was removed from the culture, and trehalose was added back, transport activity returned to maximal activity after about 2 h (Fig. 10). Resumption of trehalose transport following inhibition in the presence of glucose requires protein synthesis; addition of either cycloheximide or ethidium bromide abolished the resumption of transport activity (Fig. 10).

Two factors may provide opportunities for genetic manipulation of the transport system further: the system is inducible, and the induction appears to require transcription of mRNA and protein synthesis (Fig. 10; J.H. Crowe et al. 1991). Furthermore, mutants apparently already exist for the genes responsible for coding for the transporter, although they have not been formally characterized (Smaal, pers. commun.). Thus, it may be possible to use either these mutants or the fact that the transporter is inducible as tools in isolating and cloning the gene(s) for the transporter. We have already developed a membrane preparation from the yeasts that shows trehalose transport (so long as the cells from which the membranes were prepared were expressing the transporter), and we are using these membranes in which the transporter is present or nor to seek to identify the protein associated with the transport (Araujo et al. 1991).

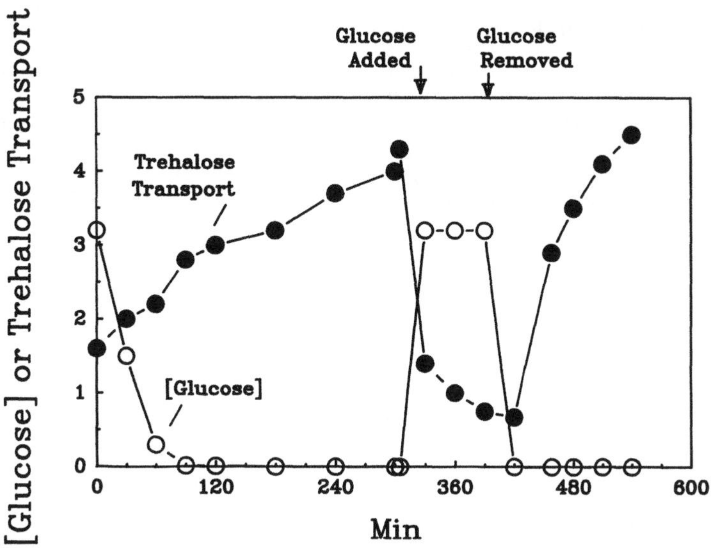


**Fig. 10.** Conditions under which the trehalose transporter is expressed in yeasts. After glucose exhaustion in the growth medium the transporter appears. It is rapidly degraded if glucose is added back to the medium, but reappears if the glucose is removed. The reappearance depends on transcription of RNA and protein synthesis, as indicated by the effects of ethidium bromide and cycloheximidine. (Data from J.H. Crowe et al. 1990)

## Summary and Prospects for Further Research

We have established that trehalose, and to a lesser extent other dissacharides, stabilize structure and function in two components of cells that are labile to drying: membranes and water-soluble proteins. We believe that we are now in a position to commence testing some of the paradigms established for these subcellular components in intact cells, and we have already begun doing so. Furthermore, we believe that with the existence of yeast mutants with defects in trehalose phosphate synthesis and for the trehalose transport system the time is ripe for isolation of the genes responsible for coding for these proteins. The potential applications of such studies are clear; if the genes can be isolated and cloned it may then become possible to introduce them, in an appropriate expression vector, into any cell that one wishes to stabilize in the dry state.

*Acknowledgments:* We gratefully acknowledge support for this project by Fulbright Senior Research Fellowships and research grants from the US National Science Foundation (DCB89-18822) and US Department of Agriculture Competetive Grants Program (88-37264-4068).

## References

Anchordoguy TJ, Rudolph AS, Carpenter JF, Crowe JH (1987) Modes of interaction of cryoprotectants with membrane phospholipids during freezing. Cryobiology 24:324–331

Arakawa T, Carpenter JF, Kita YA, Crowe JH (1990) The basis for toxicity of certain cryoprotectants: an hypothesis. Cryobiology 27:401–415

Araujo PS, Panek AC, Crowe JH, Crowe LM, Panek AD (1991) Trehalose transporting membrane vesicles prepared from yeasts. Biochem Int (in press)

Attfield PV (1987) Trehalose accumulates in *Saccharomyces cerevisiae* during exposure to agents that induce heat shock response. FEBS Lett 225:259–263

Beker MJ, Blumbergs JE, Ventina EJ, Rapoport AI (1984) Characteristics of cellular membranes at rehydration of dehydrated yeast *Saccharomyces cerevisiae*. Eur J Appl Microbiol Biotechnol 19:347–352

Bhandal IS, Hauptmann RM, Widholm JM (1985) Trehalose as cryoprotectant for the freeze preservation of carrot and tobacco cells. Plant Physiol 78:430–432

Borelli MI, Semino MC, Hernandez RE (1987) Cryopreservation of islets of Langerhans: the use of trehalose as a cryoprotective agent. Med Sci Res 15:299–300

Brana AF, Mendez C, Diaz LA, Manzanal MB, Hardisson C (1986) Glycogen and trehalose accumulation during colony development in *Streptomyces antibioticus*. J Gen Microbiol 132:1319–1326

Caffrey M, Fonseca V, Leopold AC (1988) Sugar-lipid interactions. Relevance to anhydrous biology. Plant Physiol 86:754–758

Carpenter JF, Crowe JH (1988a) Modes of stabilization of a protein by organic solutes during desiccation. Cryobiology 25:459–470

Carpenter JF, Crowe JH (1988b) The mechanisms of cryoprotection of proteins by solutes. Cryobiology 25:244–255

Carpenter JF, Crowe JH (1989) An infrared spectroscopic study of the interactions of carbohydrates with dried proteins. Biochemistry 28:3916–3922

Carpenter JF, Crowe LM, Crowe JH (1987a) Stabilization of phosphofructokinase with sugars during freeze-drying: characterization of enhanced protection in the presence of divalent cations. Biochim Biophys Acta 923:109–115

Carpenter JF, Martin B, Crowe LM, Crowe JH (1987b) Stabilization of phosphofructokinase during air-drying with sugars and sugar/transition metal mixtures. Cryobiology 24:455–464

Chandrasekhar I, Gaber BP (1988) Stabilization of the bio-membrane by small molecules: interaction of trehalose with the phospholipid bilayer. J Biomol Str Dyn 5:1163–1171
Charlab R, Oliveira DE, Panek AD (1985) Investigation of the relationship between sstl and fdp mutations in yeast and their effect on trehalose synthesis. Braz J Med Biol Res 18:447–454
Clegg JS (1986) The physical properties and metabolic status of Artemia cysts at low water contents: the "water replacement" hypothesis. In: Leopold AC (ed) Membranes, Metabolism, and Dry Organisms. Cornell Univ Press, Ithaca, NY, pp 169–187
Coutinho C, Bernardes E, Felix D, Panek AD (1988) Trehalose as cryoprotectant for preservation of yeast strains. J Biotechnol 7:23–32
Crowe JH, Crowe LM (1986) Stabilization of membranes in anhydrobiotic organisms. In: Leopold AC (ed) Membranes, Metabolism, and Dry Organisms. Cornell Univ Press, Ithaca, NY, pp 188–209
Crowe JH, Crowe LM (1988) Factors affecting the stability of dry liposomes. Biochim Biophys Acta 939:327–334
Crowe JH, Crowe LM, Jackson SA (1983) Preservation of structural and functional activity in lyophilized sarcoplasmic reticulum. Arch Biochem Biophys 220:447–484
Crowe JH, Crowe LM, Chapman D (1984) Preservation of membranes in anhydrobiotic organisms: the role of trehalose. Science 223:701–703
Crowe JH, Crowe LM, Carpenter JF, Aurell Wistrom C (1987a) Stabilization of dry phospholipid bilayers and proteins by sugars. Biochem J 242:1–10
Crowe JH, Spargo BJ, Crowe LM (1987b) Preservation of dry liposomes does not require retention of residual water. Proc Natl Acad Sci USA 84:1537–1540
Crowe JH, Crowe LM, Carpenter JF, Rudolph AS, Aurell Wistrom C, Spargo BJ, Anchordoguy TJ (1988) Interactions of sugars with membranes. Biochim Biophys Acta 947:367–384
Crowe JH, Crowe LM, Hoekstra FA (1989a) Phase transitions and permeability changes in dry membranes during rehydration. J Bioenerg Biomembr 21:77–91
Crowe JH, Hoekstra FA, Crowe LM (1989b) Membrane phase transitions are responsible for imbibitional damage in dry pollen. Proc Natl Acad Sci USA 86:520–523
Crowe JH, Hoekstra FA, Crowe LM, Anchordoguy TJ, Drobnis E (1989c) Lipid phase transitions measured in intact cells with Fourier transform infrared spectroscopy. Cryobiology 26:76–84
Crowe JH, Carpenter JF, Crowe LM, Anchordoguy TJ (1990) Are freezing and dehydration similar stress vectors? A comparison of modes of interaction stabilizing solutes with biomolecules. Cryobiology 27:219–231
Crowe JH, Panek AD, Crowe LM, Panek AC, Araujo PS (1991) Trehalose transport in yeast cell. Biochem Int (in press)
Crowe LM, Crowe JH (1988a) Lyotropic effects of water on phospholipids. In: Aloia RC (ed) Physiological Regulation of Membrane Fluidity. Alan R Liss Inc, New York, pp 75–99
Crowe LM, Crowe JH (1988b) Trehalose and dry dipalmitoylphosphatidylcholine revisited. Biochim Biophys Acta 946:193–201
Crowe LM, Crowe JH, Rudolph A, Womersley C, Appel L (1985) Preservation of freeze-dried liposomes by trehalose. Arch Biochem Biophys 242:240–247
Crowe LM, Womersley C, Crowe JH, Reid D, Appel L, Rudolph A (1986) Prevention of fusion and leakage in freeze-dried liposomes by carbohydrates. Biochim Biophys Acta 861:131–140
de Oliveria DE, Arrese M, Kidane G, Panek AD, Matoon JR (1986) Trehalose and maltose metabolism in yeast transformed by a MAL4 regulatory gene cloned from a constitutive donor strain. Curr Genet 11:97–106
Donnini C, Puglisi PP, Vecli A, Marmiroli N (1988) Germination of *Saccharomyces cerevisiae* ascospores without trehalose mobilization as revealed by in vivo. C-13 nuclear magnetic resonance spectroscopy. J Bacteriol 170:3789–3791
Gadd GM, Chalmers K, Reed RH (1987) The role of trehalose in dehydration resistance of *Saccharomyces cerevisiae*. FEMS Microbiol Lett 48:249–254
Greiner JV, Medcalf SK, Meneses P, Glonek T (1989) Trehalose maintenance of the metabolic health of the crystalline lens during severe temperature stress. Invest Opthalmol Visual Sci 27 (Suppl): 278–278
Harding TS (1923) The sources of the rare sugars. IX-History of trehalose, its discovery and methods of preparation. Sugar 25:476–478

Hino A, Mihara K, Nakashima K, Takano H (1990) Trehalose levels and survival ratio of freeze-tolerant versus freeze-sensitive yeasts. Appl Environ Microbiol 56:1386–1391

Hoekstra FA, van Roekel T (1988) Desiccation tolerance of *Papaver dubium* during its development in the anther. Possible role of phospholipid composition and sucrose content. Plant Physiol 88:626–632

Hoekstra FA, Crowe LM, Crowe JH (1989) Differential desiccation sensitivity of corn and *Pennisetum* pollen linked to their sucrose contents. Plant Cell Environ 23:83–91

Honadel TE, Killian GJ (1988) Cryopreservation of murine embryos with trehalose and glycerol. Cryobiology 25:331–337

Hottiger T, Boller T, Wiemken A (1987) Rapid changes of heat and desiccation tolerance correlated with changes of trehalose content in *Saccharomyces cerevisiae* cells subjected to temperature shifts. FEBS Lett 220:113–115

Hottiger T, Boller T, Wiemken A (1989) Correlation of trehalose content and heat resistance in yeast mutants altered in the RAS/adenylate cyclase pathway: is trehalose a thermoprotectant? FEBS Lett 255:431–434

Katohda S, Ito M, Sadaki K, Takahashi M (1987) Carbohydrate composition during germination and outgrowth of ascospores of *Saccharomyces cerevisiae*. Agric Biol Chem 51:2975–2981

Katohda S, Ito H, Takahashi H, Kikuchi H (1988) Carbohydrate metabolism during sporulation in spheroplasts of yeasts, *Saccharomyces cerevisiae*. Agric Biol Chem 52:349–355

Koster KL, Leopold AC (1988) Sugars and desiccation tolerance in seeds. Plant Physiol 88:829–832

Kotyk A, Michaljanicova D (1979) Uptake of trehalose by *Saccharomyces cerevisiae*. J Gen Microbiol 110:323–332

Kotyk A, Michaljanicova D, Struzinsky R, Baryshnikova LM, Sychrova H (1985) Absence of glucose-stimulated transport in yeast protoplasts. Folia Microbiol 30:110–116

Krag KT, Koehler I-M, Wright RW (1985) Trehalose – a nonpermeable cryoprotectant for direct freezing of early stage murine embryos. Cryobiology 22:636–636

Lee CWB, Waugh JS, Griffin RG (1986) Solid-state NMR study of trehalose/1,2-dipalmitoyl-sn-phosphatidylcholine interactions. Biochemistry 25:3737–3742

Lee CWB, Das Gupta SK, Mattai J, Shipley GG, Abdel-Mageed OH, Makriyannis A, Griffin RG (1989) Characterization of the Llambda phase in trehalose-stabilized dry membranes by solid-state NMR and X-ray diffraction. Biochemistry 28:5000–5009

Leopold AC (1986) (ed ) Membranes, Metabolism, and Dry Organisms. Cornell Univ Press, Ithaca, NY, 374 pp

Leprince O, Bronchart R, Deltour R (1990) Changes in starch and soluble sugars in relation to the acquisition of desiccation tolerance during maturation of *Brassica campestris* seed. Plant Cell Environ 13:539–546

Mackenzie KF, Singh KK, Brown AD (1988) Water stress plating hypersensitivity of yeasts: protective role of trehalose in *Saccharomyces cerevisiae*. J Gen Microbiol 134:1661–1666

Madden TD, Balley MB, Hope MJ, Cullis PR, Schieren HP, Janoff AS (1985) Protection of large unilamellar vesicles by trehalose during dehydration: retention of vesicle contents. Biochim Biophys Acta 817:67–74

Marechal LR (1984) Transport and metabolism of trehalose in *Escherichia coli* and *Salmonella typhimurium*. Arch Microbiol 137:70–73

Martin MC, Diaz LA, Manzanal MB, Hardisson C (1986) Role of trehalose in the spores of *Streptomyces*. FEMS Microbiol Lett 35:49–54

McBride MJ, Ensign JC (1987) Effects of intracellular trehalose content on *Streptomyces griseus* spores. J Bacteriol 169:4995–5001

Mouradian R, Womersley C, Crowe LM, Crowe JH (1984) Preservation of functional integrity during long term storage of a biological membrane. Biochim Biophys Acta 778:615–617

Murao S, Nagano H, Ogura S, Nishino T (1985) Enzymatic synthesis of trehalose from maltose. Agric Biol Chem 49:2113–2118

Oda Y, Uno K, Ohta S (1986) Selection of yeasts for breadmaking by the frozen-dough method. Appl Environ Microbiol 52:941–943

Ozer Y, Talsma H, Crommelin DJA, Hincal AA (1988) Influence of freezing and freeze-drying on the stability of liposomes dispersed in aqueous media. Acta Pharmacol Technol 34:129–139

Panek AD (1985) Trehalose metabolism and its role in *Saccharomyces cerevisiae*. J Biotechnol 3:121–130

Panek AD, Bernardes EJ (1983) Trehalose: its role in germination of *Saccharomyces cerevisiae*. Curr Genet 7:393–397

Panek AD, de Araujo PS, Neto VM, Panek AD (1987) Regulation of the trehalose-6-phosphate synthase complex in *Saccharomyces*. Curr Genet 11:459–465

Paschoalin VMF, Costa-Carvalho VLA, Panek AD (1986) Further evidence for the alternative pathway of trehalose synthesis linked to maltose utilization in *Saccharomyces*. Curr Genet 10:725–731

Paschoalin VMF, Panek AC, Panek AD (1987) Catabolite inactivation of trehalose synthesis during growth of yeast on maltose. Braz J Med Biol Res 20:675–683

Piper PW, Lockheart A (1988) A temperature-sensitive mutant of *Saccharomyces cerevisiae* defective in the specific phosphatase of trehalose biosynthesis. FEMS Microbiol Lett 49:245–250

Postma PW, Keizer HG, Koolwijk P (1986) Transport of trehalose in *Salmonella typhimurium*. J Bacteriol 168:1107–1111

Poy F, Jacobson GR (1990) Evidence that a low-affinity sucrose phosphotransferase activity in *Streptococcus mutans* GS-5 is a high-affinity trehalose uptake system. Infect Immun 58:1479–1480

Quinn PJ (1989) Effect of sugars on the phase behaviour of phospholipid model membranes. Biochem Soc Trans 17:953–957

Rapoport AI, Beker ME (1983) Effect of sucrose and lactose on resistance of the yeast *Saccharomyces cerevisiae* to dehydration. Mikrobiologiya 52:556–559

Reshkin SJ, Cassano G, Womersley C, Ahearn GA (1990) Preservation of glucose transport and enzyme activity in fish intestinal brush border and basolateral membrane vesicles. J Exp Biol 140:123–136

Strauss G, Hauser H (1986) Stabilization of lipid bilayer vesicles by sucrose during freezing. Proc Natl Acad Sci USA 83:2422–2426

Strauss G, Schurtenberger P, Hauser H (1986) The interaction of saccharides with lipid bilayer vesicles: stabilization during freeze-thawing and freeze-drying. Biochim Biophys Acta 858:169–180

Streeter JG (1985) Accumulation of alpha, alpha-trehalose by *Rhizobium* bacteria and bacteroids. J Bacteriol 164:78–84

Tanford C (1980) The Hydrophobic Effect. Wiley, New York

Tsvetkov TD, Tsonev LI, Tsvetkova NM, Koynova RD, Tenchov BG (1989) Effect of trehalose on the phase properties of hydrated and lyophilized dipalmitoylphosphatidylcholine multilayers. Cryobiology 26:162–169

Tsvetkova N, Tsvetkov TS, Tenchov B, Tsonev L (1988) Dependence of trehalose protective action on the initial phase state of dipalmitoylphosphatidylcholine bilayers. Cryobiology 25:256–263

Van Laere A, Slegers LK (1987) Trehalose breakdown in germinating spores of *Mucor rouxii*. FEMS Microbiol Lett 41:247–252

Van Laere A, Francois A, Overloop K, Verbeke M, Van Gerven L (1987) Relation between germination, trehalose and the status of water in *Phycomyces blakesleeanus* spores as measured by proton-NMR. J Gen Microbiol 133:239–245

Van Steveninck J, Ledeboer AM (1974) Phase transitions in the yeast cell membrane. The influence of temperature on the reconstitution of active dry yeast. Biochim Biophys Acta 352:64–70

Chapter 8

# Water Content and Metabolic Organization in Anhydrobiotic Animals

S.C. Hand

## Introduction

Fluctuation in cellular water content is a universal problem confronting both aquatic and terrestrial organisms. Subfreezing temperatures, desiccating conditions prevalent in xeric climates, and the osmotic variation seen in aqueous habitats are common environmental occurrences that can pose severe problems of water stress (Yancey et al. 1982). Both the extent and nature of energy metabolism can be profoundly influenced by changes in cell-associated water, and this topic has been a focus of our laboratory over the last several years.

One group of animals ideally suited as model systems for assessing relationships between water stress and cellular metabolism are the anhydrobiotes. These organisms, which include invertebrates like nematodes, rotifers, tardigrades, and certain crustacean embryos and insect larvae, can withstand the removal of virtually all of their cell-associated water without significant compromise of viability (Crowe and Clegg 1973, 1978; Hinton 1960). In response to dehydration these animals enter a state of cryptobiosis in which metabolism comes to a standstill in a reversible fashion (Keilin 1959; Hinton 1968; Clegg 1973); the organism at this point is reduced to nothing more than a morphological state (Hinton 1968).

The residual water in anhydrobiotes – defined by Clegg et al. (1978) as that amount of tightly bound water remaining after drying over strong desiccants and/or low pressures for long periods – can be exceedingly low. It is quite typical for certain nematodes and tardigrades to tolerate reversible dehydration down to 2% tissue water (Crowe and Madin 1974, 1975). The residual water content of anhydrobiotic brine shrimp embryos (*Artemia franciscana*) has been reduced to 0.0069 g water/g dry mass by chilling the dry embryos to liquid nitrogen temperatures under vacuum and then bombarding the embryos with nitrogen vapor (Clegg et al. 1978). Remarkably, the cellular viability was not reduced even after 11 cycles of this treatment. Considering that fully hydrated embryos have a water content on the order of 1.4 g water/g dry embryo (Clegg 1974), these encysted gastrulae can experience a 200-fold range in water content – the largest range tolerated by animal

Department of Environmental, Population and Organismic Biology, University of Colorado, Boulder, CO 80309–0334, USA

Somero et al. (Eds.)
Water and Life

cells to my knowledge. Clearly, anhydrobiotes can provide researchers with great experimental flexibility in evaluating the influence of water stress on metabolic processes.

In this essay, I will attempt to assess the current views on the nature of cellular water of anhydrobiotic organisms, correlate metabolic events with levels of hydration, evaluate the possible mechanisms for disruption of metabolic structure and function, and indicate promising techniques and experimental approaches for future research. For expanded literature reviews and historical perspectives on these topics, readers may wish to consult Keilin (1959), Cooper and Van Gundy (1971), Crowe and Clegg (1973, 1978), Clegg (1984a), Womersley (1981, 1987), and Hand (1991).

## Nature of Water in Anhydrobiotic Animals

### *Kinetic Features of Intracellular Water*

Without question, anhydrobiotic organisms are unusual in their ability to withstand water stress, but there is no reason to suggest that the nature of cell-associated water is qualitatively or quantitatively any different from other organisms when the cryptobiote is in a state of full hydration. The most extensive data set describing the properties of intracellular water in an anhydrobiotic organism comes from studies on the encysted embryo of *Artemia*. For example, J.S. Clegg and associates have employed nuclear magnetic resonance (NMR) spectroscopy (Seitz et al. 1981), quasi-elastic neutron scattering (Trantham et al. 1984), microwave dielectrics (Clegg et al. 1982), differential scanning calorimetry (Clegg 1979), and density measurements (Clegg 1984b) to probe the physical characteristics of water in these embryos. Emphasizing selected aspects of these data will help place in perspective the status of water in fully hydrated anhydrobiotic organisms.

The water in *Artemia* embryos displays kinetic properties that are clearly different from those of pure water (Table 1). As estimated by nuclear magnetic resonance (NMR) of water protons, the values for spin-lattice and spin-spin relaxation times ($T_1$, $T_2$) and the self-diffusion coefficient (D) are reduced 11-, 34-, and 6-fold, relative to those measured for pure bulk water, which suggests that some fraction of the cellular water in the embryo is restricted in movement. These reductions are very much within the range seen for numerous other cells and tissues. $T_1$ values fall roughly between 100 and 1100 ms, and $T_2$ values are generally 10–250 (Hazelwood 1979). The ratio of the self-diffusion coefficient of cellular water to that of pure water ($D/D_0$) ranges between about 0.15 to 0.7 (Hazelwood 1979; Burnell et al. 1981).

It is important to note that NMR diffusion measurements alone cannot be used to quantify the precise fraction of water molecules perturbed inside of cells (Clark et al. 1982). Explanations in the literature for these reductions in cellular $T_1$ and $T_2$ values range from a small tightly bound fraction of water that is exchanging with an unrestricted bulk phase, to a view that essentially all the cellular water has reduced motion (Clegg 1986). The self-diffusion coefficient is a simple average for the cellular water as a whole. Depending on the model chosen, a portion of the reductions in D can be explained in part by what is referred to as obstructional or compartmental

**Table 1.** Physical measurements of the motion of intracellular water in *Artemia* embryos, compared to pure water. The NMR data are from Seitz et al. (1981); QNS data are from Trantham et al. (1984)

| Technique | *Artemia* embryos[1] | Pure water |
|---|---|---|
| Nuclear magnetic resonance | | |
| $T_1$ (ms) | 275 | 3000 |
| $T_2$ (ms) | 53 | 1800 |
| D ($10^{-5}$ $cm^2/s$) | 0.4 | 2.41 |
| Quasi-elastic neutron scattering | | |
| D ($10^{-5}$ $cm^2/s$) | 0.8 | 2.4 |

[1]All values are for embryos at maximal hydration (about 1.4 g water/g dry mass cyst).

effects; since NMR measures the translational movement of water protons over relatively large distances (several microns), cellular structures can impede this movement [for further discussion see Trantham et al. (1984), Chang et al. (1973), Rorschach et al. (1973)]. So, ultimately one would like to have D values measured over short distances (i.e., short time intervals) so that water molecules do not encounter barriers.

Quasi-elastic neutron scattering (QNS) has the capability of giving translational information over periods as short as $10^{-12}$ s (corresponding to a few angstroms) (Trantham et al. 1984). However, QNS is rarely applied to biological systems because the sample must remain closely packed and sealed for several days, a procedure not tolerated by most cells/tissues (Clegg 1986). Hydrated *Artemia* embryos can easily withstand this treatment with no loss of viability, since these embryos enter a state of anaerobic dormancy (Ewing and Clegg 1969; Busa and Crowe 1983; Hand 1990, 1991) that can extend their anoxia tolerance up to months (Clegg and Jackson 1989). As seen in Table 1, the $D/D_0$ value for *Artemia* determined by QNS is 0.33, compared to 0.17 for the ratio determined by NMR. Consequently, Clegg and associates concluded that some of the reduction in D as measured by NMR was indeed due to barrier phenomena. But a $D/D_0$ value of 0.33 from QNS demonstrates that even after surface phenomena are removed the water in *Artemia* is clearly restricted compared to pure water.

### *Influence of Osmolyte Systems on Cellular Water*

One aspect of the preceding comparisons that deserves comment is that the physical values obtained for intracellular water are typically viewed in reference to those of pure water. The resulting differences between these data sets are generally ascribed to the strong organizing properties of biological surfaces inside the cell. The surface areas of cytoskeletal elements, membrane surfaces of intracellular organelles and plasma membranes, and more recently, the microtrabecular lattice (Porter 1986) represent an enormous area for water interfaces. In an excellent review, Clegg (1984a) calculated that simply to cover these surfaces with a single monolayer of water would require up to 4% of the cytoplasmic water. When one considers that hydrophobic surfaces theoretically may influence the motility of water 30 Å or more away (10 water layers or so), the percentage of cellular water perturbed could be far greater (Clegg 1984a). Thus, it can hardly be doubted that biological surfaces explain a significant portion of the restricted water movement in cells.

However, what these comparisons fail to incorporate is the strong effect that intracellular solutes have on the inter-molecular organization of water. After one extended discussion of the impact of cellular structures on $D/D_0$ ratios, which took place 18 years ago during a noted conference on water biophysics, Gilbert Ling felt compelled to remark, "I want once more to remind you that the living cell is not water alone, but ions and water also, which are very much together" (Hazelwood 1973, p. 455). Indeed, the cytoplasm is far more than water and structural surfaces. It also, among other components, contains a complex mixture of inorganic and organic solutes collectively refered to as the osmolyte system (Yancey et al. 1982; Somero 1986). To compare NMR and QNS data collected on intact cells and tissues to pure water seems to be a less than satisfactory way to dissect the role of biological surfaces in structuring cytoplasmic water. Equally important, such comparisons do not address the significant effect that osmolytes can have on the kinetic properties of water.

The data of McCall and Douglass (1965) demonstrate the effects of numerous inorganic ions on the self diffusion coefficient of water (Fig. 1). Substantial changes are generally not observed until concentrations of 0.5 M and above are reached. While such effects are not significant for a typical cell under conditions of full hydration, the perturbing effects of inorganic ions on water may become measurable when one considers the dehydration regimes experienced by anhydrobiotic animals. As will be discussed below, concentrations of inorganic ions increase markedly under these conditions of restricted water availability in *Artemia* embryos.

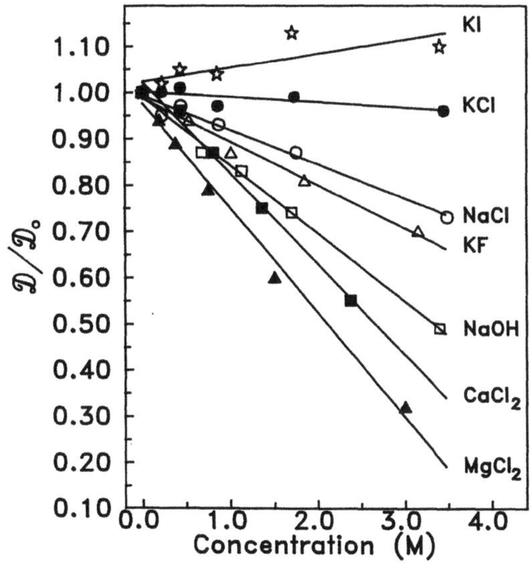


**Fig. 1.** Dependence of the relative self diffusion coefficient of water ($D/D_0$) upon the concentration of various inorganic ions. (Data plotted are from the tables of McCall and Douglass 1965)

It is now clear that the organic components of intracellular osmolyte systems, based principally on the work of Mary E. Clark and associates (also see Goldhammer and Hertz 1970), have far stronger effects on the diffusive properties of water than do inorganic ions. These classes of compounds – polyhydric alcohols (polyols), amino acids and their derivatives, other nitrogenous solutes – have pronounced influences on macromolecular structure and function, as Drs Somero and Yancey have emphasized in their contributions to this symposium. In an illuminating series of papers (Clark and Hinke 1981; Clark et al. 1981; Burnell et al. 1981; Clark et al. 1982; Clark 1985, 1987) many effects of these biological solutes on osmotic properties of the intracellular milieu have been quantified. Of particular interest for this discussion is the observation that trimethylamine-N-oxide, glycerol, and mixtures of solutes reduce significantly the self-diffusion coefficient of water at physiologically relevant concentrations (Fig. 2). Weight for weight, organic solutes have the same effect on the $D/D_0$ as soluble proteins, which also markedly reduce this value (Clark et al. 1982).

The data in Table 2 document the large contribution of organic osmolytes and soluble protein to the overall $D/D_0$ value measured for fresh, intact muscle fibers from the giant barnacle, *Balanus nubilus*. First, it is important to note that the ultracentrifugate from these muscle fibers gives a $D/D_0$ value (0.767) midway between that for the intact muscle fiber (0.551) and pure water (1.0). Considering that as much as half the reduction in $D/D_0$ seen for intact cells may potentially be explained by obstructional effects of intracellular compartments alone (see QNS data above), the ultracentrifugate value of 0.767 is far lower that might have been expected. However, Clark et al. (1982) obtained a virtually identical value for an artificial solution that mimicked the known concentrations of salt, organic solutes

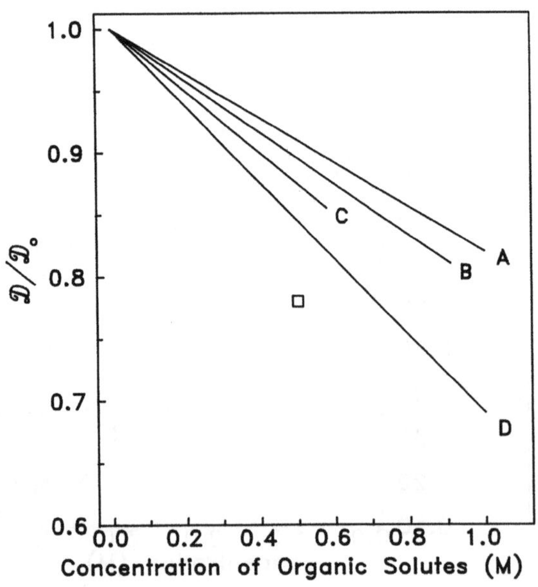


**Fig. 2.** Dependence of the relative self diffusion coefficient of water ($D/D_0$) upon the concentration of various organic solutes and mixtures of organic and inorganic solutes mimicking the concentrations in barnacle muscle. *A* mixture of glycine, asparagine, proline, arginine, taurine, alanine, valine, glutamate, betaine, trimethylamine-N-oxide; *B* glycerol; *C* mixture of propionate, KCl and NaCl; *D* trimethylamine-N-oxide; *open square* trimethylamine-N-oxide plus 170 mM propionate. (Data redrawn from Clark et al. 1982)

**Table 2.** Values for the relative self diffusion coefficient, compared to pure water, ($D/D_0$) for various barnacle muscle preparations and artificial solute solutions. (Data from Clark et al. 1982)

| Preparation | $D/D_0$ |
|---|---|
| Fresh barnacle fibers (*Balanus nubilus*) | 0.551 ± 0.021 (n = 3) |
| Fiber ultracentrifugate (corrected for any loss of protein during centrifugation) | 0.767 |
| Artificial solution mimicking the salt, organic solute and protein composition of fibers | 0.769 |
| Ultracentrifugate without protein (calculated by subtracting the contribution of protein to $D/D_0$, as estimated with values for BSA) | 0.854 |
| Artificial solution mimicking the salt and organic solute composition of fibers (no protein) | 0.863 |

and protein in these fibers (Table 2). The close agreement between these values demonstrates the striking, and generally unappreciated, contribution that these soluble components have on the translational motion of water. Similarly, when one subtracts the effect of soluble proteins from the ultracentrifugate $D/D_0$ value, good agreement is obtained between this protein-free ultracentrifugate value and the $D/D_0$ value measured for an artificial solution simulating the salt and organic solute levels in fibers (Table 2). Clearly the solute effects are additive.

Interestingly, the reduction in $D/D_0$ values caused by organic solutes is not reflected in the relaxation time $T_1$ of water protons. In other words, these components appear to strongly affect the translational movement of water molecules (and hence the solvent structure) over long distances without greatly altering the rotational freedom (Clark et al. 1982). As pointed out by these authors, an hypothesis for long-range effects of solutes was offered by Zeidler (1973) some time ago. He suggested that, at least for nonpolar alkyl groups, water molecules did not interact directly with these solutes but were constrained to form stronger or more numerous hydrogen bonds among themselves.

Organic solute levels (primarily trehalose, glycerol and TMAO) in fully hydrated *Artemia* embryos are well above the effective concentration for restricting the motion of water (Fig. 2; Table 3), and when dehydration occurs, the solute levels of course increase dramatically. The point to be made at this juncture is that the remarkable influence of intracellular osmolyte systems on the state of cellular water should not be ignored when one formulates models to explain metabolic transitions in anhydrobiotic animals, a topic to which we will return below.

**Table 3.** Internal solute concentrations of *Artemia* embryos from the Great Salt Lake as a function of NaCl concentration. (Data from Glasheen and Hand 1989)

| Solute | mosmol/ kg $H_2O$[a] | External medium (M NaCl) 0.25 | 2.0 | 3.0 | 5.0 |
|---|---|---|---|---|---|
| | | (mmol/kg $H_2O$) | | | |
| Ions | | | | | |
| Sodium | 32.8 | 36.4 ± 1.0[b] | 57.2 ± 1.1 | 69.9 ± 2.1 | 94.4 ± 1.8 |
| Potassium | 130.9 | 145.4 ± 3.9 | 228.9 ± 4.1 | 279.6 ± 8.4 | 377.4 ± 6.8 |
| Magnesium | 49.0 | 54.4 ± 1.6 | 85.7 ± 1.9 | 104.7 ± 3.4 | 141.2 ± 3.1 |
| Calcium | 5.3 | 5.9 ± 0.2 | 9.3 ± 0.3 | 11.4 ± 0.5 | 15.4 ± 0.5 |
| Chloride | 45.3 | 50.3 ± 3.2 | 79.2 ± 4.8 | 96.7 ± 6.3 | 130.5 ± 7.9 |
| Phosphate | 11.2 | 12.4 ± 1.1 | 19.6 ± 1.7 | 23.9 ± 2.1 | 32.2 ± 2.8 |
| Subtotal (ions) | 274.5 | | | | |
| Organics | | | | | |
| Trehalose | 340.5 | 340.5 ± 16.9 | 535.9 ± 24.3 | 654.7 ± 33.7 | 883.7 ± 40.1 |
| Glycerol | 319.2 | 319.2 ± 30.7 | 502.3 ± 47.4 | 613.7 ± 59.7 | 828.2 ± 78.1 |
| NPS | 62.3 | 62.3 ± 3.4 | 98.0 ± 5.0 | 119.7 ± 6.8 | 161.6 ± 8.3 |
| TMAO + betaine | 56.4 | 49.0 ± 2.0 | 77.0 ± 2.7 | 94.1 ± 4.0 | 127.0 ± 4.4 |
| Subtotal (organics) | 778.4 | | | | |
| Total osmotically active solutes | 1053 | | | | |
| Total osmotic pressure observed | 1301 | | | | |
| Unidentified solutes (by subtraction) | 248 | | | | |

[a]Values for embryos hydrated in 0.25 M NaCl; for osmotic coefficients see Glasheen and Hand (1989).
[b]Mean ± S.E.M. (N = 3).
NPS, ninhydrin-positive substances; TMAO, trimethylamine-N-oxide.

## Metabolic Arrest During Anhydrobiosis

### *Respiration Rate and Heat Dissipation Versus Water Content*

Various studies with anhydrobiotic organisms have characterized the acute metabolic transitions accompanying dehydration and rehydration. Measurements by Morris (1971) with polarographic oxygen electrodes indicated significant oxygen consumption by dried *Artemia* embryos within 10 min of immersion in 0.5 M NaCl. Later, Clegg (1976b) related the precise water content of San Francisco Bay embryos to the onset of respiration. His results showed that oxygen consumption, as measured polarographically at 25 °C, was not detectable below about 0.5–0.6 g water/g dry mass embryo, but respiration increased rapidly as hydration climbed during incubation of embryos in 0.5 M NaCl. A similar "critical hydration level" for initiation of respiration was determined in the same study by pre-equilibrating embryos in solutions of increasing ionic strength. The hydration state of embryos can be controlled in this manner because the outer cuticular membrane is completely impermeable to inorganic ions, yet fully permeable to water (Conte et al. 1977; Clegg and Conte 1980). The rapid onset of respiration was confirmed with manometric measurements, which indicated that respiration rate of embryos increased from undetectable levels of 1.0 μg $O_2$/mg dry mass/h within the first 45 min of hydration in the vapor phase of 0.5 M NaCl (Clegg 1976b). Crowe et al. (1977) showed that the respiration rate of anhydrobiotic nematodes (*Aphelenchus avenae*), after transfer from dry air to 97% relative humidity for 24 h and then to water for 10 min was about 25 nl $O_2$/mg dry mass/min at 25 °C. Oxygen consumption of fully hydrated nematodes was about 150 nl/mg dry mass/min. These values compare well to others obtained for anhydrobiotic (Bhatt and Rohde 1970) and fully hydrated (Cooper and van Gundy 1970) nematodes.

More recently, our laboratory has used open-flow microcalorimetry to follow changes in energy flow in *Artemia* embryos (Great Salt Lake, Utah) undergoing cycles of dehydration-rehydration in NaCl solutions (Glasheen and Hand 1989). Metabolic heat dissipation is a sensitive measure of total energy flow in the embryos (Hand and Gnaiger 1988, 1989), and calorimetry avoids the technical difficulties that arise when polarographic oxygen electrodes are chosen as a means to estimate aerobic metabolism in solutions of high ionic strength; electrolyte solutions within the electrolytes dehydrate under such conditions (Hale 1983). Figure 3 depicts the pattern of heat dissipation observed during a dehydration/rehydration bout. The heat signal from developing embryos rose steadily during the first 4 h in the control medium (0.25 M NaCl). When the perfusion medium was switched to 1.0 M NaCl, energy flow continued to rise, reaching 137% of control (hour 4) values. However,

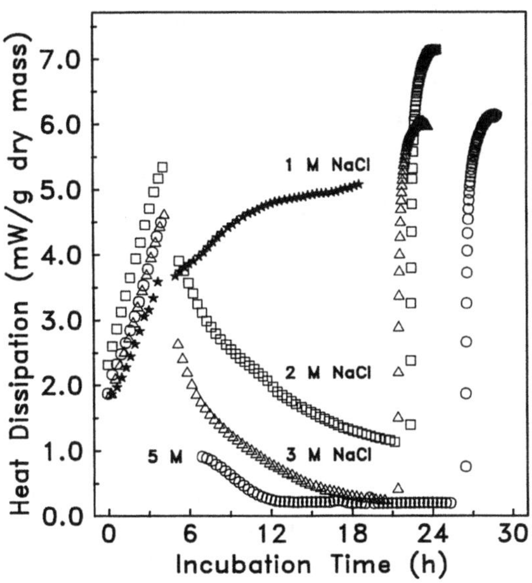


**Fig. 3.** Heat dissipation of *Artemia* embryos during incubations in various concentrations of NaCl. For all experiments, embryos were initially incubated for 4 h in 0.25 M NaCl, after which the perfusion medium was changed to either 1.0, 2.0, 3.0 or 5.0 M NaCl. After the 2.0, 3.0, and 5.0 M treatments, the perfusion medium was returned to 0.25 M NaCl, and the recovery period monitored (Data redrawn from Glasheen and Hand 1989)

when control medium was switched to 2.0 M NaCl, energy flow fell exponentially to 21% of control values after approximately 18 h. In experiments with 3 M NaCl, heat dissipation fell sharply to 6%. At higher ionic concentrations, heat dissipation declined to as low as 3%, or an absolute rate of 0.14 milliwatt/g dry mass. Thus energy flow was disrupted at the hydration level promoted by an external NaCl concentration between 2.0 and 3.0 M. When embryos were returned to control perfusion after dehydration, recovery of energy flow was rapid and complete.

These metabolic transitions were correlated with embryo hydration levels measured across the same dehydration series (Glasheen and Hand 1989). Water content of metabolically-active embryo tissue ranged from 1.34 ± 0.031 g $H_2O$/g dry mass in 0.25 M NaCl to 0.52 ± 0.006 g $H_2O$/g dry mass in 5.0 M NaCl. At the point where heat dissipation was markedly suppressed (2–3 M NaCl), cyst water content was between 0.854 ± 0.010 and 0.699 ± 0.019 g $H_2O$/g dry mass. This water content is similar to the "critical" hydration level required for the onset of carbohydrate catabolism and respiration in San Francisco Bay embryos (Clegg 1976a,b). If Clegg's hydration data are expressed as water content of metabolically active embryo tissue, a value of 0.70 g water/g dry mass was needed for the onset of respiration in the San Francisco Bay embryos.

### *Blockage of Carbohydrate Catabolism*

While these metabolic transitions appear to be a function of water content, the controlling mechanisms involved in *Artemia* embryos, as well as in other anhydrobiotic organisms, are still not understood. Because trehalose is the exclusive metabolic fuel used during development of encysted embryos (Clegg 1964), one must understand the mechanisms involved in arrest of carbohydrate metabolism if an acceptable explanation for the suppression of overall energy flow is to be attained. Consequently, we compared the changes in concentrations of trehalose, glycogen, glycolytic intermediates, and adenylate nucleotides occurring during normal aerobic development of hydrated embryos with those occuring after dehydration of embryos in concentrated salines (Glasheen and Hand 1988). The trehalose utilization and glycogen synthesis that occurred during development of fully hydrated cysts were both blocked during desiccation (Glasheen and Hand 1988; cf. Clegg 1976a). Upon return to 0.25 M NaCl solution both processes were resumed. Analysis of glycolytic intermediates suggested that the inhibition was localized at the trehalase, hexokinase and phosphofructokinase reactions. ATP levels remained constant during the 6-h period of dehydration, as did the adenylate energy charge.

It is clear that embryo $pH_i$ declines dramatically during transitions into anaerobic dormancy (Busa et al. 1982), a quiescent state brought about by anoxia. This reduction in $pH_i$ promotes suppression of various metabolic and developmental processes in these post-diapause embryos (Busa and Crowe 1983; Carpenter and Hand 1986a,b; Hand and Carpenter 1986; Utterback and Hand 1987; Hofmann and Hand 1990a,b; Hand 1990). Thus, in order to determine whether or not $pH_i$ influences the arrest of cyst metabolism during dehydration, it was important to perform experiments in which any possible reduction of $pH_i$ was prevented. A dehydration experiment was performed in saline containing a level of ammonia known to maintain an alkaline $pH_i$ in the embryos. The presence of ammonia in the external medium has been shown by $^{31}$P-NMR to promote an alkaline intracellular pH ($pH_i$) in *Artemia* cysts (Busa and Crowe 1983). The metabolic response to dehydration under these conditions was very similar to the previous dehydration series. Thus, these results were taken as strong evidence that the metabolic suppression observed during dehydration does not require cellular acidification.

### *The Vicinal-Water Network Model*

The central question to be considered in the remainder of this article is what factors *are* responsible for the metabolic arrest that is correlated with restricted availability of cellular water. The existing paradigm concerned with explaining this phenomenon comes primarily from the work of J.S. Clegg and is embodied in his "vicinal-water network model" (Clegg 1978, 1979, 1981). Clegg suggests that multiple classes or phases of water may exist in *Artemia* cysts, such that in the fully hydrated embryo, about 50% of the total water may be in a bulk (unrestricted) phase, about 40% in a vicinal phase (translationally restricted water), and about 10% may be tightly bound water (reflecting the primary hydration hydration shells of macromolecular components) (cf. Clegg 1981). These various forms of water are sequentially removed during dehydration, and Clegg has correlated the inhibition of metabolic events with the disappearance of these water classes.

A key feature of this model is that the major metabolic events are not qualitatively changed until all bulk water has been eliminated from cells, and the "vicinal water" has begun to be perturbed. As discussed earlier, the motion of vicinal-water is restricted due to its close proximity to membrane surfaces and other ultrastructural features of the cell. The vicinal-water network model ascribes extreme metabolic importance to this class of water, stating that "vicinal water is the basic cellular matrix within which most of metabolism takes place, and that the overall regulation of cellular activity is inexorably linked to this matrix" (Clegg 1979, p. 401). In *Artemia* embryos, from full hydration (about 1.4 g $H_2O$/g dry mass embryo) down to about 0.65 g/g, metabolism is qualitatively the same, and this range of hydration is referred to as the domain of conventional metabolism. Clegg refers to the hydration range from 0.65 to 0.3 g/g as the domain of restricted metabolism. Within this range, major metabolic events (respiration, carbohydrate metabolism, protein synthesis) have ceased. Finally, the ametabolic domain is the hydration range from 0.3 to 0 g/g, where no evidence exists for enzymatically catalyzed events.

As mentioned earlier, various lines of evidence clearly indicate that water in fully hydrated *Artemia* cysts exhibits motional properties different from those of pure bulk water. However, obtaining empirical evidence for three distinct classes of water in *Artemia*, and for their *sequential* removal during dehydration, has been more challenging. For example, one might expect distinct transitions to be apparent when self diffusion coefficients are plotted against cyst hydration levels. If one assumes that the bulk, vicinal, and bound water each make independent contributions to the observed value for D, and that each water "class" is removed sequentially (as per the vicinal-water network model), then the following equation could predict the expected relationship between D and hydration state of *Artemia* cysts:

$$D_{obs} = D_1\left[\frac{0.7g - X_1}{1.4g - Y}\right] + D_2\left[\frac{0.56g - X_2}{1.4g - Y}\right] + D_3\left[\frac{0.14g - X_3}{1.4g - Y}\right]$$

when $D_{obs}$ is the predicted self diffusion coefficient. $D_1$, $D_2$ and $D_3$ represent values for the independent diffusion coefficients for bulk, vicinal and bound water. The value for $D_3$, $0.4 \times 10^{-6}$ $cm^2/s$, was taken as the value experimentally determined by Clegg et al. (1982) for intracellular water in cysts at a hydration of 0.2 g/g, where the only water remaining should be "bound". $D_1$ and $D_2$ were assigned values such that $D_{obs}$, at full hydration, would equal $3.89 \times 10^{-6}$ $cm^2/s$ – approximately the value determined experimentally by NMR for fully hydrated cysts (Seitz et al. 1981).[1] Note that the bracketed terms in the equation represent the percentage of total cyst water contained in each water class. $X_1$, $X_2$ and $X_3$ represent the sequential increments of water (in grams) removed from each water class during dehydration, and Y is the total water removed from all water classes ($X_1 + X_2 + X_3$) at a given hydration state[2]. At full hydration of 1.4 g/g, the percentages of water in each water class are precisely those estimated by Clegg (1981) for bulk (0.5), vicinal (0.4), bound (0.1) water.

In Figure 4, values for $D_{obs}$ were calculated according to the above equation and plotted under the assumption of sequential linear dehydration (dashed line). The experimentally determined relationship from Seitz et al. (1981) is also shown. As cyst hydration is reduced, the above equation predicts a distinct break in the $D_{obs}$ versus hydration curve at 0.7 g $H_2O$/g dry mass, the water content at which carbohydrate metabolism is observed to be arrested. Another discontinuity is predicted at 0.14 g/g after all vicinal water has been removed. Yet, the actual experimental data measured with NMR show no abrupt change across a wide range of hydration values (solid line). From 1.4 g/g to 0.2 g/g, D is reduced in a monotonic manner; below 0.2 g/g the values tend to increase. It is possible that this latter increase could be due to the interactions of solutes with macromolecular and/or crystallization of solutes at these low water levels – both would serve to free up water of greater mobility. In any event it is clear that, at least across the 0.6–0.7 g/g "critical hydration level", there is not a noticeable transition in the measured values for the self-diffusion coefficient. Neither is there a major discontinuity in the relationship between the relaxation times for water protons ($T_1$ and $T_2$) and hydration state of cysts at 0.6 g/g (Seitz 1977).

[1]The value chosen for $D_1$ ($6.5 \times 10^{-6}$ $cm^2/sec$) is significantly constrained in that it can be no higher than $7.8 \times 10^{-6}$ (or else $D_{obs}$ would be above $3.89 \times 10^{-6}$ at full hydration), and it must be greater than the value chosen for $D_2$ ($1.5 \times 10^{-6}$). $D_2$ must fall between the values for $D_1$ and $D_3$. Altering these values within these constraints does not eliminate the discontinuities predicted in Fig. 4.

[2]This dehydration is modeled in the following manner. The removal of up to 0.7 g of water is reflected by an increase in $X_1$ from 0 to 0.7 g, at which point all bulk water has been removed (and $X_1$ is fixed at its maximum of 0.7). Beyond 0.7 g, the value of $X_2$ increases from 0 to 0.56 g ($X_1 + X_2 = 1.26$, all vicinal water removed). The removal of the final 0.14 g of bound water is reflected by an increase in $X_3$ from 0 to 0.14 g.

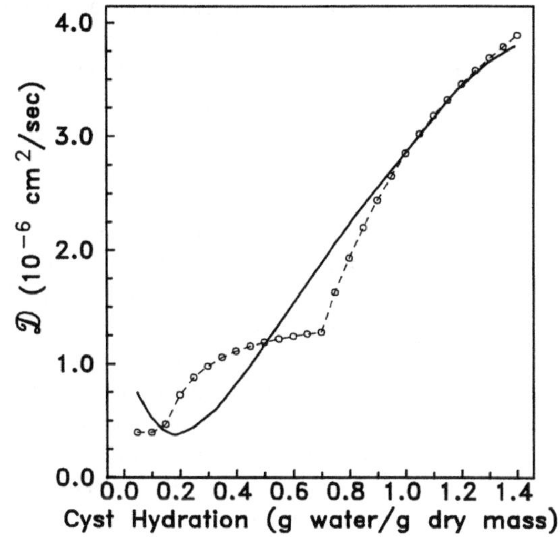


**Fig. 4.** Relationship between the self diffusion coefficient of intracellular water in *Artemia* embryos (*solid line*) determined experimentally with NMR and the predicted values for $D_{obs}$ (*dashed line*) based on the mathematical relationship presented in the text. The calculated relationship assumes that three classes of intracellular water (bulk, vicinal, and bound) exist in *Artemia* embryos, that the classes are removed sequentially during dehydration, and that the removal of one class does not influence the properties of the remaining classes. (NMR data redrawn from Seitz et al. 1981)

However, the patterns are not simply monotonic functions, as there are minima at 0.2 g/g.

A similar case was seen for the cyst dielectric constant (permittivity). While preliminary evidence suggested a break in the plot of dielectric constant versus hydration state did exist at 0.3 g $H_2O$/g dry mass (Clegg 1979), the final data set indicated a monotonic relationship across the entire hydration range (0 to 1.4 g/g) when the data were corrected for air space in the measuring chamber (Clegg et al. 1982). In contrast, the DC conductivity values showed marked increase at hydration values above 0.35 g/g (Bruni et al. 1989). While the ions giving rise to the conductivity changes remain to be identified (possibly protons), the authors tentatively proposed that at 0.35 g/g and above, a sufficient network of water was present to allow for ionic displacement over considerable distances. Thus, the coincidence of the onset of "restricted metabolism" and the ionic conductivity suggested a possible dependence between these two processes (Bruni et al. 1989). However, if the ions responsible for the increased conductance are not protons, then such a percolative process through a water lattice (i.e., a proton conduit) would be less probable.

Heat capacity measurements for cysts showed that up to 0.7 g $H_2O$/g dry mass the partial heat exceeded that of pure water; above that hydration value, the partial heat was essentially the same as bulk water. Since commencement of conventional metabolism in cysts requires approximately 0.7 g/g, the implication was that bulk water may be required for onset of conventional metabolism (Clegg 1981).

Most recently, Kasturi et al. (1990) have reported that the rotational motions of deuterium and oxygen-17 are restricted in *Artemia* embryos that were hydrated in $^2H_2O$ and $H_2{}^{17}O$, respectively. One advantage of NMR spectroscopy with these nuclei is that their relaxation is due entirely to intramolecular motions, and consequently, are viewed as better probes for rotational motion of water; complications arising from intermolecular interactions, as seen with the hydrogen nucleus of water, are minimized (Kasturi et al. 1990). The authors found striking similarities between the cyst data and the hydration dependence of deuterium relaxation rates for water in lysozyme crystals. In the case of single protein crystals (Borah and Bryant 1982), a marked change in slope was seen at a hydration level of 0.3 g $D_2O$/g, when relaxation rates ($1/T_1$ and $1/T_2$) of $^2H$ and $^{17}O$ were plotted against 1/hydration level. The slope change at this point was interpreted as the value at which the protein attained a full complement of hydration water. Likewise in *Artemia*, there was a break in the analogous hydration plot, but it occurred at a water content of about 0.6 g/g (i.e., about 1.2 g $H_2O$/g dry cyst protein). Thus there is considerable discrepancy between these systems in the hydration level required for the effect. Quantitative comparisons are problematic since *Artemia* contains organic solutes and lipid membranes that are absent from the lysozyme/$D_2O$ system. It also should be noted that complex hydration patterns are seen for *Artemia* embryos in deuterium (Bruni et al. 1989), and that deuterium reduces the development rate of the embryos two-to-threefold (Kasturi et al. 1990). The conclusion reached by Kasturi et al. (1990) was that the changed relaxation rates in cysts at 0.6 g/g were due to rotational motion of water molecules associated to the protein.

In contrast to an "onion peel" model for *Artemia* embryo dehydration, i.e., the sequential removal of static, pre-existing water classes or layers, it does not seem unreasonable to suggest that, if multiple classes exist, they may be interdependent. Stated another way, when a portion of cyst water is removed during dehydration, it may influence (draw upon) all water classes at once – bound , vicinal and bulk. Or, as suggested for some systems, one could view all water in *Artemia* cysts as possessing reduced diffusional motion, and as dehydration proceeds, the degree of this perturbation simply increases along a continuum. Such alternatives would seem more consistent, for example, with the pattern of self diffusion coefficients versus hydration state seen in Fig. 4. The simplistic formula used to model the vicinal water network idea in Fig. 4 purposely adopted a strict and literal interpretation of Clegg's model. Importantly, the values predicted by this equation fail to match the experimentally determined self-diffusion coefficients because the state of water in *Artemia* embryos during dehydration is more likely to be a continuum.

The profound influence that organic solutes have on diffusive properties of water will undeniably alter the degree of structured intracellular water as dehydration proceeds. If, as the vicinal water network model suggests, changes in the state of intracellular water are linked to perturbation of metabolic organization, then osmolytes could be key players in metabolic arrest during dehydration for this reason alone. However, I feel it is more important that both organic and inorganic osmolytes are known to alter the kinetics and structural organization of enzymes via direct binding to proteins and by altering macromolecular hydration water (see below). Thus, to fully understand the manner by which reductions is cell-associated water cause the arrest of metabolism, one cannot ignore the potential impact that changes in the intracellular osmolyte system may have.

### *A Role for Intracellular Solutes*

As reviewed above, interpretations based on solvent status have proved highly useful, but the important corollary of intracellular solute content has been relatively neglected vis a vis metabolic organization and control in *Artemia* (Glasheen and Hand 1989; Hand 1991). Until recently, one deficiency preventing investigators from evaluating the role of intracellular solutes in metabolic arrest during anhydrobiosis was the lack of a comprehensive analysis of the internal osmolyte system for *Artemia* embryos. Without such information it has been impossible to estimate the inorganic ion and organic solute concentrations present in the embryo at the onset of metabolic arrest. Yet, in over two decades of research into the *Artemia* system, many of these important constituents of the intracellular milieu have never been quantified. The few values available (amino acid, trehalose, glycerol, sodium content; Emerson 1967; Clegg 1967; Conte et al. 1977) are taken from different populations of embryos and are often expressed in ways that make it difficult to estimate internal concentrations.

Thus, we recently completed a comprehensive analysis of the internal solutes (Table 3) and macromolecular components of *Artemia* embryos (Glasheen and Hand 1989). A comprehensive inventory of the internal osmolytes indicated that inorganic ions ($Na^+$, $K^+$, $Cl^-$, $Mg^{2+}$, $Ca^{2+}$, $P_i$) accounted for 21% of the osmotic activity and 1.48% of embryo dry weight. Organic solutes (trehalose, glycerol, ninhydrin-positive substances, and TMAO + betaine) contributed 22% of the dry weight and 60% of the osmotic pressure. Macromolecular components (protein, lipids, glycogen, and DNA) were also quantified and formed the bulk of embryo mass. Taken together, 97.3% of the cyst dry mass was identified. When the cellular hydration values for these embryos were coupled to the intracellular solutes levels, it was possible to estimate the composition of the internal milieu of these embryos at any point along the graded series of dehydration (Table 3).

One interesting feature of these measurements was that the internal osmotic pressure calculated for *Artemia* embryos was substantially below that of the external medium for all incubations equal to or greater than 1.0 M NaCl. For example, the difference between external and internal osmotic pressure (the latter estimated from total solutes and total water content) for embryos incubated in 5.0 M NaCl approaches 10,000 mosm/kg water (Fig. 5). We feel very confident that the embryos were in osmotic steady state with the surrounding medium, i.e., that the total chemical potential inside the embryo matched the chemical potential existing externally. Thus, we take this trend to mean that much of the intracellular water is restricted and not osmotically available – providing indirect evidence to support the premise that the properties of intracyst water are quite different from those of the external medium. While it is probable that the osmotic coefficients of internal solutes may change with concentration, this factor is unlikely to account for the enormous differences depicted in Figure 5.

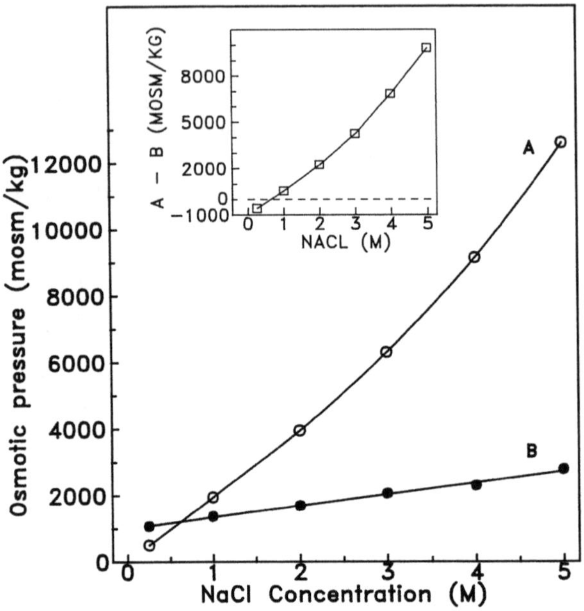


**Fig. 5.** The calculated internal osmotic pressure of *Artemia* embryos (*B, closed circles*) and the osmotic pressure of NaCl solutions (*A, open circles* from Handbook of Chemistry and Physics) plotted as a function of NaCl concentration. *Inset* curve A minus curve B. The format for this analysis was suggested by J.S. Clegg; data for *Artemia* embryos are from Glasheen and Hand (1989)

### *Disruption of Metabolon Organization via Osmolyte Effects*

At the cellular dehydration state promoting metabolic arrest, the internal concentrations of inorganic and organic osmolytes in *Artemia* embryos were about 530 mM and 1350 mmol/kg $H_2O$, respectively. In most tissues studied, levels of univalent ions above 100–200 mM are severely deleterious to enzyme function, as judged by the lowering of reaction velocities and elevation of $K_m$ values (Clark and Zounes 1977; Bowlus and Somero 1979; Yancey et al. 1982).

In addition to these direct influences on enzyme kinetics, evidence suggests that the binding of enzymes to subcellular components may be disrupted as a consequence of the elevated ionic composition of embryos at low water activities. Partitioning of enzymes into soluble and particulate fractions (with differing kinetic properties), or altering the cohesiveness of multienzyme associations (metabolons) are attractive models for metabolic regulation (Welch 1977; Wilson 1978; Srere 1987; Ovadi 1988; Somero and Hand 1990). In mammalian muscle, the adsorption of the glycolytic enzymes aldolase, lactate dehydrogenase, and pyruvate kinase to an F-actin-tropomyosin-troponin complex was significantly reduced in 150 mM KCl (Clarke and Masters 1975). Likewise, 150 mM NaCl resulted in desorption of glyceraldehyde-3-phosphate dehydrogenase from Band 3 protein in erythrocytes (Yu and Steck 1975). Thus, it is possible that ionic perturbation of enzyme-enzyme or enzyme-cytoskeletal interactions may account for at least part of the observed dehydration-induced arrest of *Artemia* metabolism.

It is appropriate to note that certain organic solutes (e.g., urea, arginine, guanidinium-based compounds) are also known to disrupt protein-protein interactions (Yancey et al. 1982; Hand and Somero 1982). Yet, the majority of organic solutes in *Artemia* embryos (trehalose, glycerol, TMAO) stabilize protein molecules and assemblies (Yancey et al. 1982; Crowe et al. 1987).

Previous experiments have indicated simultaneous inhibition of three major enzymes during glycolytic shutdown in dehydrated embryos, namely trehalase, hexokinase, and phosphofructokinase (Glasheen and Hand 1988). The catalytic properties of *Artemia* trehalase are significantly altered by modifications of macromolecular assembly state (Hand and Carpenter 1986). While *Artemia* hexokinase does not reversibly associate with isolated mitochondia (Rees et al. 1989) as does the rat brain homolog, we have preliminary evidence that phosphofrutokinase does bind to a particulate fraction in homogenates of these embryos (unpublished observations). The fact that the dehydration-induced metabolic shutdown in *Artemia* is fully reversible is consistent with the above view of metabolic organization. Such noncovalent assembly/disassembly transitions could alter enzymic function without compromising native protein structure. Indeed, it seems reasonable to suggest that conservation of macromolecular integrity is an important prerequisite for organisms capable of undergoing such drastic reductions in cell-associated water.

Consequently, we have suggested that the role of intracellular osmolytes should be more fully integrated into hypotheses explaining the dehydration-induced arrest of *Artemia* metabolism (Glasheen and Hand 1989). At low water activities, concentrations of inorganic ions rise appreciably, which could result in the disruption of protein-protein associations and the attendant loss of metabolic function. The substantial quantities of the organic solutes trehalose, glycerol, and methylamines may be involved in fostering reversibility of these anhydrobiotic bouts by preventing macromolecular denaturation during drying. Quantifying the relative influences of inorganic ions versus organic osmolyte on protein assembly state awaits further study.

## Alternative Experimental Approaches

Is it possible to test the above hypothesis that intracellular solutes have a role in the metabolic arrest seen during dehydration in *Artemia* embryos? Attempts to simulate intracellular conditions in a test tube often seem to present unending pitfalls. One critique of a manuscript submitted recently from our laboratory correctly noted that it is impossible to accurately mimic a condition (i.e., the intracellular milieu) that is at best only dimly perceived! Yet, the manipulative advantage gained through such approaches is attractive, and with sufficient care, useful information can be obtained even about a phenomenon as complex as the influence of water and solutes on metabolic organization. It is becoming increasingly clear that the intracellular microenvironment in which an enzyme functions must be simulated with the highest possible precision if the true regulatory properties of the protein are to be accurately evaluated. This type of information demands a diverse set of approaches including ultrastructural analysis, protein chemistry, and quantification of the solution chemistry in the cell of interest (Somero and Hand 1990).

### *Artificial Solutions and Cell Free Extracts*

Based on the data now available on the cellular composition of *Artemia* embryos, we are in a position to make initial attempts at separating the influences of water availability and solute perturbation on metabolism, so as to evaluate the possible contribution of internal osmolytes to metabolic arrest during anhydrobiosis. For example, one initial approach could be to use mixtures of selected enzymes, solutes of low molecular weight, and large artifical polymers to mimic the internal osmotic conditions estimated across the range of dehydration experienced by *Artemia* embryos. In a first series, the mixtures would simulate the changes in organic solute levels measured in cysts during dehydration (Table 3). This series would also contain increasing levels of hydroxyethyl starch (or polyethylene glycol) in order to adjust the self-diffusion coefficients of water in these solutions to match those measured in *Artemia* embryos during graded dehydration (Seitz et al. 1981). In a second series, the solutions would be the same except that inorganic ions would be added to simulate the levels estimated during dehydration (Table 3). [Although the addition of inorganic salts would obviously change the osmotic pressure, note that the effect on the self-diffusion coefficient of water would be negligible (Figure 1).] Then catalytic amounts of one enzyme (e.g., hexokinase, phosphofructokinase, trehalase; each shown to be inhibited during dehydration) and requisite substrates would be added to the mixtures. If inorganic ions are important for the metabolic arrest observed in vivo during dehydration, then as a function of $D/D_o$, one would predict enzyme activities to be more inhibited in the second series. Since the artificial colloid would be present in both series, any specific influence of this molecule on enzyme activity would be canceled.

One might also use embryo homogenates – dialyzed against solutions varying in solute composition – as an alternative to the mixtures employed in the previous experiments. The advantage here would be that the macromolecular components would reflect more closely the diverse array present in intact embryos. The samples would be dialyzed so as to mimic the desired self-diffusion coefficients (or the g $H_2O$/g dry mass values) measured for *Artemia* embryos across the appropriate dehydration range. Because of the viscosity/turbidity of the resulting samples, discontinuous enzyme assays or even reflectance spectrofluorometry may be required in place of traditional spectrophotometric assays. Figure 6 depicts one hypothetical pattern that could emerge from such data. Again, if increasing concentrations of inorganic osmolytes were making a significant contribution to arresting enzyme activity, then the inhibition should be more acute in the case where elevated ion levels were present.

### *Permeabilized Embryos*

We anticipate that the use of permeabilized *Artemia* embryos may represent a more biologically realistic system for estimating the in situ activity of carbohydrate regulatory enzymes as a function of solute conditions that mimic those measured during dehydration. This approach would preserve the three dimensional microenvironment of the enzymes at least to a greater extent than allowed by homogeniza-

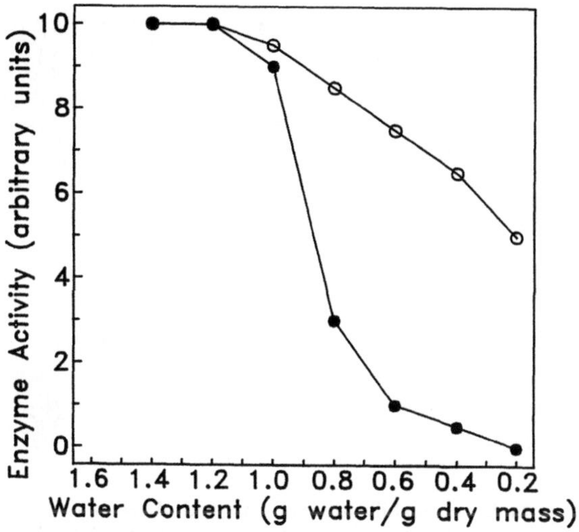


**Fig. 6.** Hypothetical results from proposed in vitro experiments designed to separate the influence of inorganic solutes from the influence of restricted water availability on the inhibition of enzymatic activity during dehydration. *Open circles* represent embryo homogenates dialyzed against solutions of high molecular weight hydroxyethyl starch (HES) and organic solutes, designed to mimic the changes in water content and organic solute changes calculated to occur during dehydration of *Artemia* embryos. *Closed circles* represent homogenates dialyzed against HES solutions where the concentrations of both inorganic and organic solutes increase as calculated to occur during dehydration. The *abscissa* represents the water content of these samples expressed as g $H_2O$/g dry mass (could also be expressed as the self-diffusion coefficients of the solutions). Selected enzyme activities in these samples would be quantified as described in text

tion. From the cell's perspective, the most "gentle" homogenation procedure undoubtedly represents cataclysmic violence (McConkey 1982; Clegg 1984a).

Although brine shrimp embryos offer numerous advantages for studies of metabolic quiescence, one feature that has limited the use of this experimental approach has been the extreme impermeability of the outer cuticular membrane. Numerous attempts to remove this permeability barrier (including exposure to concentrated acids and bases, various organic solvents, and digestive enzymes) have met with no success (reviewed by Clegg and Conte 1980). However, recently we have been successful in permeabilizing this chitinous membrane by incubating dechorionated embryos overnight in dimethylacetamide and LiCl. This chemical mixture was developed recently by organic chemists for the digestion of chitin (Rutherford and Dunson 1984). After this treatment, the outer cuticular membrane was effectively permeabilized in 70–90% of the embryos as assessed with a Hoechst dye (#33342; DNA specific, plasma membrane permeant). The dye penetrated the embryos and caused the nuclei to fluoresce brightly; untreated controls were not permeable to the dye (Fig. 7).

The treatment resulted in some temporary dehydration of the embryos that can be seen in Fig. 7c. At this stage the technique is far from fully optimized. We

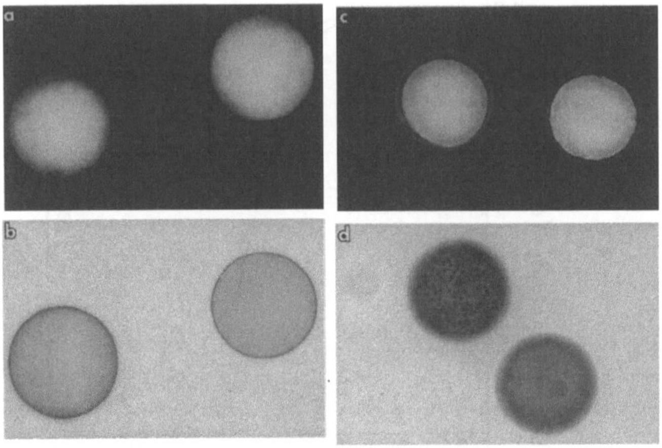

**Fig. 7a-d.** Permeabilization of encysted *Artemia* embryos. Control (untreated) embryos were photographed with transmitted light (**a**) and ultraviolet light (**b**) in the presence of Hoechst fluorescent dye. These control embryos were impermeable to this nuclear stain. Experimental embryos were treated for several hours with dimethylacetamide containing lithium chloride and viewed with transmitted light (**c**) and ultraviolet light (**d**)

anticipate that further optimization of the technique (e.g., partial digestion with dimethylacetamide, followed by chitinase or lysozyme digestion; minimizing exposure to lithium, cf. Busa and Gimlich 1989) may improve the low embryo viability (currently less than 10%). However, since enzyme activities can be readily measured in homogenates of the permeabilized embryos (e.g., malate dehydrogenase 166 units/g wet mass; phosphofructokinase 0.51 units/g wet mass), the low hatch rate may be of little concern for our primary purposes.

Permeabilizing the outer cuticular membrane, coupled with electroporation to permeabilize the plasma membranes, should allow in situ enzyme assays to be performed with *Artemia* embryos. Work with sea urchin eggs has shown that entry of enzyme substrates into the cells is facilitated with this procedure, and the activities of glutathione reductase, hexokinase, NAD kinase, glucose-6-phosphate dehydrogenase and other enzymes have been successfully measured (Swezey and Epel 1988, 1989). The approach offers the ability to retain major structural features of the cell while analyzing the catalytic features of enzymes in a microenvironment that perhaps more closely approximates the microenvironment of enzymes in living cells.

Having performed these manipulations, we would alter the external solutions bathing these permeabilized *Artemia* embryos to simulate the internal changes in solutes experienced intracellularly during dehydration. We anticipate that monitoring the activities of glycolytic enzymes across a graded series of osmolyte concentration could be quite informative. The changes in catalytic activity could be compared to our previous metabolic profiles for embryos undergoing dehydration (Glasheen and Hand 1989) with the hope of correlating metabolic arrest with catalytic inhibition.

## Concluding Remarks

The primary intent of this chapter is to call attention to the potential importance of intracellular solutes in mediating the metabolic arrest experienced by anhydrobiotic animals during dehydration. Certainly the objective is not to minimize the importance that the disposition of cellular water may have in this process, but rather to suggest that these two features of the cell are inseparable when it comes to formulating realistic models for metabolic control under conditions of restricted water availability.

The eventual goal of our ongoing research is to clarify the importance of the intracellular milieu in regulating metabolic events during dehydration, not only in terms of solvent interactions, but also in terms of the nature and quantity of critical solutes. We are optimistic that our studies with artificial solute/protein solutions and with permeabilized embryos may allow us to analyze the catalytic activity of enzymes in a more realistic microenvironment, as conditions of progressive dehydration are simulated by altering the internal milieu to which enzymes are exposed. Such enzyme studies will be technically challenging, but it is my contention that we must look beyond dilute aqueous media if we are ever to appreciate new and more meaningful catalytic features of enzymes. These studies should provide a test for the hypothesis that solute-induced disruption of organized enzyme associations are involved in the arrest of metabolism during anhydrobiosis.

*Acknowledgments.* Appreciation is extended to Professor Mary E. Clark for her helpful discussions and encouragement during preparation of this paper, and to Professor George N. Somero for stimulating my interest in the biological roles of intracellular osmolytes. I thank Professor James S. Clegg for recommending that we analyze our solute data in the format depicted in Fig. 5. Helpful comments on the manuscript were given by Steve Glasheen, Kurt Kwast, Ron Provost, and Barney Rees. Portions of the research described herein were supported by grants DCB-8316711, DCB-8702615 and BBS-8704421 from the National Science Foundation (USA).

## References

Bhatt BD, Rohde RA (1970) The influence of environmental factors on the respiration of plant-parasitic nematodes. J Nematol 2:277–285

Borah B, Bryant RG (1982) Deuterium NMR of water in immobilized protein systems. Biophys J 38:47–52

Bowlus RD, Somero GN (1979) Solute compatibility with enzyme function and structure: Rationales for the selection of osmotic agents and end-products of anaerobic metabolism in marine invertebrates. J Exp Zool 208:137–152

Bruni F, Careri G, Clegg JS (1989) Dielectric properties of *Artemia* cysts at low water contents. Evidence for a percolative transition. Biophys J 55:331–338

Burnell EE, Clark ME, Hinke JAM, Chapman NR (1981) Water in barnacle muscle. III. NMR studies of fresh fibers and membrane-damaged fibers equilibrated with selected solutes. Biophys J 33:1–26

Busa WB, Crowe JH (1983) Intracellular pH regulates transitions between dormancy and development of brine shrimp (*Artemia salina*) embryos. Science 221:366–368

Busa WB, Gimlich RL (1989) Lithium-induced teratogenesis in frog embryos prevented by a polyphosphoinositide cycle intermediate or a diacylglycerol analog. Dev Biol 132:315–324

Busa WB, Crowe JH, Matson GB (1982) Intracellular pH and the metabolic status of dormant and developing *Artemia* embryos. Arch Biochem Biophys 216:711–718
Carpenter JF, Hand SC (1986a) Arrestment of carbohydrate metabolism during anaerobic dormancy and aerobic acidosis in *Artemia* embryos: Determination of pH-sensitive control points. J Comp Physiol B 156:451–459
Carpenter JF, Hand SC (1986b) Comparison of pH-dependent allostery and dissociation for phosphofructokinases from *Artemia* embryos and rabbit muscle: nature of the enzymes acylated with diethylpyrocarbonate. Arch Biochem Biophys 248(1):1–9
Chang DC, Rorschach HE, Nichols BL, Hazelwood CF (1973) Implications of diffusion coefficient measurements for the structure of cellular water. In: Hazelwood CF (ed) Physicochemical state of ions and water in living tissues and model systems. Ann N Y Acad Sci 204:434–443
Clark FM, Masters CJ (1975) On the association of glycolytic enzymes with structural proteins of skeletal muscle. Biochem Biophys Acta 381:37–46
Clark ME (1985) The osmotic role of amino acids: discovery and function. In: Gilles R, Gilles-Baillien M (eds) Transport processes, iono- and osmoregulation. Springer, Berlin Heidelberg New York, pp 412–423
Clark ME (1987) Non-Donnan effects of organic osmolytes in cell volume changes. In: Gilles R (ed) Current topics in membranes and transport. Cell volume control: fundamental and comparative aspects in animal cells, vol 30. Springer, Berlin Heidelberg New York, pp 251–271
Clark ME, Hinke JAM (1981) Studies on water in barnacle muscle fibres. I. The intracellular dry weight components of fresh fibres. J Exp Biol 90:33–41
Clark ME, Zounes M (1977) The effects of selected cell osmolytes on the activity of lactate dehydrogenase from the euryhaline polychaete, *Nereis succinea*. Biol Bull 153:468–484
Clark ME, Hinke JAM, Todd ME (1981) Studies on water in barnacle muscle fibres. II. Role of ions and organic solutes in swelling of chemically skinned fibres. J Exp Biol 90:43–63
Clark ME, Burnell EE, Chapman NR, Hinke JAM (1982) Water in barnacle muscle. IV. Factors contributing to reduced self-diffusion. Biophys J 39:289–299
Clegg JS (1964) The control of emergence and metabolism by external osmotic pressure and the role of free glycerol in developing cysts of *Artemia salina*. J Exp Biol 41:879–892
Clegg JS (1967) Metabolic studies of cryptobiosis in encysted embryos of *Artemia salina*. Comp Biochem Physiol 20:801–809
Clegg JS (1973) Do dried cryptobiotes have a metabolism? In: Crowe JH, Clegg JS (eds) Anhydrobiosis. Dowden Hutchinson Ross, Stroudsburg, PA, pp 141–146
Clegg JS (1974) Interrelationships between water and cellular metabolism in *Artemia* cysts. I. Hydration-dehydration from the liquid and vapor phases. J Exp Biol 61:291–308
Clegg JS (1976a) Interrelationships between water and cellular metabolism in *Artemia* cysts. II. Carbohydrates. Comp Biochem Physiol 53A:83–87
Clegg JS (1976b) Interrelationships between water and cellular metabolism in *Artemia* cysts. III. Respiration. Comp Biochem Physiol 53A:89–93
Clegg JS (1978) Hydration-dependent metabolic transitions and the state of cellular water in *Artemia*. In: Crowe JH, Clegg JS (eds) Dry biological systems. Academic Press, Lond New York, pp 117–154
Clegg JS (1979) Metabolism and the intracellular environment: the vicinal-water network model. In: Drost-Hanson W, Clegg JS (eds) Cell-associated water. Academic Press, Lond, New York, pp 363–413
Clegg JS (1981) Metabolic consequences of the extent and disposition of the aqueous intracellular environment. J Exp Zool 215:303–313
Clegg JS (1984a) Properties and metabolism of the aqueous cytoplasm and its boundaries. Am J Physiol 246 (Regul Integrative Comp Physiol 15):R133–R151
Clegg JS (1984b) Interrelationships between water and cellular metabolism in *Artemia* cysts. Cell Biophys 6:153–169
Clegg JS (1986) On the physical properties and potential roles of intracellular water. In: Welch GR, Clegg JS (eds) The organization of cell metabolism. Plenum Press, New York Lond, pp 41–55
Clegg JS, Conte FP (1980) Cellular and developmental biology of *Artemia*. In: Persoone G, Sorgeloos P, Roels O, Jaspers E (eds) The brine shrimp *Artemia*, vol 2. Physiol Biochem, Mol Biol, Univ Whetteren, Belgium, pp 83–103

Clegg JS, Jackson SA (1989) Long-term anoxia in *Artemia* cysts. J Exp Biol 147:539–543
Clegg JS, Zettlemoyer AC, Hsing HH (1978) On the residual water content of dried but viable cells. Experientia 34:734–735
Clegg JS, Szwarnowski S, McClean VER, Sheppard RJ, Grant EH (1982) Interrelationships between water and cell metabolism in *Artemia* cysts. X. Microwave dielectric studies. Biochim Biophys Acta 721:458–735
Conte FP, Droukas PC, Ewing RD (1977) Development of sodium regulation and de novo synthesis of (Na + K)-activated ATPase in larval brine shrimp, *Artemia*. J Exp Zool 202:339–362
Cooper AF Jr, Van Gundy SD (1970) Metabolism of glycogen and neutral lipids by *Aphelenchus avenae* and *Caenorrhabditis* sp. in aerobic, microaerobic, and anaerobic environments. J Nematol 2:305–315
Cooper AF Jr, Van Gundy SD (1971) Senescence, quiescence, and cryptobiosis. In: Zuckerman BM, Mai WF, Rohde RA (eds) Plant parasitic nematodes, vol 2. Academic Press, Lond New York, pp 297–318
Crowe JH, Clegg JS (eds) (1973) Anhydrobiosis. Dowden Hutchinson Ross, Stroudburg, PA, 477 pp
Crowe JH, Clegg JS (eds) (1978) Dry biological systems. Academic Press, Lond New York San Francisco, 357 pp
Crowe JH, Madin KA (1974) Anhydrobiosis in tardigrades and nematodes. Trans Am Microsc Soc 93:513–523
Crowe JH, Madin KA (1975) Anydrobiosis in nematodes: evaporative water loss and survival. J Exp Zool 193:323–334
Crowe JH, Madin KAC, Loomis SH (1977) Anydrobiosis in nematodes: metabolism during resumption of activity. J Exp Zool 201:57–64
Crowe JH, Crowe LM, Carpenter JF, Wistrom A (1987) Stabilization of dry phospholipid bilayers and proteins by sugars. Biochem J 242:1–10
Emerson DN (1967) Some aspects of free amino acid metabolism in developing encysted embryos of *Artemia salina*, the brine shrimp. Comp Biochem Physiol 20:245–261
Ewing RD, Clegg JS (1969) Lactate dehydrogenase activity and anaerobic metabolism during embryonic development in *Artemia salina*. Comp Biochem Physiol 31:297–307
Glasheen JS, Hand SC (1988) Anhydrobiosis in embryos of the brine shrimp *Artemia*: Metabolic arrest during reductions in cell-associated water. J Exp Biol 135:363–380
Glasheen JS, Hand SC (1989) Metabolic heat dissipation and internal solute levels of *Artemia* embryos during changes in cell-associated water. J Exp Biol 145:263–282
Goldhammer EV, Hertz HG (1970) Molecular motion and structure of aqueous mixtures with nonelectrolytes as studied by nuclear magnetic relaxation methods. J Phys Chem 74:3734–3755
Hale JM (1983) Factors influencing the stability of polarographic oxygen sensors. In: Gnaiger E, Forstner H (eds) Polarographic oxygen sensors. Aquatic and physiological applications. Springer, Berlin Heidelberg New York, pp 3–17
Hand SC (1990) Heat dissipation during long-term anoxia in *Artemia* embryos: identification and fate of metabolic fuels. J Comp Physiol B 160:357–363
Hand SC (1991) Metabolic dormancy in aquatic invertebrates. In: Advances in comparative and environmental physiology, vol 8. Springer, Berlin Heidelberg New York, pp 1–50
Hand SC, Carpenter JF (1986) pH-Induced metabolic transitions in *Artemia* embryos mediated by a novel hysteretic trehalase. Science 232:1535–1537
Hand SC, Gnaiger E (1988) Anaerobic dormancy quantified in *Artemia* embryos: A calorimetric test of the control mechanism. Science 239:1425–1427
Hand SC, Gnaiger E (1989) Metabolic arrest in *Artemia* embryos quantified with microcalorimetric, respirometric, and biochemical measurements. In: Wieser W, Gnaiger E (eds) Energy transformations in cells and organisms. Thieme, Stuttgart New York, pp 155–162
Hand SC, Somero GN (1982) Urea and methylamine effects on rabbit muscle phosphofructokinase. Catalytic stability and aggregation state as a function of pH and temperature. J Biol Chem 257:734–741
Hazelwood CF (1973) Physicochemical state of ions and water in living tissues and model systems. Ann NY Acad Sci 204:455

Hazelwood CF (1979) A view of the significance and understanding of the physical properties of cell-associated water. In: Drost-Hanson W, Clegg J (eds) Cell-associated water. Academic Press, Lond New York, pp 165–259
Hinton HE (1960) Cryptobiosis in the larva of *Polypedilum vanderplanki* Hint. (Chironomidae). J Insect Physiol 5:286–300
Hinton HE (1968) Reversible suspension of metabolism and the origin of life. Proc R Soc B 171:43–57
Hofmann GE, Hand SC (1990a) Subcellular differentiation arrested in *Artemia* embryos under anoxia: evidence supporting a regulatory role for intracellular pH. J Exp Zool 253:287–302
Hofmann GE, Hand SC (1990b) Arrest of cytochrome *c* oxidase synthesis coordinated with catabolic arrest in dormant *Artemia* embryos. Am J Physiol 258 (Regul Integrative Comp Physiol 27):R1184–R1191
Kasturi SR, Seitz PK, Chang DC, Hazelwood CF (1990) Intracellular water in *Artemia* cysts (brine shrimp). Investigations by deuterium and oxygen-17 nuclear magnetic resonance. Biophys J 58:483–491
Keilin D (1959) The problem of anabiosis or latent life: history and current concept. Proc R Soc Lond B 150:149–191
McCall DW, Douglass DC (1965) The effect of ions on the self-diffusion of water. I. Concentration dependence. J Phys Chem 69:2001–2011
McConkey EH (1982) Molecular evolution, intracellular organization, and the quinary structure of proteins. Proc Natl Acad Sci USA 79:3236–3240
Morris JE (1971) Hydration, its reversibility, and the beginning of development in the brine shrimp, *Artemia salina*. Comp Biochem Physiol 38A:843–857
Ovadi J (1988) Old pathway – new concept: control of glycolysis by metabolite-modulated dynamic enzyme associations. Trends Biochem Sci 13:486–490
Porter KR (1986) Structural organization of the cytomatrix. In: Welch GR, Clegg JS (eds) The organization of cell metabolism. Plenum Press, New York Lond, pp 9–25
Rees BB, Ropson IJ, Hand SC (1989) Kinetic properties of hexokinase under near-physiological conditions. Relation to metabolic arrest in *Artemia* embryos under anoxia. J Biol Chem 264: 15410–15417
Rorschach HE, Chang DC, Hazelwood CF, Nichols BL (1973) The diffusion of water in striated muscle. In: Hazelwood CF (ed) Physicochemical state of ions and water in living tissues and model systems. Ann NY Acad Sci 204:444–452
Rutherford FA, Dunson WA (1984) The permeability of chitin films to water and solutes. In: Zikakis JP (ed) Chitin, chitosan, and related enzymes. Academic Press, Lond New York, pp 135–143
Seitz PK (1977) Water proton magnetic resonance of metabolic and ametabolic *Artemia* cysts. PhD Thesis, Univ Texas, Austin (Abstr 7807383)
Seitz PK, Chang DC, Hazelwood CF, Rorschach HE, Clegg JS (1981) The self-diffusion of water in *Artemia* cysts. Arch Biochem Biophys 210:517–524
Somero GN (1986) Protons, osmolytes, and fitness of internal milieu for protein function. Am J Physiol 251 (Regul Integrative Comp Physiol 20):R197–R213
Somero GN, Hand SC (1990) Protein assembly and metabolic regulation: physiological and evolutionary perspectives. Physiol Zool 63(3):443–471
Srere PA (1987) Complexes of sequential metabolic enzymes. Annu Rev Biochem 56:89–124
Swezey RR, Epel D (1988) Enzyme stimulation upon fertilization is revealed in electrically permeabilized sea urchin eggs. Proc Natl Acad Sci USA 85:812–816
Swezey RR, Epel D (1989) Stable, resealable pores formed in sea urchin eggs by electric discharge (electroporation) permit substrate loading for assays of enzymes in vivo. Cell Regul 1:65–74
Trantham EC, Rorschach HE, Clegg JS, Hazelwood CF, Nicklow RM, Wakabayashi N (1984) Diffusive properties of water in *Artemia* cysts as determined from quasi-elastic neutron scattering spectra. Biophys J 45:927–938
Utterback P, Hand SC (1987) Yolk platelet degradation in preemergence *Artemia* embryos: Response to protons in vivo and in vitro. Am J Physiol 252 (Regul Integrative Comp Physiol 21):R774–R781
Welch GR (1977) Role of organization of multienzyme systems in cellular metabolism: general synthesis. Prog Biophys Mol Biol 32:103–191

Wilson JE (1978) Ambiquitous enzymes: variation in intracellular distribution as a regulatory mechanism. Trends Biochem Sci 3:124–125

Womersley C (1981) Biochemical and physiological aspects of anhydrobiosis. Comp Biochem Physiol 70B:669–678

Womersley C (1987) A reevaluation of strategies employed by nematode anhydrobiotes in relation to their natural environment. In: Veech J, Dickson DW (eds) Vistas on Nematology. MD Soc Nematol Publ, pp 165–173

Yancey PH, Clark ME, Hand SC, Bowlus RD, Somero GN (1982) Living with water stress: Evolution of osmolyte sytems. Science 21:1214–1222

Yu J, Steck TL (1975) Associations of band 3, the predominant polypeptide of the human erythrocyte membrane. J Biol Chem 250:9176–9184

Zeidler MD (1973) NMR spectroscopic studies. In: Franks F (ed) Water: A comprehensive treatise, vol 2. Water in crystalline hydrates; aqueous solutions of simple nonelectrolytes. Plenum, New York, pp 529–584

Chapter 9

# Macroautophagy Triggered by Sucrose Starvation in Higher Plant Cells: Analysis of a Model for Prolonged Carbon Deprivation Under Water Stress

R. DOUCE, N. PASCAL, and R. BLIGNY

## Introduction

During the day almost all plant cells are amply supplied with sucrose via the phloem, provided photosynthesis is uninhibited. The immediate products of sucrose metabolism formed in the cytosolic compartment are hexose phosphates and UDP-glucose (ap Rees 1987). These, in turn, are the substrates for a large number of reactions of general metabolism and for the synthesis of storage products (sucrose and organic acids in the vacuole, starch in the plastids) as well as for respiration (Douce 1985). A notably large flux of carbon is to plant cell respiration, which accounts for the loss of over 40% of all carbon entering the cell (Farrar 1985). A long period of drought leads to stomatal closure, thus depriving most of the plant cells of sucrose. For example, there is evidence from field studies that stomatal responses correlate with soil water supply and that the total root biomass was found to decrease with progressive drought (Schulze 1986). Kaiser (1984) has reviewed data that clearly indicate that mild osmotic stress leads to a reduction of photosynthesis in leaves; probably caused by stomatal closure and direct effects on the partial pressure of $CO_2$ within the leaf. Under these conditions, restriction of $CO_2$ supply caused by stomatal closure leads to a reduction in the capacity for both starch and sucrose synthesis (Vassey and Sharkey 1989): under field conditions, the main limitation of photosynthesis appears to be stomatal closure, photosynthetic metabolism being highly resistant to dehydration under most conditions (Kaiser 1987; Eickmeier et al., this Vol.).

The work reported in this chapter deals with these aspects of sucrose starvation in heterotrophically grown sycamore cell cultures. We present evidence showing that after a long period of sucrose starvation, cytoplasmic components can be utilized as a carbon source for respiration after almost all the intracellular carbohydrate pools have disappeared. This autophagic process is distinct from senescence, i.e., the series of events subjected to direct genetic control and concerned with cellular disassembly in the leaf and the mobilization of various small molecules (amino acids, asparagine, etc.) released during this process. This has been excellently reviewed by Thomas and Stoddart (1980) and Laurière (1983). Carbohydrate starvation has been widely studied in many plant species: soybean (Kerr et al. 1985; Walsh et al. 1987),

Laboratoire de Physiologie Cellulaire Végétale, Centre d'Etudes Nucléaires, 85x F38041 GRENOBLE-CEDEX, France

Somero et al. (Eds.)
Water and Life

maize (Saglio and Pradet 1980; Pace et al. 1990), pea (Webster and Henry 1987), barley (Farrar 1981, 1985), pearl millet (Baysdorfer et al. 1988), and sycamore (Journet et al. 1986).

## Effect of Sucrose Starvation on Carbohydrate Pools

Sycamore cells have two major pools of carbohydrates sucrose: and starch. Starch is present in the stroma of amyloplasts, whereas most of the sucrose is stored in the vacuole. During the first 10 h of sucrose starvation the endogenous sucrose content decreased to 30% of the control value and starch (expressed as glucose) remained constant (Fig. 1). Sucrose efflux from the vacuole was rapid enough to maintain an optimum phosphate ester concentration in the cytosol (Journet et al. 1986; Fig. 2) for respiration. After 10 h of sucrose starvation, when a threshold of intracellular sucrose

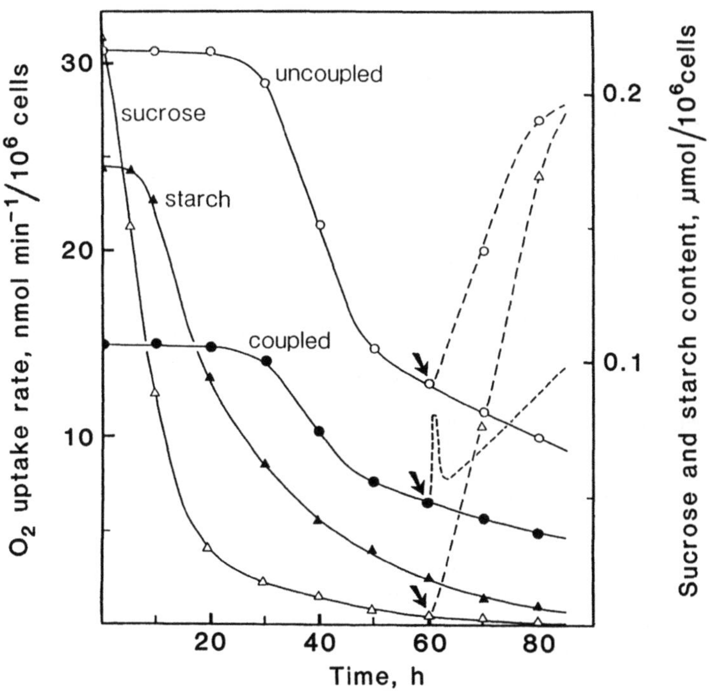


**Fig. 1.** Effect of sucrose starvation on the rate of $O_2$ consumption by sycamore cells and on carbohydrate pools (starch and sucrose). Cells harvested from the culture medium were rinsed three times by successive resuspensions in fresh culture medium devoid of sucrose and incubated at zero time (20 mg fresh weight per ml) in flasks containing sucrose-free culture medium. At each time cells were harvested and sucrose, starch and the $O_2$ consumption rate were measured. The starch content is expressed as micromole of glucose per $10^6$ cells. The *dotted lines* correspond to the enhancement of cell respiration and sucrose accumulation that are observed when 50 mM sucrose is added to the culture medium (*arrows*)

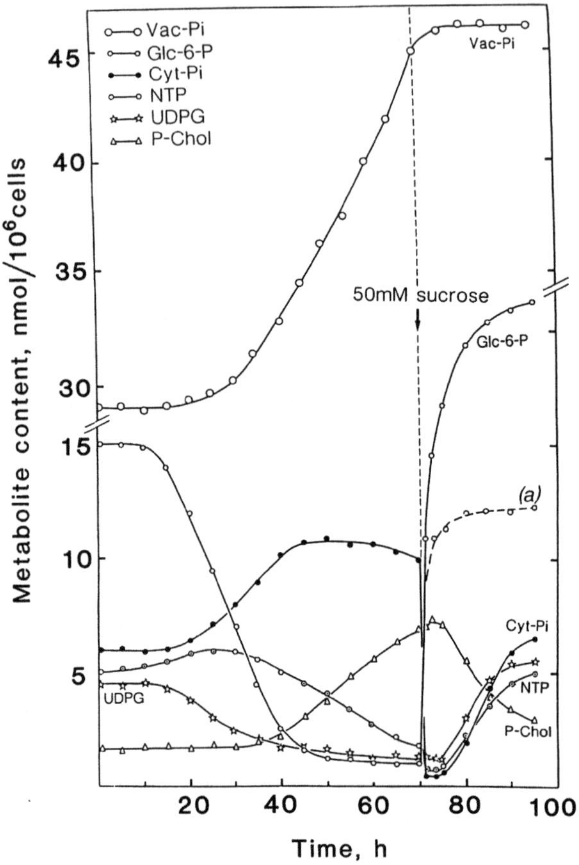


**Fig. 2.** Time-course of changes in various phosphorus compounds in sycamore cells subjected to sucrose starvation up to 70 h followed by 24 h recovery after addition of 50 mM sucrose and 50 μM $P_i$ (*arrow*) to the medium. The concentrations of organic compounds were determined biochemically. A WM 200 Bruker NMR spectrometer was used to characterize the two major intracellular pools of $P_i$ (vacuolar-$P_i$; cytoplasmic-$P_i$) known to be present in the plant cells. *NTP* nucleotide triphosphate; *Glc-6-P* glucose 6-P; *UDPG* UDP-glucose; *P-chol* phosphorylcholine. The *broken line* (*a*) shows the recovery of glucose 6-P in the absence of $P_i$ in the medium. Note in this case that the vacuolar $P_i$ pool does not fluctuate rapidly to buffer the $P_i$ in the cytoplasm. In the 70 h sucrose-starved cells the cytoplasmic $P_i$ and NTP decreased almost immediately upon addition of sucrose whereas glucose 6-P increased abruptly. Such a situation led to a transient increase of twofold in the rate of $O_2$ consumption ascribed to a marked increase in ADP delivery to the mitochondria (see Fig. 1)

concentration was attained, the concentration of glucose 6-P decreased gradually. During this period there was a marked depletion of sugar nucleotides (UDP-glucose). Pi molecules liberated from phosphate ester molecules were slowly expelled into the vacuole where they accumulated (Fig. 2). In these cells, starch breakdown, which progressively replaces sucrose hydrolysis after 10 h of sucrose starvation (Fig. 1), is not rapid enough to maintain an optimum phosphate ester concentration.

The mechanism of starch breakdown and its conversion to respiratory substrates such as malate and pyruvate is incompletely understood (Steup 1988). Plastids from soybean cultures (Macdonald and ap Rees 1983) and cauliflower buds (Journet and Douce 1985) contain α-glucan phosphorylase and all the enzyme equipment needed to convert glucose 1-phosphate to triose phosphate. These observations strongly suggest that conversion of starch to triose phosphate via glucose 1-phosphate may involve hydrolysis of starch by various amylases and/or phosphorolytic attack, followed by glycolysis to triose phosphate in the stroma and entry into the cytosol via the Pi-translocator (Heldt and Flügge 1987; Alban et al. 1988). It is possible that the immediate products of starch breakdown, hexose phosphates (glucose 6-phosphate or glucose 1-phosphate), can move via a specific carrier from the amyloplast to the cytosol. As pointed out judiciously by ap Rees (1987), it is necessary not only to establish the reactions of starch breakdown and their regulation but also which of the products of starch breakdown (hexose-phosphate, free-hexose, triose-phosphate) move from the amyloplast to the cytosol to sustain mitochondrial respiration during the course of sucrose starvation. The results presented by Journet et al. (1986) strongly suggest that, following sucrose deprivation, an increase in the cytosolic Pi level and a decrease in phosphorylated compounds (glucose 6-phosphate) apparently trigger starch breakdown in amyloplasts. Studies performed with isolated intact chloroplasts also point to a closer correlation between the extraplastidic Pi level and starch mobilization (for review see Steup 1988).

## Effect of Sucrose Starvation on the Rate of $O_2$ Consumption by Sycamore Cells

The rate of respiration of cells deprived of sucrose appeared to be constant for at least 18 h (Fig. 1). Thus by a gradual mobilization of sucrose and starch reserves, the respiratory activity was sustained and stabilized over an extended period of time during which an external carbon source was not available. Thereafter, the rate of $O_2$ consumption decreased with time. After 50 h of starvation, their capacity to utilize $O_2$ was reduced to less than 50% of that of normally growing cells. Figure 1 also indicates that the uncoupled rate of $O_2$ consumption decreased after 30 h in the same ratio as the rate of respiration without uncoupler. Furthermore, a careful examination of Fig. 1 indicates that the rate of $O_2$ consumption began to decline when the intracellular sucrose had been consumed. Under these conditions starch content was reduced to less than 30% of that of normal cells.

The progressive loss of respiratory capacity of sycamore cells during their aging in sucrose-free culture medium could be entirely attributable to a progressive decrease in the intracellular substrate levels, as suggested by several authors (for review see Yemm 1965). The respiration rate of many plant tissues falls off when they are deprived of a carbon source for a long period of time, and the reduced respiration rates resulting from such starvation can often be elevated by supplying the appropriate sugars (James 1953). It is possible, according to Lambers (1985), that in the absence of exogenous sucrose or glucose, intracellular respiratory substrates (pyruvate? malate?) are no longer "wastefully" respired via the cyanide-insensitive alternative pathway thus leading to a progressive decrease in the rate of $O_2$ consumption. Cyanide-resistant respiration has been the subject of a number of interesting reviews in recent years, and these reviews should be consulted for detailed discussions of the cyanide-resistant pathway in plant mitochondria (Lance et al. 1985; Laties 1982; Siedow 1982). In short, electrons seem to be partitioned between the cytochrome oxidase pathway and the alternative pathway according to the rate constants associated with the reactions between ubiquinol and either complex III or the alternative oxidase, respectively. It is clear, therefore, that the endogenous substrate for both pathways is the reduced form of ubiquinone-10 (Siedow 1982), and this pool of ubiquinone is kept reduced via substrate oxidation by complexes I and II. When electrons from the quinol pool flow through the alternative pathway, energy is not conserved in the form of an electrochemical gradient, and no ATP is formed. Lambers (1985) has suggested that in plant tissues in which excess carbohydrate is produced (i.e., carbohydrate that cannot be readily stored or used in growth or reproduction), the nonphosphorylating alternative pathway may contribute significantly to total respiration. In other words, the cyanide-resistant respiration may act as an overflow to drain off the energy of carbohydrates supplied in excess of demand: this pathway may be considered energetically wasteful in terms of whole-plant carbon budgets.

However, the concept of wasteful oxidation of sucrose (the level of respiratory substrate determines the degree to which the cyanide-resistant pathway contributes to respiration) is most unlikely, because in the cells the rate of mitochondrial respiration must be rigorously coordinated to meet the ATP demand of the cytoplasm (Brand and Murphy 1987; Dry et al. 1987; Douce and Neuburger 1989) and because the demonstration that wasteful respiration via a mechanism that does not involve tight allosteric control of sugar phosphate breakdown has yet to be made convincingly (ap Rees 1988). Consequently, it is difficult to imagine that the intracellular substrate level is the parameter which determines the regulation of such a branched pathway (which is probably exceedingly complex). Furthermore, if one considers that substrate levels have been depleted after a long period of sucrose starvation we should expect addition of exogenous sucrose to increase respiration immediately afterwards. In fact, this is not seen. The enhancement of uncoupled respiration that is observed when sucrose is added to the culture medium is a slow process: the full rate of uncoupled $O_2$-consumption by 70-h sucrose-starved cells was recovered several hours after the addition of sucrose (Journet et al. 1986) (Fig. 1). Likewise, Saglio and Pradet (1980) have clearly shown that $O_2$ uptake declines progressively after excision of maize root tips and that the addition of exogenous sugars induces a slow rise in the respiratory rate up to its original value while the energy charge remains constant. Direct attempts to increase the respiration of pea roots by the addition of sucrose yielded no evidence of wasteful respiration (Bryce and ap Rees 1985). The most likely explanation for the progressive loss of respiratory capacity of plant cells during their aging in sucrose-free culture medium is that the total volume of mitochondrial/cell declines progressively during the course of sucrose starvation. In support of this suggestion, studies performed with sycamore cells by Journet et al. (1986) indicated that during the course of sucrose starvation the respiration rates decreased progressively in the same ratio as the decrease in intracellular cardiolipin or cytochrome $aa_3$, two specific mitochondrial markers (Bligny and Douce 1980).

Addition of sucrose after 70 h of sucrose starvation (i.e., when the uncoupled respiration rate had decreased considerably) (Fig. 1) resulted in the disappearance of Pi from the cytoplasm and a marked increase in cytoplasmic glucose 6-P (Roby et al. 1987) (Fig. 2). Interestingly, although the vacuole had sequestered almost all of the cellular Pi liberated during the course of sucrose starvation, the Pi necessary to restart cell metabolism was utilized first from the cytoplasm, and then, if present, from the external medium (Roby et al. 1987) (Fig. 2). The consumption of vacuolar Pi was a very slow process which required several hours and occurred only in response to the depletion of Pi from the external medium: in these series of experiments the vacuolar Pi pool did not fluctuate to buffer the Pi in the cytoplasm.

## Effect of Sucrose Starvation on Polar Lipid Fatty Acids of Sycamore Cells

Figure 3 summarizes the effect of sucrose starvation of sycamore cells on intracellular fatty acid content. After 30 h of sucrose starvation, when almost all of the intracellular carbohydrate pool (starch + sucrose) (see Fig. 1) had disappeared, the cell fatty

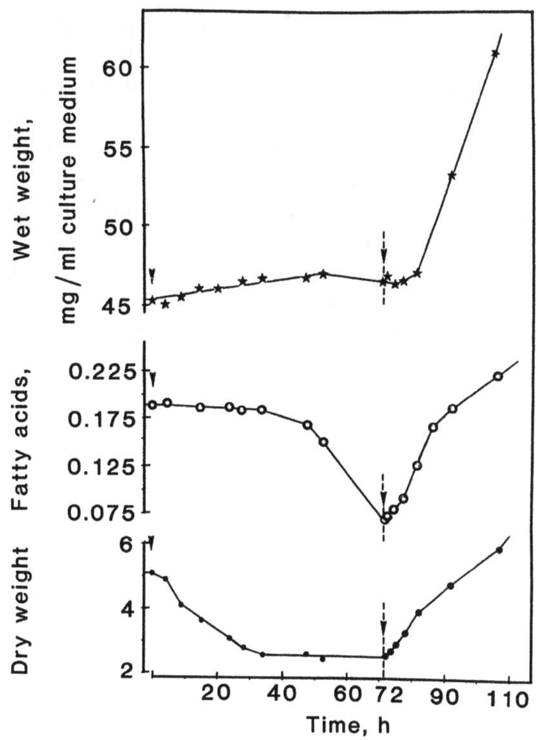


**Fig. 3.** Effect of sucrose starvation and replenishment on cell weight (fresh and dry weight) and on intracellular fatty acid content of sycamore cells. Cells harvested from the culture medium were rinsed three times by successive resuspension in fresh culture medium devoid of sucrose and incubated at zero time (45 mg fresh weight/ml) in flasks containing sucrose-free culture medium. At the time indicated by an *arrow* 50 mM sucrose was added to the culture medium

acid content declined progressively. Analysis of cell polar lipids has indicated that a long period of sucrose starvation also induced a progressive disappearance of all the cell phospholipids and galactolipids (Dorne et al. 1987). During the period of the most rapid disappearance of polar lipid fatty acids, the respiratory quotient (RQ) of the cell was as low as 0.7. Such a low RQ indicates that the fatty acids which were released during the breakdown of the endomembranes may be utilized as respiratory substrates through peroxisomal metabolic pathways (Gerhardt 1986). This observation is reminiscent of respiration in detached leaves kept in continuous darkness for several days (James 1953) and in maize root tips deprived of sucrose for several hours, where a link between the respiration of lipids and a low RQ has been shown (Pradet and Raymond 1983). Interestingly, during the course of polar lipid breakdown the total amount of sterols including steryl glucoside and acylated steryl glucoside appeared to be constant (Table 1). These observations suggest that while a large proportion of the endomembrane system disappears in starving cells, tonoplast and plasmalemma, which are enriched in sterol compounds (Hartmann and Benveniste 1987), escape the autophagic process.

After 35 h of sucrose deprivation, when the endogenous fatty acids started to decline (Fig. 3), several noticeable changes began to occur in the cells (Journet et al. 1986). Of particular interest there was the marked increase in the amount of phosphorylcholine (Fig. 2). This compound was characterized in plants for the first time by Maizel et al. (1956). It could be demonstrated that the total amount of phosphorylcholine that appeared after a long period of sucrose deprivation corresponded exactly to the total amount of phosphatidylcholine that disappeared within the same period of time (Dorne et al. 1987). The accumulation of phosphorylcholine which occurs in intact cells during the loss of fatty acids reflects, therefore, the hydrolysis of the major cell membrane polar lipid by a mechanism mediated by lypolytic acyl hydrolases. The fatty acids thus released are probably oxidized in peroxisomes and mitochondria operating in a concerted manner. In support of this suggestion it has been demonstrated that all of the plant cells studied so far contain very powerful lipolytic acyl hydrolase activities (Galliard 1980) and possess in their peroxisomes all the enzymatic equipment necessary to oxidize free fatty acids (Gerhardt 1986). Gut and Matile (1988) have found in barley leaf segments that a large proportion of polar lipids was lost during the initial phase of dark-induced senescence and that an

**Table 1.** Biochemical characterization of normal and sucrose-starved cells. These data are from a representative experiment and have been reproduced at least three times. Normal cells were harvested from the culture medium. Sucrose-starved cells: normal cell were rinsed three times by successive resuspension in fresh culture medium devoid of sucrose and incubated during 50 h in flasks containing sucrose-free culture medium. Galactolipids: monogalactosyldiacylglycerol + digalactosyldiacylglycerol. Sterols: free sterol + sterylglucoside + acylated steryl glucoside

| Cells | Protein | Phospholipids | Galactolipids | Sterols |
|---|---|---|---|---|
| | mg/$10^6$ cells | μg/$10^6$ cells | μg/$10^6$ cells | μg/$10^6$ cells |
| Normal | 0.53 | 50 | 4.5 | 3.5 |
| Sucrose-starved | 0.27 | 20 | 2.5 | 3.7 |

apparent induction of key enzymes (isocitrate lyase and malate synthase) of the glyoxylic acid cycle occurred during this period. We can conclude, therefore, that the presence of a large excess of phosphorylcholine in plant cells should be considered as a good marker of membrane utilization after a long period of sucrose starvation and is apparently related to stress.

Interestingly, Tolbert's group (Tanaka et al. 1966; Martin and Tolbert 1983) suggested that many plants accumulate large amounts of phosphorylcholine in their xylem sap during starvation in long periods of environmental stress such as drought. Thus during the course of autophagy, phosphorylcholine molecules, derived from phosphatidylcholine degradation, may escape from the cell and be exported towards the xylem vessels where they accumulate. Under these conditions, phosphorylcholine molecules would be transported in the xylem fluids to meristematic cells or fast-growing tissues where they could be used subsequently as a source of choline for membrane proliferation. In support of this hypothesis, Bligny et al. (1989) and Gout et al. (1990) have demonstrated that sycamore cells derived from the cambium, a tissue which is in close contact with the xylem vessels, possess all the enzymatic equipment (a powerful phosphorylcholine phosphatase associated with the cell wall, a unique choline carrier associated with the plasma membrane, and a cytosolic choline kinase) for trapping phosphorylcholine. Furthermore, the low half-saturation of choline transport (the apparent $K_m$ was 2–4 μM) would make possible a very active system, in conjunction with the cytosolic choline kinase, to act as a drain for sequestering choline released by the phosphorylcholine phosphatase in the cell wall.

Addition of 50 mM sucrose to the medium after 70 h of sucrose starvation resulted in a marked increase in the cell dry weight and total cell fatty acids associated with polar lipids (Fig. 3). The increase in the cell dry weight was attributable to a rapid accumulation of sucrose in the vacuole and starch in plastids, whereas the increase in total cell fatty acids was attributable to the synthesis of new cytoplasmic material such as mitochondria (Dorne et al. 1987). The burst of cell metabolism led to an increase in ATP turnover and therefore to a transient increase of 1.5 to twofold in the rate of $O_2$ consumption (Fig. 1). When the total fatty acid content had returned to near the level of normal cells (Fig. 3), growth began. The lag phase preceding growth represented the time required to synthesize new cytoplasmic material. Furthermore, 10 h after replenishment of sucrose other substantial changes were evident in the cells. Of particular interest was the marked decrease in the amount of phosphorylcholine (Dorne et al. 1987) (Fig. 2). We have shown that the linear decrease in cytosolic phosphorylcholine closely correlated with the linear increase in phosphatidylcholine (Roby et al. 1987). In these cells phosphorylcholine was readily metabolized to sustain phosphatidylcholine synthesis, the major polar lipid of all the cell membrane systems (Harwood 1980).

## Effect of Sucrose Starvation on Asparagine Accumulation

After 30 h of sucrose starvation, when almost all the intracellular carbohydrate pool (starch + sucrose) had disappeared, the cell protein content declined progressively (see Table 1). A careful analysis of free amino acids carried out during the course of protein breakdown indicated that the total amount of free amino acids which accumulated (Genix et al. 1990) did not account for all of the total protein which disappeared. These results strongly suggest that amino acids released during the course of protein breakdown could be metabolized to provide the remaining mitochondria with respiratory substrates. In general, the catabolic sequences for most amino acids lead to a compound which is capable of entering the tricarboxylic acid cycle (Mazelis 1980). Furthermore, the peroxisomes from nonfatty plant tissues are able to activate branched fatty acids which are formed by deamination of leucine, isoleucine and valine, the major amino acids of membrane proteins (Gerbling and Gerhardt 1989).

It is well established that ammonia which is released during the course of amino acid utilization is sequestered into organic compounds and that amides are important products of protein degradation (Stewart and Lahrer 1980; Sieciechowicz et al. 1988). Genix et al. (1990) have indicated that asparagine accumulated in sycamore cells deprived of sucrose. After 70 h of sucrose starvation, the total amount of asparagine present in intact sycamore cells was considerable approximately 10 µmol/g fresh weight, and the total amount of protein that had been broken down was approximately 4–5 mg/g fresh weight. From these results they calculated that the total amount of asparagine (expressed as nitrogen) and free amino acids that appeared after a long period of sucrose deprivation corresponded roughly to the total amount of protein (expressed as nitrogen), that disappeared within the same period of time.

Many questions remain to be answered regarding the synthesis of asparagine. It might be expected that aspartate aminotransferase, glutamine synthetase, glutamate synthase and asparagine synthetase play a key role in $NH_3$ reassimilation and in asparagine accumulation (Scott et al. 1976). Oxaloacetate which is necessary for asparagine biosynthesis, is very likely to be provided by the tricarboxylic acid cycle. Furthermore, in various circumstances intact plant mitochondria can export oxaloacetate (Douce and Neuburger 1989). We can conclude that the presence of a large excess of asparagine in plant cells should be considered as a good marker of protein breakdown after a long period of sucrose starvation, and like phosphorylcholine is also likely to be related to stress. Interestingly, numerous results (for a review see Stewart and Lahrer 1980 and Sieciechowicz et al. 1988) have clearly indicated that asparagine (along with other amides) is synthesized during exposure of plants to periods of environmental stress such as drought.

In the presence of sucrose, asparagine was found to be metabolized in sycamore cells and does not behave, therefore, as a dead end metabolite. Again, many questions remain to be answered regarding the fate of asparagine. It might be expected that asparagine is metabolized either by transamination or deamination by an asparaginase in agreement with what was previously observed in developing pea leaves and seeds (for review see Sieciechowicz et al. 1988). However, in mature leaves that no longer require nitrogen for growth, asparagine molecules are not readily metabolized and are re-exported towards the xylem where they accumulate (for review see Sieciechowicz et al. 1988). For example, high concentrations of this amino acid have been found in the xylem sap of peas, accounting for up to 70% of the xylem contents (Urquhart and Joy 1982), and, like phosphorylcholine, a variable proportion of this amino acid could be utilized in the developing plant at a later period.

## Summary

These data raise questions about the mechanism involved in the triggering of the autophagic process. It is possible that ubiquitin, a highly conserved protein involved in several important regulatory processes through its ATP-dependent, covalent ligation to a variety of eukaryotic target proteins or cell membranes (cell surface recognition) (Finley and Varshavsky 1985; Vierstra 1987), could play an important role in this energy-dependent autophagic process. It is also possible that tubes of smooth endoplasmic reticulum (ER) wrap themselves around portions of cytoplasm including cell organelles such as mitochondria. Once the sequestered portions of the cytoplasm have been completely enclosed by lateral fusion of the enveloping ER, various hydrolase are probably released from the surrounding exoplasmic space into the sequestered cytoplasm to form an autophagic vesicle. This phenomenon is reminiscent of the formation of vacuoles in parenchyma cells of root meristems (for review see Marty et al. 1980). Interestingly, we have observed that macroautophagy is activated early in starvation to provide the required amino acids and fatty acids. Finally, nothing is known about the mechanism by which protein degradation is controlled during the course of sucrose starvation. It is possible that the increase in amino acid levels observed when almost all the intracellular carbohydrate pools had disappeared plays a critical role in the control of cell autophagy. Caro et al. (1989) have shown that various amino acids and especially leucine at physiological concentrations strongly inhibit autophagic proteolysis in isolated rat hepatocytes. As suggested by Dice (1990), in mammalian tissues a prolonged fast might also activate a selective "lysosomal" degradation pathway that applies only to dispensable cytoplasmic proteins with specific peptide sequences.

In summary, these results demonstrate that during the course of sucrose starvation the catabolism of fatty acids and amino acids must be sufficiently rapid to supply mitochondria with respiratory substrates in order to maintain a high nucleotide energy charge in the remaining cytoplasmic fraction (Roby et al. 1987). These observations are important if one considers that ATP is a very important factor in controlling the rate of autophagy in rat hepatocytes (Caro et al. 1989). These results emphasize the extraordinary flexibility and complexity of plant cell metabolism. This flexibility is compounded by the fact that cytoplasm, in particular, can be utilized as a carbon source after a long period of sucrose starvation without significantly affecting the survival of these cells. Under these conditions, plant cells, owing to the presence of intracellular pools of carbohydrate and to their ability to control an autophagic process, can survive for a long period of time without receiving any external supply of organic carbon; a situation that may occur during exposure of plants to periods of environmental stress such as drought.

## References

Alban C, Joyard J, Douce R (1988) Preparation and characterization of envelope membranes from nongreen plastids. Plant Physiol 88:709–717

ap Rees T (1987) Compartmentation of plant metabolism. In: Davies DD (ed) The biochemistry of plants, vol 12. Physiology of metabolism. Academic Press, San Diego, pp 87–115

ap Rees T (1988) Hexose phosphate metabolism by nonphotosynthetic tissues of higher plants. In: Preiss J (ed) The biochemistry of plants. Vol 14: Carbohydrates. Academic Press, San Diego, pp 1–33

Baysdorfer C, Warmbrodt RD, Van der Woude WJ (1988) Mechanisms of starvation tolerance in pearl millet. Plant Physiol 88:1381–1387

Bligny R, Douce R (1980) A precise localisation of cardiolipin in plant cells. Biochim Biophys Acta 617:254–263

Bligny R, Foray M-F, Roby C, Douce R (1989) Transport and phosphorylation of choline in higher plant cells. Phosphorus-31 nuclear magnetic resonance studies. J Biol Chem 264:4888–4895

Brand MD, Murphy MP (1987) Control of electron flux through the respiratory chain in mitochondria and cells. Biol Rev 62:141–193

Bryce JH, ap Rees T (1985) Effects of sucrose on the rate of respiration of the roots of *Pisum sativum*. J Plant Physiol 120:363–367

Caro LHP, Plomp PJAM, Leverve XM, Meijer AJ (1989) A combination of intracellular leucine with either glutamate or aspartate inhibits autophagic proteolysis in isolated rat hepatocytes. Eur J Biochem 181:717–720

Dice JF (1990) Peptide sequence that target cytosolic proteins for lysosomal proteolysis. Trends Biochem Sci 15:305–309

Dorne A-J, Bligny R, Rébeillé F, Roby C, Douce R (1987) Fatty acid disappearance and phosphorylcholine accumulation in higher plant cells after a long period of sucrose deprivation. Plant Physiol Biochem 25:589–595

Douce R (1985) Mitochondria in higher plants. Structure, function, and biogenesis. Academic Press, Orlando

Douce R, Neuburger M (1989) The uniqueness of plant mitochondria. Annu Rev Plant Physiol Plant Mol Biol 40:371–414

Dry IB, Bryce JH, Wiskich JT (1987) Regulation of mitochondrial respiration. In: Davies DD (ed) The biochemistry of plants. Vol 11: Biochemistry of metabolism. Academic Press, San Diego, pp 213–252

Farrar JF (1981) Respiration rate of barley roots: its relation to growth, substrate supply and the illumination of the shoot. Ann Bot 48:53–63

Farrar JF (1985) Fluxes of carbon in roots of barley plants. New Phytol 99:57–69

Finley DA, Varshavsky A (1985) The ubiquitin system: functions and mechanisms. Trends Biochem Sci 117:344–347

Galliard T (1980) Degradation of acyl lipids: hydrolytic and oxidative enzymes. In: Stumpf PK (ed) The biochemistry of plants. Vol 4: Lipids: Structure and function. Academic Press, Lond New York, pp 85–116

Genix P, Bligny R, Martin JB, Douce R (1990) Transient accumulation of asparagine in sycamore cells after a long period of sucrose starvation. Plant Physiol 94:717–722

Gerbling H, Gerhardt B (1989) Peroxisomal degradation of branched-chain 2-oxo acids. Plant Physiol 91:1387–1392

Gerhardt B (1986) Basic metabolic function of the higher plant peroxisome. Physiol Vég 24:397–410

Gout E, Bligny R, Roby C, Douce R (1990) Transport of phosphocholine in higher plant cells: $^{31}P$ nuclear magnetic resonance studies. Proc Natl Acad Sci USA 87:4280–4283

Gut H, Matile Ph (1988) Breakdown of galactolipids in senescent barley leaves. Planta 176:548–550

Hartmann M-A, Benveniste P (1987) Plant membrane sterols: isolation, identification, and biosynthesis. Methods Enzymol 148:632–6501

Harwood JL (1980) Plant acyl lipids: Structure, distribution, and analysis. In: Stumpf PK (ed) The biochemistry of plants. Vol 4: Lipids: structure and function. Academic Press, Lond New York, pp 1–55

Heldt H, Flügge UI (1987) Subcellular transport of metabolites in plant cells. In: Davies DD (ed) The biochemistry of plants. Vol 12: Physiology of metabolism. Academic Press, San Diego
James WO (1953) Plant respiration. Clarendon Press, Oxford
Journet EP, Douce R (1985) Enzymic capacities of purified cauliflower bud plastids for lipid synthesis and carbohydrate metabolism. Plant Physiol 79:458–467
Journet EP, Bligny R, Douce R (1986) Biochemical changes during sucrose deprivation in higher plant cells. J Biol Chem 261:3193–3199
Kaiser WM (1984) Response of photosynthesis and dark $CO_2$-fixation to light, $CO_2$, and temperature in leaf slices under osmotic stress. J Exp Bot 35:1–11
Kaiser WM (1987) Effects of water deficit on photosynthetic capacity. Physiol Plant 71:142–149
Kerr PS, Rufty Jr TW, Huber SC (1985) Changes in nonstructural carbohydrates in different parts of soybean (*Glycine max*) plants during a light/dark cycle and in extended darkness. Plant Physiol 78:576–581
Lambers H (1985) Respiration in intact plants and tissues: its regulation and dependence on environmental factors, metabolism and invaded organisms. In: Douce R, Day D (eds) Encyclopedia of plant physiology. Vol 18: Higher plant cell respiration. Springer, Berlin Heidelberg New York, pp 418–473
Lance C, Chauveau M, Dizengremel P (1985) The cyanide-resistant pathway of plant mitochondria. In: Douce R, Day D (eds) Encyclopedia of plant physiology. Vol 18: Higher plant cell respiration. Springer, Berlin Heidelberg New York, pp 202–247
Laties G (1982) The cyanide-resistant, alternative path in higher plant respiration. Annu Rev Plant Physiol 33:519–555
Laurière C (1983) Enzymes and leaf senescence. Physiol Vég 21:1159–1177
MacDonald FD, ap Rees T (1983) Enzymic properties of amyloplasts from suspension cultures of soybean. Biochim Biophys Acta 755:81–89
Maizel JV, Benson AA, Tolbert NE (1956) Identification of phosphorylcholine as an important constituent of plant saps. Plant Physiol 31:407–408
Martin BA, Tolbert NE (1983) Factors which affect the amount of inorganic phosphate, phosphorylcholine, and phosphorylethanolamine in xylem exudate of tomato plants. Plant Physiol 73:464–470
Marty F, Branton D, Leigh RA (1980) Plant vacuoles. In: Tolbert NE (ed) The biochemistry of plants. Vol 1: The plant cell. Academic Press, Lond New York, pp 625–658
Mazelis M (1980) Amino acid catabolism. In: Miflin BJ (ed) The biochemistry of plants. Vol 5: Amino acids and derivatives. Academic Press, Lond New York, pp 541–567
Pace GGM, Volk RJ, Jacskson WA (1990) Nitrate reduction in response to $CO_2$-limited photosynthesis. Plant Physiol 92:286–292
Pardet A, Raymond P (1983) Adenine nucleotide ratios and adenylate energy charge in energy metabolism. Annu Rev Plant Physiol 34:199–224
Roby C, Martin J-B, Bligny R, Douce R (1987) Biochemical changes during sucrose deprivation in higher plant cells phosphorus-31 nuclear magnetic resonance studies. J Biol Chem 262:5000–5007
Saglio PH, Pradet A (1980) Soluble sugars, respiration, and energy charge during aging of excised maize root tips. Plant Physiol 66:516–519
Schulze E-D (1986) Carbon dioxide and water vapor exchange in response to drought in the atmosphere and in the soil. Annu Rev Plant Physiol 37:247–274
Scott DB, Farnden KJF, Robertson JG (1976) Ammonia assimilation in lupin nodules. Nature (Lond) 263:703–705
Sieciechowicz KA, Joy KW, Ireland RJ (1988) The metabolism of asparigine in plants. Phytochemistry 27:663–671
Siedow JN (1982) The nature of cyanide-resistant pathway in plant mitochondria. Rec Adv Phytochem 16:47–84
Steup M (1988) Starch degradation. In: Preiss J (ed) The biochemistry of plants. Vol 14: Carbohydrates, Academic Press. San Diego, pp 255–296
Stewart GR, Lahrer F (1980) Accumulation of amino acids and related compounds in relation to environmental stress. In: Miflin BJ (ed) The Biochemistry of plants. Vol 5: Amino acids and derivatives. Academic Press, Lond New York, pp 609–635

Tanaka K, Tolbert NE, Gohlke AF (1966) Choline kinase and phosphorylcholine phosphatase in plants. Plant Physiol 41:307–312
Thomas H, Stoddart JL (1980) Leaf senescence. Annu Rev Plant Physiol 31:83–111
Urquhart AA, Joy KW (1982) Transport, metabolism, and redistribution of xylem-borne amino acids in developing pea shoots. Plant Physiol 69:1226–1232
Vassey TL, Sharkey TD (1989) Mild water stress of *Phaseolus vulgaris* plants leads to reduced starch synthesis and extractable sucrose phosphate synthase activity. Plant Physiol 89:1066–1070
Vierstra RD (1987) Demonstration of ATP-dependent, ubiquitin-conjugating activities in higher plants. Plant Physiol 84:332–336
Walsh KB, Vessey JK, Layzell DB (1987) Carbohydrate supply and $N_2$ fixation in soybean. Plant Physiol 85:137–144
Webster PL, Henry M (1987) Sucrose regulation of protein synthesis in pea root meristem cells. Environ Exp Bot 27:253–262
Yemm EW (1965) The respiration of plants and their organs. In: Steward FC (ed) Plant physiology. A treatise, vol IV-A. Academic Press, Lond New York, pp 231–310

Chapter 10

# Desiccation Tolerance in Vegetative Plant Tissues and Seeds: Protein Synthesis in Relation to Desiccation and a Potential Role for Protection and Repair Mechanisms

J.D. BEWLEY[1] and M.J. OLIVER[2]

## Introduction

While many vascular plants produce specialized structures that are able to withstand desiccation (e.g., seeds, spores, or pollen), few can survive severe water loss in the vegetative state. The vast majority of desiccation-tolerant plants belong to the lower groups of the plant kingdom: algae, lichens, and bryophytes. They are poikilohydric in nature and exhibit a rapid equilibration of their internal water content with the water potential of the environment. A few ferns, and even fewer angiosperms are tolerant of desiccation (Davis 1972; Gaff 1980; Bewley and Krochko 1982); they exhibit modified poikilohydry, however, since most have features (e.g., stomatal complexes) that permit them to reduce their rate of water loss.

Earlier work on desiccation tolerance has led to the proposal that three critical parameters must be met for a plant, or plant tissue to survive severe water deficits. It must exhibit the ability: (1) to limit damage during drying to a repairable level, (2) to maintain its physiological integrity in the dry state, and (3) to mobilize repair mechanisms upon rehydration which effect restitution of damage suffered during desiccation (Bewley 1979). The basis for desiccation tolerance remains enigmatic, however, and very few plants have been subjected to biochemical and molecular studies to elucidate the cellular events that are involved. An appropriate approach would appear to be one which concentrates on two experimentally testable facets of tolerance – cellular protection and cellular repair. In this chapter we first consider the responses of desiccation-tolerant vegetative plants to drying and rehydration, particularly in relation to the roles of protection and repair. Such plants are subjected to drying at unpredictable intervals, and may have to accommodate water loss several to numerous times during a growing season. In contrast, the seeds of many higher plants undergo desiccation in a programmed manner at the termination of their development. Their response to drying is presented as an interesting comparison to that of vegetative tissues.

[1]Department of Botany, University of Guelph, Guelph, Ontario NIG 2 WI, Canada
[2]USDA/ARS Cropping System Lab, Route 3, Box 215, Lubbock, TX 94401, USA

Somero et al. (Eds.)
Water and Life

## Desiccation of Vegetative Plants and Tissues: Evidence for Cellular Repair

Much of the evidence for the importance of cellular repair in desiccation tolerance comes from studies of poikilohydric plants, particularly bryophytes. Upon drying, desiccation-tolerant mosses rapidly lose their capacity to conduct protein synthesis (Bewley 1972, 1973a; Siebert et al. 1976; Henckel et al. 1977). In *Tortula ruralis* this is manifested as a decline in polysomes. If water loss is rapid (desiccation in 1 h) 50% of the polysomes are retained in the dried state, whereas slow desiccation (within 4 h) results in their complete dissociation (Gwozdz et al. 1974). The retention of polysomes in rapidly dried moss is indicative that loss of water alone is not the cause of detachment of ribosomes from the mRNA template. Nor is the decline in polysomes the result of an increase in RNase activity (Dhindsa and Bewley 1976, 1977), but rather it is caused by the run-off of ribosomes from the mRNAs, concomitant with their failure to reinitiate protein synthesis (Dhindsa and Bewley 1977). The retention of polysomes in rapidly dried moss is because drying is so fast that mRNAs are trapped on polysomes before run-off is completed. Due to the speed of desiccation, and the apparent sensitivity to water loss of the initiation step of protein synthesis, it seems unlikely that "protective" proteins are synthesized during drying. This is borne out by the observation that no new mRNAs (e.g., for protective proteins) are recruited into the protein synthesis complex during drying, even when the drying rate is slow (M.J. Oliver, unpublished data). Poikilohydry, at least in this bryophyte, appears to preclude a strategy of cellular protection involving unique protein synthesis as a mechanism for desiccation tolerance. The assumption is made that this strategy demands the synthesis of proteins only as the cell is losing water. Such may not be the case, however, since it is possible that the moss synthesizes protective proteins at all times and thus does not have to undertake their production in response to water loss. Rapid desiccation of *Bryum pseudotriquetrum*, an intolerant aquatic moss, results in a complete loss of polysomes. Rehydration does not result in polysome recovery and within 24 h ribosomes themselves are degraded (Bewley et al. 1974). Rapid desiccation of the semi-aquatic moss *Cratoneuron filicinum* results in some polysome retention in the dried state, but they are absent after slow drying (L. Malek and J.D. Bewley, unpubl. data). Rehydration results in a complete inactivation of both retained polysomes and ribosomes, however, and the plant fails to recover (Gwozdz and Bewley 1975). Thus the ability to survive desiccation is not an inherent feature of all bryophytes, but is associated with those adapted to xeric and mesic environments.

For a repair-based mechanism of tolerance to be effective, it is expected that there will be some disruption of cellular integrity due to desiccation. Thus the majority of repair-related events will occur soon after the relief of desiccation, i.e., within a relatively short time of rehydration. This appears to be the case in poikilohydric mosses. Their cells sustain desiccation-induced cellular damage, which can be quite severe, especially to membranes, with the major disruptions occurring when water rushes into dried cells upon rehydration (Oliver and Bewley 1984a). Interestingly, the extent of desiccation-induced damage is related to the speed at which drying occurs; rehydration following rapid desiccation leads to a much more extensive leakage of solutes than after slow drying (Gupta 1976; Brown and Buck 1979; Bewley and Krochko 1982). This could be indicative of a physical, rather than metabolic, protection component to the desiccation-tolerance mechanism, in that slow water loss somehow allows membranes to structurally accommodate the decline in cell water status and thus restrict leakage upon rehydration.

Upon rehydration of desiccation-tolerant mosses, protein synthesis recovers rapidly, e.g., *T. ruralis* (Bewley 1973a,b; Gwozdz et al. 1974), *T. norvegica* and *T. caninervis* (M.J. Oliver, unpubl. data), *Polytrichium commune* (Siebert et al. 1976), and *Neckera crispa* (Henckel et al. 1977). The speed of desiccation affects the rate at which protein synthesis can recover upon rehydration. Slowly dried moss resumes protein synthesis at a faster rate than does rapidly dried moss (Gwozdz et al. 1974), even though no polysomes remain in the slow-dried state. This is probably because rapid desiccation leads to more cellular damage than slow drying and thus a longer recovery time is required.

RNA synthesis also recovers soon after rehydration of dried *T. ruralis* (Oliver and Bewley 1984b) and, as with protein synthesis, recovery takes longer after rapid drying. Ribosomes and ribosomal RNAs are stable during both rapid and slow desiccation. Conserved and newly synthesized ribosomes (and ribosomal RNAs) are quickly utilized in the reformation of polysomes upon rehydration (Gwozdz and Bewley 1975; Tucker and Bewley 1976; Bewley and Oliver 1983; Oliver and Bewley 1984b,c). Messenger RNAs are also stable during desiccation, although more so during slow than rapid drying, and they are quickly utilized in protein synthesis upon rehydration (Oliver 1983; Oliver and Bewley 1984c,d). Upon rehydration of the dried moss there is the de novo synthesis of mRNA to accompany the turnover of those mRNAs stored in the dry state. The rate of mRNA synthesis and its recruitment into polysomes during the first 60 min following rehydration is greater in moss that was previously dried rapidly than in that which was slowly desiccated (Oliver and Bewley 1984d). This may be in response to the more extensive loss of mRNAs as a result of rapid drying, or it may relate to the activation of cellular repair processes to overcome damage incurred under this drying regime. These observations lend credence to the contention that there is a repair-based tolerance mechanism in mosses; the greater the damage due to drying, the more substantial the response.

The relative importance of conserved and newly synthesized mRNAs in protein synthesis upon rehydration has been determined. In undesiccated control moss, the recruitment of newly synthesized mRNA is balanced by the recruitment of preexisting mRNA (Fig. 1). Upon rehydration of both slowly and rapidly dried moss the proportion of conserved mRNA soon declines and within 2 h of rehydration little is present; most of the mRNA in the protein synthetic complex at this time is newly synthesized (Fig. 1). Thus, conserved mRNAs are replaced by newly synthesized ones, even within the first hour of rehydration.

In *T. ruralis*, desiccation and rehydration lead to considerable changes in the pattern of protein synthesis, regardless of the speed of water loss. Fully hydrated moss synthesizes a complex array of proteins (Fig. 2A) but, after desiccation, the pattern of proteins synthesized during the first two hours following rehydration is substantially different (Fig. 2B). Many proteins whose synthesis is easily detectable in hydrated moss (circled on Fig. 2A) are not synthesized initially upon rehydration. Such proteins have been designated as h proteins (hydration proteins) and their

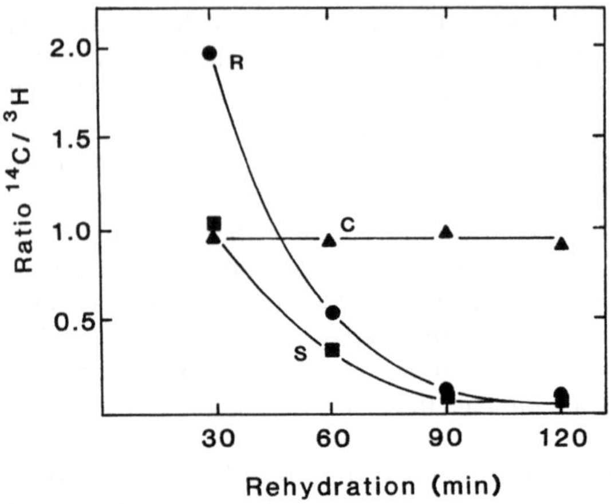


**Fig. 1.** The complement of conserved and newly synthesized mRNA (poly $A^+$ RNA) in the protein synthesis complex (polysomes) upon rehydration of rapidly (*R*) and slowly (*S*) dried *Tortula ruralis*. Hydrated moss was incubated in $^{14}C$-adenosine for 4 h before washing and desiccation. Rapidly and slowly dried moss was rehydrated for up to 2 h in $^{3}H$-adenosine. Control (*C*) moss was treated similarly, except that it was not desiccated. Ribosomal pellets were isolated and the mRNAs extracted therefrom. The ratios of conserved to newly synthesized mRNA were determined by measuring $^{14}C$ and $^{3}H$ incorporation. A high $^{14}C/^{3}H$ ratio during rehydration is indicative of a predominance of conserved mRNA, as occurs after 30 min. A low ratio shows that mRNA synthesized during rehydration is more prevalent, as occurs with increasing time after addition of water to both rapidly and slowly dried moss. Initially, there is a greater amount of conserved mRNA in rehydrated rapidly dried moss (which presumably is associated with polysomes conserved in the dry state) than in rehydrated slowly dried moss. In the undesiccated control moss there is a steady turnover of mRNA. (Oliver and Bewley 1984d)

translation obviously is suppressed in response to a desiccation-rehydration event. In contrast, there is another set of proteins whose synthesis is not substantial when the moss is in the hydrated state but is distinctly elevated during the initial phase of recovery desiccation (arrowed in Fig. 2B). These proteins have been designated as r proteins (rehydration proteins) and their translation is initiated or elevated upon rehydration after drying. A third set of proteins, (not marked), whose synthesis appears unaffected by a desiccation-rehydration event have been designated as c proteins (constitutive proteins).

The protein patterns resulting from the in vitro translation of bulk RNA extracted from hydrated, dried and rehydrated *T. ruralis* are very similar, which is indicative that the same mRNA population is present in all conditions (Oliver and Bewley 1984d). Yet the patterns of protein synthesis occurring in vivo in hydrated and rehydrated moss are distinctly different (Fig. 2A,B). Thus, the changes in gene expression in vivo due to drying and rehydration are not regulated by changes at the transcriptional level but rather at the translational level, with differential selection by the protein synthesizing complex from a constant mRNA pool.

This is somewhat of a surprising observation, since most plant stress responses involve changes in transcription (Sachs and Ho 1986). When placed in the context of

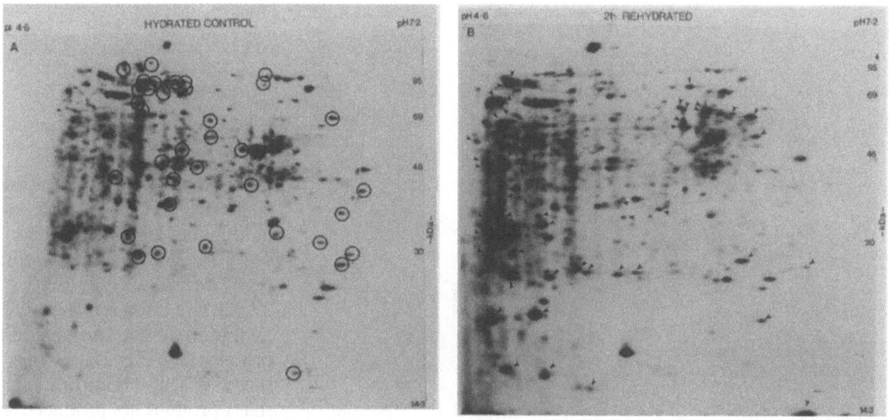


**Fig. 2A,B.** Fluorographs of two-dimensional PAGE separation (IEF/SDS PAGE) of $^{32}$S-cystine/methionine-labelled proteins synthesized in (**A**) fully hydrated *Tortula ruralis* and (**B**) following slow desiccation and rehydration for 2 h.

*Circled* are the proteins synthesized only in the moss gametophytes in the hydrated state (h proteins) and *arrowed* are those synthesized only after desiccation and rehydration (r proteins). $M_r$ markers in kDa are given on the *right-hand side*.

The pattern of protein synthesis is clearly different during the first 2 h of recovery following desiccation of the moss, compared to before desiccation. This change in protein synthesis is due to differential selection of mRNAs from the cellular pool (Oliver and Bewley 1984d), and not the result of regulation at the transcriptional level. The synthesis of r proteins is switched on as a consequence of desiccation, and that of h proteins is switched off. (Unpublished data of M.J. Oliver)

a repair-based mechanism of desiccation tolerance, however, regulation of protein synthesis at the translational level would appear to be an expedient strategy. The moss, which does not synthesize protective proteins, must initiate cellular repair immediately upon rehydration in order to quickly regain its structural and metabolic integrity. Since its metabolism is opportunistic, the moss must attempt to achieve a net gain in productivity during the brief and unpredictable periods of hydration. By reutilizing pre-existing mRNAs for repair processes upon rehydration, for example, at the expense of mRNAs for nonrepair proteins, the appropriate metabolic mode can be adopted. This, presumably, is accompanied by an up-regulation of the genes for r proteins which are already being expressed, rather than the transcription of new genes. This will result in the increased recruitment of mRNAs for these proteins into polysomes, and hence account for the new mRNAs associated therein during the first 1–2 h after rehydration (Fig. 1).

Even when *T. ruralis* is subjected to relatively mild water loss (to 85% original fresh weight) over a 30-min period, synthesis of r proteins is effected upon return to full hydration (Fig. 3), but this is accompanied by the synthesis of h proteins, an event which is inhibited only if water loss is severe (to approx. 30% fresh weight). This synthesis of h proteins upon rehydration is not due simply to run-off translation of their mRNAs. Thus, the induction of r protein synthesis, and the inhibition of h protein synthesis are not tightly linked events. If the r proteins are involved in cellular

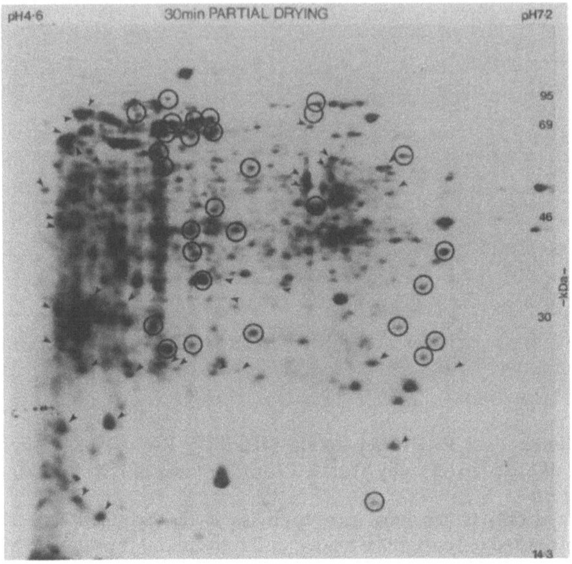


**Fig. 3.** Fluorograph of two-dimensional PAGE separation (IEF/SDS PAGE) $^{35}$S-cystine/methionine-labelled proteins synthesized after partial drying (to 85% fresh weight) for 30 min and upon return to full hydration for 2 h. Both h and r proteins (see Fig. 2) are synthesized. This demonstrates that r proteins are produced in response to relatively mild water stress, whereas h protein synthesis requires a more severe water stress to be switched off. (Unpublished data of M.J. Oliver)

repair processes it may be appropriate that their synthesis is particularly sensitive to water stress. On the other hand, it would be inefficient for h protein synthesis to cease under only mild or brief water stress conditions, and hence its lower sensitivity to this perturbation. Further studies are needed to provide evidence for these speculations, to determine the nature of the damage inflicted by desiccation, and which enzymes or structural proteins are needed to effect repair.

## Desiccation of Vegetative Plants and Tissues: Evidence for Cellular Protection

The evidence for desiccation-tolerance mechanisms based on cellular protection is equally as circumstantial as that for repair-based ones. Some indication of the former strategy can be derived from studies of the photosynthetic cyanobacterium *Nostoc commune*, which can survive for longer periods, even years, in the desiccated state. Peat et al. (1988) have demonstrated that field-collected vegetative cells and heterocysts, stored dry for 2 years before rehydration, are not structurally damaged. The functional integrity of these desiccated cells when stored for an extended period is preserved also (Scherer et al. 1984). Rapidly dried *Nostoc* fails to maintain structurally intact heterocysts (Peat and Potts 1987), however, suggesting that the rate of drying has to be slow enough to allow for protection of cellular integrity.

When *Nostoc* colonies are exposed to repeated cycles of desiccation and rehydration a group of acidic proteins, Wsp (Water Stress Proteins), are synthesized and accumulate to high concentrations (Scherer and Potts 1989). This microheterogeneous group of closely related glycoproteins may be derived from in vivo processing of a single 39 kDa Wsp; the microheterogeneity is not the result of differential glycosylation. The major Wsp39 glycoprotein occurs both intra- and extracellularly, as determined by immunocytological localization (Hill, Haludun, Scherer and Potts, pers. commun.). Water stress proteins are absent from cells that have not been exposed to desiccation or water stress. They are resistant to the proteolysis that normally occurs during prolonged desiccation and because of their persistence they may have a protective function. This conjecture is strengthened by the observation that rapidly desiccated lab-grown cells, in which structural integrity is not fully maintained, accumulate significantly smaller amounts of Wsp (Scherer and Potts 1989). A direct demonstration that the Wsp are responsible for cellular protection and desiccation tolerance in *Nostoc* has still to be achieved, but the prospect is exciting.

Desiccation-tolerance mechanisms in higher (vascular) plant species may also be based on protective mechanisms, but the evidence is largely indirect. Pretreatment of the desiccation-tolerant higher plants *Borya nitida* (Gaff and Churchill 1976) and *Coleochloa setifera* (Bartley, cited in Tymms et al. 1982) with a mild water stress leads to their ability to survive greater extremes of desiccation (faster, longer, and to lower water potentials), perhaps because of the initial induction of a protective component. In the desiccation-tolerant angiosperm *Xerophyta villosa* there is a transient increase in polysomes during the early stages of drying (Tymms et al. 1982). Concomitant with this increase is a change in the protein pattern, such that dried leaves appear to contain novel proteins. It is not known if these are involved with desiccation-tolerance because *X. villosa* undergoes many structural and morphological changes during drying, e.g., the leaves, even though they remain viable, turn yellow as a result of the destruction of chlorophyll and the disruption of the chloroplast (Gaff and McGregor 1979; Hallam and Luff 1980). Thus it is difficult to determine if the changes to proteins during drying are related to desiccation tolerance or to cellular degradation. Since chloroplasts regain their integrity upon rehydration, a repair component must be an important feature of desiccation tolerance in this plant (Eickmeier et al., this Vol.).

Novel proteins arise during dehydration of the desiccation-tolerant fern *Polypodium virginianum* (Fig. 4). These proteins are not synthesized in nonstressed (hydrated) fronds, nor within 24 h of recovery from the desiccated state, but only during slow drying. Whether these proteins are related to cellular protection, or are response proteins to dehydration stress, remains to be elucidated. Recently, a study involving a desiccation-tolerant member of the Scrophulariaceae, *Craterostigma plantagineum*, has provided more substantial evidence for a protection-based tolerance mechanism. Attached and detached leaves, and leaf-derived callus tissues of this plant can recover fully from desiccation (Gaff 1971; Bartels et al. 1990). Full dryness is achieved after exposure to an atmosphere of 60% RH for between 24 and 48 h, allowing time for the expression of a protein-based system. Interestingly, the callus cultures are only desiccation-tolerant if pretreated with 5 mg/l abscisic acid (ABA) for 4 days. Also both leaves and calli contain significantly higher concentrations of ABA (up to sevenfold) when dried, than when in the hydrated state (Bartels et al. 1990). Although long implicated in the response to water stress and in the desiccation of seeds, this is the first reported involvement of ABA in the drying response of any desiccation-tolerant vegetative tissue. Several new proteins are

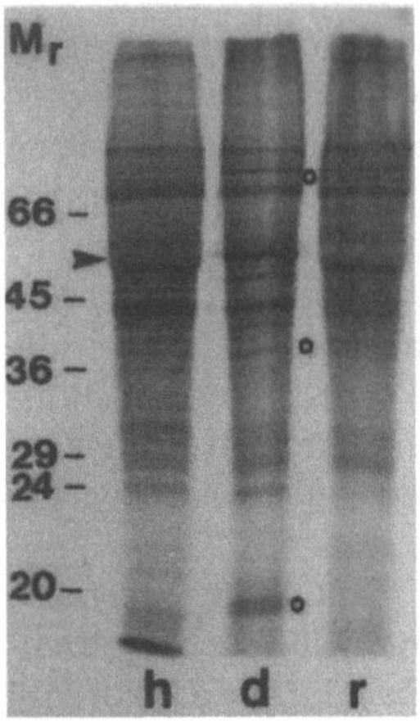


**Fig. 4.** The synthesis of proteins during slow desiccation of the tolerant fern *Polypodium virginianum*. Segments of fern fronds were labeled with $^{35}S$-methionine in the hydrated (*h*) state and then placed in a drying environment to determine protein synthesis during desiccation (*d*). Synthesis after 24 h of rehydration (*r*) was followed also. At least three proteins (*o*), identified after separation by SDS-PAGE and fluorography, are synthesized only during drying, and these may be involved in the desiccation-protection mechanism in this plant. *Arrowed* is the large subunit of Rubisco, a protein which remains stable during desiccation and rehydration, but which may exhibit increased synthesis in response to drying. $M_r$ markers in kDa are on the *left-hand side*. (Unpublished data of T.L. Reynolds and J.D. Bewley)

synthesized during drying of both leaves and calli, and many of these proteins appear when ABA is applied to nonstressed callus tissue (Bartels et al. 1990). These proteins are translated from newly synthesized mRNA, and cDNA clones for the messages of proteins synthesized during desiccation have been isolated. Messenger RNAs corresponding to these proteins are barely detectable in hydrated tissues, but when treated with ABA they increase substantially. Upon drying, the tissues synthesize mRNAs for some of the stress-related proteins within 30 min, while others are transcribed later. The early synthesized proteins may be the first steps in a cascade of responses that leads first to an increase in ABA, which then serves as a signal to induce the transcription of desiccation-tolerance genes. In the context of protection it is of note that this plant cannot survive rapid drying (Gaff 1977), and thus may require time for the full complement of protective syntheses to be completed.

## Drying in Seeds: General Considerations

Maturation drying is the normal terminal event in the development of many seeds, at which time they pass into a metabolically quiescent state. Seeds may remain in this dry state from several days to many years and retain their viability. Upon hydration under suitable conditions, the seed will commence germination, leading to the establishment of the vegetative plant. A notable feature of the life cycle of the seed is that during development its metabolism is largely anabolic, resulting in the massive deposition of polymeric reserves within the storage tissues. Following germination, these reserves are degraded and the resultant products converted to metabolites useful for early seedling development. There is now substantial evidence that the "switch" from a developmental to a germinative mode of metabolism is elicited by the intervening maturation drying event (reviewed in Kermode and Bewley 1986; Kermode et al. 1989b; Kermode 1990).

Seeds are not capable of withstanding desiccation at all stages during their development, but their acquisition of tolerance is usually substantially earlier than the drying event itself. In some species, for example, seeds may withstand desiccation even before development has reached the mid-point. It is likely that seeds undergo certain metabolic adjustments associated with the onset of tolerance, i.e., the synthesis of protective proteins, including those with hydrophilic properties, which have been implicated in maintaining the structural integrity of cellular structures during dehydration (Dure et al. 1989).

Resumption of metabolism during germination commences within minutes of the introduction of water to the dry seed. It is generally acknowledged that early metabolic events utilize and depend upon components present within the dry seed, which are replaced as normal turnover events proceed during germination. The germinating seed itself retains its tolerance of desiccation, usually until radicle emergence is completed.

In this section, changes in protein synthesis will be detailed in relation to desiccation, and the consequences of this event on the mode of metabolism.

## The Termination of Development by Desiccation and the Switch to Germination

Developing seeds acquire tolerance to desiccation prior to the time that normal maturation drying commences. Premature drying of developing seeds during their desiccation-tolerant stage leads to germination upon subsequent imbibition, although the rate at which premature drying is achieved may determine their subsequent germinability. For the castor bean seed, slow desiccation over several days is necessary for their survival during the earlier times after acquisition of desiccation tolerance, whereas closer to the time of normal maturation drying, more rapid drying can be withstood (Kermode and Bewley 1985a). It thus seems that not only is desiccation tolerance achieved at a particular stage of development, but tolerance of harsher drying regimes becomes greater as development proceeds. This is probably a consequence of morphological and physiological changes which take place with time of development, including the synthesis of specific (protective) proteins during the later stages (see next section). There are some exceptions which must be accounted for, however, since the seeds of some Gramineae appear to be able to withstand rapid desiccation relatively early during their development (see review by Kermode 1990). Perhaps these seeds undergo whatever changes are associated with the acquisition of desiccation tolerance at early times during their development. Presumably the nature of these changes is quite diverse, and they are not simply restricted to the synthesis of specific proteins.

The developing seed is usually incapable of germinating unless first dried, although the developing embryos, when isolated and placed on a liquid medium, will germinate. It is likely, therefore, that drying is overcoming an imposition on the embryo by the surrounding structures (the storage tissues, coat, and maternal environment), releasing it from these constraints. There is considerable debate at present as to the agents which prevent germination during development, whether they be osmoticum or abscisic acid (reviewed in Quatrano 1986; Kermode et al. 1989a; Skriver and Mundy 1990), but it is sufficient to surmise here that desiccation causes their decline and/or negates their effects, thus permitting germination. In some species, e.g., castor bean, the acquisition of desiccation tolerance and germinability of the embryo during development are achieved more or less simultaneously, whereas in others, e.g., barley, germinability is acquired earlier than desiccation tolerance (Bartels et al. 1988).

Studies over the past decade in several laboratories have clearly demonstrated that desiccation of developing seeds not only promotes germination upon subsequent rehydration, but also results in the cessation of developmental protein synthesis and the onset of syntheses associated with germination and post-germinative growth. This is well illustrated in the endosperm of castor bean, where premature drying leads to a change in the polypeptide profile, as demonstrated using SDS-PAGE, from one which is distinctly developmental to one which is identical to that found in germinating seeds and seedlings following normal matural drying (Fig. 5). A similar change in direction of protein synthesis as a consequence of premature drying during the desiccation-tolerant stages of development has been demonstrated in other seeds, e.g., *Phaseolus vulgaris* (Misra and Bewley 1985a). Drying of developing *P. vulgaris* seed during the desiccation-intolerant stages of development not surprisingly leads to reduced protein synthesis upon rehydration and the eventual demise of the seed. Interestingly, though, the pattern of protein synthesis exhibited upon rehydration is both developmental and germinative (Dasgupta and Bewley 1982), suggesting that the ability of the seed to completely switch the mode of its synthetic events in response to drying is acquired simultaneously with the acquisition of desiccation tolerance.

Premature drying of developing seeds also elicits the production of enzymes required for the mobilization of stored reserves following germination. In soybean and castor bean seeds, enzymes for the mobilization and conversion of lipid reserves

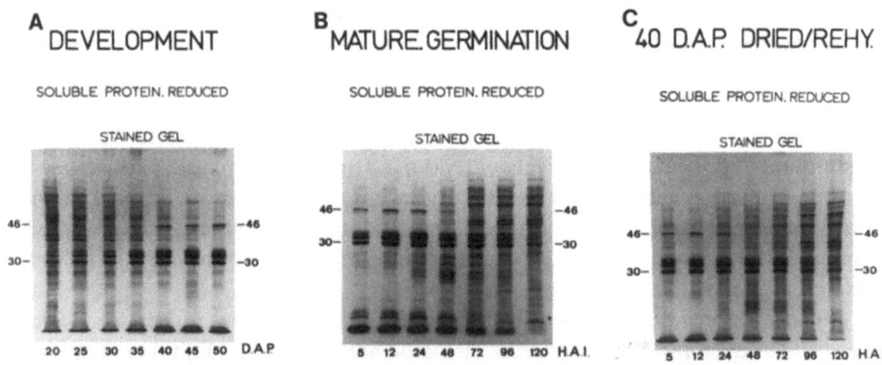


**Fig. 5A-C.** SDS-polyacrylamide stained gels of soluble proteins from castor-bean endosperms isolated during (**A**) development, (**B**) germination and growth, and (**C**) following premature desiccation and rehydration at 40 days of development.

Note that the pattern of proteins produced upon rehydration following premature drying at 40 DAP is identical to that produced in the mature dry seeds during germination and growth. Germination of the embryos is completed at 24 HAI. This demonstrates that the "switch" from a developmental to a germinative mode of metabolism is elicited by desiccation. *DAP* days after pollination; *HAI* hours after imbibition. (Kermode and Bewley 1985b) do not increase during seed development, but do so after premature drying (Adams et al. 1983; Kermode and Bewley 1985b) and to a similar extent as in the germinated mature seed.

The ability of the aleurone layer of cereals to produce α-amylase in response to GA is not normally acquired until the onset of maturation drying. However, premature drying of wheat, triticale, and maize (King et al. 1979; Armstrong et al. 1982; Oishi and Bewley 1990) all lead to synthesis of this enzyme. This has been attributed to an increase in sensitivity of the aleurone layer to GA; a mechanism requiring desiccation-induced membrane changes, and alterations to integral hormone receptors, has been invoked (Norman et al. 1982).

The changes in protein synthesis from a developmental to a germinative/growth mode are indicative of a switch in genome activity. This results in a permanent suppression of developmental protein synthesis and an induction of germination- and growth-related proteins. Drying ultimately affects transcription, i.e., mRNA synthesis, for after desiccation, be it premature or during the final stages of maturation, messages for developmental proteins are no longer synthesized, whereas those for germinative and growth proteins increase upon subsequent rehydration. Developmental messages may decline during drying itself; any residual messages present in the dry state are presumably degraded upon rehydration by the normal turnover processes, and are not replaced because their genes have been off-regulated. This is illustrated in Fig. 6 for the castor bean endosperm. Here it is evident that storage protein messages present during development decline due to drying and are not resynthesized upon subsequent rehydration. On the other hand, messages associated with post-germinative activities from 24 h onwards, and which are not abundant during development, are induced as a consequence of premature drying. Thus desiccation effects an off-regulation of developmental messages, and an on-regulation of those for germination and growth.

The ability of the seed to synthesize mRNAs for developmental proteins is retained during drying, although the rates at which their genes are transcribed declines with increasing water loss. It is upon rehydration that developmental genes are no longer transcribed, and only then is the germinative and post-germinative program of gene expression initiated (Comai and Harada 1990). However, simple on-off regulation of gene expression during development and germination may not be the norm. A residual pattern of developmental protein synthesis can continue for several to many hours during germination and growth (Kermode et al. 1985; Kermode et al. 1989b), and hence some developmental genes may retain a transient transcriptional competence even after drying. Some genes, once thought to be associated uniquely with embryogenesis, can be induced in the vegetative plant, often under stress conditions, e.g., wheat germ agglutinin (Cammue et al. 1989) and Em protein (Marcotte et al. 1989, citing unpublished work of S. Berge).

Such a switch in message population induced by premature drying has been reported in axes of *P. vulgaris*, including the off-regulation of the mRNA for the storage protein β-phaseolin (Misra and Bewley 1985a,b). In castor bean, also, premature drying results in the suppression of mRNA synthesis for the developmental protein, ricin D (Kermode et al. 1989b). In the aleurone layer cells of immature wheat grain, premature desiccation and rehydration induce mRNAs for α-amylase

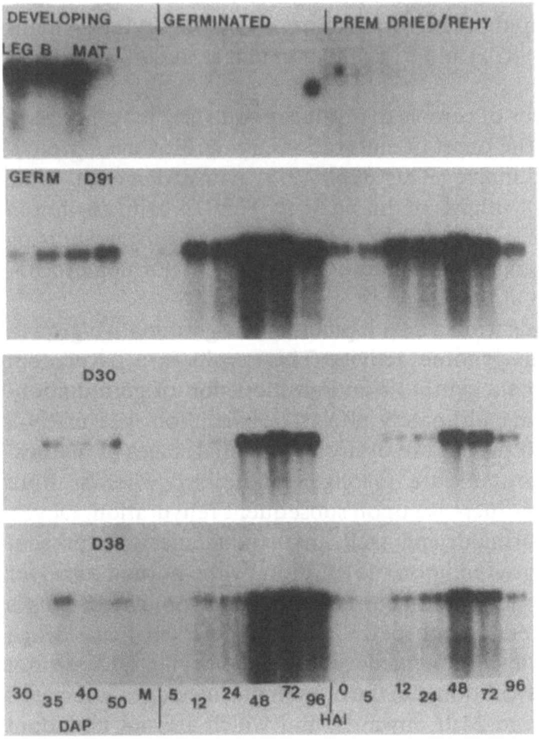


**Fig. 6.** Northern blot analysis of developmental and germination/growth-related messenger RNAs in the castor bean endosperm.

The LEG B MAT1 clone is for a developmental protein, the message for which is prevalent during the grand phase of storage protein synthesis (30–40 DAP). This mRNA is not present in the dry seed, nor upon imbibition. Premature drying of the seed at 40 DAP results in a loss of this developmental mRNA, which does not increase during germination or growth.

GERM D91, D30 and D38 clones are for mRNAs which arise post-germinatively, and predominantly from 48 h onwards. They are present in low abundance during development. After premature drying and rehydration at 40 DAP, these post-germinative mRNAs increase at the same time after the start of imbibition as in the mature seed. Thus desiccation results in the off-regulation of developmental genes and the up-regulation of germinative/post-germinative genes. (Unpublished data of D.W. Hughes, G.A. Galau, A.R. Kermode and J.D. Bewley)

(Cornford et al. 1988), although messages for some other post-germinative enzymes (e.g., acid phosphatase, ribonuclease, and proteinase) are not induced. This is suggestive of a differential sensitivity of the aleurone layer to desiccation, with sensitivity increasing with time of development, because following normal maturation drying all of these enzymes are synthesized post-germinatively.

In closing this section, it should be noted that the switch from a developmental to a germinative mode, both morphologically and synthetically can be brought about in some seeds in the absence of desiccation. Treatments such as partial drying, embryo isolation, and washing may also elicit the switch (Cornford et al. 1988;

Garcia-Maya et al. 1990), although the quantitative and qualitative response is sometimes different from that induced by drying. While maturation drying is the event normally expected to overcome the constraints on germination imposed on the developing seed, other treatments may reduce these constraints and thus trigger the switch.

## Protein Synthesis Associated with Maturation Drying

During seed development the mRNA population of the embryo and associated storage tissues changes, with messages for storage proteins being particularly prevalent from the completion of histodifferentiation until the onset of maturation drying. In cotton seeds there are at least six distinct sets of coordinately regulated mRNAs (Dure 1985) which are present in different amounts at different stages of development. One subset of messages increases in abundance as the seeds mature, may be conserved in the dry seed, but is virtually absent from the germinating seed. The possibility exists that the protein products, called late embryogenic abundant proteins (LEAs), are important to the final stages of maturation, and they have been implicated in the ability of the mature seed to withstand desiccation (Baker et al. 1988), i.e., in cellular protection.

LEA protein synthesis constitutes a large proportion of the translational activity of the embryo during late maturation (up to 25%), regulated at the level of transcription, i.e., by the abundance of *Lea* mRNAs (Hughes and Galau 1987). In the mature cotton embryos they comprise about 30% of the nonstorage protein moiety. The increase in *Lea* transcripts during cotton seed development indicates that they arise in two distinct classes. One class which is made up of six different *Lea* transcripts is first detectable soon after histodifferentiation is completed, and exhibits two transient peaks before reaching a maximum level 3 days prior to desiccation. The transcript level of the second class of 12 *Lea* mRNAs increases sharply during late maturation and peaks just prior to or at desiccation (Fig. 7). Some of the *Lea* genes may be induced by ABA, for early increases in some *Lea* transcripts can be correlated with high levels of this growth regulator (Galau et al. 1987). The activity of one particular *Lea* gene, D19 or Em protein gene, can be induced by ABA even in germinated seeds (Williamson et al. 1985). The transcription of *Lea* genes occurs even in ABA-treated and water-stressed seedlings (Mundy and Chua 1988). This has led to the suggestion that not all *Lea* genes are intrinsically developmentally regulated in their expression, but could be "on-call" genes whose expression is in response to water loss (perhaps with ABA acting as one of the signal transducers) (Baker et al. 1988; Skriver and Mundy 1990). Thus, they may be part of a general class of desiccation-protection proteins. While there is a response of some *Lea* genes to ABA during embryogenesis, it is too premature to invoke this regulators as the important or only controlling agent.

Messages homologous to *Lea* cDNA have been identified in seeds from rapeseed, barley, maize, wheat, carrot, radish, rice, and castor bean, and will no doubt be reported in many other seeds as research is continued. Certainly, in other species, changes in protein synthesis, or mRNA patterns during seed maturation have been

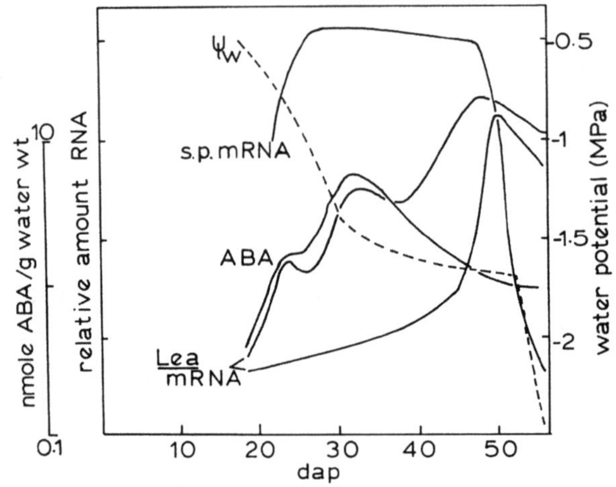


**Fig. 7.** Changes in the relative abundances of the two classes of *Lea* mRNAs during cotton seed embryogenesis in relation to changes in seed water potential ($\Psi_W$), storage protein (*s.p.*) mRNA and abscisic acid (ABA). Two distinct classes of *Lea* mRNAs increase during development. One class increases about the mid-point of development (30–35 DAP) and is coincidental with an increase in the growth regulator ABA, and the other class increases at the time of maturation drying (45–50 DAP. (After Galau et al. 1987 and Hughes and Galau 1989)

recorded, although some are unlikely to be related to LEA proteins, e.g., the 128 kDa protein which arises during late embryogenesis in soybean (Rosenburg and Rinne 1988). All of the seeds in which *Lea* transcripts have been identified are of the so-called orthodox kind which withstand drying; it will be of interest to determine if "recalcitrant" seeds, which are desiccation-intolerant, have the ability to produce LEA proteins.

*Lea* transcripts decline rapidly during germination, although the amount of LEA proteins present is not known. Perhaps there is some correlation between the loss of desiccation tolerance during germination and the inability to produce a sufficient quantity of proteins. This remains entirely speculative, given that vegetative tissues do transcribe *Lea* genes in response to stress.

The relationship between proteins synthesized during desiccation of tolerant vegetative plants and those during maturation drying of seeds remains to be tested. Homologies might be expected between at least some of the LEA proteins and those induced during drying of desiccation-tolerant tissues, especially since the structure and composition of these proteins permits them to bind to macromolecular structures in a protective manner (Dure et al. 1989). Proteins that are responsive to ABA (RAB proteins), in that they are synthesized in the presence of this regulator (Mundy and Chua 1988), might also be common to desiccation-tolerant tissues. It remains to be determined, however, if desiccation-tolerant lower plants either contain or respond to ABA.

## Desiccation Tolerance During Germination and its Loss at Radicle Emergence

Components of the protein synthesizing complex are present within the dry seed, and within minutes of imbibition polysomes are reformed as protein synthesis recommences (Bewley and Black 1985). Very few studies have been conducted on the fate of ribosomes, tRNA, and the cytoplasmic components of protein synthesis during maturation drying, but it is generally assumed that these do not decrease substantially. An obvious exception is the castor bean seed in which the total RNA component declines to a very low level with water loss (due to premature or maturation drying), then increases again after germination is completed (Roberts and Lord 1979; Kermode et al. 1989b). The reason for the loss of RNA is not known, but ribonucleases appear to be involved (Winchcombe and J.D. Bewley, unpublished data).

While the mature seed is extremely tolerant of desiccation, this property is lost during the very last stages of germination, and the growing seedling is desiccation-intolerant. Thus, while tolerance during development may be manifest for several weeks, this is lost within hours of the start of imbibition of the dry seed. The transition of the growing seedling to desiccation intolerance occurs at the time of cell division and vacuolation (Akalehiywot and Bewley 1980; McKersie and Tomes 1980; Sargent et al. 1981), and is thus associated with major morphological and physiological changes. At this time, then, protective and/or repair capabilities are lost.

During the normal course of germination, profound quantitative and qualitative changes occur to the pattern of protein synthesis. Few studies have been made on the effects of drying during germination, but it is generally accepted that the severe water loss suspends the germinative process, which resumes upon subsequent rehydration. The total time required for germination by seeds which have been desiccated and rehydrated is approximately the same as for seeds which remain in the hydrated state (Akalehiywot and Bewley 1980; Lalonde and Bewley 1985). Thus, if seeds which require 24 h to complete germination are desiccated after 12 h, they complete germination within 12–16 h upon rehydration, rather than require an additional 24 h. In other words, desiccation during germination does not cause the seeds to revert back to the mature dry state, but to resume germinative activities at a stage close to the time of drying. This is reflected in the patterns of protein synthesis (Fig. 8). In germinating pea axes, for example, upon rehydration after an interrupting desiccation treatment, the types of proteins being synthesized are recognizable as those synthesized earlier during germination: seeds imbibed for 8 h, and then dried, do not recommence synthesis of proteins typical of 8-h-imbibed axes, but rather of 4-h-imbibed axes. Thus drying of the germinating pea axes causes the synthesizing mechanism to revert to producing proteins typical of the earlier stages of imbibition (Lalonde and Bewley 1986a). Drying during germination never causes the seed to revert to the metabolic status of the initial mature dry seed, however. This change in the pattern of protein synthesis is reflected in the mRNA complement also, which is indicative that the effect of drying is on the genome. Some new proteins and messages are produced following desiccation during germination (Lalonde and Bewley 1986b; Misra and Bewley 1986). These desiccation-induced changes, which lead to the synthesis of rehydration proteins, occur at the level of transcription and

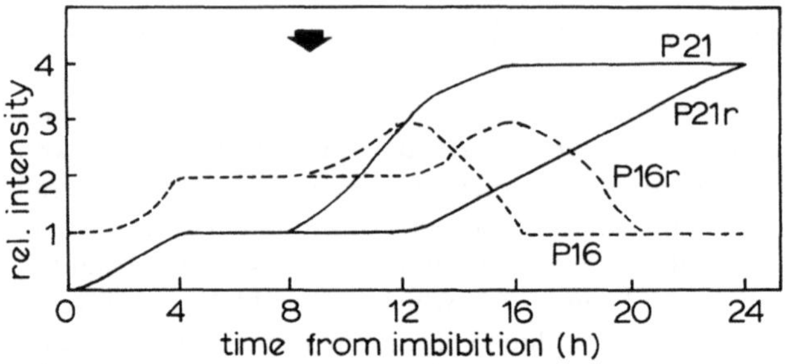


**Fig. 8.** The time course of synthesis of two specific proteins (P16 and P21) during germination of axes of pea seeds, and the effect on their synthesis of an intervening period of desiccation at 8 h from the start of imbibition. Upon rehydration following desiccation at 8 h (*arrowed*) the synthesis of P16 and P21 proteins resumes (*P16r* and *P21r*) with a delay of 4 h (*P16*) or 8 h (*P21*) before maximum synthesis is achieved. Thus the interrupting desiccation results in a delay in the synthesis of specific germination- and growth-related proteins. Desiccation does not cause the metabolism of the germinating seed to revert to that of the once-imbibed dry seed.

The relative intensity of the proteins is taken from fluorographs of $^{35}$S-methionine-labeled proteins separated by 2-D SDS-PAGE. (From original data of Lalonde and Bewley 1986b)

may result from a requirement to increase the level of certain enzymes or structural proteins to effect the repair of processes perturbed by drying.

## Conclusions

It is reasonable to assume that the morphological structure of a particular plant plays an important role in dictating which of the two processes, cellular repair or cellular protection, dominates as the desiccation-tolerance mechanism. Poikilohydric lower plants tend to lack morphological features that can act as barriers to water loss and so they desiccate when water is unavailable. Desiccation of plants such as the bryophytes can be so rapid that there is no time available for the production of proteins that could play a protective role. Thus, in these plants, desiccation-tolerance mechanisms are likely to rely more upon the activation of repair processes during rehydration. The possibility that protective proteins are synthesized constitutively, in anticipation of desiccation, cannot be discarded, however. Desiccation-tolerant higher (vascular) plants have features that can slow the rate of water loss from their vegetative tissues; in fact, none appear to be able to survive rapid drying. This reduced rate of water loss may allow the cells of these plant to prepare for the dried state. Thus protection processes are more likely to serve as the basis for the desiccation-tolerance mechanism. Both lower and higher plants may utilize a combination of repair and protection processes in achieving desiccation tolerance and is a possibility that remains to be investigated. It is difficult to assess the relative roles of protection against desiccation and repair of desiccation-induced damage in developing and germinating seeds. Certainly, upon rehydration following either maturation- or premature drying there is an extensive leakage of solutes, which ceases within a short time, suggestive of repair to membranes, but whether this involves protein synthesis is undetermined. During development, seeds pass from an intolerant stage to a tolerant one; again, whether this is due to the development of protective or repair mechanisms is debatable for morphological changes to the cells may play an important role (e.g., filling of storage vacuoles with storage materials). Developing seeds can withstand slow desiccation at stages when they are intolerant of rapid drying. This is suggestive that a protection-based mechanism is important, which requires time to be completed before the seed becomes dry. The loss of desiccation tolerance at the end of germination seems to coincide with cell division and vacuolation. The synthesis of LEA proteins might be an important prelude to desiccation, and provide the basis of an important protective mechanism. Their presence in water-stressed and desiccation-tolerant vegetative tissues opens up some interesting avenues for comparative research. Superimposed upon any effect of drying on repair or protective processes is the switch in metabolism from development to germination. This major change in direction of metabolism may mask some of the more subtle changes associated with the tolerance mechanisms.

It is tempting to speculate that, since poikilohydric plants belong to the more ancient of plant genera, repair-based tolerance mechanisms were the first to evolve and that this may have allowed for the initial colonization of dry land. During the evolution of the more advanced plant forms, the ability to survive desiccation was lost as more advanced means of conserving water evolved. For higher plants to colonize the arid regions, a re-evolution of desiccation-tolerance mechanisms must have been necessary; but this time these mechanisms were based more strongly on protection rather than repair. Perhaps these desiccation-tolerance protection systems evolved in the vegetative plant as an extension of the mechanisms already in place to protect seed tissues during embryonic maturation drying, including the synthesis of late maturation proteins.

*Acknowledgments.* This work is supported by a Natural Sciences and Engineering Research Council of Canada grant A2210 to J.D.B. and a National Science Foundation, U.S.A. grant DCB-8819019 to M.J.O. Our thanks to Tracey Reynolds for her critical reading of the manuscript.

## References

Adams CA, Fjerstad MC, Rinne RW (1983) Characteristic of soybean seed maturation: necessity for slow dehydration. Crop Sci 23:265–267

Akalehiywot T, Bewley JD (1980) Desiccation of oat grains during and following germination, and its effect on protein synthesis. Can J Bot 58:2349–2355

Armstrong C, Black M, Chapman JM, Norman HA, Angold R (1982) The induction of sensitivity to gibberellin in aleurone tissue of developing wheat grains. I The effect of dehydration. Planta 159:573–577

Baker J, Steele C, Dure L III (1988) Sequence and characterization of 6 *Lea* proteins and their genes from cotton. Plant Mol Biol 11:277–291

Bartels D, Singh M, Salamini F (1988) Onset of desiccation tolerance during development of the barley embryo. Planta 175:485–492

Bartels D, Schneider K, Terstappen G, Piatkowski D, Salamini F (1990) Molecular cloning of abscisic acid-modulated genes which are induced during desiccation of the resurrection plant *Craterostigma plantagineum*. Planta 181:27–34

Bewley JD (1972) The conservation of polyribosomes in the moss *Tortula ruralis* during total desiccation. J Exp Bot 23:692–698

Bewley JD (1973a) Polyribosomes conserved during desiccation of the moss *Tortula ruralis* are active. Plant Physiol 51:285–288

Bewley JD (1973b) Desiccation and protein synthesis in the moss *Tortula ruralis*. Can J Bot 51:203–206

Bewley JD (1979) Physiological aspects of desiccation tolerance. Annu Rev Plant Physiol 30:195–238

Bewley JD, Black M (1985) Seeds. Physiology of development and germination. Plenum, New York Lond

Bewley JD, Krochko JE (1982) Desiccation-tolerance. In: Lange O, Nobel PS, Osmond CB, Ziegler H (eds) Encyclopedia of plant physiology. Vol 12B, Physiological ecology II. Springer, Berlin Heidelberg New York, pp 325–378

Bewley JD, Oliver MJ (1983) Responses to a changing environment at the molecular level: does desiccation modulate protein synthesis at the transcriptional or translational level in a tolerant plant? In: Randall DD (ed) Current topics in plant biochemistry/physiology. Univ Missouri Press, pp 145–164

Bewley JD, Tucker EB, Gwozdz EA (1974) The effects of stress on the metabolism of *Tortula ruralis*. In: Bieleski RL, Ferguson AR, Cresswell MM (eds) Mechanisms of regulation of plant growth. R Soc NZ Bull Wellington 12:395–402

Brown DH, Buck GW (1979) Desiccation effects and cation distribution in bryophytes. New Phytol 82:115–125

Cammue BPA, Broekaert WF, Kellens JTC, Raikhel NV, Peumans WJ (1989) Stress-induced accumulation of wheat germ agglutinin and abscisic acid in roots of wheat seedlings. Plant Physiol 91:1432–1435

Comai L, Harada JJ (1990) Transcriptional activities in dry nuclei indicate the timing of the transition from embryogeny to germination. Proc Natl Acad Sci USA 87:2671–2674

Cornford CA, Black M, Chapman JM, Baulcombe DC (1988) Expression of α-amylase and other gibberellin-regulated genes in aleurone tissue of developing wheat grains. Planta 169:420–428

Dasgupta J, Bewley JD (1982) Desiccation of axes of *Phaseolus vulgaris* during development causes a switch from a developmental pattern of protein synthesis to a germination pattern. Plant Physiol 70:1224–1227

Davis JH (1972) Survival records in the algae, and the survival role of certain algal pigments, fat and mucilaginous substances. Biologist 54:52–93

Dhindsa RS, Bewley JD (1976) Plant desiccation: polyribosome loss not due to ribonuclease. Science 191:181–182

Dhindsa RS, Bewley JD (1977) Water stress and protein synthesis: V. Protein synthesis, protein stability, and membrane permeability in a drought-sensitive and a drought-tolerant moss. Plant Physiol 59:295–300

Dure L III (1985) Embryogenesis and gene expression during seed formation. In: Miflin BJ (ed) Oxford surveys in plant molecular and cell biology, vol 2. Oxford Univ Press, Oxford, p 179

Dure L III, Crouch M, Harada J, Ho THD, Mundy J, Quatrano R, Thomas T, Sung ZR (1989) Common amino acid sequence domains among the LEA proteins of higher plants. Plant Mol Biol 12:475–486

Gaff DF (1971) Desiccation-tolerant flowering plants in Southern Africa. Science 174:1033–1034

Gaff DF (1977) Desiccation-tolerant vascular plants of Southern Africa. Oecologia 31:95–109

Gaff DF (1980) Protoplasmic tolerance of extreme water stress. In: Turner NC, Kramer PJ (eds) Adaptation of plants to water and high temperature stress. Wiley, New York, pp 207–229

Gaff DF, Churchill DM (1976) *Borya nitida* Labill. – an Australian species in the Liliaceae with desiccation-tolerant leaves. Aust J Bot 24:209–224

Gaff DF, McGregor GR (1979) The effect of dehydration and rehydration on the nitrogen content of various fractions from resurrection plants. Biol Plant 21:92–99

Galau GA, Bijaisoradat N, Hughes DW (1987) Accumulation kinetics of cotton late embryogenesis-abundant mRNAs and storage protein mRNAs: Coordinate regulation during embryogenesis and the role of abscisic acid. Dev Biol 123:198–212

Garcia-Maya M, Chapman JM, Black M (1990) Regulation of α-amylase formation and gene expression in the developing wheat embryo. Role of abscisic acid, the osmotic environment and gibberellin. Planta 181:296–303

Gupta RK (1976) The physiology of desiccation resistance in bryophytes: nature of organic compounds leaked from desiccated liverwort, *Plagiochila asplenoides*. Biochem Physiol Pflanz 170:389–395

Gwozdz EA, Bewley JD (1975) Plant desiccation and protein synthesis: An in vitro system from dry and hydrated mosses using endogenous and synthetic messenger RNA. Plant Physiol 55:340–345

Gwozdz EA, Bewley JD, Tucker EB (1974) Studies on protein synthesis in *Tortula ruralis*: polyribosome reformation following desiccation. J Exp Bot 25:599–608

Hallam ND, Luff SE (1980) Fine structural changes in the mesophyll tissue of the leaves of *Xerophyta villosa* during desiccation. Bot Gaz 141:173–179

Henckel RA, Satarova NA, Shaposnikova SV (1977) Protein synthesis in poikiloxerophyte and wheat embryos during the initial period of swelling. Sov Plant Physiol 14:754–762

Hughes DW, Galau GA (1987) Translation efficiency of Lea mRNAs in cotton embryos: minor changes during embryogenesis and germination. Plant Mol Biol 9:301–313

Hughes DW, Galau GA (1989) Temporally modular gene expression during cotyledon development. Gen Dev 3:358–369

Kermode AR (1990) Regulatory mechanisms involved in the transition from seed development to germination. CRC Crit Rev Plant Sci 9:155–195

Kermode AR, Bewley JD (1985a) The role of maturation drying in the transition of seed development to germination. I Acquisition of desiccation tolerance and germinability during development of *Ricinus communis* seeds. J Exp Bot 36:1906–1915

Kermode AR, Bewley JD (1985b) The role of maturation drying in the transition from seed development to germination. II Post-germinative enzyme production and soluble protein synthetic pattern changes within the endosperm of *Ricinus communis* L. seeds. J Exp Bot 36:1916–1927

Kermode AR, Bewley JD (1986) Alteration of genetically regulated protein syntheses in seeds by desiccation. In: Leopold AC (ed) Membranes, metabolism and dry organisms. Cornell Univ Press, Ithaca, p 59

Kermode AR, Gifford DJ, Bewley JD (1985) The role of maturation drying in the transition from seed development to germination. III Insoluble protein synthetic pattern changes within the endosperm of *Ricinus communis* seeds. J Exp Bot 26:1928–1936

Kermode AR, Oishi MY, Bewley JD (1989a) Regulatory roles for desiccation and abscisic acid in seed development: a comparison of the evidence from whole seeds and isolated embryos. Crop Sci Soc Am Spec Publ 14:23–50

Kermode AR, Pramanik SJ, Bewley JD (1989b) The role of maturation drying in the transition from seed development to germination. VI Desiccation-induced changes in messenger RNA populations within the endosperm of *Ricinus communis* L. seeds. J Exp Bot 40:33–41

King RW, Salminen SO, Hill RD, Higgins TJV (1979) Abscisic-acid and gibberellin action in developing kernels of *Triticale* (cv 6A 190). Planta 146:249–255

Lalonde L, Bewley JD (1985) Desiccation of imbibed and germinating pea axes causes a partial reversal of germination events. Can J Bot 63:2248–2253

Lalonde L, Bewley JD (1986a) Pattern of protein synthesis during the germination of pea axes, and the effects of an interrupting desiccation period. Planta 167:504–510

Lalonde L, Bewley JD (1986b) Desiccation during germination of pea axes and its effects on the messenger RNA population. J Exp Bot 37:754–764

Marcotte WR, Russell SH, Quatrano RS (1989) Abscisic acid-responsive sequences from the Em gene of wheat. Plant Cell 1:969–976

McKersie BD, Tomes DT (1980) Effects of dehydration treatments on germination, seedling vigour and cytoplasmic leakage in wild oats and birdsfoot trefoil. Can J Bot 58:471–476

Misra S, Bewley JD (1985a) Reprogramming of protein synthesis from a developmental to a germinative mode induced by desiccation of the axes of *Phaseolus vulgaris*. Plant Physiol 78:876–882

Misra S, Bewley JD (1985b) The messenger RNA population in the embryonic axes of *Phaseolus vulgaris* during development and following germination. J Exp Bot 36:1644–1652

Misra S, Bewley JD (1986) Desiccation of *Phaseolus vulgaris* seeds during and following germination, and its effect upon the translatable mRNA population of the seed axes. J Exp Bot 37:364–374

Mundy J, Chua NH (1988) Abscisic acid and water stress induce the expression of a novel rice gene. EMBO J 7:2279–2286

Norman HA, Black M, Chapman JM (1982) Induction of sensitivity to gibberellic acid in aleurone tissue of developing wheat grains. II Evidence for temperature-dependent membrane transitions. Planta 154:578–586

Oishi MY, Bewley JD (1990) Distinction between the responses of developing maize kernels to fluridone and desiccation in relation to germinability, α-amylase activity and abscisic acid content. Plant Physiol 94:592–598

Oliver MJ (1983) The role of desiccation in the control of transcription and translation in the moss *Tortula ruralis*. PhD Thesis, Univ Calgary, Calgary Alberta

Oliver MJ, Bewley JD (1984a) Desiccation and ultrastructure in bryophytes. Adv Bryol 2:91–132

Oliver MJ, Bewley JD (1984b) Plant desiccation and protein synthesis: IV. RNA synthesis, stability, and recruitment of RNA into protein synthesis during desiccation and rehydration of the desiccation-tolerant moss *Tortula ruralis*. Plant Physiol 74:21–25

Oliver MJ, Bewley JD (1984c) Plant desiccation and protein synthesis: V. Stability of poly(A)$^-$ and poly(A)$^+$RNA during desiccation and their synthesis upon rehydration of the desiccation-tolerant moss *Tortula ruralis* and the intolerant moss *Cratoneuron filicinum*. Plant Physiol 74:917–922

Oliver MJ, Bewley JD (1984d) Plant desiccation and protein synthesis: VI. Changes in protein synthesis elicited by desiccation of the moss *Tortula ruralis* are effected at the translational level. Plant Physiol 74:923–927

Peat P, Potts M (1987) The ultrastructure of immobilized desiccated cells of the cyanobacterium *Nostoc commune* UTEX 584. FEMS Microbiol Lett 43:233–227

Peat P, Powell N, Potts M (1988) Ultrastructural analysis of the rehydration of desiccated *Nostoc commune* Hun. (cyanobacteria) with particular reference to the immunolabeling of NifH. Protoplasma 146:72–80

Quatrano RS (1986) Regulation of gene expression by abscisic acid during angiosperm seed development. In: Miflin BJ (ed) Oxford surveys in plant molecular and cell biology, vol 3. Oxford Univ Press, Oxford, p 467

Roberts LM, Lord JM (1979) Ribonucleic acid synthesis in germinating castor bean endosperm. J Exp Bot 30:739–749

Rosenberg LA, Rinne RW (1988) Protein synthesis during natural and precocious soybean seed (*Glycine max* [L.] Merr.) maturation. Plant Physiol 87:474–478

Sachs MM, Ho THD (1986) Alteration of gene expression during environmental stress in plants. Annu Rev Plant Physiol 37:363–376

Sargent JA, Sen Mandi S, Osborne DJ (1981) The loss of desiccation tolerance during germination: an ultrastructural and biochemical approach. Protoplasma 105:225–239

Scherer S, Potts M (1989) Novel water stress protein from a desiccation-tolerant cyanobacterium: purification and partial characterization. J Biol Chem 264:12546–12553

Scherer S, Ernst A, Chen TW, Boger P (1984) Rewetting of drought-resistant blue-green algae: time course of water uptake and reappearance of respiration, photosynthesis and nitrogen fixation. Oecologia 62:418–423

Siebert G, Loris J, Zollner B, Frenzel B, Zahn RK (1976) The conservation of poly(A) containing RNA during the dormant state of the moss *Polytrichum commune*. Nucl Acids Res 3:1997–2003

Skriver K, Mundy J (1990) Gene expression in response to abscisic acid and osmotic stress. Plant Cell 2:503–512

Tucker EA, Bewley JD (1976) Plant desiccation and protein synthesis. III. Stability of cytoplasmic RNA during dehydration and its synthesis on rehydration of the moss *Tortula ruralis*. Plant Physiol 57:564–567

Tymms MJ, Gaff DF, Hallam ND (1982) Protein sythesis in the desiccation-tolerant angiosperm *Xerophyta villosa* during dehydration. J Exp Bot 33:323–343

Williamson JD, Quatrano RS, Cuming AC (1985) $E_m$-polypeptide and its messenger RNA levels are modulated by ABA during embryogenesis in wheat. Eur J Biochem 152:501–507

Chapter 11

# Water in Dry Organisms

A.C. LEOPOLD[1], F. BRUNI[1], and R.J. WILLIAMS[2]

## Introduction

In the evolution of life systems, water has been a component of transcendent importance. In addition to its involvement in syntheses and metabolism, it has provided the cradle supporting the structures of macromolecules, and it has been the main source of the forces which generate and maintain membrane structures. The ability to survive the removal of water in anhydrous systems is especially remarkable as a departure from these structural dependencies. Because desiccation tolerance is widespread through the plant and animal kingdoms, we may hope that it might involve relatively simple biochemical modifications (Green and Angell 1989). In this chapter we will suggest that the basis for the survival of desiccation involves the accumulation of solutes which can lead to a vitrification of the cytoplasm as desiccation proceeds.

In the early stages of the evolution of life on earth, it is probable that primitive life forms were repeatedly exposed to wetting and drying events, leading to the repeated concentration of various solutes involved in metabolism and reproduction. Hinton (1968) has suggested that in view of this probability, the development of desiccation tolerance might well have occurred repeatedly in various primitive life forms. This suggestion again reinforces the concept of a relatively simple biochemical adaptation for tolerance.

The occurrence of desiccation tolerance is widespread among life forms today, including bacterial spores, fungal spores, nematodes, tardigrades, and other soil organisms, saline cysts such as *Artemia*, and of course angiospermous seeds and pollen, and occasional "resurrection plants." Can we find candidates for a simple mechanism which may account for the tolerance of these diverse organisms to the withdrawal of water?

[1]Boyce Thompson Institute, Cornell University, Ithaca NY 14853, USA
[2]American Red Cross Biomedical Research and Development Laboratory, Rockville MD 20855, USA

Somero et al. (Eds.)
Water and Life

## The Status of Water

With the progressive removal of water from tolerant organisms, the forces resisting water loss from the tissues become increasingly greater. The energies of water binding can be calculated from adsorption isotherms. It is usually possible to discriminate three regions of water binding in desiccation tolerant organisms, with binding energies increasing with successively dryer regions (Fig. 1). These regions of increased enthalpy of water binding are not observed in intolerant organisms (Vertucci and Leopold 1986). Labuza (1984) has pointed out that the three regions of water binding are characteristic of solute systems which do not crystallize. From such a suggestion, we might anticipate that the tolerance to desiccation is associated with some feature or solute combination which avoids crystallization.

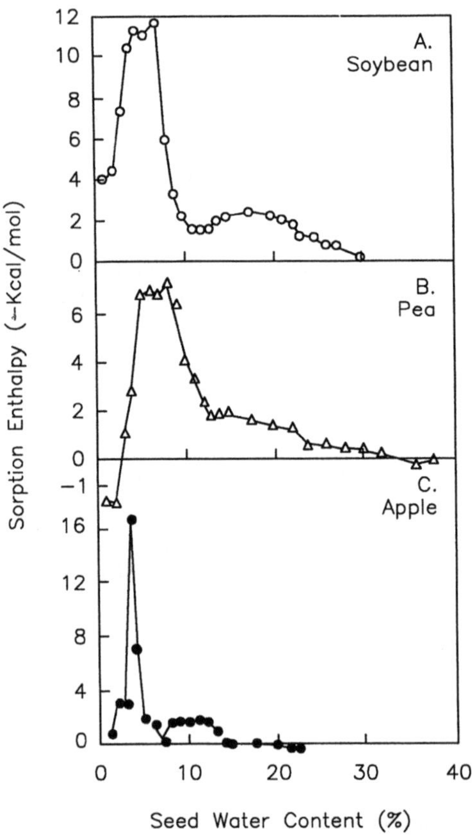


**Fig. 1A-C.** When seeds are equilibrated to different moisture levels, one can detect different binding energies at different water contents. In seeds of soybean (**A**), pea (**B**), and apple (**C**), the binding energies are evidenced in the enthalpy of water binding, as determined from moisture isotherms. (Vertucci and Leopold 1986)

## The Water Displacement Theory

In experiments with coliform bacteria, Webb (1965) made the surprising observation that desiccation treatments which were usually lethal could be withstood if the cells were suspended in an inositol solution before drying. Furthermore, the inositol could provide protection against UV and X-ray damage. This led him to propose that the inositol could provide protection by binding to the nucleic acids in the cell. He proposed, then, a "water displacement theory" by which a polyol could replace the water bound to the nucleic acid. We can interpret this idea as a stabilizing effect of the polyol on molecular structures as water was withdrawn during drying. His theory was explicitly directed to the protection of nucleic acids.

A new direction for the water displacement theory came from the work of Crowe and his coworkers (see Crowe and Crowe, this Vol.). Turning to the effects of desiccation on membrane structure, Crowe's group was able to show that numerous sugars were impressively effective in preserving the integrity of membranes during desiccation and rehydration (Crowe et al. 1986). This led them to suggest that the various sugars provided protection by replacing the water-binding sites on the membranes. Explicit experiments on this replacement concept confirmed the proposed effect (Crowe et al. 1984). It was curious, however, that inositol had little or no protective effect in Crowe's membrane experiments.

Is the protective effect by sugars distinctive to the membranes of desiccation-tolerant tissues? Crowe's group carefully examined that possibility, and found that the sugar protection effect was functional even for membranes which had come from animal species which never achieved desiccation tolerance. Furthermore, they could construct completely synthetic membranes with only phospholipids and still show the sugar protection effect upon drying (Crowe and Crowe 1986).

These experiments on sugar protection shifted the focus of the tolerance mechanism to cellular membranes, and implicated a general effect of common sugars. It is interesting, though, that there were quantitative differences in the effectiveness of different sugars.

## Sugars in Desiccation Tolerance

If sugars are an important component of desiccation tolerance, we might expect substantial accumulations of sugars to be associated with the tolerant state. A large accumulation of trehalose had been noted in the course of maturation of fungal spores by Lingappa and Sussman (1959). This finding prompted them to suggest that this sugar served as a ready substrate for germination after the passage of dormancy. Again, accumulations of trehalose have been noted in the maturing cysts of *Artemia* (Clegg 1965), as also in maturing yeast cells (Keller et al. 1982), and in nematodes and other soil organisms (Womersley 1981).

The occurrence of trehalose in dry organisms may have another meaning. It is singularly absent from angiospermous seeds and pollen. Almost all desiccation-tolerant seeds contain large amounts of sucrose, and these are usually associated with more modest amounts of oligosaccharides (Amuti and Pollard 1977). Trehalose has not been reported in any seeds. The total amounts of soluble sugars in seeds is impressively high. They constitute between 14 and 20% of the total dry weight. In the case of pollen, sucrose is again the dominant sugar, but apparently without associated oligosaccharides (Hoekstra and van Rockel 1988).

In a more general view, then, there appears to be a relationship between desiccation tolerance and the occurrence of sugars – trehalose in the cases of fungi and numerous animal species, and sucrose (usually associated with oligosaccharides) in the case of seeds.

## Desiccation Tolerance and Cytoplasmic Vitrification

Our argument so far has been that we have reason to look for a simple protective mechanism for desiccation tolerance, that it should prevent the development of crystallization in cytoplasmic components, and that it is generally correlated with the occurrence of large amounts of sugars.

At a conference on dry organisms in 1985, Michael Burke pointed out that vitrification appeared to be a promising candidate for such a protective mechanism (Burke 1986). The vitrified state would be readily formed by sugars; it would serve to limit the loss of water; and the viscous state should suppress chemical reactions that require molecular diffusion.

Using differential scanning calorimetry (DSC), we were able to detect vitrification in defatted corn embryos, as shown in Fig. 2 (Williams and Leopold 1989).

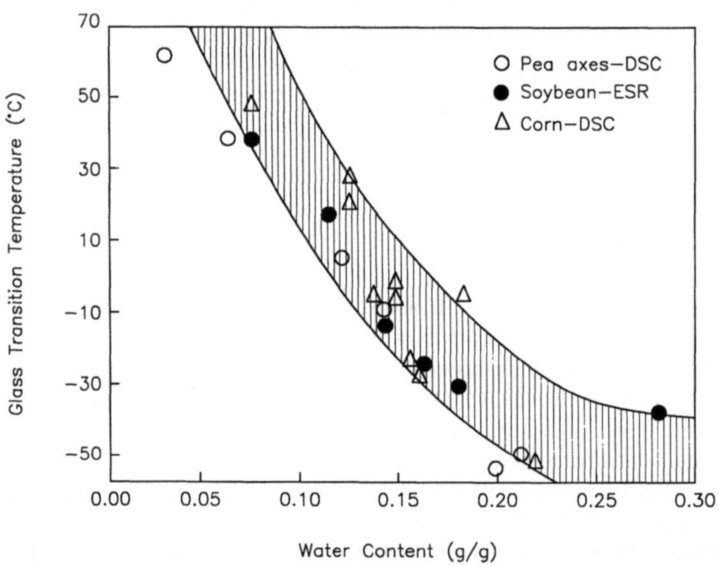


**Fig. 2.** Phase diagram of the temperature at which the glassy state is melted, as a function of the equilibrated water content in three species of seeds. Data are presented for corn embryos and pea axes determined by DSC (Williams and Leopold 1989), and for soybean axes determined by ESR (Bruni and Leopold 1991)

Working out the phase diagram showed us that at the water contents normally occurring in stored corn seeds, the aqueous components of the seed were in fact in the vitrified state, even at room temperatures.

We have subsequently combined three types of measurements to establish vitrification in three species of seeds. In addition to DSC, we have employed electron spin resonance (ESR) and thermally stimulated depolarization currents (TSDC) as techniques, and we are now able to draw phase diagrams of the vitreous condition in three different species using these three techniques (Figs. 2 and 3) (Williams and Leopold 1989; Bruni 1990; Bruni et al. 1990; Bruni and Leopold 1991).

Spin probe ESR measurements of cytoplasmic viscosity are made possible by the ability of a hydrophilic probe to diffuse into the aqueous cytoplasm. The signal generated by extracellular spin probe can be quenched with an appropriate impermeant paramagnetic agent. The ESR signal can thus be used to calculate the relative motion of the probe in the cytoplasm, as a measure of cytoplasmic viscosity.

The dielectric behavior of soybean seed axes has been studied as a function of water content by the TSDC technique (Fig. 3). Samples at a given water content are polarized by an electric field. This polarization is then "frozen in" by lowering the sample temperature sufficiently to prevent depolarization by thermal energy. The field is then switched off, and the sample is warmed at a constant rate. As the dipoles relax, the depolarization current is detected by an electrometer. This technique can provide information about the physical status of water, and particularly can identify and quantify glass transitions.

Collectively, these three types of measurements of glass transition temperatures have established that similar phase diagrams are obtained for several species of seed; and in every instance, under hydration levels at which seeds are commonly stored, the aqueous cytoplasm of the seed is seen to exist in the vitrified state. This will be true at room temperature or any temperature lower than room temperature. We

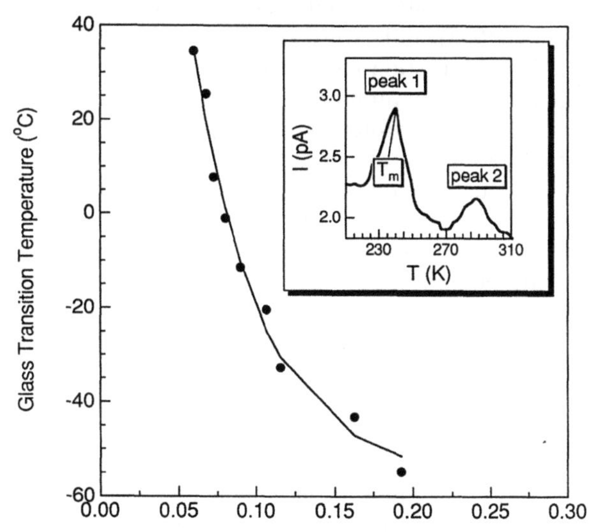


**Fig. 3.** Phase diagram for the melting temperature of the glassy state in soybean axes, determined by Thermally Stimulated Depolarization Current (Bruni 1990)

suggest, therefore, that the aqueous phase of dry seeds is ordinarily in the vitrified state.

## Functional Merits of Vitrification

Several types of physiological or physical advantages can be achieved by vitrification of dry organisms. As Burke pointed out (1986), the vitrified state would retard the loss of water, and would suppress the facility for reactions which might threaten the survivability of the organism.

The facility for spurious reactions will be suppressed by two functions of the vitrified or glassy state. First, the abilities of molecules in the cytoplasm to undergo diffusional and rotational movements will be hindered. For example, if one plots the rotational diffusion for molecules of three different radii against the viscosity of the medium (Fig. 4a), this type of movement is seen to be lowered drastically as viscosity is increased. Likewise, the translational diffusion is greatly suppressed by the glassy state (Fig. 4b). The strong increase in viscosity with vitrification would be expected to be a major barrier to spurious reactions in the dry organism, as well as a hindrance to the further loss of water.

We can illustrate this suppression of chemical reactions by vitrification with a model system involving lysozyme reacting with glucose to form Amadori products (Fig. 5). Such a reaction results in the loss of enzymatic effectiveness of the protein. We can show that in the presence of sufficient sucrose to bring the mixture into a glassy state, the loss of lysozyme activity is entirely prevented.

Another advantage to vitrification is that it can protect proteins from denaturation. Again, using a simple model system, we can show that glucose-6-phosphate dehydrogenase, which is very easily denatured in the dry state, can survive drying provided there is enough sucrose present to bring the mixture into a vitrified state (Fig. 6). A similar protection effect by trehalose has been reported by Carpenter and Crowe (1988), and Crowe and Crowe (this Vol.).

The suggestion has been made by Crowe that solutes which protect proteins against dry denaturation, viz. the sugars, have a direct interaction with the polar residues of the protein, thus keeping them in the folded condition which resists denaturation. Such an immobilization may be achieved within the vitrified cytoplasm. The immobilization holds the protein in the folded state, resulting in resistance to denaturation.

Another advantage of vitrification which we can presume to apply is that vitrification would suppress or prevent the crystallization of solutes in the cytoplasm. This suppression of crystallization is the characteristic that generated the interest in vitrification of biological materials (Takahashi et al. 1988).

## The Mechanism of Desiccation Tolerance

We have presented an argument that desiccation tolerance should be achieved through a simple mechanism. This suggestion is based on the widespread occurrence

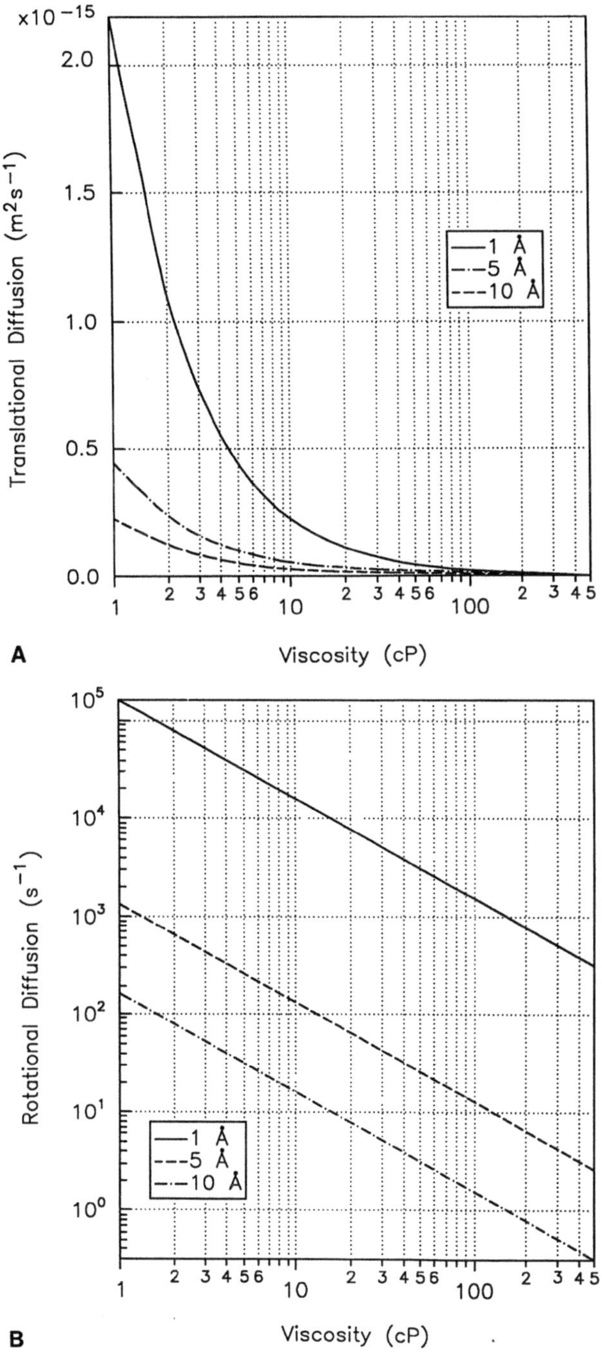


**Fig. 4A-B.** Calculated diffusion rates (**A**) and rotational rates (**B**) for molecules of three different sizes, as a function of the viscosity of the solution

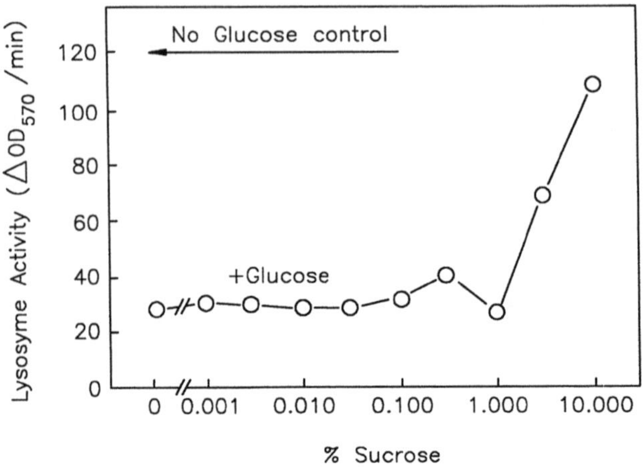


**Fig. 5.** The protection of lysozyme from destruction by the Amadori reaction, afforded by the presence of sucrose in a model system. The mixture is dried to a water content of 75% and stored at 40 °C for 5 days. In the presence of a reducing sugar (glucose), the dry mixture leads to a loss of about 85% of enzymic activity; sucrose prevents such a loss. (Wettlaufer and Leopold 1991)

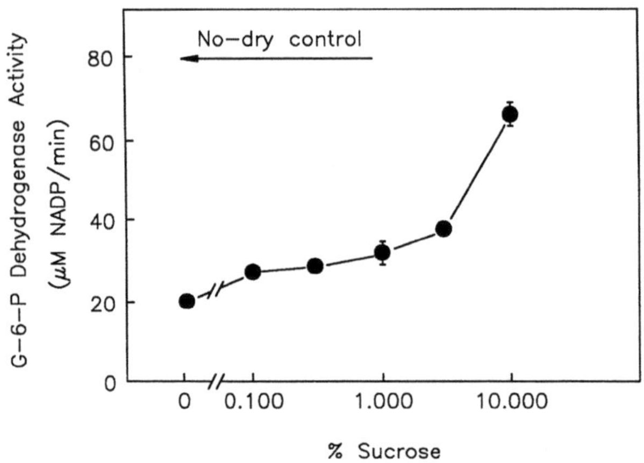


**Fig. 6.** Protection of glucose-phosphate-dehydrogenase from loss of activity (denaturation) by drying over P205, afforded by the presence of sucrose in a model system. Conditions as in Fig. 5. (Wettlaufer and Leopold 1991)

of desiccation tolerance in plant, animal, and bacterial systems. We have argued that the tolerant state would be expected to avoid crystallization and/or denaturation of cytoplasmic solutes. This expectation is based on survival experiments with model systems, and is supported by the characteristics observed for the moisture isotherms of tolerant vs. intolerant organisms. We have examined the relationship between sugars and desiccation tolerance, and noted that in the desiccated state, tolerant organisms can be shown to be in the glassy or vitrified state. We have employed model vitrifying systems to demonstrate the protective effects of a vitrifying sugar. It is our thesis that vitrification of the aqueous phase of cytoplasm is an intrinsic component of desiccation tolerance.

A curious feature of the protective effects of sugars in desiccation events is that there are substantial differences between sugars in their effectiveness (Mouradian et al. 1984). An appealing possible explanation of these differences has been proposed by Green and Angell (1989), who have compared phase diagrams of the vitrification temperatures for a diverse array of sugars. Their data show the strongest vitrification effectiveness (per sugar monomer) for trehalose, followed by maltose and then sucrose. They suggest that the protective characteristic of these sugars may be proportional to the vitrification effectiveness per sugar residue, i.e., the best protective effect is correlated with the sugars showing the highest glass-forming temperatures.

Vitrification may be the hard-core basis by which organisms survive desiccation.

## References


Amuti KS, Pollard CJ (1977) Soluble carbohydrates of dry and developing seeds. Phytochemistry 16:529–532

Bruni F (1990) Anhydrous biology and glass transitions. In: Stanley HE, Ostrowsky N (eds) Correlations and Connectivity. Kluwer Academic Publishers, pp 272–275

Bruni F, Leopold AC (1991) Glass transitions in soybean seed, relevance to anhydrous biology. Plant Physiol 96:660–663

Bruni F, Williams RJ, Leopold AC (1990) Vitrified cytoplasm in dry seeds. Plant Physiol 93:779

Burke MJ (1986) The glassy state and survival of anhydrous biological systems. In: Leopold AC (ed) Membranes, metabolism and dry organisms. Cornell Univ Press, Ithaca, pp 358–363

Carpenter JF, Crowe JH (1988) The mechanism of cryoprotection of proteins by solutes. Cryobiology 25:244–255

Clegg J (1965) The origin of trehalose and its significance during the formation of encysted dormant embryos of *Artemia*. Comp Biochem Physiol 14:135–143

Crowe JH, Crowe LM (1986) Stabilization of membranes in anhydrobiotic organisms. In: Leopold AC (ed) Membranes, metabolism and dry organisms. Cornell Univ Press, Ithaca, pp 188–209

Crowe JH, Crowe LM, Chapman D (1984) Preservation of membranes in anhydrobiotic organisms: the role of trehalose. Science 223:701–703

Crowe LM, Womersley C, Crowe JH, Reid D, Appel L, Rudolph A (1986) Prevention of fusion and leakage in freeze-dried liposomes by carbohydrates. Biochem Biophys Acta 861:131–140

Green JL, Angell CA (1989) Phase relations and vitrification in saccharide-water solutions and the trehalose anomaly. J Phys Chem 93:2880–2882

Hinton HE (1968) Reversible suspension of metabolism and the origin of life. Proc R Soc Lond B171:43–56

Hoekstra FA, van Rockel T (1988) Desiccation tolerance of *Papaver* pollen during development of the anther. Plant Physiol 88:626–632

Keller F, Schallenberg M, Wiemken A (1982) Localization of trehalose in vacuoles and of trehalose in cytosol of yeast. Arch Microbiol 131:198–301

Labuza TP (1984) Moisture Sorption. Am Assoc Cereal Chem, St Paul, 150 pp

Lingappa B, Sussman AS (1959) Endogenous substrates of dormant, activated and germinating ascospores of *Neurospora*. Plant Physiol 34:466–472

Mouradian R, Womersley C, Crowe LM, Crowe JH (1984) Preservation of functional integrity during long-term storage of a freeze-dried biological membrane. Biochem Biophys Acta 778:615

Takahashi T, Hirsh A, Erbe E, Williams RJ (1988) Cryoprotection by extracellular polymeric solutes. Biophys J 54:509–518

Vertucci CW, Leopold AC (1986) Physiological activities associated with hydration level in seeds. In: Leopold AC (ed) Membranes, metabolism and dry organisms. Cornell Univ Press, Ithaca, pp 35–49

Webb SJ (1965) Bound water in biological integrity. Thomas, Springfield, DE, 187 pp

Wettlaufer SH, Leopold AC (1991) Relevance of Amadori and Maillard reactions to seed deterioration. Plant Physiol 97:165–169

Williams RJ, Leopold AC (1989) The glassy state in corn embryos. Plant Physiol 89:977–981

Womersley C (1981) Biochemical and physiological aspects of anhydrobiosis. Comp Biochem Physiol 70B:669–678

# III Plant-Water Compartmentation and Water Stress

Chapter 12

# The Biophysics of Plant Water: Compartmentation, Coupling with Metabolic Processes, and Flow of Water in Plant Roots

E. STEUDLE

## Introduction

Water transport in living organisms is a complex phenomenon. In order to work out mechanisms and limitations, it has to be treated at the different levels of organization, i.e., at the cell, tissue, organ, and whole organism level. Usually, at the cell level irreversible thermodynamics are applied to describe water flow across membranes. In plants, the water potential concept has been successfully used to quantify the water status and driving forces for the uptake of water by roots and for water movements within the plant. Also this concept is based on solid thermodynamic grounds. Unlike solutes, water is much more mobile in cells and tissues of plants and, therefore, the compartmentation of water in plant tissue, as well as interactions between water relations (transport, water status) and active processes are often neglected. Rather, plant water relations are treated as purely hydraulic processes. However, interactions with metabolism receive considerable interest in important processes such as in the responses of plants to water shortage during drought, freezing, or high salinity. During osmoregulation and growth, active responses are also indicated by the synthesis of large amounts of different osmolytes which are produced to maintain cellular functions. Furthermore, basic phenomena such as the uptake of water and nutrients into the root or the solution flow in the phloem imply considerable coupling between metabolic and purely hydraulic processes.

In this review, basic concepts of cell, tissue, and organ water relations are presented. Emphasis will be on compartmentation and interactions with metabolic processes. Basic relations are derived which extend established concepts such as the mathematical treatments of tissue transport of water by Philip (1958a,b,c) and Molz and Ikenberry (1974). The concepts are also applied to growth. The root is treated as a good example (though still incomplete) for modeling transport at the level of cells, tissues, and an entire organ. Modern techniques such as the cell and root pressure probe are presented which allow determination of fundamental transport parameters required for modeling. Future aspects (experimental and conceptual) of plant water biophysics are discussed.

Universität Bayreuth, Lehrstuhl Pflanzenökologie, Universitätsstraβe 30, D-8580 Bayreuth, FRG

Somero et al. (Eds.)
Water and Life

## Cell Water Relations

Compared with animal cells, the field of water relations of plant cells seems to be much more developed, both with respect to the theoretical background and to the experimental techniques. This is so, because direct methods to measure essential parameters have been now available for several years. The theoretical background is based on linear force/flow relations of irreversible thermodynamics. Experiments have shown that they provide a good basis for the description of water transport. These statements hold for isolated cells such as giant-celled algae and tissue-culture cells as well as for cells in intact tissues of higher plants. Unlike animal cells, plant cells possess a rigid cell wall which allows turgor pressures of several bars to build up. The mechanics of this wall, i.e., its elastic and plastic extensibility, are also quite well understood. For technical reasons, analogous methods for determining the cell extensibilities and measuring pressure/volume relations in animal cells are lacking, since the overpressures in animal cells are usually extremely low. Thus, the existence of a cell wall, which usually complicates transport processes in plants by forming a separate compartment, turns out to be an advantage in the measurement of plant water relations.

Comprehensive overviews on the field of water relations of plant cells can be found in the literature in the reviews by Dainty (1963, 1976), Zimmermann and Steudle (1978), and Boyer (1985). House (1974) provides an excellent survey on the water relations of both plant and animal cells. More recent reviews of Tomos (1988), Steudle (1989), and Zimmermann (1989) mainly refer to applications of pressure probe techniques or to the coupling between water relations and other processes. Cosgrove (1986) focuses on growth relations of cells.

The mathematical description of water transport in plants originates from the work of Philip (1958a,b,c), who treated both the cell and tissue level. This theory has been extended by applying irreversible thermodynamics to water and solute relations (Dainty 1963, 1976; Zimmermann and Steudle 1978). The Philip model for tissue water relations has been extended considerably by Molz and Ikenberry (1974). A discussion of current mathematical models for cell and tissue water relations can be found in Molz and Ferrier (1982) and Steudle (1989).

A correct mathematical treatment of water transport across the plasma membrane of plant cells has to embrace all forces acting on water flow, i.e., couplings between water and solute flows have to be included. Active solute movement which may change osmotic gradients across the membrane also has to be incorporated. If the cell interior and the medium are treated as a two-compartment system, irreversible thermodynamics predicts for the water ($J_V$) and solute ($J_S$) flows the basic relations:

$$J_V = -\frac{1}{A}\frac{dV}{dt} = Lp\left\{P - RT\,(C^i - C^o) - \sigma_s \cdot RT \cdot (C_s^i - C_s^o)\right\}, \tag{1}$$

and

$$J_s = -\frac{1}{A}\frac{dn_s^i}{dt} = P_s\,(C_s^i - C_s^o) + (1 - \sigma_s) \cdot \bar{C}_s \cdot J_v + J_s^* . \tag{2}$$

V = cell volume; A = cell surface area; $n_s^i$ = content of permeating solutes 's' in the cell; Lp = hydraulic conductivity; P = cell turgor; $C^i$ ($C_s^i$) = concentration of non-permeating (permeating) solute in cell; $C^o$ ($C_s^o$) = concentration of nonpermeating (permeating) solute in medium; $\sigma_s$ = reflection coefficient; $\bar{C}_s$ = mean concentration of 's' in the membrane ≈ $(C_s^o + C_s^i)/2)$; $J_s^*$ = active solute flow.

In Eqs. (1) and (2), one permeating solute (denoted by subscript 's') and nonpermeating solutes are considered, but the equations can, of course, be extended to situations with more than one permeating solute. It can be seen from Eq. (1) that the volume flow ($J_V$) is related to the driving force (in brackets on the right side) by the hydraulic conductivity, Lp. The driving force represents some kind of a modified water potential gradient ($\Delta\psi = P - \Delta\pi$; $\pi = RT \cdot C$ = osmotic pressure). Since the membrane is not considered to be semi-permeable, the osmotic component due to the solutes has to be modified by the reflection coefficients which denote the selectivity of the membrane and are usually between zero and unity. The reflection coefficient can also be interpreted as a measure of the interaction between water and solutes as they cross the membrane. For most of the solutes in the sap of plant cells, $\sigma_s$ will be close to unity.

Equation (2) comprises three different components of the solute flow ($J_s$). A diffusional component [$= P_s(C_s^i - C_s^o)$] relates concentration gradients to the flow according to Fick's first law of diffusion by the permeability coefficient, $P_s$. The second term denotes the solvent drag, i.e. the amount of solute dragged along with the water flow in permeable membranes ($\sigma_s < 1$), e.g., in a membrane pore. $J_s^*$ is the active component of solute flow, i.e., it represents the interaction of solute flow with a metabolic reaction.

Equations (1) and (2) represent coupled differential equations which have to be solved to obtain changes of turgor and cell volume, i.e., of the water status of the cell. For example, if the concentration of the permeating solute is changed in the medium (i.e. $C_s^o$), the cell will undergo transient changes of turgor and volume and eventually attain a new steady value. Under these conditions, the solution of Eqs. (1) and (2) is (Dainty 1963; Steudle and Tyerman 1983):

$$\frac{V(t) - V_o}{V_o} = \frac{P(t) - P_o}{\varepsilon} = \frac{\sigma_s \cdot RT \cdot \Delta C_s^o \cdot Lp}{Lp(\varepsilon + \pi^i) - P_s} \{\exp(-k_w \cdot t) - \exp(-k_s \cdot t)\}\,, \tag{3}$$

where $\Delta C_s^o$ is the change of the concentration of solute 's' in the medium and $\varepsilon$ the elastic modulus of the cell (see below). $\pi^i$ = osmotic pressure in the cell. Eq. (3) has been derived by neglecting the solvent drag and at constant $J_s^*$. The left side represents the relative changes of cell volume following a change of $C_s^o$. $k_w$ and $k_s$ are the rate constants of water and solute exchange, respectively. Equation (3) describes a biphasic exponential process. When the solute is added to the medium, water will flow out of the cell because of the lowered water potential which the cell will tend to approach. However, since solutes enter at the same time, the process will be overlayed by an uptake of solutes and, when water flow has ceased, the process will be reverted in a following solute phase because the solute concentration in the cell will increase and the internal water potential will be lowered. Solute flow in the solute phase will cease when the same concentration of s will be attained in the cell and in the medium.

Since water flow is usually much more rapid than solute flow, two distinct phases will be obtained. If the contribution of solute flow during the water phase can be neglected, the rate constants $k_w$ and $k_s$ for the two phases will be given by:

$$k_w = \frac{\ln(2)}{T^w_{1/2}} = \frac{A}{V} Lp\,(\varepsilon + \pi^i) \ , \tag{4}$$

and

$$k_s = \frac{\ln(2)}{T^s_{1/2}} = \frac{A}{V} P_s \ . \tag{5}$$

Thus, by measuring $k_w$ and $k_s$ (or the half-times $T^w_{1/2}$ and $T^s_{1/2}$) and determining the geometry of the cell (volume, V, and surface area, A), Lp and $P_s$ can be evaluated. It has been also shown that the reflection coefficient can be determined from the maximum changes of turgor at a given change of osmotic pressure (Steudle and Tyerman 1983; Steudle 1989). For the determination of Lp according to Eq. (4), another parameter has to be known, namely, the elastic modulus of the cell, $\varepsilon$, which is an important physiological and ecological parameter. The elastic modulus relates volume and pressure of a cell. It is a measure of the rigidity of the cell wall. The inverse of $\varepsilon$ directly gives the fractional change of cell volume per unit change of turgor and, thus, represents the elastic (reversible) extensibility of the cell. The elastic modulus is defined by:

$$\varepsilon = V \frac{dP}{dV} \approx V \frac{\Delta P}{\Delta V} \ . \tag{6}$$

Many values of $\varepsilon$ are now available. Most of them have been determined by the pressure probe technique or by the pressure chamber (see reviews and below).

In an experiment, where $P_s$, $k_s = 0$ and $\sigma_s = 1$ (only non-permeating solutes present), Eq. (3) reduces to:

$$P(t) - P_o = \frac{\varepsilon}{\varepsilon + \pi^i} \Delta\pi^o \left\{\exp(-k_w \cdot t) - 1)\right\} \ , \tag{7}$$

i.e., the response becomes monophasic ($\Delta\pi^o$ = change of osmotic pressure outside the cell). There is a lot of experimental evidence that isolated plant cells do respond to changes of osmotic pressure of the medium either according to Eq. (3) or to Eq. (7) (e.g., Tyerman and Steudle 1982; Steudle and Tyerman 1983; Steudle 1989). However, for tissue cells, this type of experiment is more complicated because the tissue surrounding a test cell will function as a big unstirred layer and, since half-times of water exchange of individual cells are usually short, this layer could easily become rate-limiting. However, with techniques such as the cell pressure probe invented several years ago (Hüsken et al. 1978), these difficulties can be overcome, since changes in water potential can be also produced by manipulating cell turgor and producing "hydrostatic relaxations" which also yield $k_w(T^w_{1/2})$. This is the more common procedure of determining cell water relations parameters.

### *Stationary Responses*

For many aspects of plant water relations such as osmoregulation and growth and responses of plants to drought and water stress, stationary levels of water potential, turgor, and cell volume are important as well as the dynamics of changes. For mature cells, these responses are governed by the elastic modulus which has been, therefore, thought to be an important ecological parameter (see discussion in Steudle 1989). The way cell volume (V), turgor (P), and osmotic pressure of a cell ($\pi^i$) would change with water potential is easily verified from Eq. (6) and from the definition of water potential in the absence of matric potential ($\psi = P - \pi$). Provided that there is no active movement of solutes, it follows that

$$\frac{\Delta V}{\Delta \psi} = C_c = \frac{V}{\varepsilon + \pi^i}, \tag{8}$$

$$\frac{\Delta P}{\Delta \psi} = \frac{\varepsilon}{\varepsilon + \pi^i}, \quad \text{and} \tag{9}$$

$$\frac{\Delta \pi^i}{\Delta \psi} = -\frac{\pi^i}{\varepsilon + \pi^i}. \tag{10}$$

The storage capacity of a cell for water ($C_c$) is $\Delta V/\Delta\psi$. It can be seen that $C_c$ is not only dependent on the volume of the cell but also on its elastic extensibility ($1/\varepsilon$). Usually, $\varepsilon >> \pi^i$, for plant cells, and hence changes of water potential will be mainly reflected in changes of turgor [Eq. (9)] rather than in changes of cell volume and osmotic pressure [Eq. (10)]. However, since $\varepsilon$ will decrease with decreasing turgor, there is a possibility that, close to the plasmolytic point, changes of volume and osmotic pressure could become important as well. This has been discussed as a protective mechanism to enable plants to avoid plasmolysis and cell death under conditions of water stress. It should be noted that Eqs. (8) to (10) should also hold for wall-less animals cells. In this case, $\varepsilon$ will be virtually zero, and the changes will be predominantly in volume and osmotic pressure.

### *Active Transport*

Equations (1) and (2) simplify the real situation in that they neglect the electrical membrane potentials as a driving force in the case of ions. They also neglect (for the sake of simplicity) interactions between flows of different solutes. A direct coupling between water transport ($J_V$) and a metabolic reaction in the membrane has been omitted (active water transport), because there are no indications for an active water transport in plants. Since the water permeability (hydraulic conductivity, Lp) of plant cells is usually very high, it is evident that any active pumping of water would be rather ineffective because the pump would be short-circuited. The same applies to electro-osmotic mechanisms which have been discussed in isolated cells and also in the phloem (Steudle 1989). Nevertheless, there is a coupling between active solute movement and water transport via changes of osmotic concentrations in the cell or cell surroundings. According to Eqs. (8) and (9), changes of the content of osmotic solutes in a cell ($\Delta n_s$ in mole) result in changes of turgor and cell volume, i.e. for an isolated cell:

$$\Delta P = \frac{\varepsilon}{\varepsilon + \pi^i} RT \frac{\Delta n_s}{V} , \quad (11)$$

and

$$\Delta V = C_c \cdot RT \frac{\Delta n_s}{V} . \quad (12)$$

Of course, $\Delta n_s$ may be produced also by a degradation of osmotically inactive storage compounds in the cell (e.g., starch), rather than by an active transport process. It is worth noting that an increase of osmotic solutes in the cell would reduce its storage capacity for water as defined by Eq. (8). That means that at a high osmotic pressure in the cell, changes of $\psi$ cause rather small changes of volume because the dilution effect on osmotic pressure is large.

## Water Transport in Tissues

### *Dynamics of Water Flow*

As already mentioned, basic concepts of a mathematical description of water transport in plant tissue have been given by Philip (1958a,b,c). In Philip's concept, the apoplasmic component is neglected and changes in water potential are described in terms of a diffusion type of kinetics which results from the assumption that water flows across membranes are proportional to local differences of water potential. Molz and Ikenberry (1974) extended this concept by also considering the parallel apoplasmic path, i.e., the compartmentation of water in tissues. They assumed that, at any time, the water potential of the protoplast of a tissue cell will be very close to that of its adjacent cell wall. With this condition of local equilibrium, they arrived at somewhat extended differential equations describing the flow of water across tissues. Provided that the tissue is rather homogeneous and its geometry is known, these differential equations can be integrated for given boundary conditions to yield the changes of water potential in time and space. Today, we know from pressure probe experiments that the assumption of "local equilibrium" should be a good approximation, since half-times of water potential equilibration of cells are of the order of seconds. The rate of equilibration between protoplast and apoplast should be even faster.

In these theories, expressions are derived for the diffusivity (D) of the tissue which is a measure of the speed at which changes of water potential propagate. The diffusivity increases with the hydraulic conductance of the two possible pathways and decreases with the absolute value of its storage capacities as one would expect, since water storage damps the propagation. It has to be emphasized that the term diffusivity taken over by Philip from soil physics is only formally related to the diffusion coefficient known from Fick's laws of diffusion which describe a mass flow driven by the thermal motion of molecules, although the diffusivity has the units of a diffusion coefficient ($m^2 \cdot s^{-1}$). Rather, the transported properties in these processes are changes of the free energy of water (water potential), water content, or turgor which are all linearly related to each other.

Molz and Hornberger (1973) also incorporated the cell-to-cell (membrane-bound) diffusion of solutes into the Philip model as well as interactions between water and solutes (solvent-drag). However, their concept embraced neither the diffusion of solutes in the apoplast (which should be much more important) nor an active movement of solutes.

In the Philip theory, the diffusivity of the cell-to-cell path will be given by:

$$D_c = \frac{Lp \cdot a_{cc} \cdot \Delta x^2}{2 \cdot C_c} = \frac{a_{cc} \cdot \Delta x^2 \cdot \ln(2)}{2 \cdot A} \frac{1}{T_{1/2}^{w}}, \tag{13}$$

where $\Delta x$ is the thickness of the tissue cells in the direction of the propagation of the change of water potential and $a_{cc}$ = cross-sectional area of the cell-to-cell path (Fig. 1). The factor of 2 in Eq. (13) denotes the fact that two membranes have to be crossed per cell layer. It should be noted that the cell-to-cell path incorporates both a symplasmic component of water flow via plasmodesmata and a transcellular component which would denote flow across membranes. To date, the two components cannot be separated experimentally. It can be seen that $D_c$ is inversely proportional to the directly measurable half-time of water exchange by a factor which depends on cell dimensions and shape.

In the Molz-Ikenberry model, Eq. (13) is extended by incorporating the hydraulic conductance and storage capacity of the parallel apoplasmic path (Fig. 1). We get for the tissue diffusivity, $D_t$:

$$D_t = \frac{\Delta x\,(Lp_{cw} \cdot a_{cw} + a_{cc} \cdot \Delta x \cdot Lp/2)}{C_c + C_{cw}}. \tag{14}$$

$a_{cw}$ = cross-sectional area of apoplast; $Lp_{cw}$ = hydraulic conductivity of apoplast material; $C_{cw}$ = storage capacity of apoplast. $C_{cw}$ would be defined analogous to $C_c$ [Eq. (8)], i.e.:

$$C_{cw} = \frac{dV_{cw}}{d\psi_{cw}} \approx \frac{\Delta V_{cw}}{\Delta \psi_{cw}}, \tag{15}$$

where $V_{cw}$ is the per cell volume of water in the apoplast. Equation (14) indicates that for the two parallel pathways the conductances in the numerator and the storage capacities in the denominator are additive.

The theories of Philip and Molz and Ikenberry have been used to work out diffusivities. Since $D_c$ can be obtained from $T_{1/2}^{w}$ in pressure probe experiments and $D_t$ from following the swelling or shrinking of tissues, the comparison of values should yield some information about transport properties of the apoplast which are otherwise hard to obtain. Some of the results of $D_c$ are summarized in Table 1. It has been verified that, for plant tissues, the half-times of shrinking or swelling increase

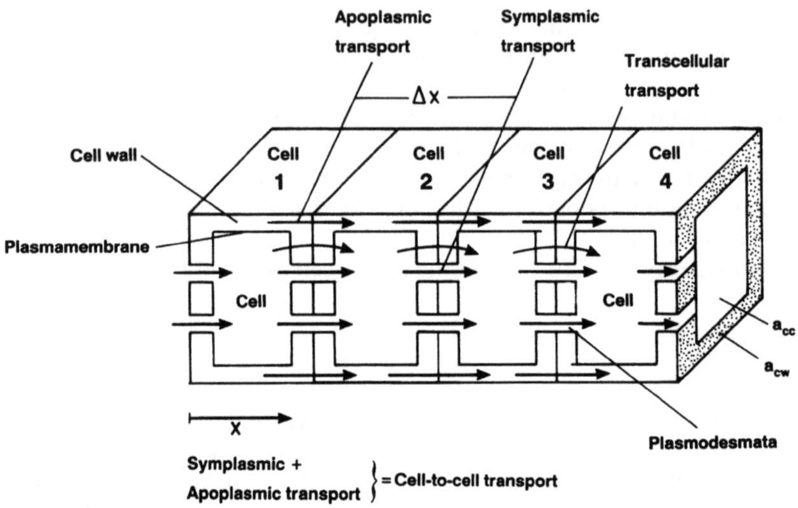


**Fig. 1.** One-dimensional cell aggregation demonstrating the different pathways of water in a tissue. The apoplasmic path is around the protoplasts. The symplasmic path is restricted to the flow across plasmodesmata. Transcellular flow denotes the path across cell membranes which also crosses the walls between adjacent cells. Transcellular and symplasmic flows are summarized as the cell-to-cell transport of water. $a_{cc}$ and $a_{cw}$ are the mean cross-sectional areas for the cell-to-cell and apoplasmic transport, respectively. $\Delta x$ is the thickness of a cell in x direction. (After Molz and Ferrier 1982)

significantly with tissue dimensions as it follows from diffusion kinetics (Westgate and Steudle 1985). Hence, there is a fairly good agreement between theoretical predictions and results. However, there are marked differences between different types of experiments, as demonstrated for the hypocotyl of soybean (Steudle and Boyer 1985) and for the midrib tissue of the maize leaf (Westgate and Steudle 1985). In experiments in which tissue was hydrated using a hydrostatic pressure gradient, $D_t$ was by an order of magnitude larger than during hydration in the absence of such a gradient. In the latter case, hydration was driven by matric or osmotic forces. Thus, the model of Molz and Ikenberry has to be extended.

The reason for the extension is that osmotic forces operating in the apoplast should have a different effect on the water movement in this path. It is easy to imagine that osmotic gradients in the apoplast should be rather ineffective, since the reflection coefficient of this structure should be close to zero (Steudle and Boyer 1985). Hence, despite a potentially high hydraulic conductance, the relative contribution of the apoplast to the overall tissue transport could be low. On the other hand, hydrostatic gradients would cause the maximum flow of water.

Solutes moving in the apoplast by diffusion, with the transpiration stream ("solvent drag"), or by permeation across cell membranes, present another problem. To date, only the latter has been incorporated into the Molz theory (see above). Apoplasmic and symplasmic solutes would exert osmotic forces which would influence the water flow equilibrium between apoplast and adjacent symplast, i.e., the "local equilibrium". Furthermore, active solutes which are taken up from the

**Table 1.** Hydraulic Conductivity (Lp), half-time of water exchange ($T^{w}_{1/2}$), and tissue diffusivity for water ($D_c$) as determined from cell pressure probe experiments[a]

| Species | Tissue/cell type | Half-time $T^{w}_{1/2}$(s) | Diffusivity, $D_c(m^2 s^{-1})$ | Hydraulic conductivity, $Lp(m\ s^{-1} MPa^{-1})$ | Reference |
|---|---|---|---|---|---|
| *Tradescantia virginiana* | Leaf epidermis | 1–35 | $(0.2–6) \times 10^{-10}$ | $(0.2–11) \times 10^{-7}$ | a,b,c |
| | Subsidiary cells | 3–34 | $10^{-11} – 10^{-10}$ | $(2–35) \times 10^{-8}$ | |
| | Mesophyll cells | 55–95 | $1 \times 10^{-12}$ | $(4–6) \times 10^{-8}$ | |
| | Isolated epidermis | 9–54 | $(0.5–3) \times 10^{-11}$ | $6 \times 10^{-8}$ | |
| *Kalanchoe daigremontiana* | CAM tissue of the leaf | 2–9 | $6 \times 10^{-10}$ | $(0.2–1.6) \times 10^{-6}$ | d |
| *Pisum sativum* | Growing epicotyl | 1–27 (epid.) | — | $(0.2–2) \times 10^{-7}$ | e,f |
| | | 0.3–1 (cort.) | $3.2 \times 10^{-10}$ | $(0.4–9) \times 10^{-6}$ | |
| *Glycine max* | Growing hypocotyl | 0.3–5.2 (epid.) | $(1–9) \times 10^{-11}$ | $(0.7–17) \times 10^{-6}$ | g |
| | | 0.4–15.1 (cort.) | $(1–55) \times 10^{-11}$ | $(0.2–10) \times 10^{-6}$ | |
| *Zea mays* | Midrib tissue of leaf | 1–8 | $(0.4–6.1) \times 10^{-10}$ | $(0.3–2.5) \times 10^{-6}$ | h |
| *Salix exigua* | Sieve elements of isolated bark strips | 110–480 | — | $5 \times 10^{-9}$ (lateral Lp) | i |
| *Hordeum distichon* | Root cortex and rhizodermis | 1–21 | $(0.5–9.5) \times 10^{-11}$ (cortex) $(1–7) \times 10^{-12}$ (rhizodermis) | $1.2 \times 10^{-7}$ | j |
| *Triticum aestivum* | Root hairs, rhizodermis, cortex | 8–12 | — | $1.2 \times 10^{-7}$ | k,l |
| *Zea mays* | Root cortex, rhizodermis | 1–28 | $(2–53) \times 10^{-12}$ | $(0.5–9) \times 10^{-7}$ | m |
| | | — | — | $(2–14) \times 10^{-7}$ | n[b] |
| | | 2–10 | — | $1.2 \times 10^{-7}$ | l |
| *Phaseolus coccineus* | Root cortex | 0.4–2.3 | $(0.3–1.7) \times 10^{-10}$ | $2 \times 10^{-6}$ | o |
| *Gossypium hirsutum* | Root cortex | — | — | $1.2 \times 10^{-7}$ | p |

[a] Dc values have been determined from $T^{w}_{1/2}$ and cell dimensions and refer to the cell-to-cell path only. For some tissues, rather large ranges are found for the parameters which indicate some inhomogenities within a tissue.

[b] Lp varied with the position (depth) in the root cortex.

(a) Tomos et al. (1981); (b) Tyerman and Steudle (1982); (c) Zimmermann et al. (1980); (d) Steudle et al. (1980); (e) Cosgrove and Cleland (1983b); (f) Cosgrove and Steudle (1981); (g) Steudle and Boyer (1985); (h) Westgate and Steudle (1985); (i) Wright and Fisher (1983); (j) Steudle and Jeschke (1983); (k) Jones et al. (1983); (l) Jones et al. (1988); (m) Steudle et al. (1987); (n) Zhu and Steudle (1991); (o) Steudle and Brinckmann (1989); (p) Radin and Matthews (1989).

apoplast into the symplast (e.g., during phloem loading) or are moving from cell to cell by different mechanisms (e.g., during the radial uptake of nutrients into the root) have to be considered.

A more general and comprehensive theory of the combined movement of water and solutes in plant tissues, which incorporates the differences in the driving forces in the apoplast (hydrostatic and matric vs. osmotic), shows that even in the absence of solute flow the Molz-Ikenberry theory has to be modified (Steudle, unpublished). For a proper description of the effects, it is useful to define a factor, F, for the apoplast which denotes the fraction of a change of water potential in the apoplast which is osmotic in nature:

$$F = \frac{d\pi_{cw}}{d\psi_{cw}} \approx \frac{\Delta\pi_{cw}}{\Delta\psi_{cw}} \quad . \tag{16}$$

Here, $\pi_{cw}$ and $\psi_{cw}$ represent the osmotic pressure and water potential in the apoplast, respectively. Since:

$$\psi_{cw} = P_{cw} - \pi_{cw} - \tau_{cw} \quad , \tag{17}$$

($\tau_{cw}$ = matric component of $\psi_{cw}$) it is evident that:

$$1 = \frac{d(P_{cw} - \tau_{cw})}{d\psi_{cw}} - \frac{d\pi_{cw}}{d\psi_{cw}} \quad . \tag{18}$$

The pressure component in the apoplast ($P_{cw}$) should be very small (if there is any), and it will hold that $0 \leq d(P_{cw} - \tau_{cw})/d\psi_{cw} \leq 1$. Hence, F varies between zero and –1. For $F = 0$, all changes of $\psi_{cw}$ will be converted into changes of the non-osmotic components of $\psi_{cw}$, and for $F = -1$, all changes will be osmotic. For example, if a tissue is subjected to a change of the osmotic pressure of the medium and is dehydrated, part of the change of the water potential in the apoplast may be matric and the rest osmotic in nature due to a reduction of the amount of apoplasmic water. Incorporating Eq. (16) into the Molz-Ikenberry theory yields:

$$D_t = \frac{Lp/2 \cdot a_{cc} \cdot \Delta x^2 + Lp_{cw} \cdot a_{cw} \cdot \Delta x \{1 + F(1 - \sigma_{cw})\}}{C_c + C_{cw}} \tag{19}$$

for the tissue diffusivity. It can be seen that for $F = -1$, the contribution of the apoplasmic conductance should be small, if $\sigma_{cw} << 1$. In fact, for $\sigma_{cw} = 0$, it will vanish. This is in agreement with the finding of low rates of equilibration of tissues in osmotic experiments (see above). On the other hand, during hydrostatic hydration $F = 0$, and it is evident that the contribution of the apoplast is optimal regardless of the absolute value of $\sigma_{cw}$. The ratio between the flows of water in the apoplast and in the cell-to-cell path ($\rho$) will be:

$$\rho = \frac{2 \cdot Lp_{cw} \cdot a_{cw} \cdot \{1 + F(1 - \sigma_{cw})\}}{Lp \cdot a_{cc} \cdot \Delta x} \quad . \tag{20}$$

Again, it can be seen that for $F = -1$, and $\sigma_{cw} = 0$, there will be no flow around the cells. For $F = 0$, there will be a maximum contribution of the apoplasmic path.

### *Influence of Passive Solute Flow*

Provided that there is no cuticle present or that the apoplast is not interrupted by Casparian bands (as in the root), the diffusion of solutes around cells may contribute to the speed at which changes of $\psi$ are propagated across a tissue. One can imagine that, if solute diffusion in the wall path is fast compared with the diffusivity for water potential propagation, this process may even dominate. For a long time, when direct measurements of the diffusivity of the cell-to-cell path were not available from pressure probe experiments, it was thought that solute diffusion could limit the process (Dainty 1976).

Analogous to the diffusion in the wall path, the passive permeation of solutes from cell to cell has to be considered which will be governed by the permeability coefficient of the cell membranes according to Molz and Hornberger (1973). Introducing these two effects into the theory presented in previous paragraphs, we obtain by a further extension of Eq. (14):

$$D_t = \frac{Lp/2 \cdot a_{cc} \cdot \Delta x^2 + Lp_{cw} \cdot a_{cw} \cdot \Delta x \,\{1 + F(1 - \sigma_{cw})\}}{C_c + C_{cw}} + \frac{a_{cw} \cdot \Delta x}{V_{cw}} D_s^{cw} + \frac{a_{cc} \cdot \Delta x}{V} \frac{P_s \cdot \Delta x}{2} \ . \qquad (21)$$

This equation differs from Eq. (19) by two additive terms on the right side which incorporate the diffusion coefficient ($D_s^{cw}$) of the given solute in the apoplast (a figure which is usually hard to measure) and the permeability coefficient of the membrane ($P_s$). Since two membranes will have to be crossed by the solute in each cell layer, a factor of two also appears in the last term on the right side of Eq. (21).

It should be noted that in Eq. (21) effects of the solvent drag on the solute flow have been neglected [see Eq. (2)]. They could be important in the wall space, but certainly not in the cell-to-cell path, since the solutes in the tissue cells usually exhibit reflection coefficients close to unity. If the solvent drag has to be taken into account, more complicated differential equations are needed which are no longer of the simple diffusion type (Steudle, in preparation).

### *Active Transport*

Two components have to be considered here, i.e., an uptake from the apoplast into the parallel symplast (cell-to-cell path) or an active transport from cell to cell. Distinct from passive transport, the active components will not necessarily be linearly related to concentration gradients and, therefore, it cannot be expected that the equations can be simply extended, as in the case of passive solute flows in the absence of solvent drag. An active movement of solutes from cell to cell may result in an increase of solute concentration in cells (protoplasts) which would then cause some re-distribution of water due to the local equilibration of water potentials. It is easy to verify that, at a given net change of solutes of $\Delta n_s'$ moles in a cell sitting at a certain position x in a tissue, the corresponding change of water potential would be:

$$\Delta\psi' = -\frac{C_c}{C_c + C_{cw}} RT \frac{\Delta n_s'}{V} . \tag{22}$$

Here, the factor of $C_c/(C_c + C_{cw})$ denotes that water will be shared in the tissue between the protoplast and apoplast according to their storage capacities.

Things becomes a little bit more complicated, if solutes are taken up from the apoplast into the adjacent symplast. In this case, the water potential in the symplast would tend to reduce whereas the water potential of the apoplast would tend to increase, it can be shown that under these conditions, the resulting change of water potential would depend on the amount of solutes ($\Delta n_s''$ in moles) transferred from the apoplast into the symplast (Steudle, in preparation):

$$\Delta\psi'' = \frac{1}{C_c + C_{cw}} RT \left\{\frac{C_{cw}}{V_{cw}} - \frac{C_c}{V}\right\} \Delta n_s'' . \tag{23}$$

It can be seen from Eq. (23) that for positive $\Delta n_s''$ (uptake into the cell), the change of water potential of the tissue can be either positive, negative, or zero depending on the absolute values of the volumes and storage capacities of the two compartments. If the absolute rates of the uptake of solutes ($\Delta n_s'$ and $\Delta n_s''$) are known, these rates may be used to calculate the effect of active transport or of other active processes on the water status incorporating the effect of tissue compartmentation. In this case, the differential equations which have to be solved numerically to obtain $\psi(x,t)$ have to be extended (Steudle, in preparation). However, the problem would be to get reliable data of $\Delta n_s'$ and $\Delta n_s''$, i.e., of the active solute fluxes within the tissue.

### *Stationary Relations of Tissues*

When subjected to changes of water potential, tissues will basically behave as isolated cells with respect to the stationary changes. However, if there are active responses of the plant, things may be different. Typical examples of such responses would be the phenomena of osmoregulation and responses of plants to drought, high salinity, and freezing. Other examples, where active processes influence the water status, are crassulacean acid metabolism (CAM) or extension growth (Steudle 1989). Under these conditions, tissues behave differently from isolated cells because of the compartmentation of water and solutes in the apoplast and symplast. Some aspects of the coupling between active processes and water status are considered in the following.

For mature tissue, two basically different situations can be distinguished: one in which active processes change the osmotic concentration in the protoplast compartment, and one in which these changes occur in the apoplast. In both cases, a redistribution of water will take place and the water potential will change. If it is assumed that a water flow equilibrium will be maintained in a tissue, changes of equilibrium values of turgor, water potential, and cell volume may be easily calculated. If the amount of nonpermeating solutes gained in the protoplasts of a tissue by active processes during a certain time interval is $\Delta n_s$, then the corresponding changes in pressure, water potential, and cell volume are (Steudle 1989):

$$P - P_o = \frac{\varepsilon}{\varepsilon + \pi^i} \frac{1}{C_c + C_{cw}} \left\{ RT \frac{\Delta n_s}{V} C_{cw} + \Delta V_{tissue} \right\} , \tag{24}$$

$$V - V_o = \frac{C_c}{C_c + C_{cw}} \left\{ RT \frac{\Delta n_s}{V} C_{cw} + \Delta V_{tissue} \right\} , \text{ and} \tag{25}$$

$$\psi_{cw} - \psi_{cwo} = \frac{1}{C_c + C_{cw}} \left\{ \Delta V_{tissue} - RT \frac{\Delta n_s}{V} C_c \right\} . \tag{26}$$

Here, reference states of turgor, cell volume, and water potential are indicated by the subscript o. The uptake of water per cell into the tissue, i.e., a supply (e.g., from the root) minus a loss (e.g., by transpiration) is denoted by $\Delta V_{tissue}$. It can be seen that changes of the osmotic pressure of the cell (= $RT \cdot \Delta n_s/V$) are not directly reflected into changes of turgor. Rather, the ratio of $C_{cw}/(C_c + C_{cw})$ plays an important role which is substantially smaller than unity. The reason for this is that, if the water potential of the protoplast becomes more negative due to an uptake of solutes, water will be shared between the protoplast and apoplast. The water potential will be maintained only if there is sufficient water supply so that $\Delta V_{tissue} = RT \cdot \Delta n_s/V \cdot C_c$. By combining Eqs. (24) and (26), it is verified that, under these conditions, there will be a change of turgor similar to that of osmotic pressure provided that $\varepsilon >> \pi^i$. Compartmentation also affects $\psi$, although the factor of $C_c/(C_c + C_{cw})$ will be usually close to unity (Steudle 1989). It is evident from the equations that a small ratio of $C_{cw}/(C_c + C_{cw})$ would strongly reduce the capability of cells to store water by active solute accumulation. This may be of importance during CAM where it has been postulated that the accumulation of large amounts of malic acid in the vacuoles during the night would prevent water losses. Estimates of $C_{cw}$ in the literature range between one third and one tenth of the total storage capacity (Molz and Ferrier 1982; Steudle 1989).

### *Apoplasmic Solutes*

Apoplasmic solutes lower the water potential of a tissue. They may be important during phloem loading and unloading and during growth. However, consequences of high solute concentrations have been merely considered quantitatively. Analogous to the symplast, stationary changes due to a per cell change of $\Delta n_s^{cw}$ are given by:

$$P - P_o = \frac{\varepsilon}{\varepsilon + \pi^i} \frac{1}{C_c + C_{cw}} \left\{ \Delta V_{tissue} - RT \cdot C_{cw} \frac{\Delta n_s^{cw}}{V} \right\} , \tag{27}$$

$$V - V_o = \frac{C_c}{C_c + C_{cw}} \left\{ \Delta V_{tissue} - RT \cdot C_{cw} \frac{\Delta n_s^{cw}}{V_{cw}} \right\} , \text{ and} \tag{28}$$

$$\psi - \psi_o = \frac{1}{C_c + C_{cw}} \left\{ \Delta V_{tissue} - RT \cdot C_{cw} \frac{\Delta n_s^{cw}}{V_{cw}} \right\} . \tag{29}$$

Apoplasmic solutes affect P, $\psi$, and cell volume in the same direction, which is different from the situation in proptoplasts (see above). Again, compartmentation plays an important role, i.e. the fractional amount of storage capacity of the protoplast. Since $V_{cw} << V$, effects of changes of the amount of solutes in the apoplast should have a larger effect than those in the cells. However, it can be seen that there will be a considerable damping of the effects according to the total storage capacity.

### *Conclusion*

The examples illustrate the importance of compartmentation for the water relations of plants. Future work has to experimentally verify the relations and has to concentrate on the determination of the parameters governing the water relations of the apoplast and symplast compartments. Data for the apoplast, such as the storage capacity and hydraulic conductance, are lacking. Direct methods to determine these parameters are badly needed.

## Water Transport Coupled with Growth

So far, we have applied concepts of compartmentation and active solute accumulation to mature tissue exhibiting perfect elastic (reversible) extensibility of cell walls. In this section, we should extend the procedure to growing tissue and plastic (irreversible) extension of cell walls. During expansion growth, a large uptake of water is required in order to maintain turgor and to drive the plastic (viscous) deformation of cell walls. According to Lockhart's (1965) theory, two different processes have to be considered to describe the change in cell volume with time, i.e., the uptake of water according to a modified Eq. (1) and denoted by a change of cell water volume $dV_w/dt$, and a change of volume due to the plastic deformation, $dV_m/dt$, i.e.:

$$\frac{dV_w}{dt} = -Lp \cdot A\left\{P(t) - \Delta\pi(t)\right\}, \text{ and} \tag{30}$$

$$\frac{dV_m}{dt} = V \cdot m\left\{P(t) - P_c\right\}, \tag{31}$$

where $\Delta\pi(t)$ is the difference of osmotic pressure between cell and surroundings in Eq. (31), the plastic deformation is proportional to a driving force $(P(t) - P_c)$ and an extensibility factor, m. $P_c$ is a threshold of turgor which has to be overcome to induce a plastic flow of the wall material. Hence, cell walls are considered to behave like Bingham bodies. It should be noted that stretching of the walls is proportional to turgor, because the tension within the walls (which is the force acting) is proportional to turgor. Thus, the extensibility factor would also incorporate factors due to the size (diameter) of the cells, cell wall thickness, etc. (Steudle et al. 1982a). It should be also noted that the application of Eq. (31) in tissues or organs of higher plants may cause problems, because certain tissues may limit extension (Kutschera 1988), and some caution is necessary, when interpreting absolute values of m. As far as the cell level is concerned, Eq. (31) has been applied only to *Nitella* internodes, where it has been shown that the parameters m and $P_c$ are not real constants, but are under physical and metabolic control (Green et al. 1971).

Usually, Eqs. (30) and (31) are used under stationary or quasi-stationary conditions, i.e., when the mechanical expansion is just compensated by an equivalent uptake of water so that turgor remains constant. Under these conditions, the combination of the equations yields:

$$\left(\frac{1}{Lp \cdot A} + \frac{1}{V \cdot m}\right) \frac{dV}{dt} = (\Delta\pi - P_c). \tag{32}$$

This relation shows that there are two "growth resistances" which limit cell expansion, a hydraulic ($1/L_p \cdot A$) and a mechanical ($1/m \cdot V$) resistance. In principle, both resistances may affect the growth rate. For isolated cells, it is certainly true that the mechanical rather than hydraulic resistance is limiting, since the hydraulic conductance of plant cells is usually high. However, in tissues or organs of higher plants, water has to cross several layers to reach the growing tissue, and, therefore, it has been proposed that water transport could become limiting or co-limiting. This hypothesis was supported by a comprehensive theory for water movement in growing tissue by Molz and Boyer (1978). However, measurements of Cosgrove and co-workers indicated that this may be not true, because in growing tissue of pea, no gradients of turgor could be detected using the pressure probe technique (Cosgrove and Cieland 1983b). Cosgrove (1986) concluded from this and other work that the growth-induced water potential which indicated considerable hydraulic resistances in the work of Boyer and co-workers, were due to wall relaxation processes during the measurement and due to apoplasmic solutes causing a low water potential (Cosgrove and Cleland 1983a). Cosgrove (1985, 1987) measured very small rates of cell wall relaxation in growing pea seedlings (i.e., a small "m") which suggested to him a rate-limitation by mechanical extensibility. On the other hand, Boyer et al. (1985) have shown in excision experiments with soybean hypocotyl that wall relaxation was fast (i.e., "m" was high). They suggested that long half-times of relaxation could be due to large quantities of mature tissue attached to the growing tissue which could serve as a water reservoir and damp the relaxation of turgor (Matyssek et al. 1988).

Besides mechanical or hydraulic effects, solute relations and compartmentation should be considered in the context of a limitation of growth. Steudle (1985) has pointed out that the dilution effect could become serious in cells growing at rates of 10 to 20%/h. Since the osmotic pressure in the growing cell is the driving force for growth [Eq. (32)], dilution may soon cancel growth, if there is no supply. To my knowledge, there has been no serious attempt to quantify these effects, i.e., to measure changes of the content of osmotic solutes in growing cells due to active processes.

Effects due to compartmentation need to be considered in growing tissue, for which there should be equations analogous to Eqs. (24) to (29). In expanding cells, as distinct from mature cells, a time-dependent extension has to be taken into account, i.e., the extensibility would not simply be given by $1/\varepsilon$, but would also incoporate the time and the extensibility factor m. We will consider an immature cell which is not growing at a reference state where $P = P_c$, $V = V_c$, and $\pi^i = \pi^i_c$. If we then induce growth by increasing the water potential and cell turgor, we may look at the cell after a certain time interval $\Delta t$. According to Eqs. (6) and (32), the total (elastic plus plastic) extension per unit change of pressure (Ex) would be:

$$Ex = m \cdot \Delta t + 1/\varepsilon \,. \tag{33}$$

It is evident that, if there are concomitant elastic and plastic extensions, the storage capacity of the cell would be larger than in the case of a mature cell. If this is taken into account, we obtain [analogous to Eq. (8)] for a given time interval a storage capacity $C^g_c$ of the growing cell of:

$$C^g_c = \frac{V_c}{1/Ex + \pi^i_c} \,. \tag{34}$$

It is evident that $C^g_c > C_c$. Using the relation given in Eq. (34) and the same procedure as for mature cells, we can also evaluate changes of turgor, water potential, and cell volume when also allowing for some uptake of solutes into the symplast ($\Delta n_s$) and apoplast ($\Delta n^{cw}_s$). By neglecting possible changes of the storage capacity of the apoplast during growth, we get:

$$P - P_c = \frac{1/Ex}{1/Ex + \pi^i_c} \, \frac{1}{C^g_c + C_{cw}} \left\{ \Delta V_{tissue} + RT \cdot C_{cw} \left( \frac{\Delta n_s}{V} - \frac{\Delta n^{cw}_s}{V_{cw}} \right) \right\} , \tag{35}$$

$$V - V_c = \frac{C^g_c}{C^g_c + C_{cw}} \left\{ \Delta V_{tissue} + RT \cdot C_{cw} \left( \frac{\Delta n_s}{V} - \frac{\Delta n^{cw}_s}{V_{cw}} \right) \right\} , \text{ and} \tag{36}$$

$$\psi - \psi_c = \frac{1}{C^g_c + C_{cw}} \left\{ \Delta V_{tissue} - RT \cdot \left( \frac{\Delta n_s}{V} C_c + \frac{\Delta n^{cw}_s}{V_{cw}} C_{cw} \right) \right\} . \tag{37}$$

These equations are analogous to those for mature cells [Eqs. (24) to (29)]. They incorporate effects of solutes in the apoplast as well as in the symplast. In fact, by setting $I/Ex = \varepsilon$ and $C^g_c = C_c$ (i.e., by ignoring plastic extension), the appropriate relations for mature cells are obtained from Eqs. (35) to (37). The difference between mature and growing cells is that, in the growing cells, the extension term (Ex) and the per cell storage capacity will be time-dependent, and, therefore, the equations will be only valid for a certain time interval. For longer periods of time, they would have to be integrated to yield P(t), V(t), etc. The factor of $Ex^{-1}/(EX^{-1} + \pi^i_c)$ is a dilution factor which tends to decrease $(P - P_c)$ and, hence, the growth rate. It can be seen that an increase of $\Delta n_s$ will increase $P - P_c$ and the growth rate, whereas an increase of $\Delta n^{cw}_s$ will decrease it. It can be also seen that, as for a mature cell, compartmentation will strongly influence changes of turgor. The fraction of $C_{cw}$ of the total storage capacity will be also important during growth. At steady growth, the dilution effect expressed by the first factor on the right side of Eq. (35) will be compensated by an uptake of solutes into the protoplast to maintain a constant P – $P_c$. It is also evident that at that time the ratio between $\Delta n_s$ and $\Delta V_{tissue}$ must be constant (Steudle 1985).

Equations (35) to (37) also indicate effects of withholding water from growing tissue. This means that $\Delta V_{tissue} = 0$ which would tend to make $\psi$ more negative at $\Delta n_s^c$, $\Delta n_s^{cw} > 0$. The equations predict possible rate limitations of growth by different factors. The rate of any of the flows, i.e., $\Delta V_{tissue}/\Delta t$, $\Delta n_s/\Delta t$, or $\Delta n_s^{cw}/\Delta t$ could become rate-limiting. The limitation via the mechanical and hydraulic properties (m, Lp) would be incorporated in the total water flow $\Delta V_{tissue}/\Delta t$.

The use of $\Delta n_s$ and $\Delta n_s^{cw}$ needs some explanation, since sometimes it has been mistakenly assumed that apoplasmic solutes may terminate growth in a static sense (Cosgrove and Cleland 1983a). It is clear that for a control of growth rates, a continuous supply of water together with solutes would be necessary to maintain a gradient of $\psi$, e.g., by a solute flow from the phloem. In fact, the ratio between solute and water flow to the growing tissue will be critical. Any change of this parameter may cause large changes of the growth rate. These changes could be misunderstood as a limitation by water supply or by mechanical properties, if solutes are not carefully excluded as the limiting factor.

## Techniques for Measuring Plant Water Relations Parameters

### *Plant Cell Level*

Before the pressure probe for higher plant cells was introduced in 1978 (Hüsken et al. 1978), data on water relations parameters of individual plant cells were rather rare. They were mostly determined from tracer flow (for references, see Steudle 1989) or plasmolysis experiments (for references, see Stadelmann and Lee-Stadelmann 1989). Since then the pressure probe has been applied to many plant species and tissues. The technique is explained in Fig. 2. Some typical data of Lp, $T_{1/2}^w$, and $D_c$ obtained so far with the pressure probe and other techniques are listed in Tables 1 and 2.

It can be seen that the Lp values range over several orders of magnitude. This large variation is due to technical problems in determining the parameters. The tracer derived data are strongly influenced by unstirred layers and by uncertainties of using the proper models for evaluationg Lp. The NMR data are surprisingly high, although they may still suffer from effects of unstirred layers.

Data from plasmolysis are of the same order of magnitude as those from the cell pressure probe, even though unstirred layer effects may be a problem in some cases. With this technique, problems may also occur from a possible pressure dependence of Lp, although this has been shown to date for only one species (Steudle et al. 1982b). The cell pressure probe data range between $10^{-8}$ and $10^{-6}$ $m \cdot s^{-1} \cdot MPa^{-1}$ which seems to be the range for higher plant tissue cells. These values are fairly large and cause the half-time of water exchange to be rather short ($T_{1/2}^w \approx 10$ s). Hence, the

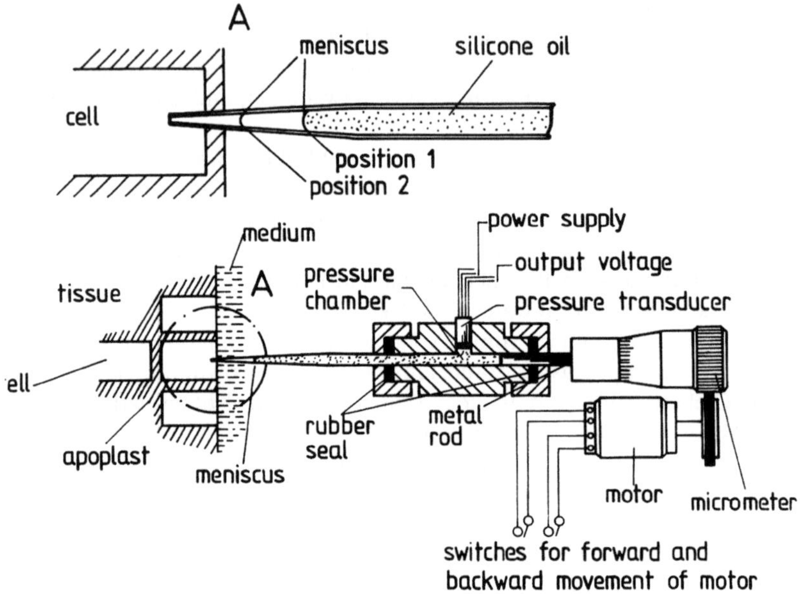


**Fig. 2.** Pressure probe for measuring water relations parameters of tissue cells of higher plants (cell pressure probe). The tip of the probe (diameter: 3–6 μm) is introduced into a cell under the microscope so that the cell turgor is transmitted to the oil in the probe and measured by a transducer. The meniscus between oil and cell sap in the tip of the microcapillary serves as a point of reference. It can be moved in order to quantitatively change the cell volume for measuring elastic coefficients of cells [Eq. (6)] and to induce water flows across the cell membrane. For example, if cell turgor was first stabilized at a position 1, and the meniscus is then moved to a position 2 and is fixed there, pressure relaxations are obtained which yield $T_{1/2}^{w}$ (Lp)

diffusivity of tissues is also rather large and plant tissues tend to be close to water flow equilibrium at dimensions of the order of a millimeter, i.e., of about 20 cell layers. In this case, the half-times of the equilibration of water flows will be of the order of minutes (e.g., Steudle and Boyer 1985). However, if growth occurs, this may cause gradients within the tissue, provided that the extensibility of the cells is large, compared with the hydraulic conductance of the tissue (Molz and Boyer 1978; see above). The relative importance of the growth resistances [Eq. (32)] is still under debate.

### *Techniques for Tissue Water Relations*

Techniques such as the plasmolytic method or the pressure probe measure the hydraulic conductivity. Exchange rates of water in turgid tissue can be followed and elastic properties of cells can be measured with the pressure probe. There are other techniques which measure the elasticity and water potentials of tissues or of entire plant organs such as psychrometry and the Scholander bomb (for references, see Kramer 1983; Boyer 1985). In all these techniques, water potentials are measured

**Table 2.** Hydraulic conductivity (Lp) and half-time of water exchange ($T_{1/2}^{w}$) of cells of higher plants determined by methods different from the cell pressure probe technique[a]

| Species | Tissue cell type | Technique | Half-time $T_{1/2}^{w}$ (s) | Hydraulic conductivity Lp (m $s^{-1}$ $MPa^{-1}$) | Reference |
|---|---|---|---|---|---|
| *Allium cepa* | Bulb scale epidermis | Plasmolysis | — | $4.3 \times 10^{-8}$ | a |
| | | External force | 9–55 | $(2–5) \times 10^{-7}$ | b |
| *Pisum sativum* | Stembase subepidermis | Plasmolysis | — | $1.9 \times 10^{-7}$ | a |
| *Quercus rubrum* | Root cortex | Plasmolysis | — | $2.6 \times 10^{-7}$ | a |
| *Helianthus tuberosus* | Storage tissue | External force | 50 | $(6–36) \times 10^{-8}$ | c |
| *Beta vulgaris* | Storage tissue | External force | 55–690 | $(2–6) \times 10^{-8}$ | d |
| *Hedera helix* | Bark cells | NMR | — | $2 \times 10^{-6}$ | e |
| *Zea mays* | Root cells | NMR | — | $5 \times 10^{-7}$ | f |
| | | Tracer flow | — | $(3–4) \times 10^{-7}$ (cell-to-cell model); $1 \times 10^{-9}$ (isolated cell model) | g |

[a]Only a few typical examples are given for the plasmolytic, tracer flow, external force, and NMR techniques. All data have been transformed into units of hydraulic conductivities (Lp). For the tracer flow data, Lp values strongly depend on the model used for the calculation of Lp, i.e. on whether the tissues would behave like isolated cells or whether water moves from cell to cell within the tissues. (a) see Stadelmann and Lee-Stadelmann (1989); (b) Ferrier and Dainty (1977); (c) Ferrier and Dainty (1978); (d) Green et al. (1979); (e) Stout et al. (1978); (f) Bacic and Ratkovic (1984); (g) Woolley (1965).

and, if water potential is determined as a function of water content, the data also yield pressure-volume curves and some measure of the elastic modulus of the tissue. Osmotic pressures and turgor can be estimated as well. However, the latter parameters represent averaged values for a sample in which different tissues contribute to the overall value according to their volume. This has to be taken into account when these methods are used (Zimmermann and Steudle 1978; Steudle 1989).

The absolute value of $\varepsilon$ determined from individual cells as well as from overall measurements has been discussed in terms of strategies of plants to withstand drought and other types of water stress. According to Eqs. (9) and (10), large absolute values of $\varepsilon$ would tend to reduce turgor instead of osmotic pressure upon a change of water potential. Thus, there would be a danger that the cell would be plasmolyzed and the buffering capacity of the system would be low. On the other hand, a low absolute value of $\varepsilon$ would also cause changes of the osmotic pressure and of the cell volume depending on the absolute value of $\pi^i$ compared with that of $\varepsilon$. However, literature data do not show clear evidence for such a protective mechanism, and therefore it is doubtful whether it really exists (see discussion in Steudle 1989).

### *Open Questions in the Field of Cell and Tissue Water Relations*

Application of the cell pressure probe in different laboratories over the past 13 years has given considerable information about hydraulic properties of plant cells, namely, about the hydraulics of protoplasts, and the mechanical properties of cell walls. Complete modeling of tissues and testing of mathematical models would require further data about the hydraulic properties of the apoplast. Another problem is the solute relations of the apoplast, i.e., data about diffusion and reflection coefficients of this compartment. Direct measurement of the parameters would be the best, but the technical problems are enormous. The more pragmatic way is, perhaps, to measure entire tissues and individual cells and obtain information about the apoplast by difference, as has been already done. In some tissues such as the root, water and solute relations parameters of the apoplast may change during development, due the deposition of hydrophobic material. These changes may be studied using modern fluorescent staining techniques and apoplasmic dyes (Peterson 1988). In any studies of apoplasmic water and solute relations, detailed structural investigations are necessary in order to work out the sizes of compartments which determine storage capacities and cross-sectional areas available for transport.

Another point which will deserve more interest in future is the coupling between water and active and passive solute relations. In isolated cells such as the *Chara* internode or the isolated epidermis, interactions between water and solute flows have been studied intensively (Steudle and Tyerman 1983; Tyerman and Steudle 1982). However, detailed studies on the interaction between active solute flow and water relations are still lacking. This does not mean that basic concepts of water relations will change or are threatened by change as has been recently suspected by Kramer (1988). Rather, these phenomena have to be integrated into the theory which has a solid thermodynamic basis (Schulze et al. 1988). Interactions between water relations and active processes are most obvious in phloem transport, in the movement of guard cells, and in the coupled flow of water and nutrients in the root. These interactions also deserve attention in the context of mechanisms of the communication between root and shoot and of growth.

The root has become a promising object not only to work out the mechanisms of the coupled uptake of water and solutes, but also to elucidate basic questions of water and solute relations of tissues as discussed above. The reason for this is that the root has a fairly regular cylindrical geometry and its developmental state can be properly defined. Furthermore, with a modified pressure probe, water transport across roots can be directly measured. The technique can be combined with measurements at the cell level. The results obtained with these new methods may be compared with those obtained with other techniques. Some recent results are summarized in the next sections.

**Water and Solute Transport in Roots**

In the soil-plant-atmosphere continuum, water flow is driven through a series arrangement of hydraulic resistances which can be identified as the soil, the root, the xylem, the leaf with the stomata, and a boundary layer adjacent to the leaf surface. Usually, the variable resistance of the stomata, including the boundary layer, is the most important. The next resistance in size is that of the root or the soil, depending on the water content of the soil, and these resistances are also variable. Hydraulic properties of roots have been adequately dealt with in several reviews and textbooks (Anderson 1976; Pitman 1982; Weatherley 1982; Kramer 1983; Dainty 1985; Passioura 1988; Steudle 1989). Some recent aspects which may be of interest for the understanding of basic water relations of tissues and for future research are discussed here.

The difference in water potential between the soil solution adjacent to the root and the root xylem is the driving force for the uptake of water into the root. As for cells, it is very unlikely that there is an active uptake of water into the root. In the transpiring plant, the driving force will be mainly set up by the tension (negative pressure) produced in the root xylem. However, there will be also a component related to the active uptake of solutes (nutrients) into the root xylem which results in an osmotic gradient across the root cylinder. This component should become important at very low or zero rates of transpiration. Under these conditions, positive root pressures are developed in the xylem which result in phenomena such as root exudation and guttation.

In the picture depicted so far, the root acts as an osmometer which requires some kind of a semipermeable membrane. This root membrane or osmotic barrier is usually identified as the endodermis, a cell layer surrounding the stele in which the apoplasmic path is interrupted by the Casparian band. It is thought that nutrients are taken up actively by cells of the rhizodermis or cortex and then cross the endodermis in the symplast to be actively secreted into the xylem by xylem parenchyma cells.

Recently, this model of the root has been further refined by Katou and coworkers, who applied the standing gradient model of Diamond and Bossert (1967) to the root (Katou and Taura 1989; Taura et al. 1988). According to these authors, the active uptake of solutes from the canal system of the cortical apoplast provides an osmotic gradient for water uptake. In the stele, the process is reversed, which results in a coupled movement of water and solutes across the root. Katou and coworkers assume that this mechanism of a coupling between water and active processes also works in the transpiring plant. They argue that it could explain findings such as low reflection coefficients of roots.

In an alternative model, the root is treated as a simple two-compartment system (like a plant cell). As the internal compartment, the xylem is separated from the soil solution or root medium by a membrane-like structure. Linear irreversible thermodynamics have been applied to this system to describe water and solute relations (Fiscus 1975; Dalton et al. 1975). As pointed out by Dainty (1985), this approach has the advantage that it is fairly flexible. Deviations from linearity can be easily expressed by changes in the absolute values of the coefficients, although the meaning of the coefficients is then unclear. The dilution of the xylem sap during root exudation could be accounted for in the two-compartment model (Fiscus 1975). However, deviations may also occur as soon as more than one barrier is considered (Newman 1976). There is evidence from experiments on cortical sleeves prepared from roots that the cortex without the endodermis also exhibits some selective properties (Ginsburg and Ginzburg 1970a,b; Table 4) and that there could be an effective barrier to ion flow in the exodermis, which may also develop Casparian bands (Peterson 1988).

The results obtained from measurements with the root pressure probe have led to the conclusion that the root may be considered as a composite membrane system made up by a series and parallel arrangement of barriers (Steudle et al. 1987; Steudle 1989). Series elements could be the exodermis, cortex, endodermis, etc. Parallel elements could be represented by the different pathways, i.e., the cell-to-cell and apoplasmic path (Fig. 1). Furthermore, arrays of the root at different stages of development could act as parallel elements. Composite membrane systems are known to exhibit characteristic deviations from those of the isolated elements. The contributions of the elements to the overall properties are not simply additive (Kedem and Katchalsky 1963a,b). If the elements could be measured individually, this approach could be useful to provide quantitative models for water and solute transport in roots. By considering the apoplast and cell-to-cell path as separate pathways, compartmentation in the root is taken into account by the model. Alternatively, the extended model of Molz and Ikenberry can be applied (see above).

In models of the root, the fact that the longitudinal hydraulic resistance (xylem resitance) could be also important, is a further complication. This component may be substantial near the root apex, as long as the xylem is not fully developed, and also in plants having tracheids. The complication has been also considered in models (Landsberg and Fowkes 1978; Frensch and Steudle 1989). The axial component can be also determined experimentally during root development (see below).

To be realistic, all models have to take into account the coupling between water and solute flows in the root, the variation of transport properties along the root during development of the xylem and of the endo- and exodermis, as well as during secondary thickening and suberization. For a proper, quantitative evaluation of water and solute relations, transport and morphological studies are required at the cell, root zone, and entire root level. In principle, the data which are required for the mathematical modeling of the root water and solute relations are now available by the root pressure probe technique.

### *Experimental Techniques*

In intact plants, the hydraulic resistance of the root system can be measured by determining the flow in the stem directly (stem flow techniques; e.g., Schulze et al. 1985) or from transpiration (Pearcy et al. 1989) and measuring the stationary difference in water potential between leaves and soil. However, this resistance will incorporate that of the soil and of the xylem besides the root resistance. Additional experiments would be required to separate resistances. In other experiments, profiles of the water uptake into the roots of intact plants have been measured using potetometers (Häussling et al. 1988).

Information about the hydraulic conductivity of roots ($Lp_r$ per $m^2$ of the surface area of the root) has been obtained from exudation experiments by measuring the exudation rate in $m^3 \cdot s^{-1}$ of excised roots, as well as the root surface area and the difference of osmotic pressure between exudate and root medium. Root exudation is varied by applying hydrostatic pressure to the root system sitting in a pressure chamber. Steady flow rates have been measured as a function of applied pressure (e.g., Fiscus 1986). In these techniques, the local forces driving radial water flow may be different from the overall osmotic pressure measured in the exudate, and the reflection coefficient of the solutes may differ from unity. This will be also true for stop-flow techniques, where excised roots are allowed to exude at a constant rate, and exudation is then stopped by applying a certain amount of a nonpermeating solute to the root medium (Pitman and Wellfare 1978) or by applying a step change of hydrostatic pressure to the cut surface of the root just sufficient to stop exudation (Miller 1980).

These techniques yielded important information on the hydraulic conductivity ($Lp_r$) and root reflection coefficients ($\sigma_{sr}$), but a more detailed resolution down to the root zone or cell level was not possible. However, this is badly needed for a precise modeling and a characterization of the contribution of pathways to the overall transport.

### *Root Pressure Probe Technique*

In extending the cell pressure probe technique described earlier in this review, a root pressure probe has been developed to determine water and solute relations of excised roots (Steudle and Jeschke 1983; Steudle et al. 1987; Steudle 1989). Root systems, parts of root system, or root end segments may be used. The technique permits the measurement of transport coefficients such as $Lp_r$ and $\sigma_{sr}$, but also of permeability coefficients ($P_{sr}$) of roots (Steudle and Frensch 1989; Steudle and Brinckmann 1989). The technique has been also employed for the measurement of longitudinal hydraulic resistances of the root xylem (Frensch and Steudle 1989) and has been combined with simultaneous measurements at the cell level (Zhu and Steudle 1991). The technique is briefly reviewed here. Results are compared with those from other techniques and are discussed with respect to basic mechanisms of the water and solute relations of higher plant tissue.

In the root pressure probe technique, an excised root is attached to the probe using silicone seals so that the root pressure can be built up in the system (Fig. 3). As with the cell pressure probe, a meniscus is formed in a measuring capillary between the silicone oil in a pressure chamber and the solution placed on top of the excised root (usually 0.1 mM $CaSO_4$ solution). This meniscus serves as a point of reference in the measurements. When a stationary root pressure has been attained (usually between 0.1 and 0.5 MPa), this pressure can be changed hydrostatically by changing the pressure in the measuring system or osmotically by changing the osmotic pres-

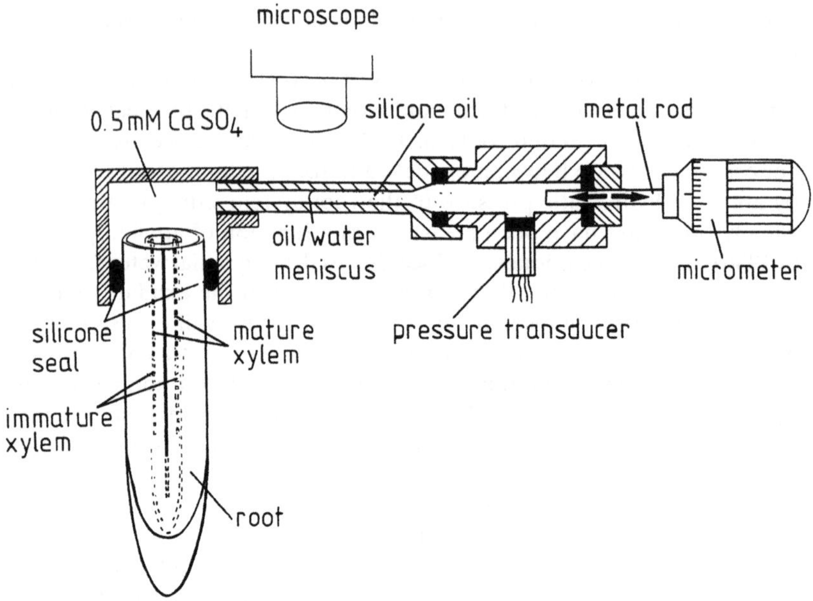


**Fig. 3.** Pressure probe for measuring water and solute flows in plant roots (root pressure probe; schematic). The excised root is connected with the probe by a silicone seal so that root pressure can build up in the system which is filled with 0.1 mM $CaSO_4$ solution and silicone oil. Root pressures are recorded by a pressure transducer. They can be manipulated with the aid of a metal rod connected to a micrometer screw (hydrostatic experiments). Changes of the osmotic pressure of the root medium will also cause water flows (osmotic experiments). From the different responses, the hydraulic conductivity ($Lp_r$) as well as the root permeability ($P_{sr}$) and reflection ($\sigma_{sr}$) coefficients are evaluated

sure of the root medium. The procedures are analogous to those used with plant cells. From the responses of root pressure, the transport parameters are evaluated, thus giving the hydraulic conductivity ($Lp_r$) as well as the permeability ($P_{sr}$) and reflection coefficient ($\sigma_{sr}$) for a given test solute. Basically, the theory used for these evaluations is the same as for cells, but there are important differences which mainly result from the fact that the root attached to the probe is an open system, since the cut xylem is in contact with the solution sitting on top of the cut surface (Fig. 3).

It is also possible to measure water flows and hydraulic conductivity in pressure clamp experiments by producing a step change in root pressure and estimating the corresponding water flow across the root from the shift of the meniscus in the capillary with time (Steudle and Frensch 1989). Under these conditions, the hydraulic conductivity of the root can be determined under nearly stationary conditions and at high flow rates for both hydrostatic and osmotic types of experiments.

Typical examples of root pressure probe experiments are shown in Fig. 4. Some typical data of transport coefficients are presented in Tables 3 and 4 and are compared with data obtained by other techniques. The data show that in some species such as maize, the hydraulic conductivity was strongly dependent on the

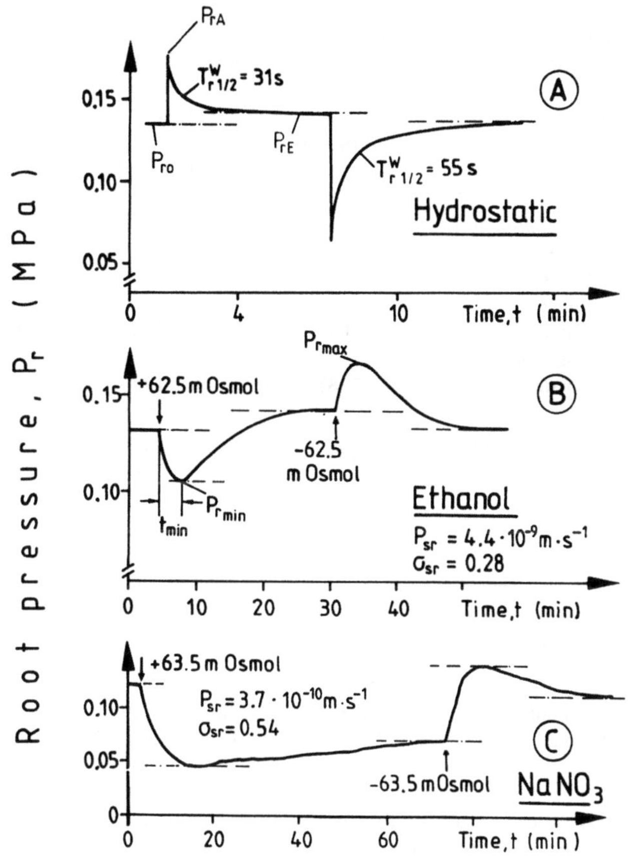


**Fig. 4A-C.** Typical relaxations of root pressure as measured with the root pressure probe (Fig. 3) on roots of *Phaseolus coccineus*. In (**A**), hydrostatic experiments are shown in which root pressure was manipulated with the aid of the probe. In (**B**) and (**C**), typical osmotic experiments are shown using permeating solutes (ethanol, $NaNO_3$). In these experiments, solute permeability ($P_{sr}$) and reflection ($\sigma_{sr}$) coefficients are determined from the curves besides the hydraulic conductivity, $Lp_r$

nature of the driving force, i.e., osmotic gradients produced a much smaller water flow than hydrostatic gradients. Furthermore, a comparison of the root $Lp_r$ with the cell Lp of cortical cells showed for maize that $Lp_r$ was of the same order of magnitude as the cell Lp. This led to the conclusion that the radial water transport across the root cylinder was apoplasmic in the cortex of this species and, at least to some extent, also in the endodermis. The apoplast had a potentially high hydraulic conductance. In the osmotic experiments, the driving force in the apoplast was $\sigma_{sr}^{cw} \cdot \Delta\pi_s^{cw}$ and, since the reflection coefficient of the wall should be close to zero, this resulted in a substantial cell-to-cell transport. Thus, at least for some species, the situation in the root appeared to be similar to that found for other tissue where large differences in water transport were found depending on the driving forces used (soybean hypocotyl, midrib tissue of maize leaf; see above).

Other species, however, revealed similar osmotic and hydrostatic values of $Lp_r$. This was explained in terms of much tighter Casparian bands in the endodermis (which should have interrupted an apoplasmic by-pass) or by a much larger cell Lp, as in the bean root (Fig. 4). In these cases, the comparison between root and cell

**Table 3.** Hydraulic conductivity of entire roots ($Lp_r$) and of individual root cells ($Lp$). Water flows were either induced by hydrostatic or osmotic gradients

| Species | Root $Lp_r \times 10^8$ | Cell $Lp \times 10^8$ ($ms^{-1}$ $MPa^{-1}$) | Techniques used ($ms^{-1}$ $MPa^{-1}$) | Reference |
|---|---|---|---|---|
| *Hordeum distichon* | | | | |
| osm.flow | 0.5–4.3 | — | | |
| hydr.flow | 0.3–4.0 | 12 | | |
| *Zea mays* | | | Root and cell | a,b,c |
| osm./hydr. | 1.4/10 | —/24 | pressure probe | |
| *Phaseolus coccineus* | | | | |
| osm./hydr. | (2–8)/(3–7) | —/190 | | |
| *Triticum aestivum* osm./hydr. | (1.6–5.5)/— | —/12 | Osmotic stop flow and cell pressure probe | d |
| *Zea mays* | | | | |
| osm./hydr. | (0.9–5)/— | —/12 | | |
| *Gossypium hirsutum* osm./hydr. | —/23 | —/12 | Pressure-flow and cell pressure probe | e |

(a) Steudle and Jeschke (1983); (b) Steudle et al. (1987); (c) Steudle and Brinckmann (1989); (d) Jones et al. (1988); (e) Radin and Matthews (1989).

hydraulic conductivity indicated a predominant cell-to-cell transport. Thus, it can be concluded that the mechanisms of water transport across the root cylinder may differ depending on the species and on the nature of the driving force. The latter may be also related to the finding of variable root resistances, i.e., to the fact that root conductances increase with increasing water flow in intact plants (cf. Weatherley 1982). At high transpirational demands and flow rates, one can imagine that the flow would be predominantly driven by the tension in the root xylem (hydrostatic flow), whereas at low flow rates the contribution of the osmotic component could become important.

Transport parameters such as $Lp_r$, $P_{sr}$, and $\sigma_{sr}$ may vary along the root as a consequence of root development and different root zones may contribute differently to the overall values. This has been found in stop-flow experiments (Jones et al. 1988) and with the root pressure probe (J. Frensch and E. Steudle, unpublished results). With the latter technique, a detailed analysis of transport parameters was possible. The analysis showed that $P_{sr}$ decreased and $\sigma_{sr}$ increased in young maize roots with increasing distance from the apex as one would expect. However, $Lp_r$ was constant along the root (Frensch and Steudle 1989).

In young maize roots, profiles of longitudinal resistance along the root also were determined using the root pressure probe (Frensch and Steudle1989). The data showed that the longitudinal resistance decreased by several orders of magnitude in

**Table 4.** Reflection ($\sigma_{sr}$) and permeability ($P_{sr}$) coefficients of roots determined by the root pressure probe and other techniques

| Root | Solute | Reflection coefficient, $\sigma_{sr}$ | Permeability coefficient, $P_{sr} \cdot 10^{10}$ (m $s^{-1}$) | Reference |
|---|---|---|---|---|
| *Glycine max* | Nutrients | 0.9 | — | a |
| *Zea mays*, cortical sleeves | Sucrose | 0.9–1.0 | — | b,c |
| | Urea | 0.85 | — | b,c |
| *Zea mays* | Nutrients | 0.85 | — | d |
| | Ethanol | 0.27 | 60–190 | e,f |
| | Mannitol | 0.74 | — | e,f |
| | Sucrose | 0.54 | 30 | e,f |
| | PEG 1000 | 0.82 | — | e,f |
| | NaCl | 0.5–0.6 | 60–140 | e,f |
| | $KNO_3$ | 0.5–0.7 | 10–80 | e,f |
| *Phaseolus coccineus* | Methanol | 0.6–0.34 | 27–62 | g |
| | Ethanol | 0.15–0.47 | 44–73 | g |
| | Urea | 0.41–0.51 | 11 | g |
| | Mannitol | 0.68 | 1.5 | g |
| | KCl | 0.43–0.54 | 7–9 | g |
| | NaCl | 0.59 | 2 | g |
| | $NaNO_3$ | 0.59 | 4 | g |

(a) Fiscus (1977); (b) Ginsburg and Ginzburg (1970a); (c) Ginsburg and Ginzburg (1970b); (d) Miller (1985); (e) Steudle and Frensch (1989); (f) Steudle et al. (1987); (g) Steudle and Brinckmann (1989).

the tip region as the xylem developed. Except for this region, the longitudinal resistances were small compared with the radial. Experimental values have been compared also with calculated values, according to Poiseuille's law, using cross-sections to estimate the number and size of conducting vessels. From the detailed analysis of radial and longitudinal resistances, profiles of water potential and the cumulative water flow could be calculated for the end segments of the maize roots assuming a steady transpirational drag at the basal end. These data allowed to quantify the efficiency of roots in collecting water.

### *Solute Parameters*

Solute parameters ($P_{sr}$, $\sigma_{sr}$) measured with the root pressure probe are summarized in Table 4. The data indicate fairly low $P_{sr}$ values, i.e., solute permeabilities which are low enough to prevent considerable leakage of nutrients from the xylem. The pump rates for the uptake of nutrients (which are available from the literature) are thus sufficient to compensate and overcompensate the leak, and permit proper function of the root. However, at the same time, the absolute values of the reflection coefficients are significantly smaller than unity for solutes for which cells exhibit a $\sigma_s \approx 1$. This means that the root differs from the behavior of an ideal osmometer. The reason is probably found in the complex structure of the root as a composite membrane system (see above). This model would also explain differences between osmotically and hydrostatically measured values of $Lp_r$. To date, problems with the model occur when comparing the permeability coefficients of $K^+$ salts with those of sodium or ammonium salts. Potassium salts exhibit a much smaller $P_{sr}$ than the other salts, which is not readily understood in terms of a passive by-pass in the apoplast. Perhaps the differences are due to high concentrations of $K^+$ salts in the xylem, or in the root apoplast, which reduce the rate of diffusion into the root. Unstirred layers have been also discussed in the context of low $\sigma_{sr}$ and $P_{sr}$. However, a detailed analysis showed that the effects and, namely, that of low $\sigma_{sr}$ could not be completely explained by unstirred layers. Effects of unstirred layers also do not explain the differences between different types of salts.

The composite structure model of the root is supported by combined measurements at the cell and root level using the double pressure probe technique (Zhu and Steudle 1991). In this technique, measurements of turgor in cortical cells of roots are performed simultaneously with measurements of root pressure. The results (again obtained on young maize roots) showed that the apparent $\sigma_s$ of root cells, measured at different depths in the cortex, decreased towards the endodermis. In the rhizodermis, $\sigma_s$ was close to unity as expected, whereas in cell layers close to the endodermis, the absolute value was close to that for the root which was determined in the same experiment. In cells sitting deeper in the tissue, the decrease of $\sigma_s$ could be interpreted in terms of the composite membrane model, since these cells are surrounded by more osmotic barriers (cell layers) exhibiting low reflection coefficients. The initial radial profiles of water potential across the root as well as the water flows across each layer could be evaluated from the initial water flows measured during double pressure probe experiments. Since hydraulic conductance of the cellular path was also known from cell pressure probe experiments, the contribution of the wall path could be evaluated. It turned out that $Lp_{cw}$ was quite high, and that during osmotic dehydration most of the water from the root cortex left the tissue via the apoplast. Nevertheless, the trans-root hydraulic conductivity was low, because of the low reflection coefficient of the wall path, and a substantial amount of trans-root water flow took place from cell to cell.

It may be argued that the root pressure probe experiments take place at positive root pressure, whereas in the intact plant there are usually tensions that drive the uptake of water. Thus, experiments have been performed in the range of negative pressures in the measuring system and xylem (Heydt and Steudle 1991). These experiments are difficult to perform because cavitation problems occur as soon as there are gas seeds of sufficient size left in the system. To date, experiments have been performed in a pressure range of down to –0.3 MPa (–3 bar) of absolute pressure (vacuum = 0 MPa). However, it was only possible to perform the experiments when the young roots of maize used were first infiltrated to remove gas-filled intercellular spaces of the apoplast. Furthermore, only osmotic experiments could be performed, since the movement of the metal rod in the root pressure probe immediately caused cavitations. It could be shown that the transport coefficients ($Lp_r$, $P_{sr}$, and $\sigma_{sr}$) obtained in this way for the range of negative pressure were similar to those obtained for the range of positive root pressure. Hence, the use of the data for the range of tensions in the xylem seems to be justified, at least for maize.

In conclusion, the root work demonstrates that the water and solute relations of tissues are not straightforwardly the sum of those of the cells they consist of. Compartmentation and the composite structure of tissues result in unexpected and important differences between the cell and the tissue level. Furthermore, the development of the xylem and of the endo- and exodermis complicates the situation.

### *Future Work on Roots*

In future work on root water and solute transport, extended mathematical models for the combined transport of water and solutes have to be applied in order to work out alternatives to the two-compartment model which assumes a single root membrane. This requires further knowledge of transport coefficients and of their variation along the root. Effects of active and passive solute flow have to be integrated to describe transport in root systems. Additional information about transport coefficients of secondary roots, effects of secondary root thickening, and suberization will be necessary. The effects of the exodermis, which has been proposed to be an important barrier to water and solute flow, have to be quantified. The root pressure probe has to be applied to root systems in soil. Effects of mycorrhiza, which are said to alter the hydraulic conductivity of roots, have to be measured. An important problem to be addressed is measurement of longitudinal resistances of plants having tracheids, such as conifers. Here, the effects of the longitudinal component of water transport should be relatively large.

In context with basic aspects of tissue water relations such as compartmentation and active process, the root work will yield valuable information about water relation parameters of the apoplast, i.e., of the hydraulic conductivity and storage capacity of this structure. The nature of apoplasmic water flow should be worked out with roots which have the advantage that there are modifications of the apoplast permeability to water. Couplings between active solute transport and root water relations which can be studied in different zones during root development will be of major interest.

*Acknowledgment.* This work was supported by a grant from the Deutsche Forschungsgemeinschaft, Sonderforschungsbereich 137.

## References

Anderson WP (1976) Transport through roots. In: Lüttge U, Pitman MG (eds) Encyclopedia of plant physiology, vol 2, Part B, Transport in plants. Springer, Berlin Heidelberg New York, pp 129–156

Bacic G, Ratkovic S (1984) Water exchange in plant tissue studied by proton NMR in the presence of paramagnetic centers. Biophys J 45:767–776

Boyer JS (1985) Water transport. Annu Rev Plant Physiol 36:473–516

Boyer JS, Cavalieri AJ, Schulze ED (1985) Control of the rate of cell enlargement: excision, wall relaxation, and growth-induced water potentials. Planta 163:527–543

Cosgrove DJ (1985) Cell wall yield properties of growing tissues. Evaluation by in vivo stress relaxation. Plant Physiol 78:347–356
Cosgrove DJ (1986) Biophysical control of plant cell growth. Annu Rev Plant Physiol 37:377–405
Cosgrove DJ (1987) Wall relaxation in growing stems: comparison of four species and assessment of measurement techniques. Planta 171:266–278
Cosgrove DJ, Cleland RE (1983a) Solutes in the free space of growing stem tissues. Plant Physiol 72:326–331
Cosgrove DJ, Cleland RE (1983b) Osmotic properties of pea internodes in relation to growth and auxin action. Plant Physiol 72:332–338
Cosgrove DJ, Steudle E (1981) Water relations of growing pea epicotyl segments. Planta 153:343–350
Dainty J (1963) Water relations of plant cells. Adv Bot Res 1:279–326
Dainty J (1976) Water relations of plant cells. In: Lüttge U, Pitman MG (eds) Encyclopedia of plant physiology, vol 2, Part A. Springer, Berlin Heidelberg New York, pp 12–35
Dainty J (1985) Water transport through the root. Acta Hortic 171:21–31
Dalton FN, Raats PAC, Gardner WR (1975) Simultaneous uptake of water and solutes by plant roots. Agron J 67:334–339
Diamond JM, Bossert WH (1967) Standing-gradient osmotic flow. A mechanism for coupling of water and solute transport in epithelia. J Gen Physiol 50:2061–2083
Ferrier JM, Dainty J (1977) A new method for measurement of hydraulic conductivity and elastic coefficients in higher plant cells using an external force. Can J Bot 55:858–866
Ferrier JM, Dainty J (1978) The external force method for measuring hydraulic conductivity and elastic coefficients in higher plant cells: application to multicellular tissue sections and further theoretical development. Can J Bot 56:22–26
Fiscus EL (1975) The interaction between osmotic and pressure induced water flow in plant roots. Plant Physiol 59:1013–1020
Fiscus EL (1977) Determination of hydraulic and osmotic properties of soybean root systems. Plant Physiol 59:1013–1020
Fiscus EL (1986) Diurnal changes in volume and solute transport coefficients of *Phaseolus* roots. Plant Physiol 80:752–759
Frensch J, Steudle E (1989) Axial and radial hydraulic resistance to roots of maize (*Zea mays* L.). Plant Physiol 91:719–726
Ginsburg H, Ginzburg BZ (1970a) Radial water and solute flows in the roots of *Zea mays*. I. Water flow. J Exp Bot 21:580–592
Ginsburg H, Ginzburg BZ (1970b) Radial water and solute flows in roots of *Zea mays*. II. Ion fluxes across root cortex. J Exp Bot 21:593–604
Green PB, Erickson RO, Buggy J (1971) Metabolic and physical control of cell elongation rate. In vivo studies in *Nitella*. Plant Physiol 47:423–430
Green WN, Ferrier JM, Dainty J (1979) Direct measurement of water capacity of *Beta vulgaris* storage tissue sections using a displacement transducer and resulting values for cell membrane hydraulic conductivity. Can J Bot 57:981–985
Häussling M, Jorns CA, Lehmbecker G, Hecht-Buchholz CH, Marschner H (1988) Ion and water uptake in relation to root development in Norway spruce [*Picea abies* (L.) Karst.]. J Plant Physiol 133:486–491
Heydt H, Steudle E (1991) Measurement of negative pressure in the xylem of excised roots: effects on water and solute relations. Planta 184:389–396
House CR (1974) Water transport in cells and tissues. Edward Arnold, London
Hüsken D, Steudle E, Zimmermann U (1978) Pressure probe technique for measuring water relations of cells in higher plants. Plant Physiol 61:158–163
Jones H, Tomos AD, Leigh RA, Wyn Jones RG (1983) Water-relation parameters of epidermal and cortical cells in the primary root of *Triticum aestivum* L. Planta 158:230–236
Jones H, Leigh RA, Wyn Jones RG, Tomos AD (1988) The integration of whole root and cellular hydraulic conductivities in cereal roots. Planta 174:1–7
Katou K, Taura T (1989) Mechanism of pressure-induced water flow across plant roots. Protoplasma 150:124–130
Kedem O, Katchalsky A (1963a) Permeability of composite membranes. Part 2: Parallel elements. Trans Far Soc Lond 59:1931–1940

Kedem O, Katchalsky A (1963b) Permeability of composite membranes. Part 3: Series array of elements. Trans Far Soc Lond 59:1941–1953
Kramer PJ (1983) Water relations of plants. Academic press, Orlando
Kramer PJ (1988) Changing concepts regarding plant water relations. Plant Cell Environ 11:565–568
Kutschera U (1988) Cooperation between outer and inner tissues in auxin-mediated plant organ growth. In: Cosgrove DJ, Knievel DP (eds) Physiology of cell expansion during plant growth. ASPP, Rockville, MA, USA, pp 215–226
Landsberg JJ, Fowkes ND (1978) Water movement through plant roots. Ann Bot 42:493–508
Lockhart JA (1965) An analysis of irreversible plant cell elongation. J Theor Biol 8:264–276
Matyssek R, Maruyama S, Boyer JS (1988) Rapid wall relaxation in elongating tissues. Plant Physiol 86:1163–1167
Miller DM (1980) Studies on root function of *Zea mays*. I. Apparatus and methods. Can J Bot 59:351–360
Miller DM (1985) Studies on root function of *Zea mays*. III. Xylem sap composition at maximum root pressure provides evidence of active transport into the xylem and a measurement of reflection coefficient of the root. Plant Physiol 77:162–167
Molz FJ, Boyer JS (1978) Growth-induced water potentials in plant cells and tissues. Plant Physiol 62:423–429
Molz FJ, Ferrier JM (1982) Mathematical treatment of water movement in plant cell and tissue: a review. Plant Cell Environ 5:191–206
Molz FJ, Hornberger GM (1973) Water transport through plant tissues in the presence of a diffusable solute. Soil Sci Soc Am Proc 37:833–837
Molz FJ, Ikenberry E (1974) Water transport through plant cells and cell walls: theoretical development. Soil Sci Soc Am J 38:699–704
Newman EI (1976) Interaction between osmotic- and pressure-induced water flows in plants. Plant Physiol 57:738–739
Passioura JB (1988) Water transport in and to roots. Annu Rev Plant Physiol Mol Biol 39:245–265
Pearcy RW, Schulze ED, Zimmermann R (1989) Measurement of transpiration and leaf conductance. In: Pearcy RW, Ehleringer J, Mooney HA, Rundel PW (eds) Plant physiological ecology, field methods and instrumentation. Chapman and Hall, New York, pp 137–160
Peterson CA (1988) Exodermal Casparian bands: their significance for ion uptake by roots. Physiol Plant 72:204–208
Philip JR (1958a) The osmotic cell, solute diffusability, and the plant water economy. Plant Physiol 33:264–271
Philip JR (1958b) Propagation of turgor and other properties through cell aggregations. Plant Physiol 33:271–274
Philip JR (1958c) Osmosis and diffusion in tissue: Half-times and internal gradients. Plant Physiol 33:275–278
Pitman MG (1982) Transport across plant roots. Q Rev Biophys 15:481–554
Pitman MG, Wellfare D (1978) Inhibition of ion transport in excised barley roots by abscisic acid: relation to water permeability of the roots. J Exp Bot 29:1125–1138
Radin J, Matthews M (1989) Water transport properties of cells in the root cortex of nitrogen- and phosphorus-deficient cotton seedlings. Plant Physiol 89:264–268
Schulze ED, Cermak J, Matyssek R, Penka M, Zimmermann R, Vasicek F, Gries W, Kucera J (1985) Canopy transpiration and water fluxes in the xylem of the trunk of *Larix* and *Picea* trees – a comparison of xylem flow, porometer and cuvette measurements. Oecologia 66:475–483
Schulze ED, Steudle E, Gollan T, Schurr U (1988) Responses to Dr. P.J. Kramer's article 'Changing concepts regarding plant water relations', volume 11, Number 7, pp 565–568. Plant Cell Environ 11:573–576
Stadelmann EJ, Lee-Stadelmann OY (1989) Passive permeability. In: Fleischer S, Fleischer R (eds) Methods in enzymology, vol 174, Biomembranes, Part U: Cellular and subcellular transport: Eukaryotic (nonepithelial) cells. Academic Press, Lond New York, pp 246–266
Steudle E (1985) Water transport as a limiting factor in extension growth. In: Baker NR, Davies WD, Ong C (eds) Control of leaf growth, SEB Seminar, vol 27. Cambridge Univ Press, Cambridge, pp 35–55

Steudle E (1989) Water flow in plants and its coupling to other processes: an overview. In: Fleischer S, Fleischer R (eds) Methods in enzymology, vol 174, Biomembranes, Part U: Cellular and subcellular transport: Eukaryotic (nonepithelial) cells. Academic Press, Lond, New York, pp 183–225

Steudle E, Boyer JS (1985) Hydraulic resistance to radial water flow in growing hypocotyl of soybean measured by a new pressure-perfusion technique. Planta 164:189–200

Steudle E, Brinckmann E (1989) The osmometer model of the root: water and solute relations of roots of *Phaseolus coccineus*. Bot Acta 102:85–95

Steudle E, Frensch J (1989) Osmotic responses of maize roots: water and solute relations. Planta 177:281–295

Steudle E, Jeschke WD (1983) Water transport in barley roots. Planta 158:237–248

Steudle E, Tyerman SD (1983) Determination of permeability coefficients, reflection coefficients, and hydraulic conductivity of *Chara corallina* using the pressure probe: effects of solute concentrations. J Membr Biol 75:85–96

Steudle E, Smith JAC, Lüttge U (1980) Water relations parameters of individual mesophyll cells of the CAM plant *Kalanchoe daigrmontiana*. Plant Physiol 66:1155–1163

Steudle E, Ferrier JM, Dainty J (1982a) Measurement of the volumetric and transverse elastic extensibilities of *Chara corallina* internodes by combining the external force and pressure probe technique. Can J Bot 60:1503–1511

Steudle E, Zimmermann U, Zillikens J (1982b) Effect of cell turgor on hydraulic conductivity and elastic modulus of *Elodea* leaf cells. Planta 154:371–380

Steudle E, Oren R, Schulze D (1987) Water transport in maize roots. Measurement of hydraulic conductivity, solute permeability, and of reflection coefficients of excised roots using the root pressure probe. Plant Physiol 84:1220–1232

Stout DG, Steponkus PL, Cotts RM (1978) Nuclear magnetic resonance relaxation times and plasmalemma water exchange in ivy bark. Plant Physiol 62:636–641

Taura T, Iwaikawa Y, Furumoto M, Katou K (1988) A model for radial water transport across plant roots. Protoplasma 144:170–179

Tomos AD (1988) Cellular water relations of plants. In: Franks F (ed) Water science reviews, vol 3. Cambridge Univ Press, Cambridge, pp 186–277

Tomos AD, Steudle E, Zimmermann U, Schulze ED (1981) Water relations of leaf epidermal cells of *Tradescantia virginiana*. Plant Physiol 68:1135–1143

Tyerman SD, Steudle E (1982) Comparison between osmotic and hydrostatic water flows in a higher plant cell: determination of hydraulic conductivities and reflection coefficients in isolated epidermis of *Tradescantia virginiana*. Austr J Plant Physiol 9:461–479

Weatherley PJ (1982) Water uptake and flow in roots. In: Lange OL, Osmond CB, Ziegler H (eds) Encyclopedia of plant physiology, vol 12B: Physiological plant ecology. Springer, Berlin Heidelberg New York, pp 79–109

Westgate ME, Steudle E (1985) Water transport in the midrib tissue of maize leaves. Direct measurement of the propagation of changes in cell turgor across a plant tissue. Plant Physiol 78:183–191

Woolley JT (1965) Radial exchange of labelled water in intact maize roots. Plant Physiol 40:711–717

Wright JP, Fisher DB (1983) Estimation of the volumetric elastic modulus and membrane hydraulic conductivity of willow sieve tubes. Plant Physiol 73:1042–1047

Zhu GL, Steudle E (1991) Water transport across maize roots: Simultaneous measurement of flows at the cell and root level by double pressure probe technique. Plant Physiol 95:305–315

Zimmermann U (1989) Water relations of plant cells: pressure probe technique. In: Fleischer S, Fleischer R (eds) Methods in enzymology, vol 174. Biomembranes, Part U: cellular and subcellular transport: eukaryotic (nonepithelial) cells. Academic Press, Lond New York, pp 338–366

Zimmermann U, Steudle E (1978) Physical aspects of water relations of plant cells. Adv Bot Res 6:45–117

Zimmermann U, Hüsken D, Schulze ED (1980) Direct turgor pressure measurements in individual leaf cells of *Tradescantia virginiana*. Planta 149:445–453

Chapter 13

# Water Compartmentation in Plant Tissue: Isotopic Evidence

D. YAKIR

## Introduction

The fast diffusion of water molecules in an aqueous solution (self diffusion coefficient $2.3 \times 10^{-5}$ cm$^2$ s$^{-1}$), the high permeability of the biological membrane to water (permeability coefficient of ca. $10 \times 10^{-5}$ cm s$^{-1}$), and the small distances involved at the cellular level (typical cell diameter $3 \times 10^{-5}$ cm) provide a strong basis for the generally accepted assumption that water in plant cells should be a well mixed pool. Thus, if isotopically labeled water, e.g., tritiated water, is introduced into leaves it would be expected that a complete equilibrium and isotopic homogeneity would be reached within a few seconds. It has long been recognized that this is not so, and this paper presents experimental evidence in conflict with this expectation and explores the biophysical basis for these observations.

Any form of water compartmentation in plant tissue has wide-ranging implications. Water is the solvent in which most biochemical reactions take place and its compartmentation is bound to influence the conditions under which these reactions occur. Water compartmentation implies that restricted "communication" between water compartments, or pools, exists. Such restricted communication can, in turn, influence the way changes in external environmental parameters, such as water availability and ambient humidity, are transduced. It can also influence the intrinsic intracellular conditions at the sites of key biochemical processes such as water oxidation in photosynthesis.

Most of the evidence for water compartmentation in plant tissue comes from isotopic studies. In some of these studies a radioactive isotope, tritium, has been used and the movement of the radioactive label was followed by sampling the tissue water and measuring its specific activity. There is, however, an increasing number of studies that employ stable, nonradioactive, isotopes of oxygen and hydrogen. There are two major advantages in using this approach. First, no artificial labeling is necessary and experiments can be performed under natural conditions by following changes in the natural abundance of the isotopes. Second, artifacts stemming from large artificial gradients associated with introduction of radioactive isotopes can be avoided.

The use of stable isotopes in the study of water status and water relations in plants has developed from the fact that during evapotranspiration there is a preferen-

Botany Department, Duke University, Durham, NC 27706, USA.
Present address: Department of Environmental Sciences and Energy Research, Weizmann Institute of Science, Rehovot 76100, Israel

Somero et al. (Eds.)
Water and Life

tial loss of water species containing the lighter isotopes. Specifically, $H_2{}^{16}O$ has a higher vapor pressure and will evaporate faster than $H_2{}^{18}O$ or deuterated water ($D_2O$), resulting in enrichment of D and $^{18}O$ in the water remaining in the leaf. This isotopic signal, measured as the $^{18}O/^{16}O$ and D/H ratios in water, is produced at the evaporating surfaces and must then propagate into the rest of tissue water by diffusion, convection or circulation.

Whether radioactive or stable isotopes are used, the long-term persistence of isotopic nonhomogeneity in cell or tissue water, under steady state conditions, is a strong indication for restricted mixing that is termed water compartmentation. It should be emphasized in this context that nonhomogeneity in the *structure* of water in the living cell is an intensively studied but still controversial issue. However, it is widely recognized that water structure, including various types of "bound" water, has a relatively small effect on the actual mobility of the individual water molecules (reducing the lifetime of a water molecule at a certain site from ca $10^{-11}$ sec to ca. $10^{-7}$ s, e.g., Tait and Frank 1971; Rupley et al. 1983; Shibata and Iwayanagi 1989). Evidently, the existence of any form of structural nonhomogeneity of water does not in itself provide a basis for incomplete mixing and equilibrium between water species in the cell or tissue. There is no other known basis for explanation of water compartmentation and, therefore, discussion of the experimental evidence which follows is clearly speculative.

There is enough evidence to demonstrate "abnormality" in the patterns of water mixing in the plant tissue. Recognition of "abnormality" in the behavior of a natural system is an infallible indication of our lack of understanding, and therefore should serve as a clear mark of an area in need of further research.

## Evidence for Leaf Water Compartmentation from Stable Isotope Ratios

The use of isotope ratios, at the natural abundance level, as a research tool became practical with the development of the mass spectrometer in the 1940s (Nier 1947). However, the observation that leaf water is "heavy" relative to ground or stem water was noted earlier (Washburn and Smith 1934). This was later established in a more precise manner by Gonfiantini et al. (1965) for oxygen isotopic ratios and by Warshaw et al. (1970) for hydrogen isotopic ratios, and evapotranspiration was recognized as the driving force for these enrichments (for recent reviews see White 1988; Sternberg 1988). The enrichment of heavy isotopes in leaf water does not proceed indefinitely, and as the concentration of the heavier isotopes in the liquid phase increases, the proportion in the evaporating water will also increase until it is equal to that in the source water. At this point, provided ambient conditions do not change, a steady state isotopic composition will be established.

The factors that determine the steady state isotopic composition of a constantly fed water body were first described numerically by hydrologists (Craig and Gordon 1965) and the model was subsequently applied successfully to hydrological systems (Allison and Leaney 1982). Similarly, leaves were regarded as constantly fed water bodies exposed to evapotranspiration, and it was expected that the Craig and Gordon model could be applied to the study of plant water relations (Dongmann et al. 1974; Farris and Strain 1978). It was quickly realized, however, that although there might be a qualitative agreement between changes in the model-predicted isotopic ratios and those actually observe, a significant quantitative discrepancy persisted (Farris and Strain 1978; White 1988; Walker et al. 1989; Yakir et al. 1990). Moreover, it was noted in practically all cases that the measured isotopic composition was consistently less enriched in the heavy isotopes than the model-predicted values.

A number of hypotheses have been offered to explain the discrepancy. Some suggest that whereas the model predicts isotopic composition at steady state, leaf water never reaches this state (Flanagan and Ehleringer 1991). Strictly speaking, this hypothesis is likely to be true under highly variable natural conditions. However, several studies have shown that even when the isotopic composition of leaf water remains constant for extended periods, i.e., when steady-state conditions are established, the discrepancy between predicted and observed values persists (Farris and Strain 1978; Walker et al. 1989; Yakir et al. 1990). The attainment of isotopic steady-state conditions in leaf water can be demonstrated in a more rigourous way by measuring identical isotopic ratios in the transpiration water leaving the leaf and in the stem water entering it (Fig. 1). Under these strict steady-state conditions evidence for water comparmentation (see below) is still observed.

Isotopic nonhomogeneity in leaf water has been confirmed by recent experiments. In one case (Yakir et al. 1989), pressure was applied to the exterior of a sunflower or an ivy leaf in a pressure chamber, slowly forcing tissue water out of the leaf through the petiole that extended out of the chamber. Each 5-μl water fraction obtained in this way was then analyzed for its isotopic composition and a large

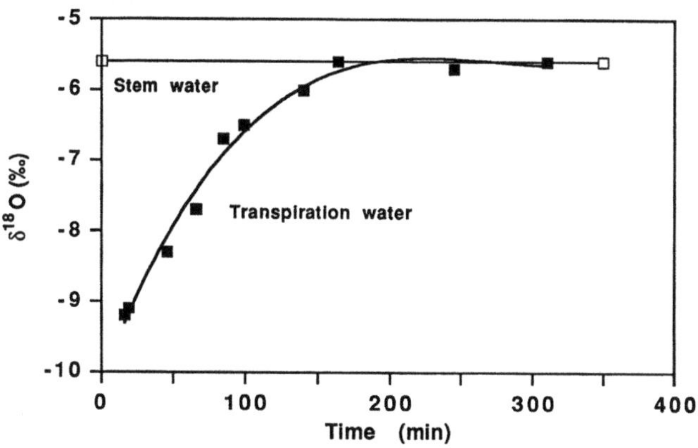


**Fig. 1.** Establishment of isotopic steady state in leaf water. Intact leaf of a sunflower plant, taken from a growth chamber at about 90% relative humidity, was enclosed in a leaf gas exchange chamber at 78% relative humidity. Transpiration water was trapped at the times indicated (*closed symbols*) and analyzed mass spectrometrically for its $^{18}O/^{16}O$ ratio. The constant isotopic composition of stem (input) water is indicated (*open symbols*) and the equality with transpiration (output) water, reached within 3 h, indicates the attainment of isotopic steady state in the water that remains in the leaf. Isotopic ratios are expressed in the delta notation where $\delta x = [R_{sample}/R_{standard}) - 1]1000$, x is $^{18}O$ or D, R is $^{18}O/^{16}O$ or D/H and the standard is SMOW). (Yakir, Berry and Osmond, unpubl.)

isotopic nonhomogeneity among water fractions could be demonstrated. Additional ion analyses of the water fractions helped in distinguishing between extra- and intracellular water sources (see below). In a second study using a different approach, the physical separation between two tissue layers of a *Peperomia* leaf (about 1.5 mm thick) enabled separate isotopic analyses of each section. Sampling the two layers over 24 h revealed a completely different set of isotopic ratios for each tissue layer (Fig. 2). In the lower tissue, in which stomata are embeded, a typical diel cycle in D/H (e.g., Förstel 1978) could be observed that reflects changes in ambient relative humidity. In the upper tissue, with no stomata, the diel cycle was practically unrecognizable and lower values of D/H ratios were maintained. It is important to note that microscopic examination of the leaf tissue revealed no distinct barrier to water movement throughout the leaf cross section.

These confirmations of isotopic nonhomogeneity in leaf water may indicate that water mixing in the leaf is somehow restricted and therefore some form of compartmentation must exist. It does not, however, provide much information as to the identity of the possible compartments or the basis for such a phenomenon.

## Tissue Level Explanations for Water Compartmentation

A number of suggestions and hypotheses have been advanced to explain the isotopic nonhomogeneity of leaf water. All the models thus far offered have serious difficulties in explaining the observed data. However, taken together with the evidence from other approaches, a compartmentation of water at the cellular level cannot be ruled out. It should be noted that some of the unpublished ideas presented here have been subjected to re-interpretations by the present author.

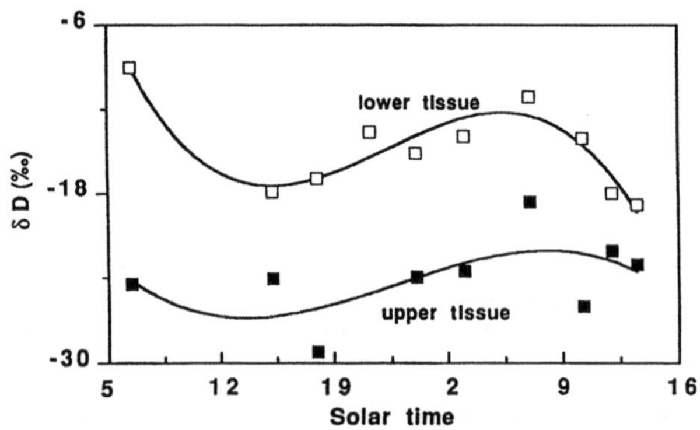


**Fig. 2.** Diurnal changes in the hydrogen isotopic composition of tissue water in a *Peperomia* leaf. Leaves from the same plant were sectioned to lower (chlorenchyma) and upper (water storage parenchyma) (about 0.5 mm thick) before water was extracted for isotopic analysis. Stomata are present only at the lower surface of the leaf and when open expose leaf water to the effects of ambient humidity. Water in the two parts of the leaf was not in equilibrium at any time during the diurnal cycle. For definition of the δ notation see Fig. 1. (Yakir, Ting and DeNiro, unpubl.)

The simplest possibility to explain isotopic nonhomogeneity in leaf water proposes that total leaf water consists of two pools. One constitutes tissue water that becomes enriched in $^{18}O$ and deuterium, and the second is composed of unaltered vein water (Leaney et al. 1985; Fig. 3A). Mixing the two water fractions during extraction should yield, in accordance with observations, a lower D/H and $^{18}O/^{16}O$ isotopic ratios than expected. However, for this possibility to reconcile the differences between observed and measured isotopic ratios, one must assume that the fraction of vein water is in the order of 25–50% of total leaf water (Leaney et al. 1985; Walker et al. 1989). Very little information is available on the actual proportion of vein water in leaves. Emmerst (1961) has estimated that xylem water in stems and petioles of spinach plants was about 2% of total tissue water. A detailed anatomical analysis in barley leaves was carried out by Rayan and Matsuda (1988) that led to an estimate of 0.8% xylem water in the leaf blade. We also estimated (Yakir et al. 1989) the proportion of vein and cell wall water in leaves of sunflower and ivy to be 3–5%, based on an isotopic mass balance calculation. A wide range of studies employing pressure-volume analyses indicate that total apoplastic water (veins and cell walls combined) in crop plants is in the order of 5–30% of total leaf water (Tyree and Jarvis 1982). However, these results should be considered together with those of Boyer (1967), who showed that cell-wall water alone may constitute 9–28% of total leaf water in several crop plants. Taken together, these studies indicate that only a small proportion of total leaf water is contained in the veins.

Another two-pool model was developed independently by White (1988). This author speculated that part of the large discrepancy between the model predictions and the measured values may arise from the existence of an intracellular "crystalline" water fraction that does not readily interact with the rest of the tissue water, although it may be affected isotopically by cell metabolism.

Walker et al. (1989), in supporting, and extending, the two-pool hypothesis of Leaney et al. (1985), also realized that "the physiological interpretation of the two pool model has yet to be resolved". They go on to speculate that, in fact, the unaltered water fraction could be associated with cellular water, rather than that in

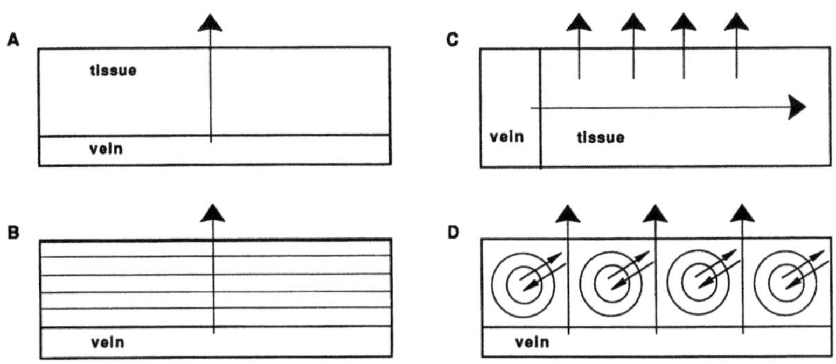


**Fig. 3A-D.** Schematic representation of the various models proposed to explain isotopic nonhomogeneity of leaf water as discussed in the text. *Large arrows* indicate the direction of the transpiration stream. *Small arrows* indicate possible slow exchange

the veins. It should be mentioned that both Walker et al. (1989) and White (1988) recognized that the discrepancies between the model-predicted and the observed data cannot be reconciled by simply adjusting variables such as leaf water content, boundary layer characteristics, etc. (see also Flanagan and Ehrlinger 1991) that are used as inputs in the numerical models.

A somewhat different approach to leaf water isotopic composition was offered by Farquhar (unpublished; see White 1988). This model relates to leaf water, or part of it, as an aqueous body supplied with source water on one side which evaporates on the other (Fig. 3B). In this model the isotopic enrichment taking place at the surface decays exponentially as it diffuses back toward the source water. Obviously, the measured isotopic composition of this water body will be an averaged intermediate. This model, however, has a number of difficulties. First, in most leaves the fine mesh of veins (Esau 1965), leads to very short water pathways to reach the evaporating site. Rayan and Matsuda (1988) show that any xylem element in the leaf blade of barley is only three cells away from the nearest stoma. The type of changes in isotopic composition suggested by the Farquhar model occur only at a microscopic scale. In fact, since the numerical model includes a parameter describing the average distance between the source and the surface, one can obtain an estimate of it by fitting the model with experimental data. By doing so, one obtains distances of many millimeters, representing values that are much greater than anatomical realities. The fit could be improved if the water pathway was restricted to narrow passages with much higher flux rates than indicated by average leaf transpiration rate. For example, if most of the transpiration flux was restricted to a narrow passage between a xylem element and the substomatal cavity. Even accepting such restrictions, the model applies to only part of leaf water, and some other restrictions to water mixing within the tissue must exist to maintain the isotopic nonhomogeneity.

In a system such as portrayed by the above model, a proportional decay in both the $^{18}O$ and deuterium enrichment in leaf water would be expected. Thus, plots of the isotopic composition of oxygen vs that of hydrogen, for various fractions of leaf water, should yield a linear relationship that extrapolates to that of the source water. In at least two cases such as linear relationship was not observed. First, in the leaf pressurization experiments in which successive fractions of water were gradually forced out of the leaf (Yakir et al. 1989), a clear lack of linearity, as compared with the tight relationships always observed in open water, was apparent (Fig. 4). Second, it was shown that source water, total leaf water and the predicted steady-state value corresponding to surface water in the Farquhar model, as well as estimates corresponding to the intermediate layers, do not lie on the same line when plotted in D/H vs. $^{18}O/^{16}O$ space (Yakir et al. 1990). The importance of such multi-isotope analysis is obvious and more study is clearly needed to verify the question of linearity between the oxygen and hydrogen isotopic composition in leaf water.

In a variation of the above model suggested by J.A. Berry, it is proposed that water entering leaf tissue at the base will be continuously subjected to evaporation, and therefore enrichment, as it moves to the top of the leaf (Fig. 3C). In such a system an imaginary series of pools is formed of which only the first is directly fed by the source water and only this pool would be expected to behave according to the Craig and Gordon model. The rest of the "pools" in the chain will be fed by ever more

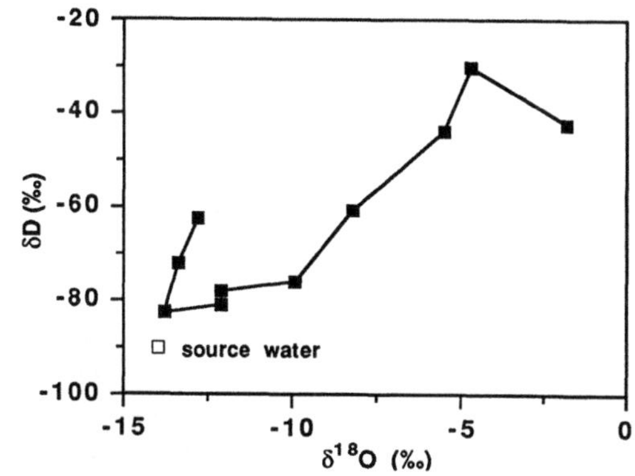


**Fig. 4.** Nonlinearity in the relationship between hydrogen and oxygen isotopic composition of leaf water. The oxygen and hydrogen isotopic composition was determined for successive fractions of water gradually forced out through the petiole of an ivy leaf enclosed in a pressure chamber with its petiole extending out. The isotopic composition of the irrigation (source) water is indicated. For definition of the δ notation see Fig. 1 and for details of the experimental approach see Yakir et al. (1989)

enriched water and will therefore reach an increasingly higher level of steady-state isotopic enrichment. Clearly, it can be intuitively expected, as well as calculated, that total leaf water in such a leaf would have an isotopic composition significantly more enriched than predicted by the Craig and Gordon model. However, this prediction is in sharp contrast with observations that isotopic composition of leaf water is consistently less enriched than predicted by the model (White 1988; Walker et al. 1989; Yakir et al. 1990). Furthermore, the difficulties discussed above with respect to anatomical realities and linearity of the $^{18}O/^{16}O$ vs. D/H line also apply here.

The above model may be applied also in another form. That is, if the water in the minor veins has a significant degree of lateral exchange with tissue water (e.g., Van Bel 1976; Cataldo et al. 1972; Biddulph and Cory 1957), then vein water itself will become slightly, but progressively, more enriched as it moves along the leaf. It is possible that this is the situation reflected in the isotopic composition of a corn leaf (Fig. 5 insert). However, it seems that although this effect may occur in some leaves with a large number of extremely long, parallel veins with a few cells in between, the effect is relatively small when compared with the large overall enrichment observed in the leaf above that of stem water (Fig. 5). It is also important to remember that in most dicot leaves the situation is markedly different from that in a corn leaf.

Another variation of the hypotheses discussed above suggests that at least three distinct water compartments should be recognized in leaf water (Yakir et al. 1989; 1990). These are, water in the veins with isotopic composition similar to that of the source, intercellular and cell-wall water that is directly exposed to evaporation and attains the isotopic composition predicted by the conventional model, and a large intracellular water pool that interacts relatively slowly with other pools (Fig. 3D). As mentioned above, several authors speculated that the answer to the persistent discrepancy between predicted and measured isotopic values may be found in some type of water compartmentation at the cellular level. The models, nevertheless, invariably treat tissue water as a single parcel of water, though a nonhomogeneous one.

The version of the model given in Fig. 3D identifies intracellular water as a specific isotopic entity and argues that restricted mixing between the water compartments is the basis for their persistence. The primary objections to this model were

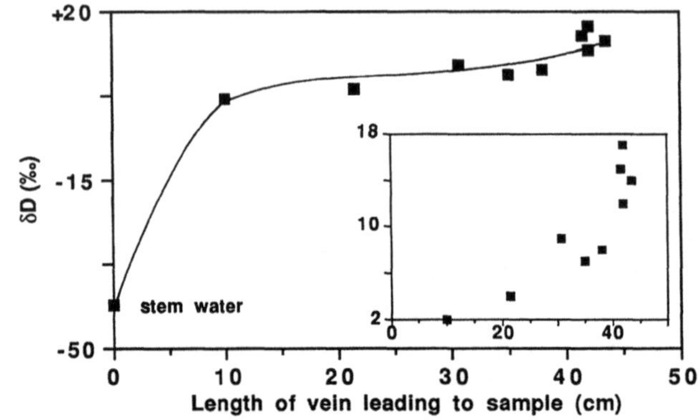


**Fig. 5.** Changes in the hydrogen isotopic composition of leaf water of a corn plant as a function of the distance of water movement in the veins leading to the sample. Leaf tissue samples ( 5 × 1 cm) were punched out of the leaf and the vein leading to the sampled area was traced down the leaf to the point of convergence with the main midrib vein. Water samples (3 µl) for isotopic analysis were obtained by crushing the samples in a vice. *Insert* shows an expansion of the upper portion of the main graph. For definition of the δ notation, see Fig. 1. (Yakir, Berry and Osmond unpubl.)

stated earlier in the introduction. Nevertheless, it seems that recent experiments such as those reported below, as well as evidence in the literature that goes back at least 25 years, support the probable existence of water compartmentation at the cellular level.

In a recent experiment we took advantage of the fact that oxygen in $CO_2$ and that of water come to almost instantaneous isotopic equilibrium in the presence of the enzyme carbonic anhydrase (the spontaneous hydration of $CO_2$ is several orders of magnitude slower.) It was possible to demonstrate that respired $CO_2$ that flows out of a sunflower leaf is fully equilibrated with cellular water (where the enzyme resides). In addition, the isotopic relationship between $CO_2$ and water at equilibrium is precisely known (Brenninkmeijer et al. 1983). Therefore, by collecting $CO_2$ respired into $CO_2$-free air in a leaf chamber we obtained an isotopic signal, $^{18}O/^{16}O$, that is easily translated into the $^{18}O/^{16}O$ ratio of the water in the metabolic compartment in which the carbonic anhydrase-catalyzed equilibrium took place. By monitoring the conditions in the leaf chamber and sampling the input and transpired water, it was also possible to obtain a precise estimate of the expected isotopic ratio at steady state based on the conventional model (Craig and Gordon 1965). In addition, the establishment of an isotopic steady state at the time of measurements was confirmed by demonstrating that the isotopic compositions of the transpiration and the source waters were identical (Fig. 1). At the end of each experiment, each of which was conducted at different air humidity, total leaf water was vacuum distilled from the leaf and the isotopic composition of total leaf water was determined.

The results (Fig. 6) clearly show that the intracellular water pool had a distinct $^{18}O/^{16}O$ ratio that was considerably less enriched in $^{18}O$ than total leaf water. Clearly, total leaf water must be a mixture of this "intracellular pool" and another, significant-

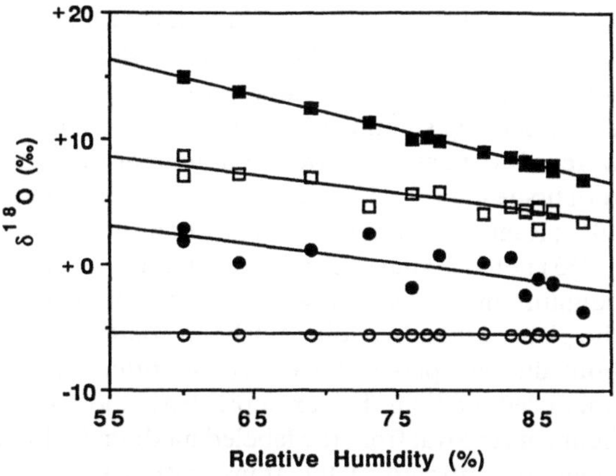


**Fig. 6.** Water compartmentation in a sunflower leaf. An attached sunflower leaf was enclosed in a gas exchange chamber at each of the indicated relative humidities and constant conditions were maintained until isotopic steady state was attained (as shown in Fig. 1). The chamber was then flushed with $CO_2$ free air and the respired $CO_2$ was trapped and analyzed for its oxygen isotopic composition. This isotopic composition reflects that of the intracellular water pool in which carbonic anhydrase catalyzes the extremely rapid hydration of $CO_2$ (*closed circles*). Total leaf water (*open squares*) and stem water (*open circles*) was then extracted for isotopic analysis. The isotopic composition of the water at the evaporating surfaces (*closed squares*) was calculated using the conventional model of Craig and Gordon (1965). For definition of the δ notation see Fig. 1. (Yakir, Berry and Osmond, unpubl.)

ly more enriched, water fraction. It was assumed that this enriched water pool most likely corresponds to the water at or near the evaporating surfaces that directly interact with the ambient air and must, therefore, behave as predicted by the conventional model. Taking the fraction of vein water to be 2% and using simple mass balance calculations one can obtain a first estimate of the size of the intracellular compartment of ca. 65% of total leaf water. This estimate is in agreement with a previous estimate using a completely different approach (Yakir et al. 1989) as well as other estimates in the literature (e.g., 60–90%, Tyree and Jarvis 1982). It should be noted also that, as would be expected, the distinct inner compartment was less responsive to changes in ambient relative humidity than predicted (and tested, Allison and Leaney 1982) for surface water. Thus, these results not only support the idea of water compartmentation but also help in identifying the major component as the "metabolic" pool.

## Supporting Evidence for Water Compartmentation at the Cellular Level

Evidence in support of water compartmentation at the cellular level comes mostly from tracer (tritiated water) and NMR studies. The former evidence dates as far back as the early 1950s; the latter methods are only now beginning to realize their potential.

Cline (1953), in a noteworthy study with bean plants, demonstrated that tissue water in plants fed with labeled water did not reach equilibrium with the nutrient solution even after three days (maximum specific activity of tissue water ca 60% of that in feeding solution). This was true for either stem or leaves and similar results were subsequently reported in a number of studies for leaves and stems as well as root tissue immersed in the feeding solution (Biddulph et al. 1961; Ordin and Gairon 1961; Raney and Vaadia 1965a,b; Shone et al. 1980; Rayan and Matsuda 1988). Cline (1953) noted, as have all subsequent authors mentioned above, that this incomplete equilibrium is consistent with restricted mixing between different water fractions in the plant tissues. In addition, labeling patterns shown in these studies were biphasic, both during uptake and efflux measurements after plants were moved back to unlabeled medium. For example, leaves lost some 90% of their label within a few hours of removal from the labeled medium (Cline 1953), as would be expected if the label concentrated in the transpiration stream. In contrast, the remaining label persisted until the end of the experiment 2 days later, indicative of another, only slightly labeled, water fraction that was not in good communication with the rapidly washed out transpiration stream.

Another intriguing aspect of the results reported by Cline (1953) is that organic matter in the leaves was only slightly labeled within the first 24 h, with no appreciable change for the remaining 46 h of the experiment. Since tissue water is the only source for hydrogen in autotrophically produced organic matter, this again indicated that labeled water did not participate in, or simply was not available for, leaf metabolism, even though the experimental conditions were favorable for photosynthesis. This notion was reinforced by the results of another experiment in the same study in which a leaf was labeled by exposing it to tritium-labeled air. After 8 h in the light, very high label concentration was detected in leaf water but only traces in its organic matter.

Interestingly, almost 30 years after this study, another study using stable isotope ratio of hydrogen in microalgae (Estep and Hoering 1981) provided strikingly similar results. It was shown that microalgae grown in medium with constant D/H maintained a constant D/H value in their organic matter. The medium was then enriched with deuterium, but only an extremely slow response could be detected during the next 3 days in the D/H of the algal organic matter. This was in spite of the observation that dry weight tripled about every 24 h. Obviously, complications due to transpiration and water transport, such as encountered in a study with whole plants, are avoided here and the results could be interpreted as restricted communication between extracellular water and water at the metabolic sites.

Two important points relevant to the discussion above were addressed by Vartapetyan and Kursanov (1959). First, they moved plants into nutrient solutions containing water labeled with $^{18}O$, rather than tritium or deuterium, to test for possible artifacts associated with the use of hydrogen isotopes. As in the studies cited above, plant tissue water reached a maximum of only about 60% of the specific activity in the nutrient solution. Second, plants were also grown in $H_2{}^{18}O$ from seed, to test for any major isotope effects. In this case, label concentration reached almost complete equality with the nutrient solution.

Other studies using tritiated water in barley leaves (Matsuda and Riazi 1981; Rayan and Matsuda 1988) showed that when the tracer is added to the root medium it first appears in the transpiration water and only much later (more than 3 h in one treatment) in the leaf blade tissue water. Moreover, these studies also indicate that label concentration is significantly greater in water extracted from cell walls than it is in water extracted from the rest of the tissue. In addition, these studies show that mild water stress induced a reduction in water potential in the xylem while no change in water status of the mesophyll was observed. These results led the authors to conclude that water in leaf mesophyll cells does not readily exchange with water in the "transpiration stream".

The idea mentioned above that a water-potential gradient reflects water compartmentation (Rayan and Matsuda 1988; see also Rygol et al. 1989) is supported by another study (Roy and Berger 1983) in which the authors measured, by using a number of instruments (i.e., pressure chamber, dew point hygrometer, and in-situ hygrometer), different water potentials on the same tissue of a transpiring leaf. It was argued that the various instruments reflected the water potential in the xylem elements, in the vacuole and in the evaporating sites, respectively, and it must therefore be concluded that these compartments are separated by some resistance to water flow.

The pulse NMR method is based on the relaxation response of protons in a strong magnetic field after absorbing a pulse of energy. In one such study (Shibata and Iwayanagi 1989) the relaxation response in a spinach leaf was found to change sigmoidally with the time interval between pulses. It was argued that a slow water exchange between two sites must be responsible for this observation. The rate of this exchange, which is evident only in intact leaves, is some three orders of magnitude slower than that, for example, between water bound to protein and the bulk free water. A three water-species model was suggested by these authors consisting of extracellular water, cytoplasmic and vesicular water compartments. It is important to note that molecular mobility at the NMR scale and diffusion accessibility have only rarely been related. However, at least in one case, where NMR relaxation times were compared with rates of replacement of water with $D_2O$ (Katona and Vasilescu 1989), a direct correlation has been observed.

A major advance in techniques is being made with the introduction of imaging-NMR with continuously increasing resolution (see Brown et al. 1988 for an introduction to the field). In a recent study employing this method Walter et al. (1989) found variations in the amounts of "free" and "bound" water (see Kramer 1955 for definitions) in different types of mesophyll cells in the leaves of a CAM plant. The authors observed that the variation in "bound" water was associated with asymmetric water movement in the leaf and it was suggested that water movement through cells that strongly "bind" water was hindered. As will be discussed below, these results and their interpretations introduce a new perspective to the idea of water compartmentation. It is important to reiterate, however, that the existence of "bound" water in itself does not represent compartmentation, in the sense used here.

## Some Explorations of the Possible Biophysical Basis for Water Compartmentation

Since most of the evidence for water compartmentation involves isotopes, it is appropriate to ask whether any isotope effects are associated with water movement, particularly its movement through biological membranes. Unfortunately, very little information on this topic is available. Mills (1973) determined the self diffusion coefficients of different isotopic species of water, i.e., HTO, DTO, HDO, using either normal or heavy water as the solvent and found variations that cannot be explained by considering only differences in mass. Nonetheless, the differences in self diffusion coefficients were rather small (2.236, 2.272 and 2.299 × $10^{-5}$ $cm^2$ $s^{-1}$ for HTO, HDO, and HHO, respectively, in normal water at 25 °C) and no evidence for its physiological significance was provided. Wang et al. (1954) obtained somewhat different values for self-diffusion coefficients (2.34, 2.44, and 2.66 × $10^{-5}$ $cm^2$ $s^{-1}$ for HDO, HTO, and $H_2{}^{18}O$, respectively, in normal water), again indicating that more than relative mass is involved in the isotope effect, but that the effects are small.

Samuilov and Nikiforova (1974) have demonstrated that the rate of water absorption by plants decreases with increasing concentrations (from 0 to 96%) of $D_2O$ in the medium. Other workers also observed that deuterated water penetrated cells more slowly than ordinary water (Lawaczeck 1984; Owen et al. 1970). The effect of deuterium substitution in water can be substantial, i.e., more than 30% reduction in rate of water transport (Owen et al. 1970), but only at very high deuterium concentrations and it is not clear to what extent, if at all, this applies to natural abundance levels.

Allison et al. (1985) examined isotopic fractionation, at the natural abundance level, during evaporation of water through various types of artificial membranes and found practically no effect. Compressing water through a colloidal system, on the other hand, resulted in a small isotopic fractionation for both oxygen and hydrogen which was observed at the natural abundance level (Coplen and Hanshaw 1973), but these results were obtained under extremely high pressure. Crocker and Deniro (1985) reported that water of *Nautilus* eggs had an oxygen isotopic ratio consistently different from that of the seawater in which they were immersed, and speculated that these differences were due to membrane discrimination during water uptake. Isotopic discrimination by biological membranes was observed during water uptake by mangroves trees growing in sea water (Sternberg, pers. commun.).

One theory of water transport through biological membranes provides a possible mechanism for membrane-associated isotope effects. According to the hypothesis of "single-file transport" (Levitt 1974; Dani and Levitt 1981; Stein 1986) the bulk of water transport through a biological membrane may be accomplished through channels with a diameter so small that water molecules can not pass each other. A labeled water molecule, therefore, will traverse the channel only if all the water molecules lying in front of it in the path of diffusion (6–9 molecules) leave the channel first. Thus, if isotopic enrichment is induced in the water on one side of the membrane, water on the other side of the membrane will remain depleted for a much longer time than expected. This is because the diffusion of the labeled water molecules through the channels in the membrane will be slowed down by the movement of the nonlabeled molecules for which no gradient was induced. In fact, Stein (1986) estimated that diffusion permeability coefficients for water transport through membranes are likely to be underestimated by a factor of about six (the number of water molecules that form a "file" in the channel) when determined with labeled water.

Such membrane effects could be a significant factor under natural conditions where changes in "label" concentration are relatively rapid (e.g., during the diel cycle in leaf water isotopic composition). Changes in isotopic composition of an intracellular compartment are likely to lag behind changes that are induced in an extracellular water fraction that is directly exposed to ambient conditions. However, this membrane effect is probably not sufficient to account for the long-term (days) steady-state isotopic disequilibrium between medium and tissue water as found in the tracer experiments discussed above. Clearly, addressing the question of possible isotope effects associated with biological membranes by direct experimental methods is absolutely essential before a coherent picture of the isotopic patterns in tissue water can be obtained.

There are other possible isotope effects that have not yet been considered in a biological context. For example, it is known from geochemical and hydrological studies that solute concentration can significantly alter the equilibrium isotopic fractionation factor between water in solution and water vapor (for reviews see Kyser 1987; Hoefs 1987). Strong dependence on the type of solute as well as differential effects on oxygen or hydrogen isotopes in water have been observed. Practically all of the work in this area has considered only the effects of salts or total dissolved matter, and there has been no attempt to look for similar effects associated with sugars or proteins. Also, geochemists have demonstrated that large isotope effect are associated with the partitioning of water between hydration water and bulk water in minerals, soils, and other geochemical systems at physiological temperatures (Kyser 1987; Hoefs 1987). Hydration water can comprise a large fraction of cellular water in any biological system but the question of isotope effects during water exchange with this fraction have not been considered. Moreover, it seems possible that the isotope effects associated with coexisting phases, such as mentioned above, can also take place between any two liquid phases, or compartments, in the cytoplasm if solute concentrations, density, water activity, etc. are distinct. In other words, there may be isotope effects in water that are associated with solute compartmentation.

The isotope effects discussed above do not necessarily involve any restriction to net water movement. In fact, there is a growing contention that the cellular enrivonment in plants, including the cell membranes, do not pose a major barrier to water movement (Boyer 1985; Cosgrove 1986). This question is important in determining the water pathway in the soil-plant-atmosphere continuum. Whether or not the large transpirational water flux will bypass the cytoplasm, or be restricted mostly to the low-resistance cell walls is determined by the symplastic resistance. In fact, it is quite possible that the pathways in a leaf may change with changing transpirational demands (Nobel 1988; Vogt et al. 1983; Proctor 1982). Even if the major water flow is restricted to the cell wall pathway, a very rapid lateral exchange between water on the two sides of the cytoplasmic membrane still needs to be considered (for some information regarding lateral water exchange in plants see: Biddulph and Cory 1957; Biddulph et al. 1961; Cataldo et al. 1972; Van Bel 1976). In most of the discussions concerning water pathway in plants the intracellular environment is treated as a homogeneous aqueous solution. This is, of course, not the case (see, for example, Hand, this Vol.) and it is conceivable, as will be discussed below, that water compartmentation may arise from restriction to water flow by parts of the intracellular environment, regardless of the effects of the cytoplasmic membrane.

When the question of intracellular water mobility is raised it is normally coupled with the phenomenon of "bound" water (Kramer 1955; Cameron et al. 1988; Vertucci and Leopold 1984; Rupley et al. 1983; Walter et al. 1989; Katona and Vasilescu 1989). Although the term "bound" is misleading and such water molecules have a very short life time, it may form some type of boundary layer and affect the flux of water through it. For example, Walter et al. (1989) reported that in leaves of a CAM plant, "pockets" of mesophyll cells show a higher proportion of "bound" water as compared with cells in the surrounding tissue. Furthermore, this coincided with observations of asymmetric water flow within the tissue that seems to avoid passage through these "pockets". The authors concluded that "... because the presence of mesophyll cells which strongly bind water, ... water transport may be hindered through that tissue". Unfortunately, the resolution of the immaging-NMR technique employed in this study is still limited. It is possible, however, that these, or similar, "pockets" and therefore restrictions to water flow occur at the subcellular level.

Interestingly, Dick (1966) observed a linear relationship between a cell's apparent permeability and its surface to volume ratio. He then rejected the possibility that membrane permeability is dramatically different among the cells (which have been shown to be physically similar) and concluded that this relationship must mean that "water transfer ... is not being solely restricted to membrane resistance but that the diffusion coefficient of water in the cell interior is sufficiently low to have a significant effect in retarding water flow." It should be emphasized that such retardation can be more severe in parts of the cells since the observed effect, as is the case for most observations cited above, is only a macroscopic average.

These observations support the idea that water flux through a cell can encounter boundary-layer-type resistance and may even bypass a large portion of the cytoplasmic solution if it contains high proportions of "bound" water. The cytoplasm in most cells is dense with proteins, organelles and other structures which "bind" or otherwise affect water mobility (Clegg 1979; Dorst-Hansen 1971; Peschel 1976). Indeed, a reduction in intracellular water self diffusion by a factor of 2 to 10 can be observed (Dorst-Hansen 1971; Leyton 1975; Wang 1954; Blum et al. 1986; Dick 1966; Tanner 1983; Ordin and Gairon 1961; Beall et al. 1979; Beall et al. 1982). Leopold (see elsewhere in this Vol.) envisages the formation of vitrified water, with viscosity some 40 times that of pure water, in dehydrated seeds. Whether "pockets" of vitrified water are preserved in fully hydrated cells has not yet been considered. It should perhaps be noted that movement of many cell components in the gelatinous cytoplasm is only maintained by active cytoplasmic streaming. This streaming, however, may be regarded more as a moving conveyor belt than as a mixing process (Hayashi 1961). Plant cell organelles, especially the chloroplasts and mitochondria of leaves, contain extraordinarily high concentrations of proteins such as ribulose-1, 5-biphosphate carboxylase and glycine decarboxylase, which could significantly modify water activity. If this protein environment significantly influence water movement or modified isotopic exchange of $H_2O$ species, then some of the isotopic fractionation of photosynthesis (e.g., about –150‰ for deuterium; Yakir et al. 1989) may be due to these physical, rather than biochemical, factors.

The definition of "bound" water is continuously changing and recent definitions consider "bound" water to be a part a larger fraction of "hydration water." For example, Cameron et al. (1988) recently defined hydration water as "all water molecules for which motion is perturbed from that of bulk water". By this definition hydration water includes "bound" water that is defined as "water molecules which are bound by two or three hydrogen bonds to fixed polar sites and those bound to ionic sites on the protein." Water that is not covered by the above definitions is termed "bulk" water. It is possible that cells, and certainly some subcellular compartments, do not have "bulk" water at all (Ling 1988; Cameron et al. 1988; Clegg 1979). Other studies have also shown that normal metabolic activity does not require the presence of "bulk" water (Clegg 1978; Vertucci and Leopold 1987; Ling and Walton 1976; see also Srivastava and Bernhard 1987).

## Concluding Remarks

The isotopic evidence for water compartmentation in plants is strong, but we have only the most crudely developed explanations of these phenomena. A large body of experimental evidence indicates that the patterns of water flow and mixing in plant cells and tissues do not obey the expected rules. The fact that the evidence comes almost exclusively from isotope studies is perhaps unfortunate, but most likely unavoidable. Under these circumstances, much further work is needed to elucidate many possible isotope effects. Understanding such effects can help not only in clearing up the "noise" in our observations, but also in developing new tools for the study of water activity at the cellular level and its interactions with the external environment.

Plants, even under the best conditions, are continuously exposed to environmental changes that strongly influence their water status. The dynamics of water movement at the tissue and cellular level is clearly a fundamental factor in understanding how favorable conditions are constantly maintained at the metabolic sites. It seems that in this fundamental area we have just noticed that, perhaps, a new set of rules may have to be developed.

*Acknowledgments.* Some of the results reported in this paper are from a collaborative work with C.B. Osmond and J.A. Berry supported by DE-FG05-89ER14005. I thank Drs. W.G. Eickmeier and C.B. Osmond for critically reading the manuscript, and Dr. Ld.S.L. Sternberg for providing some of the references.

## References

Allison GB, Leaney FW (1982) Estimation of isotopic exchanged parameters, using constant-feed pans. J Hydrol 55:151–161

Allison GB, Gat JR, Leaney FW (1985) The relationship between deuterium and oxygen-18 delta values in leaf water. Chem Geol (Isotope Geosci Sec) 58:145–156

Beall PT, Chang CD, Hazlewood CF (1979) Temperature dependent shifts in microtubule structure and the diffusion coefficient of water in cells. J Cell Biol 83:332a

Beall PT, Hazlewood CF, Chang DC (1982) Microtubule organization and the self-diffusion coefficient of water in baby hamster kidney cells as a function of temperature. J Cell Biol 95:334a
Biddulph O, Cory R (1957) An analysis of translocation in the phloem of the bean plant using THO, $^{32}P$ and $^{14}C$. Plant Physiol 32:608–619
Biddulph O, Nakayama FS, Cory R (1961) Transpiration stream and ascension of calcium. Plant Physiol 36:429–436
Blum FD, Pickup S, Foster KR (1986) Solvent self-diffusion in polymer solutions. J Colloid Interface Sci 113:336–361
Boyer JS (1967) Matric potentials of leaves. Plant Physiol 42:213–217
Boyer JS (1985) Water transport. Annu Rev Plant Physiol 36:437–516
Brenninkmeijer CAM, Kraft P, Mook WG (1983) Oxygen isotope fractionation between $CO_2$ and $H_2O$. Isotope Geosci 1:181–190
Brown JM, Thomas JF, Cofer GP, Johnson GA (1988) Magnetic resonance microscopy of stem tissue of *Oekagronium hortum*. Bot Gaz 149:253–259
Cameron IL, Hunter KE, Fullerton GD (1988) Quench cooled ice crystal imprint size: a micromethod for study of macromolecular hydration. Scanning Microsc 2:885–898
Cataldo DA, Christy AL, Carlson CL (1972) Solution-flow in the phloem. II Phloem transport THO in *Beta vulgaris*. Plant Physiol 49:690–695
Clegg JS (1978) Interrelationships between water and cellular metabolism in *Artemia* cysts. J Cell Physiol 94:123–138
Clegg JS (1979) Metabolism and the intracellular environment: The vicinal-water network model. In: Dorst-Hansen W, Clegg JS (eds) Cell-associated water. Academic Press, Lond New York
Cline JF (1953) Absorption and metabolism of tritium oxide and tritium gas by bean plants. Plant Physiol 28:717–723
Coplen TB, Hanshaw BB (1973) Ultrafiltration by a compacted clay membrane. I. Oxygen and hydrogen isotopic fractionation. Geochim Cosmochim Acta 37:2295–2310
Cosgrove D (1986) Biophysical control of plant cell growth. Annu Rev Plant Physiol 37:377–405
Craig H, Gordon L (1965) Deuterium and oxygen-18 variation in the ocean and marine atmosphere. Proceedings of conference on stable isotopes in oceanography studies and paleotemperatures. Lab Geol Nucl Sci, Pisa, pp 9–130
Crocker KC, DeNiro MJ (1985) Stable isotopic investigations of early development in extent and fossil chambered cepholopods I. Oxygen isotopic composition of eggwater and carbon isotopic composition of siphuncle organic matter in *Naitulus*. Geochim Cosmochim Acta 49:2527–2532
Dani JA, Levitt DG (1981) Water transport and ion-water interaction in the gramicidin channel. Biophys J 35:501–508
Dick DAT (1966) Cell water. Butterworth, Washington
Dongmann G, Nurnberg HW, Förstel H, Wagener K (1974) On the enrichment of $H_2{}^{18}O$ in leaves of transpiring plants. Rad Environ Biophys 11:41–52
Dorst-Hanson W (1971) Structure and properties of water at biological interfaces. In: Brown HD (ed) Chemistry of the cell interface part B. Academic Press, Lond New York, pp 1–184
Emmerst FH (1961) Volume determination of xylem conducts in stem and petioles of *Phaseolus vulgaris* using radioactive phosphorus. Physiol Plant 14:470–477
Esau K (1965) Plant anatomy, 2nd edn. Wiley, New York Lond
Estep MF, Hoering TC (1981) Stable hydrogen isotope fractionation during autotrophic and mixotrophic growth of microalgae. Plant Physiol 67:474–477
Farris F, Strain BR (1978) The effects of water stress on leaf $H_2{}^{18}O$ enrichment. Rad Environ Biophys 15:167–202
Flanagan LB, Ehleringer JR (1991) Stable isotope composition of stem and leaf water: applications to the study of plant water-use. Funct Ecol (in press)
Förstel H (1978) The enrichment of $^{18}O$ in leaf water under natural conditions. Rad Environ Biophys 15:323–344
Gonfiantini R, Gratsin S, Tonqiori E (1965) Oxygen isotopic composition of water in leaves. In: Isotopes and radiation in soil. Plant Nutr Stud. Int Atomic Energy Agency, Vienna, pp 405–410
Hayashi T (1961) How cells move. Sci Am 205:184–204

Herbst MD, Goldstein JH (1989) A review of water diffusion measurement by NMR in human red blood cells. Am J Physiol 256:c1097–c1104
Hoefs J (1987) Stable isotope geochemistry. Springer, Berlin Heidelberg New York
Katona E, Vasilescu V (1989) Dynamics and compartmentation of water in certain biosystems. Rev Roum Morphol Embriol Physiol 26:285–296
Kramer PJ (1955) Bound water. In: Ruhland W (ed) Encyclopedia of plant physiology, vol I. Springer, Berlin Heidelberg New York, pp 223–242
Kyser TK (ed) (1987) Short course in stable isotope geochemistry of low temperature fluids, vol 13. Mineral Assoc Can, Ontario
Lawaczeck R (1984) Water permeability through biological membranes of isotopic effects of fluorescence and light scattering. Biophys J 45:491–496
Leaney FW, Osmond CB, Allison GB, Ziegler H (1985) Hydrogen-isotope composition of leaf water in $C_3$ and $C_4$ plants: its relationship to the hydrogen-isotope composition of dry matter. Planta 164:215–220
Levit DG (1974) A new theory of transport for cell membrane pores I. General theory and application to red cell. Biochem Biophys Acta 373:115–131
Leyton L (1975) Fluid behavior in biological systems. Clarendon Press, Oxford
Ling GN (1988) Solute exclusion by polymer and protein-dominated water: correlation with results of nuclear magnetic resonance (NMR) and calorimetric studies and their significance for the understanding of the physical state of water in living cells. Scanning Microsc 2:871–884
Ling GN, Walton CL (1976) What retains water in living cells? Science 191:293–295
Matsuda K, Riazi A (1981) Stress-induced osmotic adjustment in growing regions of barley leaves. Plant Physiol 68:571–576
Mills R (1973) Self-diffusion in normal and heavy water in the range 1–45°. J Phys Chem 77:685–688
Nier AO (1947) A mass spectrometer for isotope and gas analysis. Rev Sci Instrum 18:398–411
Nobel PS (1988) Environmental biology of agaves and cacti. Cambridge Univ Press, Cambridge
Ordin L, Gairon S (1961) Diffusion of tritiated water into roots as influenced by water status of tissue. Plant Physiol 36:331–335
Owen JD, Bennion BC, Holmes LP, Eyring EM, Berg MW, Lords JL (1970) Temperature jump relaxation in aqueous saline suspensions of human erythrocytes. Biochem Biophys Acta 203:77–82
Peschel G (1976) The structure of water in the biological cell. In: Lange OL, Kappen L, Schulze ED (eds) Water and plant life. Problems and modern approaches. Ecological studies, vol 19. Springer, Berlin Heidelberg New York, pp 6–18
Proctor NCF (1982) Physiological Ecology: water relations, light and temperature responses, carbon balance. In: Smith AJE (ed) Bryophyte ecology. Chapman and Hall, Lond New York, pp 333–381
Raney F, Vaasia Y (1965a) Movement and distribution of THO in tissue water and vapor transpired by shoots of *Helianthus* and *Nicatrana*. Plant Physiol 40:383–388
Raney F, Vaadia Y (1965b) Movement of tritiated water in the root system of *Helianthus annus* in the presence and absence of transpiration. Plant Physiol 40:378–382
Rayan A, Matsuda K (1988) The relation of anatomy to water movement and cellular response in young barley leaves. Plant Physiol 87:853–858
Roy J, Berger A (1983) Water potential measurement, water compartmentation and water flow in *Dactylis Glomerata* L. leaves. New Phytol 93:43–52
Rupley JA, Gratton E, Careri G (1983) Water and globular proteins. Trends Biochem Sci 8:18–22
Rygol J, Zimmermann V, Balling A (1989) Water relations of individual leaf cells of *Mesembryanthemum crystallinum* plants grown at low and high salinity. J Membr Biol 107:203–212
Samuilov FD, Nikiforova VI (1974) The study of the state of water in plants using heavy water ($D_2O$) and the NMR method. Dokl Bot Sci 214:26–28
Shibata T, Iwayagani S (1989) Pulse proton magnetic resonance study of water exchange in spinach leaf. Sci Pap Inst Phys Chem Res 82:52–58
Shone MGT, Bartlett BO, Flood AV (1980) A model of diffusion and mass flow of water in cylindrical membrane systems with application to plant roots. J Membr Biol 53:171–177
Srivastava DK, Bernhard SA (1987) Biophysical chemistry of metabolic reaction sequences in concentrated enzyme solution and in the cell. Ann Rev Biophys Chem 16:175–204

Stein WD (1986) Transport and diffusion across cell membranes. Academic Press, Orlando London

Sternberg LDSL (1988) Oxygen and hydrogen isotope ratios in plant cellulose: mechanisms and applications. In: Rundel PW, Ehleringer JR, Nagy KA (eds) Stable isotope in ecological research. Ecological Studies, vol 68. Springer, Berlin Heidelberg New York, pp 106–141

Tait MJ, Frank F (1971) Water in biological system. Nature (Lond) 230:91–94

Tanner JE (1983) Intracellular diffusion of water. Arch Biochem Biophys 224:416–428

Tyree MT, Jarvis PG (1982) Water in tissue and cells. In: Lange OL, Nobel PS, Osmond CB, Ziegler H (eds) Encyclopedia of plant physiology, vol 12B. Springer, Berlin Heidelberg New York, pp 35–78

Vartapetyan BB, Kursanov AL (1959) A study of water metabolism of plants using water containing heavy oxygen, $H_2{}^{18}O$. Fiziol Rastennii 6:154–159

Van Bel AJE (1976) Different mass transfer rates of labeled sugars and tritiated water in xylem vessels and their dependency on metabolism. Planta Physiol 57:911–914

Vertucci CW, Leopold AC (1984) Bound water in soybean seed and its relation to respiration and imbibitional damage. Plant Physiol 25:114–117

Vertucci CW, Leopold AC (1987) Oxidative processes in soybean and pea seeds. Plant Physiol 84:1038–1043

Vogt E, Schönherr T, Schmidt HW (1983) Water permeability of periderm membranes isolated enzymatically from potato tubers (*Solanum tuberasum* L.) Planta 158:294–301

Walker CD, Leaney FW, Dighton JC, Allison GB (1989) The influence of transpiration on the equilibration of leaf water with atmospheric water vapor. Plant Cell Environ 12:221–234

Walter L, Balling A, Zimmermann V, Haase A, Kuhn W (1989) Nuclear magnetic resonance imaging of leaves of *Mesembryanthemum crystallinum* L. plants grown at high salinity. Planta 178:524–530

Wang JH (1954) Theory of the self-diffusion of water in protein solutions. A new method for studying the hydration and shape of protein molecules. J Am Chem Soc 76:4755–4763

Wang JH, Robinson CV, Edelman IS (1953) Self-diffusion and structure of liquid water III. Measurement of the self-diffusion of liquid water with $H^2$, $H^3$ and $O^{18}$ as tracers. J Am Chem Soc 75:466–470

Warshaw RL, Friedman I, Hellen SJ, Frank PA (1970) Hydrogen isotopic fractionation of water passing through trees. In: Hobson GD (ed) Advances in oceanic geochemistry. Pergamon Press, Oxford, pp 55–67

Washburn EW, Smith ER (1934) The isotopic fractionation of water by physiological processes. Science 79:188–189

White JWC (1988) Stable hydrogen isotope ratios in plants: a review of current theory and some potential applications. In: Runden PW, Ehleringer JR, Nagy KA (eds) Stable isotope in ecological research. Ecological Studies, vol 68. Springer, Berlin Heidelberg New York, pp 142–162

Yakir D, DeNiro MJ, Rundel PW (1989) Isotopic inhomogeneity of leaf water: evidence and implications for the use of isotopic signals transduced by plants. Geochim Cosmochim Acta 53:2769–2773

Yakir D, DeNiro MJ, Gat JR (1990) Natural deuterium and oxygen-18 enrichment in leaf water of cotton plants grown under wet and dry conditions: evidence for water compartmentation and its dynamics. Plant Cell Environ 13:49–56

Chapter 14

# Photosynthetic Water Oxidation and Water Stress in Plants

W.G. EICKMEIER,[1,2] J.G. LEBKUECHER[2], and C.B. OSMOND[1,3]

## Introduction

Water oxidation, or the photolysis of water in oxygenic photosynthesis, is the foundation of autotrophy. It is the source of $O_2$, which makes the Earth habitable for higher life forms. It is the source of electrons, which generate reductants and ATP, which, in turn, sustain all other biological activities. About two thirds of global water oxidation occurs in leaves of land plants and the effectiveness of water oxidation in leaves is governed by access to $CO_2$, the terminal electron acceptor of photosynthesis. Access to atmospheric $CO_2$ in most leaves is limited by stomata. These pores are variable valves which effect a fundamental compromise between $CO_2$ uptake and $H_2O$ loss, described by Stocker (1952) as the path plants must steer on their Homeric journey between the Scylla of death by hunger, and the Carybdis of death by thirst (see Steudle, this Vol.).

Water oxidation is a trivial component of total water flux, even in plants in which stomata exert a strong control over water loss. Stomatal closure to restrict water loss has to be most effective during the brightest period of the day, the period when the potential for water oxidation is greatest. This is also the period in which plants first experience water stress, in which they become desiccated most rapidly, and in which the potential for light-dependent damage, for photoinhibition, is greatest. Although some plants display water stress-dependent leaf movements which minimize light interception under stress (Ludlow and Björkman 1984), most land plants are unable to draw the blinds in bright light. We must conclude that the complex reactions of photosynthetic water oxidation are capable of down-regulation and maintenance in bright light in the face of decreasing plant water potential.

This chapter deals briefly with land plants in which stomatal control of water loss and $CO_2$ access is effective over a narrow range of leaf water potential (about –0.1 to –1.5 MPa) so that down-regulation of water oxidation occurs over a range of relatively high water activity. The chapter deals in more detail with xeric and desiccation-tolerant plants which are responsible for a trivial part of global water

[1] Botany Department, Duke University, Durham, NC 27706, USA
[2] Department of Biology, Vanderbilt University, Nashville, TN 37235, USA
[3]Present address: Research School of Biological Sciences, Australian National University, Canberra ACT 2601, Australia

Somero et al. (Eds.)
Water and Life

oxidation, yet in which the stability of the water oxidation apparatus is preserved, in a nonfunctional state, to very low water potentials (to –150 MPa or more).

## Structure and Function of the Water Oxidation Complex of Photosynthesis

The water-oxidizing, $O_2$-evolving complex of photosynthesis is an association of three major extrinsic polypeptides (33, 23 and 17 kDa) associated with the transmembrane core (D1 and D2 polypeptides) of photosystem II (PSII) on the inner, lumen side of stacked (granal) thylakoid membranes in the chloroplast. Much evidence indicates that the functional water oxidation system contains a polynuclear complex of four Mn atoms, and that $Cl^-$ and $Ca^{2+}$ ions also have important roles in water oxidation (Ghanotakis and Yocum 1990; Hansson and Wydrzynski 1990). The overall function of PSII can be summarized as the photoinduced production of electrons and protons from $H_2O$ in the lumen, and vectorial electron transfer to the stromal side of the thylakoid where quinone reduction initiates electron transport to photosystem I (PSI) and NADP reduction. Proton generation from $H_2O$ in the lumen also contributes about 50% of the transthylakoid pH gradient ($\Delta$ pH) to drive photophosphorylation of ADP.

The primary processes of photosynthesis involve the transfer of excitation absorbed by the chlorophyll protein complexes of the antennae in the thylakoid membranes to the special chlorophylls (P680) of the PSII reaction center. The single photon charge separation events which yield oxidized $P680^+$ in the PSII reaction center are individually inadequate to extract electrons from water. Water oxidation releases four electrons simultaneously so the biological catalyst of water oxidation must stabilize water molecules during a gradual oxidation process which extracts electrons one at a time. Using single-turnover flashes, a four-flash periodicity in the maximum flash yield of $O_2$, and a similar periodicity in flash induced proton yield, were found. The $O_2$ flash yield periodicity of four, was explained by Kok et al. (1970) in terms of a five-state water oxidation system. The so-called Kok-clock undergoes a transition in four steps from one state to another (to the next highest oxidation state), with each photon trapped by the conversion of P680 to $P680^+$. After the absorption of four photons by P680, the catalytic complex has released four electrons to the reaction center and is in a position to remove four electrons from water.

The states of the water-oxidation complex are currently thought to be discrete oxidation states of the Mn complex at the interface between the largest extrinsic polypeptide of the complex, and the D-1 and D-2 transmembrane polypeptides. A special tyrosine residue on the D-1 polypeptide is involved in electron transfer from state 4 of the manganese complex to P680 (Brudvig et al. 1989). The water-oxidation complex is thus a transthylakoid photocell connected to a manganese battery on the lumen side of the membrane.

In spite of enormous advances in this field, some of them recognized in the Nobel prize for chemistry in 1988, the molecular mechanisms are scarcely understood. Sauer (1990) notes "Information is currently lacking about mechanism by which water becomes involved. Specifically, it is not known when the oxygen atoms become oxidized, when the O-H bonds are broken, when the O-O bond is formed, and how the final oxidation to $O_2$ is accomplished. . . . The actual mechanism of water oxidation is even less well understood at present than is the involvement of Mn in the storage of oxidizing equivalents. . . ."

Whether particular states of water are involved in the environment of the water-oxidation complex remains unclear. Radmer and Ollinger (1986) used $^{18}O_2$ flash yield studies to conclude that the higher oxidation states of the system contained no appreciable non-exchangeable $H_2O$. That is, the oxidation takes place by an all-or-nothing reaction with free water as follows:

$$2H_2O + S_4 \rightarrow S_0 + O_2 + 4H^+$$

On the other hand, others postulate tightly controlled access of water to the water oxidation catalyst (Hansson and Wydrzynski 1990). Nevertheless, Kawamoto and Asada (1990) found that $O_2$-evolving particles isolated from spinach leaves retained 70% of their activity when resuspended in aqueous media after freeze drying. However, freeze drying apparently converted the particles to the $S_2$ state (six flashes were required to elicit maximum $O_2$ evolution per flash). Pre-illumination for 60s in air was enough to restore the freeze dried particles to the $S_0$-$S_1$ state and normal four-flash maximum $O_2$ yield. Pre-illumination in a desiccator with silica gel for up to 90 min failed to restore the $S_0$-$S_1$ state, indicating that strongly bound water and two photons are required to recover $S_0$ from the $S_2$ state. These observations may be interpreted in other ways, but they are consistent with evidence given below, that the water oxidation complex in vivo is exceedingly robust in the face of very low water potential.

## Water Oxidation as a Function of Photon Flux

As shown in Fig. 1, in most plants, in most light environments, most of the time, most of the photons absorbed in the light harvesting antennae chlorophyll-protein complexes in the thylakoid are wasted as heat. Theoretically, photosynthesis at $CO_2$ saturation has a maximum efficiency of 8–9 mol photons $mol^{-1}$ $O_2$ released, or a quantum yield of 0.111–0.125 mol $O_2$ $mol^{-1}$ absorbed photons, and measurements with intact leaves approach these values (Björkman and Demmig 1987). However, as shown in region A of Fig. 1, this efficiency is retained only over the initial narrow range of photon flux, which corresponds to naturally shaded habitats. Plants grown in the sun are able to extend the range of high efficiency photosynthesis to some extent, but in general terms photosynthetic $O_2$ evolution is an unreliable photon detector. Photosynthesis does not function as a photomultiplier because the maximum rate of electron flow which can be achieved under weak light is down-regulated by most of the reactions subsequent to the primary processes. Thus the maximum rate at light and $CO_2$ saturation (region B; Fig. 1) is susceptible to reduction by a host of environmental stress factors, especially water stress. In the simplest case, in plants intolerant of water stress, stomata close when the environment and plant hydraulics conspire to lower leaf water potential by as little as 1.0 MPa. Restricted access of $CO_2$ leads to down-regulation of photosynthetic efficiency, a decrease in $O_2$ evolution, and an increase in the proportion of excitation which must be wasted as heat.

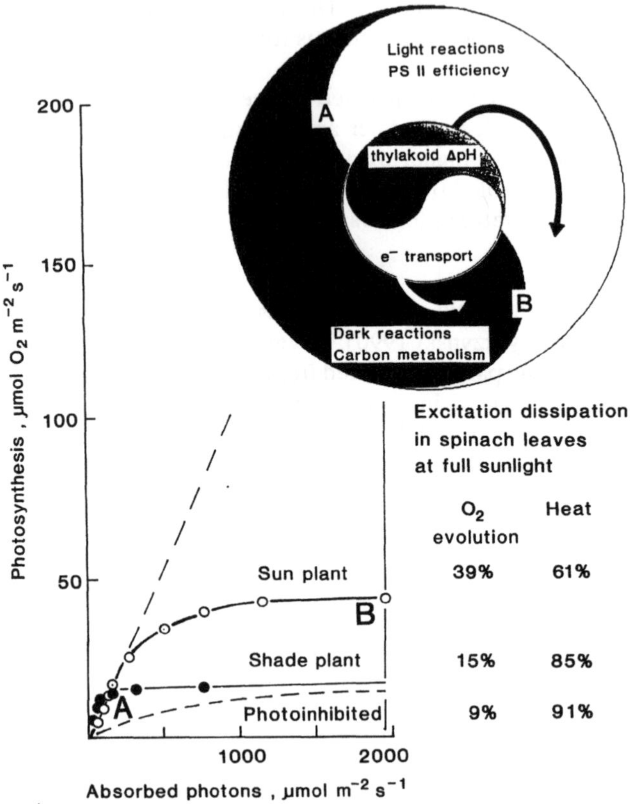


**Fig. 1A,B.** Schematic representation of the light dependence of photosynthetic $O_2$ evolution during water oxidation. High efficiency of photon use is confined to an initial, narrow range of photon flux (**A**). Extrapolation of this efficiency to full sunlight (2000 µmol photons $m^{-2}s^{-1}$) and comparisons to the actual rate of photon use (**B**) of sun plants, shade plants, and shade plants showing photoinhibitory damage after prolonged exposure to bright light, permits calculation of the extent to which absorbed excitation is wasted as heat. Down-regulation (feedback) of photochemical efficiency, coincident with up-regulation (feedforward) of the rate of carbon metabolism, is currently thought to depend on regulatory events associated with the Δ pH gradient across thylakoid membranes

Photoinhibitory damage, due to chronic exposure to high irradiance, leads to a significant lowering of photosynthetic efficiency in both parts of the light response curve (Fig. 1).

The biophysics of the processes leading to down-regulation of PS-II efficiency in vivo is not well understood. However, most evidence indicates a central role for transthylakoid Δ pH as a sensor, or driver, of most of the mechanisms linking light reactions and dark reactions of photosynthesis (Fig. 1). A small proportion of absorbed excitation is re-emitted as fluorescence, and this has proved to be a useful indicator of events in the PSII reaction center itself. Indeed a simple arithmetic relationship can be constructed from the maximum fluorescence yield ($F_m$) elicited from a saturating flash in a dark-adapted system, and that which can be elicited during steady state photosynthesis ($F_v$) at any point on the water-oxidation vs. photon flux density curve. This relationship permits calculation of photosynthetic rate from fluorescence data and photon flux alone (Weis and Berry 1987; Genty et al. 1989). It also correlates well with the efficiency of PSII photochemistry.

Reduction in the efficiency of photosynthetic water oxidation in bright light may be diagnosed in terms of two main processes (Krause 1988):

- photoprotection, in which excess excitation is deflected as heat before it is transferred to the reaction center of PSII. These processes are associated with relatively rapidly reversible decrease in both the initial level of chlorophyll fluorescence ($F_o$), and in the variable yield of fluorescence ($F_v$). These processes seem to explain the reversible changes in PSII efficiency which occur in response to the naturally occurring variations in photon flux throughout the day.
- photoinhibitory damage, which occurs when the capacity of photoprotective processes is exceeded, allowing excess excitation to the PSII reaction centers. This damage is associated with slowly reversible increases in $F_o$ and decreases in $F_v$. These fluorescence responses, and other criteria, can be used to assess the well-being of the water-oxidation system in the face of water stress.

There is now much evidence that a change in the composition of the xanthophyll pigments, presumably those associated with the light-harvesting chlorophyll-protein complexes of the antennae, plays a significant role in the dissipation of excess excitation (Demmig et al. 1987; Bilger et al. 1989a). In many plants, it has been found that whenever, and for whatever reason, excess excitation is transferred to PSII, there is an increase in the pool size of zeaxanthin, usually at the expense of violaxanthin. This change in zeaxanthin pool size is thought to be driven by $\Delta$ pH, and it correlates with an increase in the rate of excitation dissipation as deduced from fluorescence quenching (Demmig-Adams 1990). The mechanisms involved in enhanced dissipation of excitation when zeaxanthin is present are not yet known. One possibility, which needs to be evaluated, is that the much higher hydrophobicity of zeaxanthin, (which is obvious during chromatographic separation) may simply enable this potential quencher greater access to chlorophylls in the antennae pigment bed. Whatever the mechanism, these processes seem to be a nearly universal property of excitation dissipation in green plants, and may be a major factor responsible for the down-regulation of photosynthetic efficiency with increasing photon flux.

Several other processes may also contribute to this down-regulation in the efficiency of water oxidation in high light. Weis and Berry (1987) proposed that with increasing photon fluxes, a proportion of PSII reaction centers was transformed into a low efficiency state by the high transmembrane pH gradient thought to prevail at high light. Subsequent studies also implicate the redox state of PSII in these changes, and point to the possibility of electron cycling around PSII (Foyer et al. 1990). Just how these reaction center-based processes relate to the antennae based processes, remains to be determined, but all fit within the $\Delta$ pH driven scheme of Fig. 1.

## Photoinhibitory Damage to the Water Oxidation Complex

The capacity of the above processes to bring about a down-regulation of water-oxidation efficiency at sustained high photon fluxes can be exceeded in many natural circumstances. When this happens, the maximum efficiency and maximum rate of photosynthesis declines, as shown in Fig. 1. This condition, described as photoinhibitory damage, is also readily identified from fluorescence criteria (Krause 1988). The mechanisms of photoinhibitory damage in vivo are complex, and interact with $O_2/CO_2$ availability, temperature, water status, and nutrition. Photoinhibitory damage stems from loss of PSII reaction center function, and can be associated with proteolytic removal of the D-1 polypeptide, one of the two major structural transmembrane proteins of the reaction center (Chow et al. 1989).

In vitro studies show that the ionic environment of the water oxidation complex is an important component of photoinhibition, and the complex is readily photoinactivated following chaotropic interactions with anions (Miyao and Murata 1987). The effects of anions can be mitigated by glycine betaine (Papageorgiou et al. 1990), a compatible solute which accumulates in chloroplasts (Robinson and Jones 1986). Interestingly, there is no evidence that plants which differ markedly in salinity tolerance show differences in the stability of the PSII water-oxidation system in responses to anions (Ball and Anderson 1986). On the other hand, $Cl^-$-depleted thylakoids are also very sensitive to photoinhibition, because the $S_2$-$S_3$ transition is blocked, and illumination leads to the accumulation of oxidizing species on the donor side of P680 (Styring et al. 1990). An increase in the lifetime of the P680 triplet state, and its quenching by $O_2$, leading to the formation of singlet $O_2$ (Durrant et al. 1990), seems to explain some forms of photoinhibition of photosynthesis in vivo (Richter et al. 1990a,b).

Following photoinhibitory damage to the reaction center of PSII in vivo, the D-1 polypeptide may be degraded by proteases, with the 32 kDa protein being cleaved into 10 kDa and 22 kDa residues. The extent of D-1 turnover is dependent on many factors, especially temperature (Chow et al. 1989), and there is increasing evidence for lateral movement of the damaged PSII centers from granal to stromal thylakoids for repair (Adir et al. 1990). There is evidence that the polypeptides of the water-splitting complex itself remain as a mobile pool in the lumen, available for reassembly. Repair of photoinhibition is sensitive to the inhibitor of chloroplast-directed protein synthesis chloramphenicol. It requires weak light, and interpretation of responses in vivo is complicated by concurrent damage and repair.

## Effects of Water Stress on Photosynthetic Water Oxidation in Vivo

Although many generalizations are possible, the physiology of water relations in different plant life forms is so different that we must consider these responses under several categories.

### *Marine Macroalgae*

These plants, native to coastal and estuarine habitats, are exposed to water stress at low tide. The extent of stress experienced depends very much on the timing of the tidal cycle in relation to light intensity, but stress is always accompanied by large increases in ionic activity. Satoh et al. (1983) speculated that low water activity at high salinity in the desiccated red alga *Porphyra perforata* might itself inhibit electron flow from water in the water oxidation system of PSII. However, they and other authors (Öquist and Fork 1982; Smith et al. 1986; Herbert 1990) believe that photoprotection can be effective during dehydration of these organisms. It is thought that changes in the efficiency of excitation transfer to PSII, and increased "spillover" to PSI, are important features of photoprotective processes in water stress tolerance in the red algae.

Other green algae, such as *Ulva rotundata*, do not fare as well following emersion. If low tide coincides with solar noon, these plants experience significant photoinhibitory damage, possibly because they have limited ability to transform violaxanthin to zeaxanthin (S. Lindley, W.L. Henley, G. Levavasseur, L. Franklin and C.B. Osmond, unpubl. data). Dring and Brown (1982) concluded that zonation of inter-tidal brown algae was best understood in terms of recovery of photosynthesis after desiccation. The extent to which recovery is dependent on repair of photoinhibitory damage in the brown algae is not known (cf. *Porphyra*, Herbert 1990).

### *Mesophytic Terrestrial Angiosperms*

There is substantial literature on the water relations of land plants, but little which deals explicitly with the effects of water stress on primary photosynthetic processes such as water oxidation. Interpretations of the impairment of photosynthesis during water stress have ranged from those in which the limitations are primarily stomatal (i.e., $CO_2$ access), to those in which limitations are primarily biochemical, and back to the currently favored view of the dominant role of stomata (Farquhar et al. 1989). New techniques, in particular the ability to measure photosynthetic $O_2$ exchange at saturating $CO_2$ (Ben et al. 1987; Graan and Boyer 1990), and to evaluate the state of the photosynthetic apparatus via fluorescence measurements, have permitted significant progress. Photoprotective processes are of paramount importance in mesophytic plants such as cotton which are grown under high-light, high-temperature conditions with irrigation. The proportion of excitation wasted as heat by photoprotective processes increases with water stress of irrigated cotton (Björkman 1989).

Ben et al. (1987) showed that, although light- and $CO_2$-saturated photosynthesis in *Helianthus* and *Xanthium* declined at low water potential, the quantum efficiency of photosynthesis did not decline until severe water stress (–2.0 MPa) was experienced (Fig. 2). Fluorescence parameters indicated that primary photochemical functions were sustained until excised leaves were practically crisp. Only under these extreme conditions were irreversible, light-dependent changes in fluorescence observed, suggestive of photoinhibitory damage. It was concluded that death following

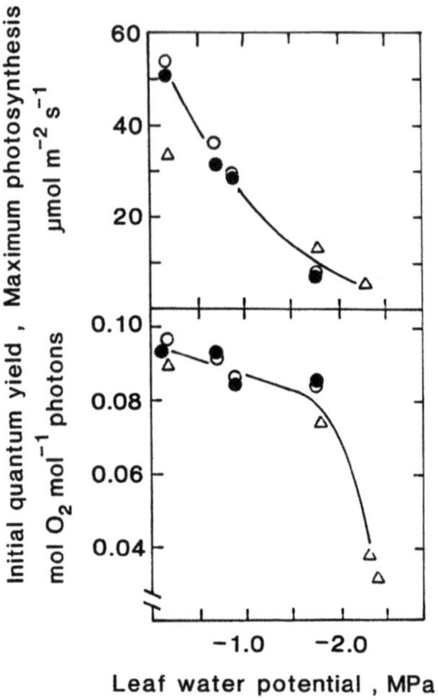


**Fig. 2.** Effects of chronic, severe water stress on photosynthetic $O_2$ evolution in mesophytic *Xanthium strumarium* in the sun (○) and shade (●) and in *Helianthus annus* in the sun (Δ). Plants did not recover from –2.0 mPa when rewatered. Maximum photosynthetic rate at $CO_2$ saturation was much more sensitive to water stress than initial quantum yield, indicating that effects of water stress on carbon metabolism and the down-regulation of PSII predominated over those on primary photochemistry and water oxidation mechanisms. (After Ben et al. 1987)

severe water stress in these plants was due to irreversible damage to processes other than photosynthetic water oxidation.

Desert plants with high water storage capacity (Nobel, this Vol.) avoid drought, but endure a complex of high-temperature stress at relatively high water potential, with little net $CO_2$ exchange, under exceptionally high irradiance. The $CO_2$ recycling of crassulacean acid metabolism is crucial to photosynthetic stability of these succulents, but marked differences in photosynthetic stability on the sun-exposed and shaded side of *Opuntia* cladodes have been observed. Subsequent studies highlight the role of photoprotection in preventing photoinhibitory damage in these plants (Adams et al. 1989).

### *Xerophytic Terrestrial Angiosperms*

Although water storage is common among some taxa in some deserts and saline environments, most species in these habitats do not display the drought-avoiding physiology of desert succulents. Rather, they display xermorphic properties which are associated with osmotic adjustment (via salt accumulation or compatible solute synthesis), and maintain physiological function at low water potentials (–2.0 to –8.0 MPa). It is increasingly clear that preservation of the primary photochemistry, including water oxidation in PSII, in these plants depends on the amplification of the capacity for excitation dissipation as heat. Björkman (1989) showed that changes in 77 K fluorescence parameters were consistent with a 10- to 15-fold increase in the rate constant for heat dissipation in *Nerium oleander* as water potential declined

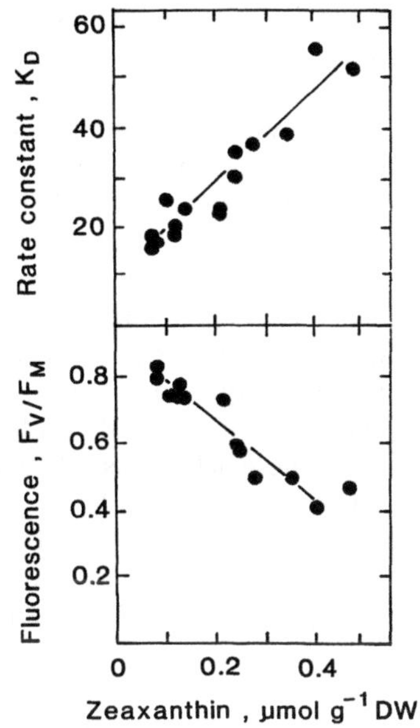


**Fig. 3.** Effects of water stress on photochemical parameters of photosynthesis in the xerophyte *Nerium oleander*. Data points relate to plants from control (–0.5 MPa) to extreme water stressed treatments (–4.5 MPa), from which the plants recovered. The down-regulation of PSII and water oxidation efficiency is directly correlated with a decline in the ratio $F_v/F_m$ of chlorophyll fluorescence. This, in turn, is highly correlated with the enhanced synthesis of the xanthophyll, zeaxanthin, which is believed to facilitate the increased dissipation of excitation as heat, indicated by increase in the calculated rate constant $K_D$. (After Demmig et al. 1988)

from –1.0 to –6.2 MPa. Demmig et al. (1988) showed similar responses in this organism which were correlated with zeathanthin content (Fig. 3). Mid-day depression of photosynthesis in the xerophyte *Artubus unedo* under natural conditions in Portugal also correlates with reduced efficiency of PSII photochemistry, due to increased dissipation of excitation as heat (Demmig-Adams et al. 1989). In mangroves in which compartmentation of salt and synthesis of compatible solutes appear to mitigate water and ionic stress at low leaf water potentials, rapidly reversible down-regulation of PSII efficiency is also observed (Björkman et al. 1988). There is thus little evidence of photoinhibitory damage in these water stress-tolerant xerophytes in their natural habitats, but like the mesophytes, one might expect photoinhibitory damage under extreme stress conditions.

## The Special Case of Poikilohydric Photosynthetic Systems (Lichens and Resurrection Plants)

Few higher plants withstand desiccation of photosynthetic tissues to an air-dry state, but poikilohydry is well represented among lower plants. The lichens are of particular interest because of the distinctly different water absorption processes in species with green-, and with blue-green algal phycobionts. Lange et al. (1986) have surveyed many species, to show that whereas green algal phycobiont lichen species (e.g., *Ramalina* spp.) absorb sufficient water from humid atmospheres to commence photosynthetic activity, those lichens with blue-green algal phycobionts (e.g., *Peltigera* spp.) only become photosynthetically active after exposure to liquid water

(Fig. 4). The biophysics of these processes have yet to be explained, but almost certainly, the hygroscopic properties of extracellular materials surrounding the algae need to be taken into account.

Analysis of the status of the photosynthetic apparatus in dry lichens shows functional detachment of the antennae pigments from PSII reaction centers (Bilger et al. 1989b). In green algal phycobiont lichens water uptake from humid air leads to restoration of connections between light harvesting chlorophyll and PSII and recovery of normal fluorescence quenching (Fig. 4). In blue-green algal phycobiont lichens the detachment of phycobilisomes was only restored following exposure to liquid water. Subsequent studies (Lange et al. 1989) confirm these responses using other techniques, but do not, as yet, indicate mechanisms of water uptake or activity which could explain different responses in the reactivation of water oxidation.

In these lichens, as in higher plants, significant correlations between the stability of primary photosynthetic processes and violaxathanin-zeaxanthin content and tolerance of desiccation in bright or dim light have been recorded. The blue-green algal phycobiont lichens have very limited capacity to synthesize zeaxanthin and consequently, when hydrated and exposed to bright light, suffer photoinhibitory damage. The green-algal phycobiont lichens show significantly enhanced zeaxanthin levels and excitation dissipation as heat in bright light, even after absorption of water from water vapor (Demmig-Adams et al. 1990). These differences in water uptake processes and in pigment compositions are major factors in restricting blue-green algal lichens to moist, shaded habitats, and in allowing green algal lichens to occupy bright, dry habitats with seasonal fogs.

Resurrection plants, the angiosperms, ferns, and club mosses which can tolerate desiccation to the air-dried state, are not very numerous in the plant kingdom, but can be locally dominant in some regions, especially in southwestern USA, Africa, and Australia (Eickmeier 1980; Gaff 1981). Of particular interest in the present context are those species in which chlorophyll is fully degraded during desiccation (e.g., the poikilochlorophyllous species *Borya nitida*), and those in which it is retained (e.g., most resurrection grasses from Australia and all resurrection dicotyledons and pteridophytes.) In the former case, the stability of the photosynthetic apparatus has been studied during dehydration and rehydration using chlorophyll fluorescence analysis. The rate of dehydration is crucial. Rapid dehydration of *Borya nitida* leads to retention of chlorophyll which appears to be capable of charge separation on excitation and some water oxidation (Hetherington et al. 1982). However, these rapidly dehydrated plants do not recover photosynthesis on rehydration, and PSII reaction centers appear to be detached from the antennae. Slow dehydration, which led to loss of chlorophyll and of water oxidation function, was fully reversible and accompanied by resynthesis of chlorophyll and normal function. The return to normal function was slow (Bewley and Oliver, this Vol.).

Resurrection plants which retain chlorophyll in the dry state may regain photosynthetic function more quickly when rewetted. This rapid recovery from compete desiccation is possible because of the conservation of constituents of the photochemical apparatus during dehydration. A particularly noteworthy example that illustrates this capacity is the Chihuahuan desert spikemoss *Selaginella lepidophylla* (Eickmeier 1979). Preliminary studies reported here show that PSII

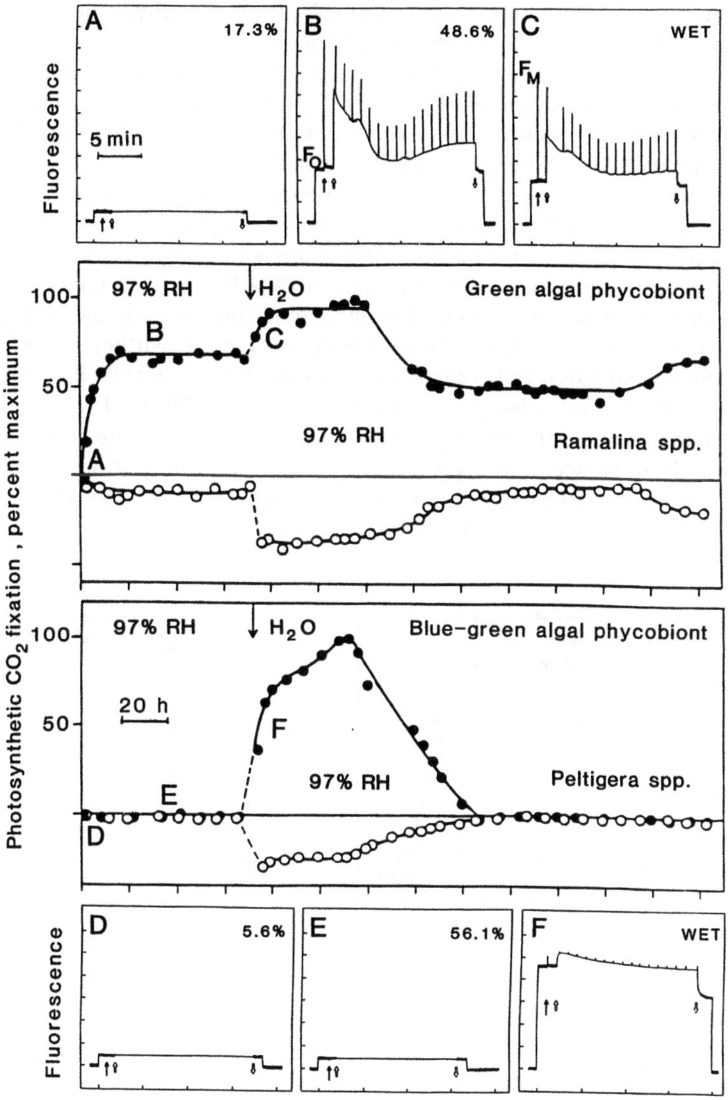


**Fig. 4A-F.** Differences in recovery from desiccation in lichens containing green algal, or blue-green algal phycobionts, as shown by rates of $CO_2$ fixation and chlorophyll fluorescence responses. Air-dry lichens were first allowed to rehydrate in air with 96% RH at 15 °C for 52 h. This treatment leads to 70% recovery of photosynthesis (●) and significant respiration (○) in *Ramalina*, but no recovery in *Peltigera*. Fluorescence analysis of *Ramalina* shows normal $F_v/F_m$ ($F_v = F_m - F_o$) and fluorescence quenching during an induction curve (compare **B** with **A**), but no change in *Peltigera* (compare **E** with **D**), in spite of the slightly higher water content in the blue algal lichen. Spraying with liquid water, and subsequent maintenance at 97% RH, leads to a transient increase in photosynthesis in *Ramalina* and qualitatively no change in fluorescence (compare **C** and **B**). In *Peltigera*, spraying results in a rapid increase in photosynthesis which subsequently declines again to zero in 97% RH. This is accompanied, within 30 min, by the onset of some variable fluorescence and fluorescence quenching (compare **E** with **F**). (After Lange et al. 1986, 1989)

function is quite robust both during desiccation and rehydration, so long as the plants are not subjected to high PFD simultaneously.

Net photosynthetic activity can be maintained to low relative water content (RWC) as *S. lepidophylla* dries (Fig. 5) at near optimal values to about 40% RWC (which corresponds to a tissue water potential of about –10 MPa). This suggests that, compared with other plants, photosynthetic water oxidation and electron transport is tolerant of very low tissue water potential. Direct measurement of room-temperature chlorophyll fluorescence confirms this suggestion. Both PSII photochemical efficiency ($F_v/F_m$) and photochemical quenching ($q_P$) at low PFD remain at near maximal values to between 30 and 40% RWC and decrease rapidly (Fig. 5b). Nonphotochemical fluorescence quenching ($q_{NP}$) rises as the plants dry and is high when $q_P$ falls in conjunction with decreasing photosynthesis. The intrinsic fluorescence yield when PSII reaction centers are fully oxidized ($F_o$) dips slightly at near 40% RWC and subsequently increases near 0%/RWC (Fig. 5c). Thus, both PSII electron transport and net photosynthetic capacity are quite desiccation-tolerant and the fluorescence data suggest that PSII efficiency may be well down-regulated and photoprotected by the time desiccation-induced limitations of $CO_2$ fixation occur. Continued illumination in this condition can lead to photoinhibitory damage.

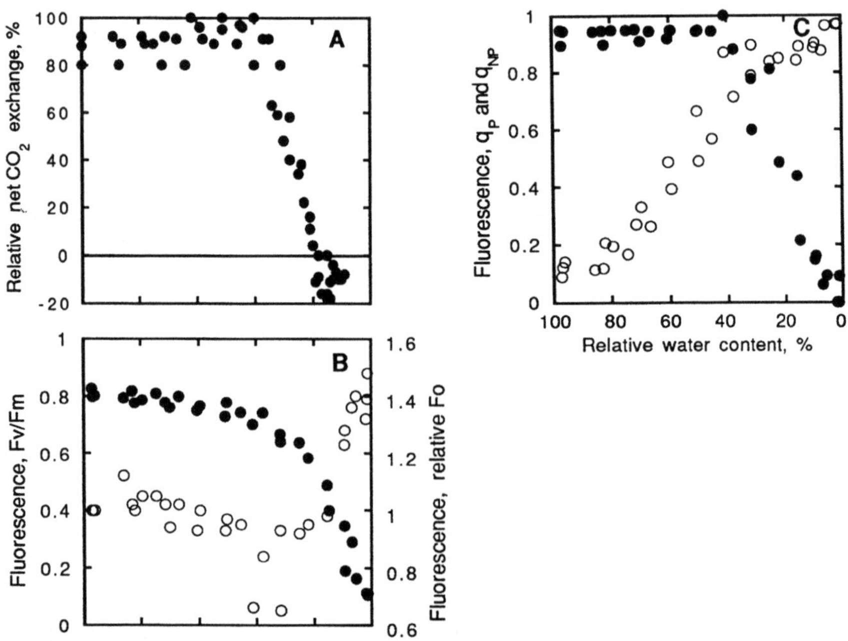


**Fig. 5A-C.** Net gas exchange and fluorescence characteristics of the resurrection plant *Selaginella lepidophylla* as functions of relative water content during desiccation. **A** Relative net gas exchange with positive and negative values on the ordinate representing net $CO_2$ uptake and loss, respectively. **B** The variable to maximum fluorescence yield ratio ($F_v/F_m$) and the intrinsic photosynthetic ($F_o$ normalized to initial value). **C** Photochemical fluorescence quenching ($q_P$) and nonphotochemical quenching ($q_{NP}$). All panels include the data from three replicate experiments. (Data of W.G. Eickmeier and C.B. Osmond, unpubl.)

The potential resiliency of carbon reduction and PSII function is also evident during rehydration of dry *Selaginella*. The redevelopment of photosynthetic capacity may occur even in the presence of the chloroplast-dependent protein synthesis inhibitor chloramphenicol (Fig. 6A). After an initial 24-h period of rehydration in the dark, both control and chloramphenicol-treated plants reach comparable rates of net photosynthesis soon after exposure to saturating light. Only with continued exposure to high light did the effects of chloramphenicol lead to an exponential decline in photosynthesis. The redevelopment of PSII function may also occur in the presence of chloramphenicol at very low light (Fig. 6B). Both the variable fluorescence yield ($F_v$) and the ratio $F_v/F_m$ rise rapidly in control and chloramphenicol-treated plants reaching near maximal values after 5 h of rehydration. This is long before substantial chloroplast protein synthesis, which only commenced at least 12 h after rehydration (Eickmeier 1982). However, if the plants are exposed to moderate light during rehydration in the presence of chloramphenicol, both $F_v$ and $F_v/F_m$ are significantly reduced, consistent with $CO_2$ exchange experiments. These observations suggest that the components of PSII may reintegrate quickly, and independently of protein synthesis in *S. lepidophylla*, but that the maintenance of PSII in high light requires chloroplast-dependent protein synthesis.

It is clear that water stress alone is not a serious challenge to stability of water oxidation in resurrection plants. On the other hand, water stress coupled with high-light stress can present a serious environmental stress. There may be several periods of heightened PSII sensitivity to high-light exposure during dehydration, during the fully desiccated state, and/or again during rehydration. Photoinhibitory damage could disrupt the disassembly and reassembly of PSII that seems to occur with change in plant hydration level. Like a very few mesophytic plants (Ludlow and Björkman 1984), resurrection plants may effectively close these windows of sen-

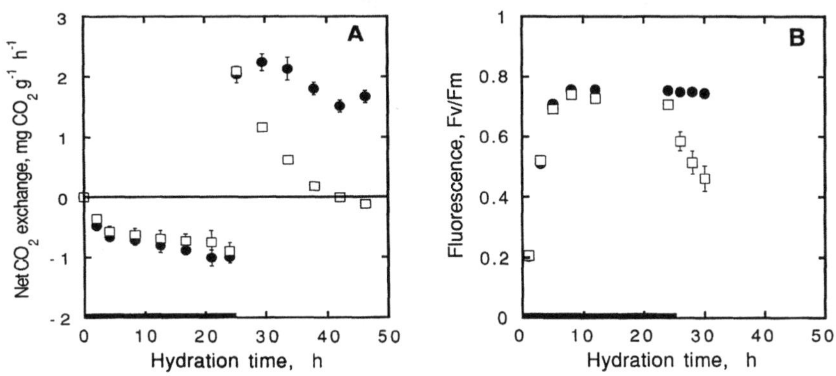


**Fig. 6A,B.** Net $CO_2$ exchange and $F_v/F_m$ ratio of the resurrection plant *S. lepidophylla* as functions of hydration time in the presence (*open symbols*) and absence (*closed symbols*) of chloramphenicol (0.31 mM). **A** Net $CO_2$ exchange was first measured in the dark for 24 h followed by an additional 24-h period under light-saturating conditions at a 25 °C plant temperature throughout. *Points* are means (n = 3) ± 1 SE. **B** The $F_v/F_m$ ratio was first measured for 24 h at ca. 10 μmol photons $m^{-2}s^{-1}$ at a 22 °C plant temperature. *Points* are means (n = 8) ± 1 SE (Data of W.G. Eickmeier and C.B. Osmond, unpubl.)

sitivity by reducing light absorption at critical stages in desiccation and hydration. This is accomplished via the dramatic curling and uncurling of stems during desiccation and hydration, respectively, which is characteristic of this and many other resurrection plant species.

The significance of stem curling and uncurling can be demonstrated by restraining plants as they dry under various temperature and light conditions, followed by subsequent measurement of PSII electron transport and fluorescence yields after 24 h of rehydration (Table 1; Fig. 7). High irradiance during desiccation significantly reduced PSII electron transport and photochemical efficiency ($F_v/F_m$), as well as chlorophyll content, relative to plants desiccated at low irradiance. However, plants which were allowed to curl normally during the bright light desiccation treatments suffered significantly less PSII damage and chlorophyll photooxidation than those that were artificially restrained. Not surprisingly, restraint was unimportant during desiccation in low light. Collectively, these studies show that the photosynthetic water oxidizing system of chlorophyll-retaining resurrection plants can cope successfully with extreme water stress. The stability of the water-oxidizing system in *Selaginella* in the face of water stress can be ameliorated via a suite of molecular, physiological, and morphological mechanisms that deserve closer scrutiny.

## Summary

As the above experiments with resurrection plants show, the water oxidation apparatus of photosynthesis is exceptionally robust in the face of water stress. Yet we know very little about the biophysics of the involvement of water in this process. Isotopic studies show that no net discrimination occurs during water oxidation

**Table 1.** PSII electron transport capacity, chlorophyll content, and PSII photochemical efficiency ($F_V/F_M$) of resurrection plants after rehydration following desiccation under various irradiance, temperature, and restraint conditions. (Data of J.G. Lebkuecher and W.G. Eickmeier, unpubl.).

| Photosynthetic parameter | °C | PPFD[a] | Plant treatment | |
|---|---|---|---|---|
| | | | Unrestrained | Restrained |
| PSII electron transport | 25 | 175 | 94 ± 10 | 106 ± 11 |
| (μmol DCPIP reduced mg $chl^{-1}$ $h^{-1}$) | 25 | 2000 | 77 ± 19 | 36 ± 6 |
| | 37 | 175 | 115 ± 18 | 100 ± 22 |
| | 37 | 2000 | 68 ± 11 | 54 ± 4 |
| Chlorophyll content | 25 | 175 | 0.14 ± 0.01 | 0.14 ± 0.01 |
| [mg $g(FW)^{-1}$] | 25 | 2000 | 0.11 ± 0.01 | 0.09 ± 0.01 |
| | 37 | 175 | 0.13 ± 0.01 | 0.12 ± 0.01 |
| | 37 | 2000 | 0.07 ± 0.01 | 0.05 ± 0.01 |
| PSII photochemical | 25 | 175 | 0.811 ± 0.006 | 0.758 ± 0.022 |
| efficiency | 25 | 2000 | 0.698 ± 0.028 | 0.323 ± 0.052 |
| ($F_v/F_m$) | 37 | 175 | 0.820 ± 0.002 | 0.803 ± 0.003 |
| | 37 | 2000 | 0.605 ± 0.071 | 0.414 ± 0.058 |

[a]Photosynthetic photon flux density; μmol photons $m^{-2}s$ -1.

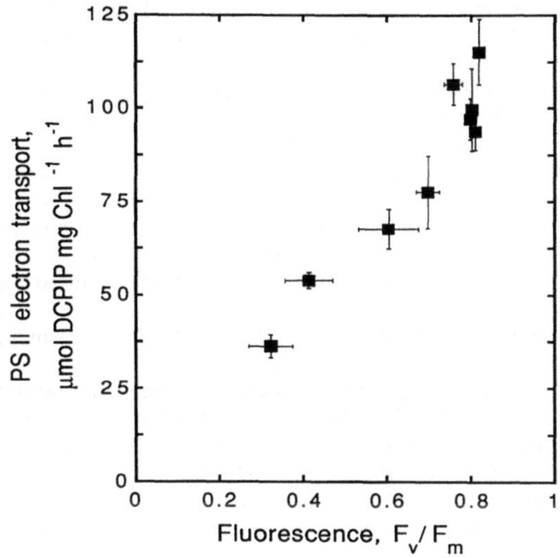


**Fig. 7.** The correlation between the $F_v/F_m$ ratio and the photosystem (*PS*) II electron transport capacity of *S. lepidophylla* plants desiccated under the various light, temperature, and plant restraint treatment combinations (plus an unmanipulated control treatment) given in Table 1. The plants were rehydrated for a 24-h recovery period before assay. The regression equation ($Y = -6.9 + 134.2\,x$) yields an $r^2$ of 0.905. (Data of J.G. Lebkuecher and W.G. Eickmeier, unpubl.)

(Yakir, this Vol.), yet controlled access of water to the biological catalyst itself has been proposed recently (Hansson and Wydrzynski 1990). We are learning more about the ways in which the efficiency of the water oxidation system can be down-regulated and photoprotected in the light so as to avoid photoinhibitory damage, but we know little about the repair systems which must be deployed to restore this damage when it occurs. In all of these studies, investigation of the comparative physiology of plants from diverse habitats is providing new insights. Some of them, such as the reconstruction of the photosynthetic apparatus upon addition of water to desiccated *Selaginella*, which is apparently independent of chloroplast protein synthesis, simply defy our present understanding (Bewley and Oliver, this Vol.) and encourage further research.

*Acknowledgments.* The authors thank Drs. Adams, Demmig-Adams, and Lange for advice and access to preprints during preparation of this manuscript. We also thank John Skillman for help in execution of Fig. 1A.

## References

Adams III WW, Diaz M, Winter K (1989) Duirnal changes in photochemical efficiency, the reduction state of Q, radiationless energy dissipation, and non-photochemical fluorescence quenching in cacti exposed to natural sunlight in northern Venezuela. Oecologia 80:553–561

Adir N, Shocat S, Insue Y, Ohad I (1990) Mechanism of the light dependent turnover of the D-1 protein. In: Baltscheffsky M (ed) Current research in photosynthesis. Kluwer, Dordrecht, 2:409–413

Ball MC, Anderson JM (1986) Sensitivity of photosystem II to NaCl in relation to salinity tolerance. Comparative studies with thylakoids of the salt-tolerant mangrove, *Avicennia marina*, and the salt-sensitive pea, *Pisium sativum*. Aust J Plant Physiol 13:689–698

Ben G-Y, Osmond CB, Sharkey TD (1987) Comparisons of photosynthetic responses of *Xanthium strumarium* and *Helianthus annuus* to chronic and acute water stress in sun and shade. Plant Physiol 84:476–482

Bilger W, Björkman O, Thayer S (1989a) Light-induced spectral absorbance changes in relation to photosynthesis and epoxidation state of xanthophyll cycle components in cotton leaves. Plant Physiol 91:542–551

Bilger W, Rinke S, Schreiber U, Lange OL (1989b) Inhibition of energy transfer to photosystem II in lichens by dehydration: different properties of reversibility with green and blue-green phycobionts. J Plant Physiol 134:261–268

Björkman O (1989) Some viewpoints on photosynthetic response and adaptation to environmental stress. In: Briggs WR (ed) Photosynthesis. Liss, New York, pp 45–58

Björkman O, Demmig B (1987) Photon yield of $O_2$ evolution and chlorophyll fluorescence characteristics at 77K among vascular plants of diverse origins. Planta 170:489–504

Björkman O, Demmig B, Andrews TJ (1988) Mangrove photosynthesis: response to high irradiance stress. Aust J Plant Physiol 15:43–61

Brudvig GW, Beck WF, de Paula JC (1989) Mechanism of photosynthetic water oxidation. Annu Rev Biophys Biophys Chem 18:25–46

Chow WS, Osmond CB, Huang LK (1989) Photosystem II function and herbicide binding sites during photoinhibition of spinach chloroplasts in vivo and in vitro. Photosynth Res 21:17–26

Demmig B, Winter K, Kruger A, Czygan F-C (1987) Photoinhibition and zeaxanthin formation in intact leaves. Plant Physiol 84:218–224

Demmig B, Winter K, Kruger A, Czygan F-C (1988) Zeaxanthin and the heat dissipation of excess light energy in *Nerium oleander* exposed to a combination of high light and water stress. Plant Physiol 87:17–24

Demmig-Adams B (1990) Carotenoids and photoprotection in plants: a role for the xanthophyll zeaxanthin. Biochim Biophys Acta 1020:1–24

Demmig-Adams B, Adams III WW, Winterk, Meyer A, Schreiber U, Pereira JS, Krüger A, Czygan FC, Lange O (1989) Photochemical efficiency of photosystem II, photon yield of $O_2$ evolution, photosynthetic capacity, and carotenoid composition during the mid-day depression of net $CO_2$ uptake in *Arbutus unedo* growing in Portugal. Planta 177:377–387

Demmig-Adams B, Adams III WW, Czygan FC, Schreiber U, Lange OL (1990) Differences in the capacity for radiationless energy dissipation in the photochemical apparatus of green and blue-green algal lichens associated with differences in carotenoid composition. Planta 180:582–589

Dring MJ, Brown FA (1982) Photosynthesis of intertidal brown algae during and after periods of emersion: a renewed search for physiological causes of zonation. Mar Ecol Prog Ser 8:301–308

Durrant JR, Giorgi LB, Barber J, Klug DR, Porter G (1990) Characterization of triplet states in isolated photosystem II reaction center: oxygen quenching as a mechanism for photodamage. Biochim Biophys Acta 1017:167–175

Eickmeier WG (1979) Photosynthetic recovery in the resurrection plant *Selaginella lepidophylla* after wetting. Oecologia 39:93–106

Eickmeier WG (1980) Photosynthetic recovery of resurrection spikemosses from different hydration regimes. Oecologia 46:380–385

Eickmeier WG (1982) Protein synthesis and photosynthetic recovery in the resurrection plant, *Selaginella lepidophylla*. Plant Physiol 69:135–138

Farquhar GD, Wong SC, Evans JRE, Hubick KT (1989) Photosynthesis and gas exchange. In: Jones HG, Flowers TJ, Jones MB (eds) Plants under stress. SEB Sem Ser 39:47–69, Cambridge Univ Press, Cambridge

Foyer C, Furbank R, Harbinson J, Horton P (1990) The mechanisms contributing to photosynthetic control of electron transport by carbon assimilation in leaves. Photosynth Res 25:83–100

Gaff D (1981) The biology of resurrection plants. In: Pate JS, McComb AJ (eds) The biology of Australian plants. Univ Western Australia Press, Nedlands, pp 114–146

Genty B, Briantis JM, Baker NR (1989) The relationship between the quantum yield of photosynthetic electron transport and quenching of chlorophyll fluorescence. Biochim Biophys Acta 990:87–92

Ghanotakis DF, Yocum CF (1990) Photosystem II and the oxygen-evolving complex. Annu Rev Plant Physiol Plant Mol Biol 41:255–276

Graan T, Boyer JS (1990) Very high $CO_2$ partially restores photosynthesis in sunflower at low water potentials. Planta 181:378–384

Hansson O, Wydrzynski T (1990) Current perceptions of photosystem II. Photosynth Res 23:131–162
Herbert SH (1990) Photoinhibition resistence in the red alga *Porphyra perforata*. Plant Physiol 92:514–519
Hetherington SE, Smillie RM, Hallam ND (1982) In vivo changes in chloroplast thylakoid membrane activity during viable and non-viable dehydration of a drought-tolerant plant, *Borya nitida*. Aust J Plant Physiol 9:611–621
Kawamoto K, Asada K (1990) Oxygen evolution of lyophilized photosystem II membranes. In: Baltscheffsky M (ed) Current research in photosynthesis. Kluwer, Dordrecht, 1:889–892
Kok B, Forbush B, McGloin M (1970) Cooperation of changes in photosynthetic $O_2$ evolution-1. A linear four-step mechanism. Photochem Photobiol 11:457–475
Krause GH (1988) Photoinhibition of photosynthesis. An evaluation of damaging and protective mechanisms. Physiol Plant 74:566–574
Lange OL, Kilian E, Ziegler H (1986) Water vapor uptake and photosynthesis of lichens: performance differences with green and blue-green algae as phycobionts. Oecologia 71:104–110
Lange OL, Bilger W, Rinke S, Schreiber U (1989) Chlorophyll fluorescence of lichens containing green and blue-green algae during hydration by water vapor uptake and by addition of liquid water. Bot Acta 102:306–313
Ludlow MM, Björkman O (1984) Paraheliotropic leaf movement in Siratro as a protective mechanism against drought-induced damage to primary photosynthetic reactions: damage by excessive light and heat. Planta 161:505–518
Miyao M, Murata M (1987) Photoinactivation of the oxygen-evolving complex of photosystem II. In: Kyle DJ, Osmond CB, Arntzen CJ (eds) Photoinhibition. Elsevier, Amsterdam, pp 289–307
Öquist G, Fork DC (1982) Effects of desiccation on the excitation energy distribution from phycoerythrin to the two photosystems in the red alga *Porphyra perforata*. Physiol Plant 56:56–62
Papageorgiou GC, Fujimura Y, Murata N (1990) On the mechanisms of betaine protection of photosynthetic structures in high salt environment. In: Baltcheffsky M (ed) Current research in photosynthesis, vol 1. Kluwer, Dordrecht, pp 957–960
Radmer R, Ollinger O (1986) Do the higher oxidation states of the photosynthetic $O_2$-evolving system contain bound $H_2O$? FEBS Lett 195:285–289
Richter M, Rühle W, Wild A (1990a) Studies on the mechanism of photosystem II photoinhibition I. A two-step degradation of D-1 protein. Photosynth Res 24:229–235
Richter M, Rühle W, Wild A (1990b) Studies on the mechanism of photosystem II photoinhibition II. The involvement of toxic oxygen species. Photosynth Res 24:237–243
Robinson SP, Jones GP (1986) Accumulation of glycine-betaine in chloroplasts provides osmotic adjustment during salt stress. Aust J Plant Physiol 13:659–668
Satoh K, Smith CM, Fork DC (1983) Effects of salinity on primary processes of photosynthesis in the red alga *Porphyra perforata*. Plant Physiol 73:643–647
Sauer K (1990) Photosystem II and water oxidation. In: Baltcheffsky M (ed) Current research in photosynthesis, vol 1. Kluwer, Dordrecht, pp 675–684
Smith CM, Satoh K, Fork DC (1986) The effects of osmotic tissue dehydration and air drying on morphology and energy transfer in two species of *Porphyra*. Plant Physiol 80:843–847
Stocker O (1952) Grundriss der Botanik. Springer, Berlin Heidelberg New York
Styring S, Jegenschöld C, Virgin I, Ehrenberg A, Andersson B (1990) On the mechanisms for the photoinhibition of the electron transfer and the light-induced degradation of the D-1 protein in photosystem II. In: Baltscheffsky M (ed) Current research in photosynthesis, vol 2. Kluwer, Dordrecht, pp 349–356
Weis E, Berry J (1987) Quantum efficiency of photosystem II in relation to energy-dependent quenching of chlorophyll fluorescence. Biochim Biophys Acta 894:198–208

Chapter 15

# Desiccation and Freezing Phenomena for Plants with Large Water Capacitance – Cacti and Espeletias

P.S. NOBEL and G. GOLDSTEIN

## Introduction

Cacti and espeletias are native to dissimilar habitats, although the two groups must cope with several similar environmental constraints. Cacti, which are succulent plants that utilize crassulacean acid metabolism (CAM), grow in habitats characterized by long periods of drought (Ting 1985; Nobel 1988a). Over 80% of the aboveground plant biomass in parts of the Sonoran Desert can be cacti (Ting and Jennings 1976), and epiphytic CAM species occupying arid microhabitats are common in many lowland tropical forests (Medina et al. 1989). As for cacti, espeletias also have conspicuous stem water storage reservoirs, but they utilize the $C_3$ pathway and grow in tropical high-altitude (alpine) habitats. These habitats are characterized by diurnal rather than seasonal temperature variations that frequently involve nocturnal freezing (Troll 1968; Meinzer and Goldstein 1986). In the American tropics the giant rosette species belonging to the genus *Espeletia* (Compositae) experience air temperatures that even during the daytime tend to be suboptimal for many physiological processes (Goldstein et al. 1989). Studies on plants from cold temperate zones, particularly conifers (Kaufmann 1975, 1977; Running and Reid 1980), as well as on cacti (Lopez and Nobel 1991), indicate that water uptake by roots may be severely impaired by low soil temperatures (0 to 5 °C). In the tropical alpine habitats, such physiological drought is particularly likely during the early morning when soil temperatures in the root zone of espeletias are near freezing and potential transpiration is high due to high solar irradiation leading to relatively high leaf temperatures. Thus, both cacti and espeletias are exposed to drought episodes, the former on a seasonal basis and the latter predominantly on a daily basis.

Another environmental constraint on cacti growing in temperate deserts and espeletias growing in tropical alpine habitats is the occurrence of subzero temperatures. Approximately 35% of the cactus species in the United States and Canada grow at latitudes or elevations at which they experience subzero temperatures during the winter season. For *Opuntia fragilis* and *O. humifusa*, such periods can last for months and the temperature can be continuously below 0 °C for weeks (Nobel 1988a). Espeletias, particularly those growing at higher elevations, may be exposed

Department of Biology and Laboratory of Biomedical and Environmental Sciences, University of California, Los Angeles, CA 90024, USA

Somero et al. (Eds.)
Water and Life

to subzero temperatures for a few hours at night for about half of the nights, although daytime temperatures are above 0 °C. Thus, both groups of plants are exposed to temperatures that are not only lower than the equilibrium freezing point of shoot tissues but also sufficiently low to cause metabolic disturbances, even if tissue freezing does not occur.

This chapter focuses on the physiological and morphological adaptations that have evolved in cacti and espeletias, allowing these plants to survive drought and freezing temperatures. First, the biophysical characteristics of substantial internal water storage (large capacitance) for cacti and espeletias and its significance in the water economy of these plants will be addressed. Because of the large amounts of stored water, the possible responses to freezing temperatures, such as tolerance of low subzero temperatures, are constrained. Second, supercooling mechanisms and acclimation to low temperatures are discussed from both a phenomenological and a mechanistic point of view. Responses to drought and to freezing will be addressed at organismic and cellular levels.

## Gas Exchange, Morphology

Several morphological and physiological characteristics of cacti are advantageous in environments with a limited water supply, such as stomatal opening at night when evaporative demand is low. CAM plants incorporate $CO_2$ predominantly by dark fixation and store it mainly in the form of malic acid; decarboxylation of malic acid releases $CO_2$ internally during the daytime, thereby eliminating the need for daytime stomatal opening (Nobel 1988a). Transpiration at night (Fig. 1), when the water vapor concentration differences between the shoot and the ambient air are lower, allows cacti to achieve a high water-use efficiency. Espeletias, on the other hand, have stomatal opening during the daytime and fix $CO_2$ by $C_3$ photosynthesis. Stomatal conductances for espeletias tend to be maximal early in the morning, when evaporative demand is relatively high, and usually exhibit a midday decline associated with reduced interception of irradiance because of the erect position of the leaves (Goldstein et al. 1989). Higher transpiration rates tend to occur in the early morning (Fig. 1), although soil water is not as readily available then because of the low soil temperatures, resulting in potential water stress.

The Cactaceae consist of about 122 genera and just over 1600 species (Gibson and Nobel 1986). Most of the leafless (evolutionarily advanced) cacti have either cylindrical, globular, or flattened photosynthetic stems. Even though cacti exhibit great diversity in stem form, which affects the plant energy budgets and thermal relations (Lewis and Nobel 1977; Nobel 1978), cacti have a similar internal stem organization. The photosynthetic tissue, or chlorenchyma, is composed of 10 to 20 layers of chlorophyll-containing cells. Interior to the chlorenchyma is the water-storage parenchyma (WSP). Vascular bundles terminate within or immediately beneath the chlorenchyma, and mucilage-containing cells are commonly interspersed in the water-storage parenchyma, particularly for platyopuntias (McGarvie and Parolis 1981; Trachtenberg and Mayer 1982). Because of low stem surface to volume ratios, cacti can store large amounts of water, mainly in the WSP. Further-

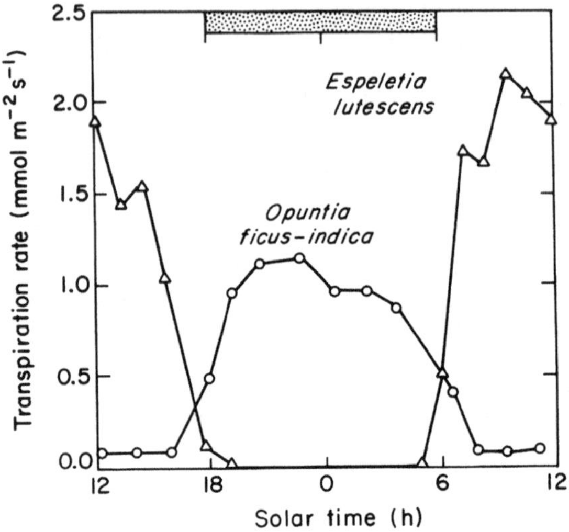


**Fig. 1.** Diel transpiration rates of the cactus *Opuntia ficus-indica* (O) and of the giant rosette species *Espeletia lutescens* (Δ). Day/night air temperatures on clear days averaged 6/0 °C for *E. lutescens* and 22/15 °C for *O. ficus-indica*. Unpublished observations of G. Goldstein and P.S. Nobel

more, the distance for water movement into the chlorenchyma is reduced by its close proximity to the WSP, which is particularly important during prolonged drought when soil water is not available to replace the water lost by nocturnal transpiration.

The genus *Espeletia* consists of over 60 species, all growing in the Andes of South America, that typically have an erect unbranched stem (up to 3.5 m tall) supporting a single parabolic rosette of large pubescent leaves (Meinzer et al. 1985). The rosette is formed by a dense mass of spirally arranged leaves separated by very short internodes, the newly expanding leaves surrounding a massive apical bud located in the focus of the parabola. Although the espeletias growing at higher elevations (up to 4400 m) have a prominent erect stem (they are caulescent), many espeletias at lower elevations (2600 to 3300 m) tend to be acaulescent, but still have a single terminal rosette. Inside the stem of both caulescent and acaulescent species, a central pith is made up of parenchymatous tissue (here called the WSP) about 7 to 12 cm in diameter that can provide water to the transpiring leaves (Goldstein and Meinzer 1983). The WSP in the stem is surrounded by conducting tissue (xylem and phloem) and then a 20- to 30-cm thick cylinder of dead leaves that remain attached to the plants after leaf senescence (Smith 1979; Goldstein et al. 1984).

Despite morphological and physiological differences between the two plant groups, a key aspect of their functional morphology related to water economy is similar – both have very large internal water storage reservoirs. The parenchyma apparently functions mainly as a water reservoir and in both cases is closely linked hydraulically with the photosynthetic tissue. Table 1 summarizes some morphological and physiological characteristics of the two groups.

**Table 1.** Summary of morphological and physiological characteristics of cacti and espeletias

| | Cacti | Espeletias |
|---|---|---|
| Photosynthesis | CAM | $C_3$ |
| Transpiration | Nocturnal, low | Diurnal (particularly morning), high |
| Root system | Shallow | Shallow |
| Stem capacitance | Large | Large |
| Type of stem water storage | Parenchyma (elastic) | Parenchyma (elastic) |
| Gas exchange surface/water storage volume | Low | High |
| Distance between water storage parenchyma and photosynthetic tissue | Short | Long |

## Capacitance and Water Flow – Whole Plant

The continuum of water flow from the soil through the roots, stems, and leaves to the atmosphere is conventionally viewed as a series of resistances through which water moves in response to gradients in water potential (van den Honert 1948). In this steady-state view of a plant, the amounts of water moving through each portion of the transpiration stream are equal to that extracted from the soil and lost by evaporation through the surface of the shoot. In reality, however, time lags occur between water uptake and water loss, indicating internal water storage, or capacitance (Nobel 1991). In this more realistic nonsteady-state representation, capacitance of various organs can have a significant effect on the water relations of plants.

Extensive water-storage tissue, such as the WSP of cacti and espeletias, can supply a considerable amount of water to the transpiration stream, thus buffering the decline in water potential of the photosynthetic tissue. Adult individuals of *Espeletia lutescens* at 4200 m in the Venezuelan Andes usually exhibit little diurnal fluctuation in leaf water potential, whereas the fluctuation can be pronounced for the sympatric species, *E. moritziana*, despite similar daily transpiration (Fig. 2). The species showing substantial daytime water potential fluctuations exhibits a lower capacitance than the species with buffered water potential fluctuations. A quantitative index of relative capacitance for the espeletias is the WSP volume per unit of actively transpiring leaf area, which is $11.3 \times 10^{-4}$ m for *E. lutescens* and $5.5 \times 10^{-4}$ m for *E. moritziana* (Table 2). Moreover, this index, which is significantly correlated with the extent of daytime water potential fluctuations for seven species of *Espeletia* (Goldstein et al. 1984), tends to increase with elevation (Table 2).

The volume of water in the WSP among species of espeletias also tends to increase with elevation (Table 2), as does plant height (Goldstein et al. 1984). From a hydraulic standpoint, long narrow cylindrical water reservoirs that maximize the area of contact between the xylem and the WSP are more effective for water exchange than are short wide ones of similar volume. This may help explain the pattern of increasing plant height with increasing elevation and the accompanying lower temperatures, as low temperatures can reduce water movement to the leaves in the same way that water uptake from the soil is impaired by low temperatures. In

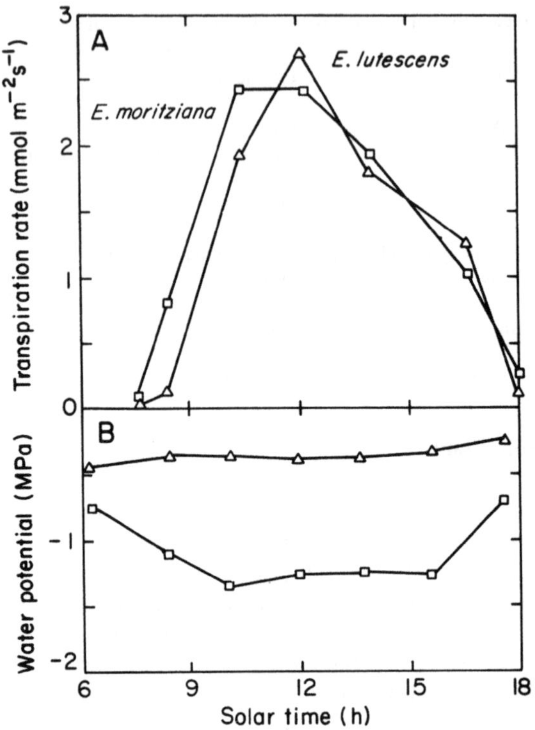


**Fig. 2A,B.** Daytime variations in transpiration rate (**A**) and leaf water potential (**B**) for *Espeletia lutescens* (Δ) and *E. moritziana* (□) at 4200 m in Piedras Blancas, Venezuela. Unpublished observations of G. Goldstein and F. Meinzer for days with some morning cloudiness and with average day/night air temperatures of +6/–1 °C

**Table 2.** Summary of volume of the water-storage parenchyma (WSP), leaf area of the plant's rosette, total hourly transpiration rate, and possible transpiration duration using water from WSP capacitance for seven espeletia species growing at different elevations in Venezuela. (Goldstein et al. 1984)

| Species | Elevation (m) | Volume of wsp ($\times 10^{-4}$ $m^3$) | Leaf area ($m^2$) | Transpiration rate ($g\ h^{-1}$) | Transpiration of stored water (h) |
|---|---|---|---|---|---|
| *E. lutescens* | 4200 | 9.68 | 0.860 | 71 | 2.50 |
| *E. moritziana* | 4200 | 2.68 | 0.485 | 40 | 1.44 |
| *E. spicata* | 4200 | 5.66 | 0.996 | 82 | 1.96 |
| *E. schultzii* | 3560 | 4.66 | 1.158 | 95 | 1.04 |
| *E. floccosa* | 3560 | 0.72 | 0.545 | 45 | 0.59 |
| *E. marcana* | 3100 | 2.70 | 0.673 | 55 | 1.56 |
| *E. atropurpurea* | 2850 | 0.29 | 0.198 | 16 | 0.57 |

the early morning water thus moves radially from the WSP into the xylem and then axially into the leaves; after soil temperatures rise as the daytime progresses, the leaves are supplied primarily by water moving from the soil along the low-resistance xylem pathway. At night, the WSP is recharged as water potential equilibrium is approached between the plant and the soil (Goldstein and Meinzer 1983). The dead leaves surrounding the stem thermally insulate the WSP, helping prevent freezing injury and thus facilitating water delivery in the early morning (Goldstein and Meinzer 1983).

The driving force for water movement in cacti is not only the transpirational water loss (as is the case for the $C_3$ espeletias) but also the diel fluctuations in osmotic pressure of the chlorenchyma induced by CAM (Nobel 1988a; Schulte et al. 1989). Transpiration during nocturnal stomatal opening decreases the water potential near the sites of water loss at the stem surface. As the water potential in the xylem then decreases, water enters the transpiration stream from the storage tissues, thus reducing the demand for water from the soil. A decrease in water potential can also be produced by the nocturnal increase in osmotic pressure of the chlorenchyma cells characteristic of CAM plants (Lüttge and Nobel 1984; Smith and Lüttge 1985). Such osmotic pressure changes provide a driving force for the movement of water from the WSP to the chlorenchyma as well as uptake from the soil (Lüttge 1986). Relatively large diel fluctuations in the osmotic pressure, and hence in the water potential, occur in the photosynthetic tissue, whereas the variations of the water potential components in the adjacent water-storage parenchyma are small. During the daytime, the water potential gradient between the chlorenchyma and the WSP is actually reversed as a consequence of a decrease in the malic acid concentration, and thus in the osmotic pressure, in the chlorenchyma cells. Substantial movement of water from the chlorenchyma to the WSP during the daytime is therefore expected for cacti such as *Ferocactus acanthodes* (Schulte et al. 1989).

In both cacti and espeletias, large capacitance provides water to the photosynthetic tissues when soil water is not readily available. Adult individuals of *E. lutescens*, for example, can have a WSP that occupies $9.7 \times 10^{-4}$ $m^3$ of the stem. Even though a large fraction of this volume represents water (0.86), only 21% of this water is readily available for transpiration (Goldstein et al. 1984). For an average transpiration rate (82 g $m^{-2}$ $h^{-1}$) of a typical rosette with an exchange surface area of 0.86 $m^2$, about 2.5 h of transpirational losses can be sustained by the internal water storage (Table 2). This is approximately the amount of time in the early morning during which water is not readily available from the soil.

Cacti also have large internal water-storage compartments. However, the high water-use efficiency due to CAM and the tight stomatal closure that can occur once soil water potential has decreased below shoot water potential (Nobel 1988a) make the WSP a source of water for a much longer period of time than for espeletias. For example, a representative individual of *F. acanthodes* with a stem surface area of 0.23 $m^2$ and a stem volume of $1.03 \times 10^{-2}$ $m^3$ can continue nocturnal stomatal opening for about 46 days after the plant is no longer able to obtain water from the soil, during which time it loses about 33% of the water present at full hydration (Nobel 1977). During the subsequent 2.5 and 6 months of drought, it can lose an additional 15% and 19%, respectively, of its maximum water content (Nobel 1977). Moreover, the water loss comes predominantly from the WSP, with a much lower fractional loss from the chlorenchyma (Barcikowski and Nobel 1984). This differential water loss helps maintain a positive turgor pressure in the chlorenchyma and prolongs its photosynthetic activity by recycling $CO_2$ released internally by respiration (Kluge and Ting 1978; Barcikowski and Nobel 1984).

**Water Relations Characteristics of Capacitance Tissues**

An expression for capacitance (C) incorporating cellular parameters is

$$C = dV/d\psi = V/(\varepsilon + \pi),$$

where V is the water volume in the cells, $\psi$ is the water potential, $\varepsilon$ is the bulk elastic modulus of the cells, and $\pi$ is the cellular osmotic pressure (Molz and Ferrier 1982; Nobel 1991). This equation is obtained by differentiating an expression for the total water potential ($\psi = P - \pi$, where P is the hydrostatic pressure) with respect to volume, incorporating the definition for bulk elastic modulus (namely, $\varepsilon = V\Delta P/\Delta V$), and assuming that no changes occur for osmotically active solutes in the cells. Because $\varepsilon$ indicates the relationship between symplastic volume and hydrostatic (turgor) pressure, it is influenced primarily by the elastic properties of the cell walls. At high water content, $\varepsilon$ is usually the dominant influence on $\psi$, because a given change in V leads to a much larger fractional change in P than in $\pi$ (Nobel 1991). The above equation indicates that capacitance is directly proportional to the total water volume in a tissue and inversely proportional to $\varepsilon$ of its constituent cells (usually, $\varepsilon \gg \pi$).

The ideal cellular properties of tissues functioning as plant water reservoirs can be predicted. For example, a large decline in water content should not cause a large drop in water potential, as tends to be the case for large cells ($\varepsilon$ is inversely proportional to the radius of spherical cells; Tyree and Jarvis 1982). Because $\varepsilon$ is directly proportional to the cell wall thickness (Tyree and Jarvis 1982), elastic cell walls (low $\varepsilon$) can also limit the decline in tissue water potential resulting from a decrease in cellular water content. Cells of the water-storage parenchyma of *Opuntia ficus-indica* have extremely elastic cell walls ($\varepsilon$ = 0.38 MPa; G. Goldstein, J.L. Andrade, and P.S. Nobel, unpublished observations). Its $\varepsilon$ is in the lower end of the range of values measured for other plants, including the CAM species *Kalanchoë daigremontiana* (Zimmermann and Steudle 1978; Steudle et al. 1980). Furthermore, the WSP cells of *O. ficus-indica* can dehydrate beyond the point of zero turgor and still fully recover after rehydration. This ability to withstand a high degree of dehydration without irreversible damage is another desirable characteristic of water-storage tissues. The WSP of espeletias can also recover after reaching tissue water potentials below turgor loss point. However, $\varepsilon$ of their WSP cells (10.6 MPa for *E. lutescens*) is much higher than $\varepsilon$ for the WSP of *O. ficus-indica*, consistent with the relatively short period (less than 3 h) during which transpiration by espeletias is supplied by stored water (Goldstein et al. 1984).

Another desirable feature of a water-storage tissue is the presence of a material with colloid-like hygroscopic characteristics, which could occur in the intercellular spaces. The biophysical characteristics of this material should allow a large decline in its water content with only a small drop in its water potential (Morse 1990). Apparently the fibrous, highly branched polysaccharide mucilage, which exists in many platyopuntias (Gibson and Nobel 1986; Nobel 1988a), has this desired property. Extracellular (apoplastic) mucilage could passively regulate tissue water content during an initial phase of plant dehydration. Only when the mucilage water content is decreased substantially during drought, causing the water potential of the mucilage-water complex to decline appreciably, would the regulation of tissue water content shift to intracellular processes requiring metabolic energy, such as the production of more osmotically-active solutes or changes in cell wall properties that lead to a lower $\varepsilon$. Extracellular polysaccharides have also been implicated in the water relations of the $C_3$ plants *Argyroxiphium grayanum* (Robichaux and Morse 1990) and *Hemizonia luzulifolia* (Morse 1990).

## Shoot Temperatures

Subzero temperatures, occurring during winter for high latitude and high altitude populations of cacti and nightly in the tropical alpine areas where espeletias grow, make the measurement of thermal properties a prerequisite for understanding the ecophysiological responses of both groups. In the 1970s, a computer model was developed that can predict the surface temperatures of cacti based on their morphology and the prevailing environmental conditions (Lewis and Nobel 1977; Nobel 1978, 1988b). The cactus stem is modeled as a set of subvolumes and the temperature is predicted for each subvolume using an energy budget analysis (Nobel 1988a). Using such a model, the influence of morphological features, such as stem diameter and stem shortwave absorptance, on tissue temperature can be quantitatively evaluated.

Modeling attention has been focused on the stem shading by the spines and the pubescence that covers the apical meristem responsible for primary stem growth of columnar and barrel cacti (Nobel 1978). For instance, under the same environmental conditions, the minimum surface temperatures were the highest for *Carnegiea gigantea* (the giant saguaro), 1.8 °C lower for *Stenocereus thurberi*, and 3.8 °C lower for *Lophocereus schottii*, which is the same relative order as the northernmost limits of these three columnar cacti in the Sonoran Desert (Nobel 1980a). Based on stem height, stem diameter, spine properties, apical pubescence, and other morphological features of the four *Ferocactus* species occurring in the southwestern United States, their predicted minimum stem surface temperatures and latitudinal ranges agreed with field observations (Nobel 1980a). (These predictions involve the assumption that tissue sensitivities to subzero temperatures are similar among the species.) The 10-mm-thick apical pubescence at the apex of *C. gigantea* leads to a simulated 2.4 °C higher nocturnal temperature compared with a plant without pubescence, which extends the distribution of this species into regions of lower temperature (Nobel 1980a). Nurse plants, which affect longwave radiation exchange by partially replacing the potentially cold nighttime sky with a radiator (the branches and the leaves of the nurse plant) close to air temperature, can also increase the nocturnal stem temperatures of seedlings and thereby extend the range of *C. gigantea* to colder regions (Nobel 1980a). The model also closely predicts the upper elevational limits of populations of *Eriosyce ceratistes* and *Trichocereus chilensis* over a 500-km transect in Chile. Taking into consideration ecotypic variations in the morphological features of the stems and topographical features such as cold-air drainage, the upper elevational limits of the populations are sensitive to a change of 0.1 °C, which corresponds to only 20 m in elevation (Nobel 1980b).

Manipulative experiments and energy balance models have also helped explain certain morphological features of espeletias, such as the thick leaf pubescence and the stem insulation by dead leaves. The pubescence of espeletia leaves primarily influences leaf temperature through increased boundary-later resistance to heat transfer, rather than through reduced absorptance of solar irradiation (Meinzer and Goldstein 1985). The decrease in sensible heat loss under high solar irradiation because of pubescence can result in leaf temperatures up to 10 °C above ambient air temperature, which can lead to higher photosynthetic rates by allowing metabolic processes to operate closer to optimal temperatures, especially at higher elevations. In this regard, the pubescence of leaves of *Espeletia schultzii* increases in thickness from 1.1 mm at 2600 m to 2.6 mm at 4200 m (Meinzer et al. 1985). However, energy-balance simulations and field data indicate that pubescent leaves are further below air temperature at night than glabrous leaves because of the reduced convective heat transfer (Meinzer and Goldstein 1986). The inward bending of the expanded leaves at night (nyctinasty) reduces heat conduction and convection, thus preventing the bud tissues and small expanding leaves from freezing (Smith 1974).

## Freezing Process and Supercooling

Cacti have stem cellular osmotic pressures of 0.5 to 1.2 MPa (Soule and Lowe 1980; Nobel 1988a), and the higher elevation espeletias have leaf osmotic pressures of 0.7 to 1.4 MPa (Rada et al. 1985). Because the freezing point depression is only 0.4 to 1.2 °C for such osmotic pressures (at equilibrium, the freezing point depression of a 1 molal solution is 1.86 °C; Nobel 1991), the shoot tissues of both groups cannot avoid freezing by equilibrium responses. However, supercooling has been observed for these tissues, as the stem or leaf temperature can decrease below the temperature that at equilibrium leads to freezing, yet the tissue water remains in the liquid phase. The lowest subzero temperature attained before ice formation occurs does not result in tissue injury. Moreover, decreasing water content or increasing osmotic pressure leads to a lowering of the equilibrium freezing temperature and, in addition, may result in a proportional increase in the extent of supercooling possible (Sakai and Larcher 1987). Because cacti can exhibit a nocturnal increase of osmotic pressure in the chlorenchyma, supercooling may be greater at night, although the magnitude of the effect would be small. For certain plants, especially those with high supercooling capacity such as espeletias, freezing occurs in the extracellular spaces and essentially simultaneously inside the cells, resulting in irreversible tissue damage (Modlibowska 1957; Rada et al. 1985; Goldstein et al. 1985b).

Many cacti can be transiently supercooled to −4 to −12 °C (Nobel 1988a). Such supercooling has been proposed to protect young *C. gigantea* from freezing damage on cold slopes (Steenbergh and Lowe 1976). On the other hand, *Coryphantha vivipara* can survive temperatures far below the temperature to which it will supercool (about −6 °C), so supercooling is not the means by which it avoids freezing damage (Nobel 1981). Yet the freezing behavior of supercooled water can help interpret some of the cellular events that take place (Fig. 3).

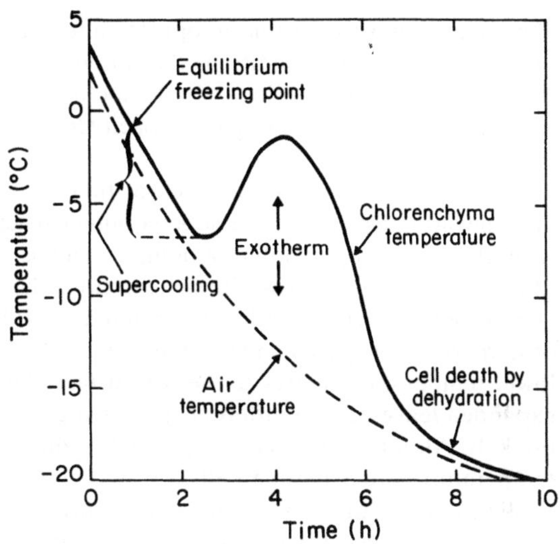


**Fig. 3.** Generalized representation of the lowering of chlorenchyma temperature leading to supercooling. After cooling below the equilibrium freezing temperature (supercooling), an exothermic reaction occurs as ice crystals are nucleated and form extracellularly. As cooling continues, water is distilled out of the cells and incorporated into the growing ice crystals, leading to the concentration of cellular solutes and eventual death by dehydration

During cooling of *C. vivipara*, an exothermic reaction occurs at –6 °C and the tissue temperature increases due to the release of the heat of fusion (the heat required to melt unit mass of ice; Nobel 1981). Based on this heat release and the specific heat of water, the extent of the temperature rise corresponds to the freezing of about 10% of the stem water, which is identified as extracellular water based on microscopic observations of *C. vivipara* (Nobel 1981) and results with other cacti (Uphof 1916). The ice crystals formed during the exotherm grow in size as water is distilled from within the protoplast, thereby concentrating the cellular solutes left behind and raising the cellular osmotic pressure. Such dehydration results in the increasingly shrunken appearance of the protoplasts of *C. vivipara* at different stages of cooling (Nobel 1981) and is known as "frost plasmolysis" (Levitt 1980). The growing ice crystals do not enter the cells and are not the cause of cellular death. Rather, intracellular dehydration leads to the death of the tissue, so freezing damage is thus essentially a water-stress injury.

For *C. vivipara* under well-watered conditions and moderate growth temperatures, death occurs near –15 °C. The cellular dehydration leading to a freezing point depression of –15 °C corresponds to an osmotic pressure of 7 MPa. If the water content varies inversely with the osmotic presure, then a loss of 94% of the cellular water would occur at the temperature leading to death of *C. vivipara*, in close agreement with the 91% water loss that leads to death by desiccation (Nobel 1981). Similarly, for seedlings of *Ferocactus acanthodes*, 91% of the water content is lost for death by freezing dehydration and 96% when they succumb to drought (Jordan and Nobel 1981).

For plants in general, supercooling is favored by small cell size, small intercellular air spaces, the absence of ice nucleators (or the presence of antinucleators), hydrophobic stem or leaf surfaces acting as barriers against external nucleators, and relatively low tissue water content (Sakai and Larcher 1987). Also, the accumulation of certain solutes that act as cryoprotectants can lead to a greater degree of supercooling. The climatic conditions that prevail in the high elevation tropical Andes, characterized by brief periods of mild nocturnal freezing temperatures, should favor selection for plants exhibiting the capacity for deep, repeated supercooling (Larcher 1981).

If supercooling is the mechanism that allows avoidance of ice formation for high-elevation espeletias, the supercooling capacity of their leaves should increase with elevation. Supercooling among espeletias not only increases with elevation (Table 3), but also it follows the same trend as the temperature lapse (approximately −0.6 °C per 100 m of elevation; Goldstein et al. 1985b). Leaves of shorter individuals of *E. timotensis* supercool more than leaves of taller individuals, consistent with the lower temperatures closer to the soil surface (Goldstein et al. 1985a). Because the expanded leaves of the different species supercool to −6 to −11 °C, leaf freezing is unlikely. Although leaves damaged by freezing are not observed in the field, if leaf temperatures are experimentally lowered in the laboratory to the maximum supercooling temperature, extracellular and intracellular freezing occurs more or less simultaneously, resulting in tissue injury (Table 3). Although tissue death essentially always accompanies intracellular freezing (Levitt 1980), cellular dehydration concomitant with the growth of extracellular ice crystals may actually be responsible for the death of leaves of espeletias.

The leaf apoplastic water content at full saturation (an indirect indication of intercellular spaces) in the higher elevation espeletia species is lower (2 to 4%) than in espeletias from lower elevation (7 to 36%; Table 3). Similarly, higher altitude populations of *E. schultzii* have relatively low apoplastic water content and high leaf supercooling capacity compared with populations from lower elevations (Rada et al. 1987). The small intercellular spaces apparently help to enhance supercooling by reducing the size of apoplastic spaces suitable for equilibrium freezing. Specifically, a reduction in apoplastic water available for nucleation reduces the production of ice

**Table 3.** Summary of maximum supercooling temperature, lowest tolerable temperature, and apoplastic water content for seven espeletia species growing at different elevations in Venezuela. (Goldstein et al. 1985b)

| Species | Elevation (m) | Maximum supercooling temperature (°C) | Lowest tolerable temperature (°C) | Apoplastic water at full saturation (% of total water) |
|---|---|---|---|---|
| *E. lutescens* | 4200 | −10.5 | −10.2 | 2.2 |
| *E. schultzii* | 4200 | −10.0 | −11.2 | 3.9 |
| *E. floccosa* | 3560 | −8.5 | −9.3 | 7.3 |
| *E. schultzii* | 3560 | −10.8 | −10.0 | 16.0 |
| *E. jahnii* | 3100 | −5.7 | −5.6 | 25.1 |
| *E. atropurpurea* | 3100 | −7.3 | −8.1 | 19.9 |
| *E. marcana* | 3100 | −9.1 | −8.0 | 20.5 |
| *E. angustifolia* | 2850 | −6.6 | −6.1 | 35.8 |

crystals that lead to cellular dehydration and that may also disrupt plasmalemmas and initiate lethal intracellular nucleation.

## Freezing Tolerance and Acclimation

Besides avoiding freezing damage by supercooling or morphological adaptations, increased tolerance to subzero temperatures is generally observed following previous exposure to low temperatures. Such acclimation is widespread among cacti, particularly those species occurring at relatively high latitudes or altitudes (Nobel 1988a). Cacti such as *Opuntia humifusa* and *O. erinacea* begin to dehydrate in late summer and become curled and prostrate by late autumn, having lost 45% of their tissue water despite relatively constant levels of soil water (Koch and Kennedy 1980; Littlejohn and Williams 1983). CAM activity of *O. humifusa*, measured by gas exchange and by acid accumulation, decreases in parallel with tissue water content, is minimal in winter, and increases in the spring (Koch and Kennedy 1980; Littlejohn and Williams 1983). Snow cover can provide thermal insulation as air temperatures fall below –40 °C, but even so stem temperatures of about –17 °C have been recorded for *O. humifusa* under these conditions (Koch and Kennedy 1980). Studies based on the sensitivity of plant tissues to freezing temperatures indicate that *O. polycantha* can tolerate –24 °C (Rajeshekar et al. 1979) and that *O. erinacea, O. humifusa, O. phaeacantha, O. spinosior,* and *O. stricta* can tolerate –17 or –18 °C but not –20 °C (Nardina and Mukhammedov 1973).

Low-temperature acclimation occurs in *Coryphantha vivipara*, which is distributed from northern Mexico (30°N) to southern Canada (50°N; Nobel 1981). Lowering the ambient air temperature from +30 to –10 °C in 10 °C steps at weekly intervals causes the subzero temperature that eliminates cellular uptake of the vital stain neutral red (indicating cellular death) to decrease from –15 to –22 °C. Since the study with *C. vivipara*, low-temperature tolerance and low-temperature acclimation have been determined for 22 other species of cacti (Table 4). The lowest temperature tolerated by cacti for a subzero treatment of 1 h at an ambient day/night air temperature of 10/0 °C ranges from –8 to –26 °C, with an average of –13 °C. The low-temperature acclimation per 10 °C decrease in ambient day/night air temperatures ranges from 0.0 to 3.7 °C, with an average of 1.0 °C (Table 4). The half-time for the low-temperature acclimation for two species of cacti from north-central Argentina, *Denmoza rhodacantha* and *Trichocereus candicans*, is about 3 days (Nobel 1982). Low-temperature acclimation is important for tolerating subzero temperatures in the field, as species with the greatest freezing tolerance when fully acclimated (–25 °C and lower), such as *C. vivipara* and *O. humifusa*, are also those showing substantial acclimation ability.

Because cacti are apparently less susceptible to a particular freezing episode when they are partially dehydrated, cultivators of cacti often withhold water when subzero temperatures are expected. Under natural conditions the osmotic pressure of *O. erinacea* and *O. humifusa* increases during the autumn (Koch and Kennedy 1980; Littlejohn and Williams 1983), which may reflect a decreased soil and root hydraulic conductivity at the lower temperatures, leading to less water uptake by the

**Table 4.** Low-temperature tolerance and low-temperature acclimation of cacti. Low-temperature tolerance is the subzero temperature (determined graphically) at which the uptake of the vital stain neutral red is just eliminated following a 1-h treatment. The low-temperature acclimation indicates the lowering of such a temperature as the day/night air temperatures are lowered from 20/10 °C to 10/0 °C over a 1-week period. (Data adapted from Nobel 1982, 1984, 1990 and Nobel and Loik 1990)

| Species | Temperature just leading to elimination of stain uptake by chlorenchyma cells for plants maintained at 10/0 °C (°C) | Low-temperature acclimation per 10 ° decrease in day/night air temperatures (°C) |
|---|---|---|
| *Carnegiea gigantea* | –13 | 0.5 |
| *Coryphantha vivipara* | –24 | 1.7 |
| var. *deserti* | –23 | – |
| var. *rosea* | –26 | – |
| *Denmoza rhodacantha* | –14 | 1.0 |
| *Eriosyce ceratistes* | –14 | 0.8 |
| *Ferocactus acanthodes* | –13 | 0.3 |
| *F. covillei* | –11 | 0.0 |
| *F. viridescens* | –10 | 0.3 |
| *F. wislizenii* | –12 | 0.3 |
| *Lophocereus schottii* | –11 | 0.5 |
| *Opuntia bigelovii* | –11 | 0.8 |
| *O. ficus-indica* (five accessions) | –9 to –12 | 0.2 to 1.4 |
| *O. fusicaulis* | –10 | 1.7 |
| *O. humifusa* (two populations) | –13 to –15 | 3.6 to 3.8 |
| *O. hyptiacantha* | –9 | 0.2 |
| *O. megacantha* | –9 | 1.1 |
| *O. polyacantha* | –21 | – |
| *O. ramosissima* | –8 | – |
| *O. rastrera* | –13 | 2.6 |
| *O. streptacantha* (five accessions) | –8 to –14 | 0.6 to 2.7 |
| *Pediocactus simpsonii* | –22 | – |
| *Stenocereus thurberi* | –13 | 0.3 |
| *Trichocereus candicans* | –11 | 1.0 |
| *T. chilensis* | –12 | 0.9 |

plants. The accompanying cellular changes that enhance freezing tolerance are not known. Less apoplastic water is probably available for freezing under the dehydrated condition, which could postpone the growth of the extracellular ice crystals that may eventually become large enough to disrupt the plasmalemmas, although the primary cause of death for cacti appears to be frost plasmolysis. A lower water content may also result in slower water movement out of the cells during freeze dehydration, resulting in less irreversible damage (Nobel and Loik 1990). A related matter is the increase as winter approaches in soluble sugars and other solutes that might act as cellular cryoprotectants.

Freezing tolerance has also been observed for *Draba chionophila* (Brassicaceae), a small rosette with dead leaves surrounding the stem and a stem WSP as for espeletias. It reaches the highest altitudes among vascular plants in the Venezuelan Andes, approximately 4700 m (Azocar et al. 1988). Sudden 1 to 2 °C increases in its leaf temperatures at night (from about –6 °C) result from the release of the heat of fusion (Fig. 3) during the freezing of extracellular water (Azocar et al. 1988). In laboratory experiments, exothermic reactions indicating ice formation occurred at approximately the same temperature, and freezing injury occurred at about –15 °C. The difference between freezing temperatures and the temperatures at which injury occurs indicates that this species is capable of tolerating and recovering from the freezing of extracellular water in its intercellular spaces, just as do cacti (Nobel 1988a).

*Draba chionophila* shows little supercooling compared with espeletias, whose exposed leaves can supercool from –6 to –11 °C (Table 3) and even to –16 °C for some species (Goldstein et al. 1985b). Espeletias, however, reach lower maximal altitudes and very infrequently are exposed to the low temperatures (–10 to –12 °C) that *D. chionophila* experiences at higher altitudes. The osmotic pressure and soluble carbohydrate levels increase at night in leaves of *D. chionophila* by an average of 0.8 MPa and 100%, respectively (Azocar et al. 1988), which may have an adaptive advantage, because soluble carbohydrates in general and sucrose in particular can be membrane cryoprotectants (Levitt 1980). High elevation tropical plants can thus use two mechanisms to cope with subzero temperatures at increasing altitude: (1) an increase in the ability to supercool (i.e., espeletias); and (2) tolerance of substantial intracellular dehydration (i.e., *D. chionophila*), as is also the case for cacti.

## Conclusions and Future Research

Basic principles concerning capacitance effects on water relations and concerning freezing avoidance as well as tolerance have been examined for cacti and espeletias. The cacti considered are CAM plants with a large water-storage parenchyma in close proximity to the photosynthetic tissue. A small gas exchange surface per unit shoot water volume and nocturnal stomatal opening help to explain their success in the most extreme desert environments, where net $CO_2$ uptake can proceed for many weeks after the onset of drought. The $C_3$ espeletias have conspicuous stem water storage that is distant from the rosette leaves but well connected hydraulically. The thermally insulated water-storage parenchyma, which is larger for the higher elevation species, can also help explain the success of espeletias in tropical alpine environments where stored tissue water is needed to sustain transpiration for a few hours each morning.

The WSP of cacti and espeletias differ in some cellular properties. The WSP of cacti has cells with very elastic cell walls and various amounts of extracellular mucilage. This leads to a relatively large volume of water that can be released per unit decrease in water potential, an advantage for a water-storage tissue that experiences a large decline in water content over a long period of time. The WSP of the espeletias, on the other hand, does not experience a large decline in water content,

as the drought period may last for only a few hours; their WSP is composed of cells with relatively inelastic walls and lacks apoplastic mucilage. The hydraulic connections with the photosynthetic tissues are tight in both cases, although the predominant mechanisms of water transport between storage regions and the sites of evaporation are different – long distance mass flow in the xylem in the espeletias and diffusion over cellular distances in the cacti. Thus, the water-storage parenchyma of the espeletias helps buffer water deficits of a few hours duration, and the WSP of the cacti helps buffer water deficits lasting up to several months.

The processes driving water movement are somewhat different in the two groups of plants. Whereas daytime transpiration is the major process responsible for water movement along the soil-plant-atmosphere continuum in espeletias, nocturnal transpiration as well as diel fluctuations in osmotic pressure of the chlorenchyma are both important in the water movement of well-watered cacti. When stomata remain closed at night after several weeks of drought for cacti, the fluctuations in osmotic pressure of the chlorenchyma cells remain the only driving force for water redistribution between the water-storage parenchyma and the photosynthetic tissue.

Although various aspects of the role of capacitance in the water relations of cacti and espeletias are well understood, the water circulation pathways from the soil into storage as well as between storage and the photosynthetic tissues remain mainly hypothetical. Experiments with radioactive tracers, such as tritiated water, and fluctuations in stable isotope composition of water ($^{2}H$ and $^{18}O$) in different tissues and at different times can provide direct evidence of water movement and its internal redistribution throughout a 24-h cycle. Additional studies of the water relations properties of the water-storage and photosynthetic tissues and the variation in all water potential components of these tissues on a 24-h and seasonal basis are necessary. More mechanistic models can then be developed that integrate this information at a whole-plant level.

Cacti and espeletias can both be subjected to freezing episodes. Supercooling appears to be an important mechanism to avoid freezing damage for espeletias, where such episodes can occur nightly for a few hours in the high tropics. However, leaf tissues experience irreversible damage once temperatures are experimentally lowered below the supercooling temperature. Small intercellular air spaces, nyctinastic inward bending of the leaves, and hydrophobic leaf surfaces appear to be important characteristics related to the supercooling behavior of espeletia leaves. In many cacti native to high elevations or high latitudes, on the other hand, freezing after supercooling does not necessarily result in injury to the tissue. Damage appears to be caused by the movement of a large amount of water across the cell membranes once ice crystals start to grow in the intercellular spaces, leading to cellular dehydration and water-stress injury. Indeed, tolerance of temperatures below –25 °C has been observed in acclimated cacti. Research is necessary to identify the nucleating agents for ice formation in both cacti and espeletias, and the possible role that antinucleating substances may play in the freezing process.

The increase in the ability to tolerate extracellular freezing as ambient temperatures decrease apparently is a widespread phenomenon for cacti, especially species native to relatively high latitudes or high elevations. This acclimation may be related to the partial dehydration accompanying low day/night temperatures, which some-

how allows the plants to tolerate further dehydration. The biophysical and biochemical changes occurring at cellular and tissue levels during acclimation are not known. Particular attention should be paid to which osmotically active solutes change, such as cryoprotectants, and the changes in the apoplastic water fraction that can affect the growth of extracellular ice crystals.

Water relations characteristics and the mechanisms of tolerating low temperatures are strongly interrelated for both cacti and espeletias. The large stem capacitances and the duration of the freezing episodes have undoubtedly exerted constraints on the evolution of physiological and morphological adaptations to cope with freezing temperatures. Certain species of cacti can tolerate equilibrium freezing at lower temperatures if, during the acclimation process, the shoot water available for crystallization decreases. Apparently some of the changes in the tissues and cells during acclimation to low temperatures are similar to changes occurring during drought. Although espeletia leaves cannot tolerate experimentally induced tissue freezing, in their native habitats they can span the nocturnal subzero period in an unfrozen, supercooled state. Their stem water-storage parenchyma, which does not exhibit the supercooling capacity of the leaves, is protected during the short period of nocturnal frosts by an insulating layer of dead leaves. Future research on the mechanisms of freezing, including assessment of nucleators, cryoprotectants, and interfacial phenomena, should provide further insights into the interrelationships of water and low temperature relations for cacti and espeletias in particular and for vascular plants in general.

*Acknowledgments.* Financial support was provided by the Ecological Research Division of the Office of Health and Environmental Research, U.S. Department of Energy contract DE-FCO3-87-ER60615.

## References


Azocar A, Rada F, Goldstein G (1988) Freezing tolerance in *Draba chionophila*, a "miniature" caulescent rosette species. Oecologia 75:156–160

Barcikowski W, Nobel PS (1984) Water relations of cacti during desiccation: distribution of water in tissues. Bot Gaz 145:110–115

Gibson AC, Nobel PS (1986) The cactus primer. Harvard Univ Press, Cambridge

Goldstein G, Meinzer F (1983) Influence of insulating dead leaves and low temperatures on water balance in an Andean giant rosette plant. Plant Cell Environ 6:649–656

Goldstein G, Meinzer F, Monasterio M (1984) The role of capacitance in the water balance of Andean giant rosette species. Plant Cell Environ 7:179–186

Goldstein G, Meinzer F, Monasterio M (1985a) Physiological and mechanical factors in relation to size-dependent mortality in an Andean giant rosette species. Oecol Plant 6:263–275

Goldstein G, Rada F, Azocar A (1985b) Cold hardiness and supercooling along an altitudinal gradient in Andean giant rosette species. Oecologia 68:147–152

Goldstein G, Rada F, Canales MO, Zabala O (1989) Leaf gas exchange of two giant caulescent rosette species. Oecol Plant 10:359–370

Jordan PW, Nobel PS (1981) Seedling establishment of *Ferocactus acanthodes* in relation to drought. Ecology 62:901–906

Kaufmann MR (1975) Leaf water stress in Engelmann spruce. Plant Physiol 56:841–844

Kaufmann MR (1977) Soil temperature and drying cycle effects on water relations of *Pinus radiata*. Can J Bot 55:2413–2418

Kluge M, Ting IP (1978) Crassulacean acid metabolism. Analysis of an ecological adaptation. Ecol Stud Ser 30, Springer, Berlin Heidelberg New York

Koch KE, Kennedy RA (1980) Effects of seasonal changes in the Midwest on Crassulacean acid metabolism (CAM) in *Opuntia humifusa* Raf. Oecologia 45:390–395

Larcher W (1981) Resistenzphysiologische Grundlagen der evolutiven Kalteakklimatisation von Sprosspflanzen. Plant Syst Evol 137:145–180

Levitt J (1980) Responses of plants to environmental stresses, 2nd edn, vol 1, Chilling, freezing, and high temperature stresses. Academic Press, Lond New York

Lewis DA, Nobel PS (1977) Thermal energy exchange model and water loss of a barrel cactus, *Ferocactus acanthodes*. Plant Physiol 60:609–616

Littlejohn RO, Williams GJ (1983) Diurnal and seasonal variations in activity of Crassulacean acid metabolism and plant water status in a northern latitude population of *Opuntia erinacea*. Oecologia 59:83–87

Lopez FB, Nobel PS (1991) Root hydraulic conductivity of two cactus species in relation to root age, temperature, and soil water status. J Exp Bot 42:143–149

Lüttge U (1986) Nocturnal water storage in plants having Crassulacean acid metabolism. Planta 168:287–289

Lüttge U, Nobel PS (1984) Day-night variations in malate concentration, osmotic pressure, and hydrostatic pressure in *Cereus validus*. Plant Physiol 75:195–200

McGarvie D, Parolis H (1981) The acid-labile peripheral chains of the mucilage of *Opuntia ficus-indica*. Carbohydr Res 94:57–65

Medina E, Cram WJ, Lee HSJ, Lüttge U, Popp M, Smith JAC, Diaz M (1989) Ecophysiology of xerophytic and halophytic vegetation of a coastal alluvial plain in northern Venezuela I. Site description and plant communities. New Phytol 111:233–243

Meinzer FC, Goldstein G (1985) Some consequences of leaf pubescence in the Andean giant rosette plant *Espeletia timotensis*. Ecology 66:512–520

Meinzer FC, Goldstein G (1986) Adaptations for water and thermal balance in Andean giant rosette plants. In: Givnish T (ed) On the economy of plant form and function. Cambridge Univ Press, New York, pp 381–411

Meinzer FC, Goldstein G, Rundel PW (1985) Morphological changes along an altitude gradient and their consequences for an Andean giant rosette plant. Oecologia 65:278–283

Modlibowska I (1956) Le problème des gelées printanières et la fruitierie. Rapp Gen Congr Pomol Int, Namur 1956, pp 83–111

Molz FJ, Ferrier JM (1982) Mathematical treatment of water movement in plant cells and tissue: a review. Plant Cell Environ 5:191–206

Morse SR (1990) Water balance in *Hemizonia luzulifolia*: the role of extracellular polysaccharides. Plant Cell Environ 13:39–48

Nardina NS, Mukhammedov GM (1973) Kul'tura vidov Opuntia Mill. v tsentral'nykh Karakumakh. Probl Osvo Pustyn' 5:60–61

Nobel PS (1977) Water relations and photosynthesis of a barrel cactus, *Ferocactus acanthodes*, in the Colorado Desert. Oecologia 27:117–133

Nobel PS (1978) Surface temperatures of cacti – influences of environmental and morphological factors. Ecology 59:986–996

Nobel PS (1980a) Morphology, surface temperatures, and northern limits of columnar cacti in the Sonoran Desert. Ecology 61:1–7

Nobel PS (1980b) Influences of minimum stem temperatures on ranges of cacti in southwestern United States and central Chile. Oecologia 47:10–15

Nobel PS (1981) Influence of freezing temperatures on a cactus, *Coryphantha vivipara*. Oecologia 48:194–198

Nobel PS (1982) Low-temperature tolerance and cold hardening of cacti. Ecology 63:1650–1656

Nobel PS (1984) PAR and temperature influences on $CO_2$ uptake by desert CAM plants. Proc IV Int Congr Photosynth, Adv Photosynth Res IV. 3:193–200

Nobel PS (1988a) Environmental biology of agaves and cacti. Cambridge Univ Press, New York

Nobel PS (1988b) Principles underlying the prediction of temperature in plants, with special reference to desert succulents. In: Long SP, Woodward FI (eds) Plants and temperature. Soc Exp Biol, Company Biol, Cambridge, pp 1–23

Nobel PS (1990) Low-temperature tolerance and $CO_2$ uptake for platyopuntias – a laboratory assessment. J Arid Environ 18:313–324
Nobel PS (1991) Physicochemical and environmental plant physiology. Academic Press, San Diego
Nobel PS, Loik ME (1990) Thermal analysis, cell viability, and $CO_2$ uptake for a widely distributed North American cactus, *Opuntia humifusa*, at subzero temperatures. Plant Physiol Biochem 28:429–436
Rada F, Goldstein G, Azocar A, Meinzer F (1985) Freezing avoidance in Andean giant rosette plants. Plant Cell Environ 8:501–507
Rada F, Goldstein G, Azocar A, Torres F (1987) Supercooling along an altitudinal gradient in *Espeletia schultzii*, a caulescent giant rosette species. J Exp Bot 38:491–497
Rajeshekar C, Gusta LV, Burke MJ (1979) Membrane structural transitions: probable relation to frost damage in hardy herbaceous species. In: Lyons JM, Graham D, Raison JK (eds) Low temperature stress in crop plants. Academic Press, Lond New York, pp 255–274
Robichaux RH, Morse SR (1990) Extracellular polysaccharide and leaf capacitance in a Hawaiian bog species, *Argyroxiphium grayanum* (Compositae-Madiinae). Am J Bot 77:134–138
Running SW, Reid PC (1980) Soil temperature influences of *Pinus contorta* seedlings. Plant Physiol 65:635–640
Sakai A, Larcher W (1987) Frost survival of plants: responses and adaptation to freezing stress. Springer, Berlin Heidelberg New York
Schulte PJ, Smith JAC, Nobel PS (1989) Water storage and osmotic pressure influences on the water relations of a dicotyledonous desert succulent. Plant Cell Environ 10:639–648
Smith AP (1974) Bud temperature in relation to nyctinastic leaf movement in an Andean giant rosette plant. Biotropica 6:163–266
Smith AP (1979) The function of dead leaves in *Espeletia schultzii* (Compositae), an Andean giant rosette plant. Biotropica 11:43–47
Smith JAC, Lüttge U (1985) Day-night changes in leaf water relations associated with the rhythm of crassulacean acid metabolism in *Kalanchoë daigremontiana*. Planta 163:272–282
Soule OH, Lowe CH (1980) Osmotic characteristics of tissue fluids in the sahuaro giant cactus (*Cereus giganteus*). Ann MO Bot Garden 57:265–351
Steenbergh WF, Lowe CH (1976) Ecology of the saguaro: I. The role of freezing weather in a warm-desert plant population. In: Research in the parks. National Park Service Monograph Series, Number 1. US Gov Printing Office, Washington DC, pp 49–92
Steudle E, Smith JAC, Lüttge U (1980) Water relation parameters of individual mesophyll cells of the crassulacean acid metabolism plant *Kalanchoë daigremontiana*. Plant Physiol 66:1155–1163
Ting IP (1985) Crassulacean acid metabolism. Annu Rev Plant Physiol 36:595–622
Ting IP, Jennings W (1976) Deep canyon, a desert wilderness for science. Boyd Deep Canyon Desert Res Cent, Univ California, Riverside
Trachtenberg S, Mayer AM (1982) Composition and properties of *Opuntia ficus-indica* (L.) Mill. mucilage. Phytochemistry 21:2835–2843
Troll C (1968) The cordilleras of the tropical Americas. Aspects of climate, phytogeography and agrarian ecology. In: Troll C (ed) Geo-ecology of the mountain regions of the tropical Americas. UNESCO, New York, pp 13–56
Tyree MT, Jarvis PG (1982) Water in tissues and cells. In: Lange OL, Nobel PS, Osmond CB, Ziegler H (eds) Encyclopedia of plant physiology, New Series, vol 12B, Physiological plant ecology II. Springer, Berlin Heidelberg New York, pp 36–77
Uphof JCTh (1916) Cold-resistance in spineless cacti. Bull 79, Univ Arizona Agric Exp Stat, Tucson
van den Honert TH (1948) Water transport as a catenary process. Disc Faraday Soc 3:146–153
Zimmermann U, Steudle E (1978) Physical aspects of water relations of plant cells. Adv Bot Res 6:45–117

# IV Freezing Stress

Chapter 16

# Ice Nucleating Agents in Cold-Hardy Insects

K.E. Zachariassen

## Introduction

Water is the solvent of all biological systems. Whenever insects and other poikilothermic animals are exposed to subfreezing temperatures, there is a possibility that their body water will freeze. Organismal freezing is usually fatal. The problems of survival at subzero temperatures are thus to a great extent the problems of how to deal with the body water at subzero temperatures.

Insects may use two fundamentally different strategies to handle the problems of freezing of body water at subzero temperatures. They may seek to avoid freezing by allowing their body fluids to remain in the liquid state at temperatures far below zero. This strategy is used by a great number of insects, and must involve removal or inactivation of all components which may trigger freezing. Alternatively, insects may establish a controlled, tolerable freezing, which must be restricted to the fluid surrounding closed compartments such as cells and intestine. In many insects which tolerate freezing this is achieved by means of potent ice nucleating agents (INAs), which are present in the extracellular fluid (hemolymph) during the cold season. Thus, both strategies used by cold-hardy insects to control the freezing behavior of their body water involve manipulation of ice nucleating agents.

The physiological mechanisms involved in insect cold-hardiness have previously been dealt with in a number of review articles (Baust 1973; Duman and Horwath 1983; Zachariassen 1985; Storey and Storey 1988; Block 1990). This chapter is a review of the present knowledge on the physicochemical properties and the physiological functions of the biological ice nucleating agents involved in the cold-hardening of insects.

## Definitions

The *melting point* or equilibrium freezing point of a solution is the temperature where the last tiny ice crystal disappears when a frozen solution is slowly heated. Pure water has a melting point of 0 °C. Addition of solutes to the water causes a depression of the melting point. This depression is one of the so-called colligative

Department of Zoology, University of Trondheim, AVH, 7055 Dragvoll, Norway

Somero et al. (Eds.)
Water and Life

properties of the solution. The osmolality is proportional with the melting point depression, and thus, the osmolality can be calculated from melting point via the osmolal melting point depression, which is 1.86 °C per Osmol.

When a system exists in an unfrozen state at a temperature below its melting point, the system is said to be *supercooled*. The supercooled state is physically metastable, and sooner or later, depending on factors such as temperature, solute concentration, and sample volume, the system will enter the physically stable frozen state. The temperature where spontaneous freezing occurs in a supercooled system is called the *supercooling point* of the system. The difference between the melting point and the supercooling point is the *supercooling capacity* of the system.

## Freezing of Physical Solutions

### *Initiation of Freezing, Ice Nucleation*

Freezing of a supercooled solution may be initiated spontaneously by the water molecules themselves (homogeneous nucleation) or by foreign components, so-called ice nucleating agents (heterogeneous nucleation). Homogeneous nucleation is probably caused by water molecules forming aggregates, with an ice-like structure fit to seed ice formation. The aggregates increase in size as the temperature is reduced. When an aggregate reaches a certain critical size, with a critical surface curvature, further addition of water molecules will no longer increase but reduce the free energy of the system (Knight 1967). This favors further growth of the crystal, and ice grows from the seeding crystal and spreads throughout the system.

### *The Freezing Process in Solutions*

Ice structures contain only water, and therefore, solutes are excluded from the growing ice crystals and left behind in the fluid fraction surrounding the ice, where they become gradually more concentrated as the freezing proceeds. There must always be vapor pressure equilibrium between the fluid fraction and the ice, i.e., the concentration of the fluid fraction must represent a melting point equal to the temperature of the system. If the temperature of the system changes, water will freeze out of the fluid fraction or melt away from the ice until the fluid fraction has attained a melting point equal to the new temperature.

When a supercooled solution freezes, the released heat of fusion of freezing water will cause an immediate temperature increase of the system (Fig. 1). When the temperature has increased to the melting point of the unfrozen fluid fraction, a further temperature increase would cause the ice to melt, and thus, the arrival at this temperature marks the end of the initial freezing phase.

Since the fluid fraction becomes concentrated as ice forms, the temperature reached at the end of the initial freezing phase will not be the melting point of the unfrozen body fluid, but the melting point of the unfrozen fraction when the initially frozen solvent water has been transferred to ice (Fig. 1). Due to the high heat of

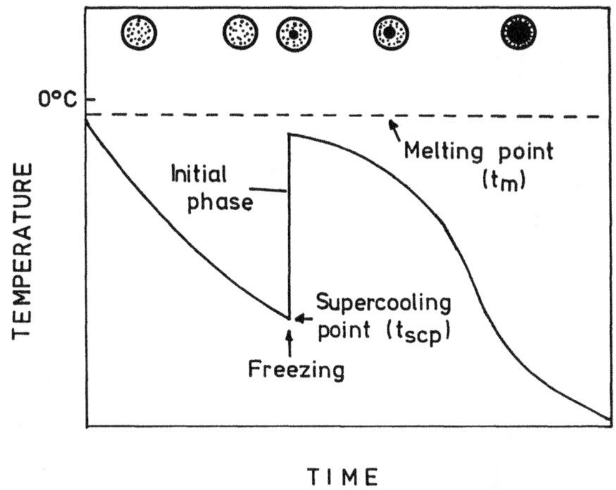


**Fig. 1.** Temperature curve before and during freezing of a supercooled solution. *Upper circles* represent concomitant changes in fraction of ice (*dark central circle*) and the concentration of solutes (*small points*)

fusion of water (79.71 cal/g), only a small fraction of the water will usually freeze during the initial phase.

After the initial freezing phase, the freezing can proceed only as fast as the heat of fusion released by further freezing can be removed from the system.

## *Freezing of Biological Systems/Freezing Injuries*

Biological fluids contain a wide variety of dissolved ionic and molecular solutes. Normal physiological body fluids are relatively dilute and display only moderate melting point depressions, in most cases less than 1 °C. Since there is osmotic equilibrium across animal cell membranes (Schmidt-Nielsen 1983), these melting point values apply to intracellular as well as extracellular fluids.

When ice formation is initiated inside a closed compartment such as a cell or the intestine, the increased concentration of the fluid fraction inside the compartment leads to an osmotic influx of water from the extracellular fluid. The compartment subsequently swells. If the swelling becomes too comprehensive, the membrane surrounding the compartment may rupture (Fig. 2). The destruction of the intestinal wall would cause digestive enzymes to enter the hemolymph, an event which is likely to cause the death of the organism. If the membrane surrounding a cell ruptures in the same manner, the cell will be destroyed.

Cooling rates in nature are normally very moderate, and too low to make nucleation likely to occur at more than one nucleation site. If the nucleation site is intracellular, the rupture of the cell membrane will cause ice to spread in the extracellular fluid and cause an osmotic shrinking of other closed compartments (cells and intestine) which must stay in osmotic equilibrium with the extracellular fluid (Fig. 2). Osmotic shrinkage is generally more tolerable than swelling, and the osmotic outflux of water will cause all closed compartments to reach vapor equilibrium with the extracellular ice. This will leave no compartment supercooled and

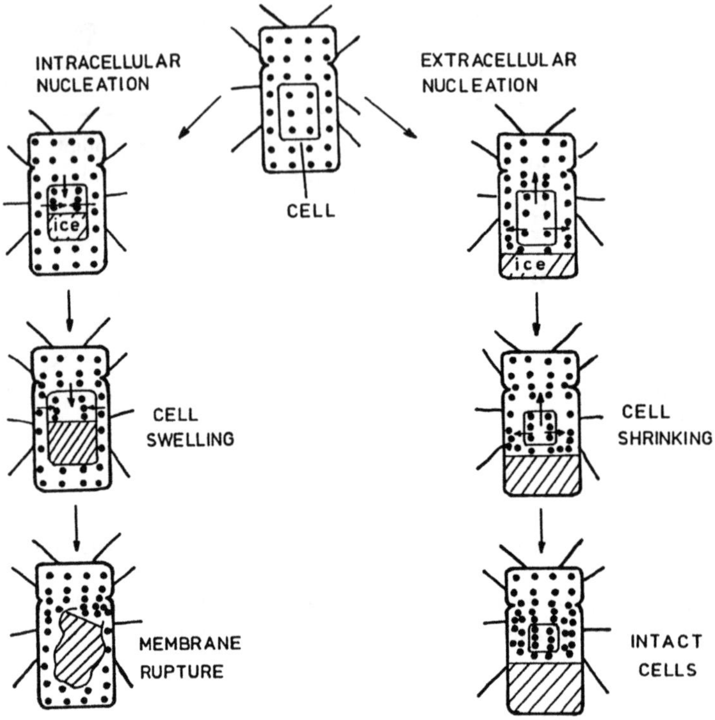


**Fig. 2.** Freezing of insect with extracellular and intracellular ice nucleation. The *inner "box"* represents cellular compartment. Solutes in the body fluid compartments are represented by the *black points*, whereas the *shaded area* represents ice. *Small arrows* indicate the movement of water across cell membranes. (After Zachariassen 1989)

eliminate the possibility of nucleation in other closed compartments. Thus, intracellular ice nucleation is likely to destroy only the cell in which nucleation took place, and leave the other cells protected against new nucleation events.

One might argue that the loss of one cell should be tolerable to an organism, particularly if the nucleation leads to a general protection of the other cells by the subsequent extracellular spreading of the ice. However, many cold-exposed organisms freeze and thaw over and over again during the cold season, and nucleation may predominantly take place in tissues where even the destruction of few cells is important (as may be the case with epithelia and nervous cells.) Under such conditions serious injury may develop as a consequence of intracellular ice nucleation (Zachariassen 1980).

If the cooling rate is too high to allow water to leave the cells as the freezing proceeds and the extracellular fluid becomes concentrated, the fluid of other cells may become sufficiently supercooled to cause intracellular nucleation in a great number of cells, leading to the destruction of a substantial number of cells (Mazur 1963). There is also a possibility that freezing at a high degree of supercooling allows ice spears to exist with a radius sufficiently small to penetrate membrane pores and thus seed intracellular freezing.

Freezing may also lead to other categories of injury. As the temperature drops and the solute concentrations of the fluid fraction increases, salts and other noncompatible solutes may reach such high concentrations that they affect the structure and function of proteins (see Hand, Somero, both this Vol.). The lyotropic effects of salts in frozen erythrocytes have been studied by Lovelock (1953), and later also by other investigators (Farrant 1969; Moiseyev et al. 1982), who ascribed freezing injuries to salts reaching toxic concentrations in the fluid fraction. Meryman (1971) ascribed freezing injuries to excessive cellular shrinkage. The membranes and organelles present within the cells are likely to form a matrix which represents the minimum volume the cells can reach upon shrinkage. If the cooling proceeds beyond the point where the cell membrane rests upon the intracellular matrix, the cells can no longer respond to increasing extracellular osmolality by shrinkage, and increasing osmotic stress may eventually cause the cell membrane to rupture.

Lyotropic injury as well as injuries caused by osmotic stress may be counteracted by the colligative action of polyols, which will act to reduce the concentrations of other solutes in the fluid fraction of a frozen system. The cryoprotective effects of polyols are not the topic of this chapter, but are dealt with by Lovelock (1953), Meryman (1971), Storey and Storey (1988) and Zachariassen (1990).

### *Ice Nucleating Agents*

Ice nucleating agents (INAs) responsible for so-called heterogeneous nucleation are substances which cause water to freeze at temperatures higher than the nucleation temperature of the INA-free system. They appear to have the capacity to organize water molecules in an ice-like configuration, so that there is an increased probability of the formation of an ice nucleus of a critical size at a high subzero temperature. While samples of INA-free distilled water and solutions may supercool to –20 °C, the presence of potent INAs may cause the samples to freeze at a few degrees below zero (Fig. 3).

### *Chemical Structure of Ice Nucleating Agents*

A number of inorganic and organic compounds display an ice nucleating activity. All known biological ice nucleators are organic molecules, most frequently proteins or lipoproteins.

Bacterial INAs have been subject to detailed structural studies, which have revealed that they are complex proteins (Green and Warren 1985). Less information is available regarding the chemical structure of insect INAs. Duman et al. (1984) reported that the hemolymph of the freeze-tolerant hornet *Vespula maculata* contains a proteinaceous ice nucleator with a molecular weight of 74 000. The protein had a remarkably high content of hydrophilic amino acids. Neven et al. (1986) found that the hemolymph of nonhardy *Ceruchus piceus* beetles contains INAs which are lipoproteins. These lipoproteins are assumed to function in lipid transport, and they are removed during the cold-hardening, which is based on the strategy of freeze-avoidance. The freeze-tolerant tipulid *Tipula trivittata* has lipoprotein as well as protein INAs in the hemolymph in winter (Neven et al. 1989). The lipoprotein INAs of the latter species contain phosphatidyl-inositol amounting to as much as 11% of the total phospholipids. The structure of insect INAs is dealt with by Duman in this volume.

### *Ice Nucleating Activity and Ice Nucleator Concentration*

Since the capacity of INAs to initiate freezing varies substantially from one nucleator species to another, it is important to distinguish between the activity and the concentration of INAs. This section is dealing with methods to determine these parameters.

#### *Ice Nucleating Activity*

The most obvious, but somewhat inexact, measure of ice nucleating activity is the temperature at which ice nucleation takes place, i.e., the supercooling point of the system. It is inexact because the activity should be measured not as the temperature of nucleation, but as the degree of supercooling when nucleation occurs. Thus, the difference between the supercooling point and the melting point of the system is a more adequate measure of nucleating activity.

The supercooling point of a system is normally measured by following the temperature of the system during cooling at a slow rate. The cooling rate should be so slow that there is a minimum of temperature gradients between the system and the temperature probe, which is usually situated outside the sample. When freezing is initiated, the temperature will immediately increase as the heat of fusion of water is released. The lowest temperature observed prior to the temperature increase is conventionally taken as the supercooling point (Fig. 1).

The nucleating activity of hemolymph may be measured on samples of hemolymph taken from the insects and transferred to glass capillaries. 5 μL is a convenient sample size, which has been used by several investigators. However, when measurements are to be carried out on hemolymph from small insects, it may be impossible to obtain samples of this size. In such cases smaller volumes may be used, but if the samples are too small, it may be difficult to detect the temperature inflection at the supercooling point with the externally situated temperature probe. This can be compensated for by diluting the sample to 5 μL by means of isosmotic saline. As will be outlined below, dilution with an INA free isosmotic solution will not affect the INA activity of the sample, but the increased water volume will amplify the thermal expression of the freezing.

#### *Specific Ice Nucleating Activity*

During winter the hemolymph of freeze-tolerant insects can be diluted by a factor of up to $10^6$ (Fig. 3; Hanzal and Zachariassen, unpubl. results) without any significant reduction in nucleating activity. Thus, within this concentration range the nucleating

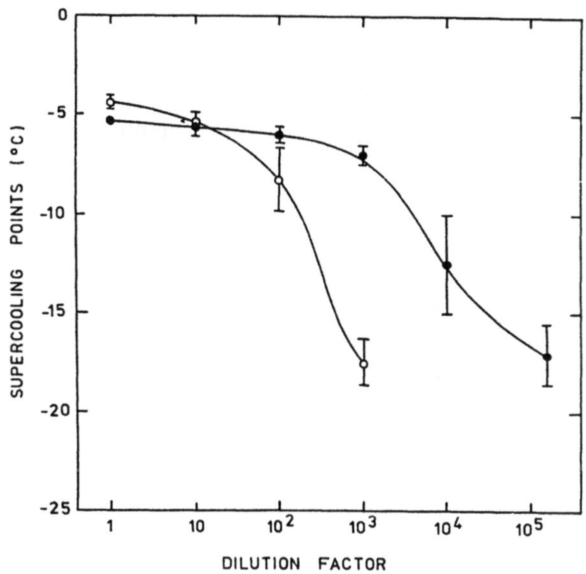


**Fig. 3.** Effect of dilution on the nucleation temperature of 5-μl samples of fluid from the afro-alpine plant *Lobelia telekii* (○) and hemolymph from the freeze-tolerant beetle *Eleodes blanchardi* (●). (Zachariassen and Hammel 1988)

activity forms a plateau where it is not significantly affected by variations in the nucleator level. Accordingly, extrapolation of the dilution curve in the direction of higher INA levels does not give nucleation activities above this plateau, which thus represents the highest nucleating potential of the particular INA species.

Figure 3 shows that the plateau of INAs from the freeze-tolerant beetle *Eleodes blanchardi* is lower than the plateau of INAs from the afro-alpine plant *Lobelia telekii*. Thus, the plateaus appear to be specific for the various types of INAs. The nucleating activities represented by the plateaus can be termed the specific activities of the respective INAs (Zachariassen and Hammel 1988).

*Ice Nucleator Concentration*

A method to determine ice nucleator concentration was described by Zachariassen et al. (1982). The method was based on an isovolumetric dilution technique, in which a 5-μl sample of hemolymph is diluted repeatedly by a factor of 10 with 0.9% saline, and the supercooling point determined for 5-μl samples representing the different degrees of dilution. The nucleator concentration is determined as the dilution factor that causes a 50% reduction in nucleator activity. Accordingly, the nucleator concentration is expressed in arbitrary logarithmic units.

To make it a true concentration measure, the logarithmic values should be related to the volume of hemolymph contained in the sample that forms the initial undiluted step in the dilution series. In this way the nucleator concentration is expressed in a standardized manner, which allows comparisons to be made between nucleator concentrations also in cases when different volumes of undiluted samples are used.

### *Mechanisms of Action of Ice Nucleating Agents*

Heterogeneous ice nucleation is likely to be accomplished by structures on the INA molecules with the ability to favor an ice-like organization of water molecules. When an ice-like water molecule aggregate attains a certain critical size or surface curvature, the probability that new water molecules will join the aggregate will increase rapidly and ice will grow and spread throughout the sample.

Mueller et al. (1990) have recently made observations indicating that bacterial INAs occur in clusters on the bacterial membrane, and that the clustering of INAs is necessary to obtain a high nucleating activity. These results appear to contradict the observation of Zachariassen and Hammel (1988) (Fig. 4), who found that nucleating activity was independent of INA concentration, i.e., on the average distance between ice nucleators.

Also the results of Zachariassen and Hammel (1988) were obtained with bacterial INA. Since the INA molecules of bacteria are situated in fixed positions on the bacterial surface (Mueller et al. 1990), the concentration of ice nucleating bacteria is not likely to have any effect on the interaction between different INA molecules or clusters of INA molecules. However, the results in Fig. 4 reveal that for INA clusters with a given specific ice nucleating activity, the amount of INA clusters is the determinant of the nucleating activity.

This implies that the results obtained by Zachariassen and Hammel (1988) do not necessarily apply to INAs in insect hemolymph, which, in contrast to bacterial INA clusters, may have an increasing opportunity to interact when their concentration in the hemolymph increases. However, the fact that the ice nucleating activity of insect hemolymph does not vary significantly when the INA concentration varies by a factor of 1000 (Fig. 3) suggests that within this concentration range interaction

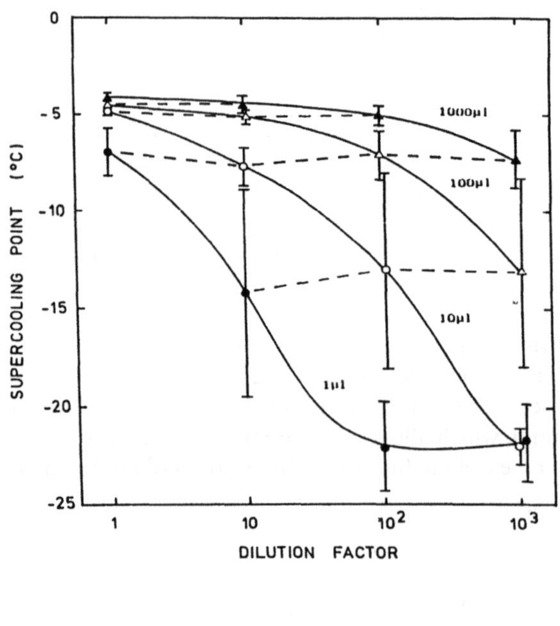


**Fig. 4.** Effect of dilution on the nucleation temperature of different volumes of fluid from the afro-alpine plant *Lobelia telekii*. The volumes were 1 (●), 10 (○), 100 (Δ) and 1000 (▲) µl. *Vertical bars* represent SD. *Solid lines* are drawn between points representing the same volumes, whereas *dashed lines* are drawn between samples with the same nucleator content. (Zachariassen and Hammel 1988)

between different INA molecules is not involved in the ice nucleation caused by insect INAs. This problem requires further experimental elaboration.

The narrow temperature range within which samples with potent biological INAs freeze (Figs. 3 and 5) indicates that within this range there is a rapid increase in the probability of nucleation. The fact that increasing solute concentrations depress the narrow temperature range of nucleation equivalent to their colligative effects indicates that the increased probability is due to phenomena involving the physical state of the water molecules, i.e., the readiness of the water molecules to be organized in an ice-like configuration under the influence of the nucleating structures on the INA molecules.

The relationship between the chemical structure and the nucleating activity of INAs has been studied by several investigators. Duman et al. (1984) reported that hemolymph INAs from the hornet *Vespula maculata* have a high content of hydrophilic amino acids, and suggested that the hydrophilic groups may be involved in organizing water molecules into embryo ice crystals. One may speculate whether also hydrophobic groups, situated at optimal locations on the INA molecules, and acting in combination with the hydrophilic groups, may be of importance to the ice nucleating activity.

Studies on bacterial INAs have shown that phosphatidylinositol (PI), attached to the nucleator protein by covalent bonds, is essential to their nucleating activity (Kozloff et al. 1987). Neven et al. (1989) found that PI was present also in hemolymph INAs from the freeze-tolerant cranefly *Tipula trivittata*, but absent in the other hemolymph lipoproteins. This strongly suggests that PI is essential also to the nucleating activity of insect INAs. They also obtained evidence indicating that the apoprotein of the lipoprotein INA is essential for the nucleating effect.

### *Factors that Influence Ice Nucleating Activity*

The activity of ice nucleating agents is influenced by several factors. Insight into the relationship between these factors and the nucleating activity is important both in order to understand the mechanisms by which INAs afford nucleation and to predict nucleating activity under different conditions.

#### *Effect of Ice Nucleating Agents*

Ice nucleating activity is a function of the content of INAs or INA units of a system. This is revealed in Fig. 4, which shows that although sample volumes and INA concentrations differ by factors of up to 1000, all samples which have the same content of INAs have the same supercooling points. This implies that INA induced supercooling points can be predicted from the nucleator content of a sample, and that the water volume in which the nucleators are distributed is of no importance (Zachariassen and Hammel 1988).

#### *Effect of Solute Concentration*

Increasing solute concentrations depress the temperature at which biological INAs induce freezing. This depression seems to be equivalent with the colligative depression of melting points (Zachariassen and Hammel 1976). This is illustrated in Fig. 5,

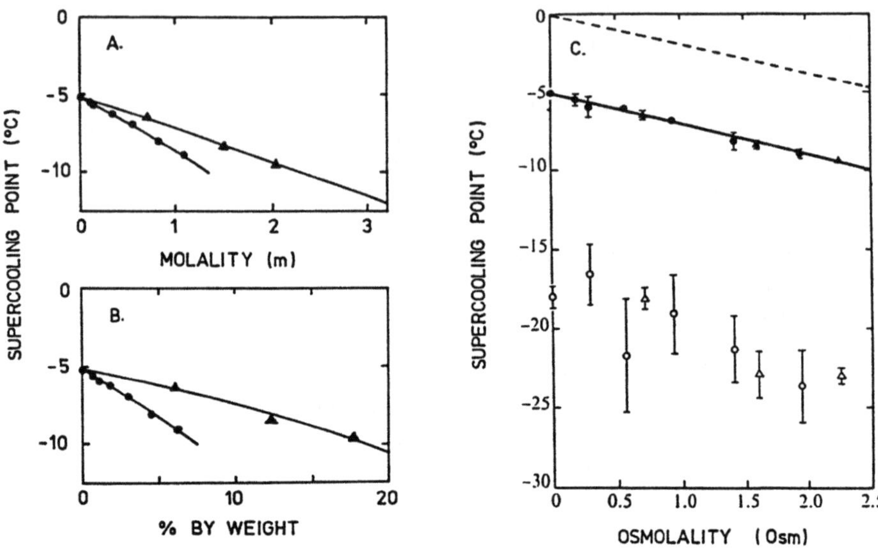


**Fig. 5A-C.** Supercooling points of 5-µl samples of solutions of NaCl (*circles*) and glycerol (*triangles*) with (*closed symbols*) and without (*open symbols*) 5 vol % of hemolymph from freeze-tolerant *Eleodes blanchardi* beetles, plotted as a function of solute concentration expressed in terms of molality (**A**), % by weight (**B**) and osmolality (**C**). *Vertical bars* represent SD. *Broken line* melting points. (**C** from Zachariassen and Hammel 1976)

which shows how different concentrations of NaCl and glycerol affect the nucleation temperature of hemolymph INAs when the solute concentrations are expressed in terms of % by weight, molality, and osmolality. Figure 5 shows that when the concentrations are expressed in terms of % by weight and molality, the two solutes appear to depress nucleation temperatures to different degrees, whereas when plotted as a function of solution osmolality, NaCl and glycerol display the same depressive effect. Furthermore, the depression of the nucleating temperature is not only related to the colligative properties of a solution, it is equivalent to the colligatively based depression of the melting point.

The colligative properties of a solution express the physical status of the water in a solution, i.e., they comprehend parameters such as vapor pressure, melting point, boiling point, and osmotic pressure. Since the process of ice nucleation involves the organization of water molecules, the physical status of the water molecules is likely to be an important parameter for the process, and, as indicated in Fig. 5, the colligative properties seem to reflect the readiness of water molecules to be organized in an ice-like pattern and form an active crystallization nucleus.

## Ice Nucleating Agents in Insect Cold Hardiness

The transitions between summer-adapted and winter-adapted states of insects involve marked changes in the supercooling points of the insects. This is illustrated in Fig. 6, which shows that while summer insects have supercooling points in the range from –8 to –12 °C, the supercooling points of freeze-avoiding insects drop to about –20 °C in the fall. In contrast to these insects, insects tolerant to freezing display an increase in supercooling points to the range from –4 to –6 °C as they enter the winter adapted state. As shown in Fig. 7, these changes take place even in the absence of a concomitant change in polyol levels and body fluid osmolality.

Zachariassen (1980) presented evidence indicating that the changes in supercooling points reflect changes in the levels and distribution of INAs as illustrated in Fig. 8. The physiological importance of this redistribution and the mechanisms involved are outlined in this section.

### *Ice Nucleating Activity in Active Nonhardy Insects*

Cells and intestine of active organisms contain components which give the organisms supercooling points in the range from –7 to –12 °C. This supercooling limit appears to be almost universal in active insects, and is seen in tropical desert beetles (Zachariassen 1980) as well as in temperate winter active insects which live on the surface of snow (Husby and Zachariassen 1980), where they are active in a supercooled state, very close to the temperature where the components may cause a lethal freezing.

Evidence that food particles present in the gut may have a strong nucleating activity, limiting the supercooling capacity of active insects, has been obtained by

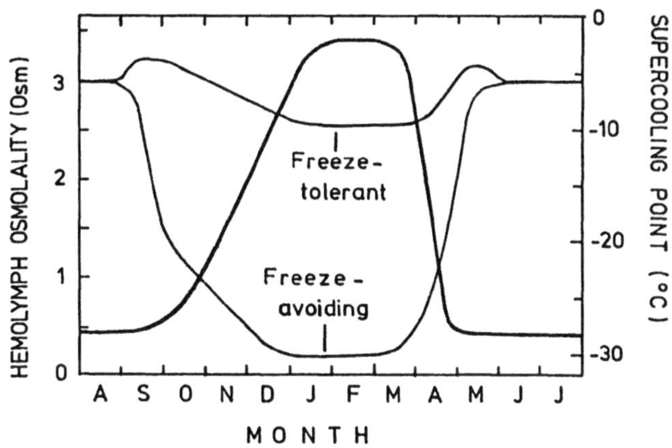


**Fig. 6.** Seasonal variations in hemolymph osmolality (*heavy line*) and supercooling points of freeze-tolerant and freeze-avoiding insects (*light lines*). (Zachariassen 1989)

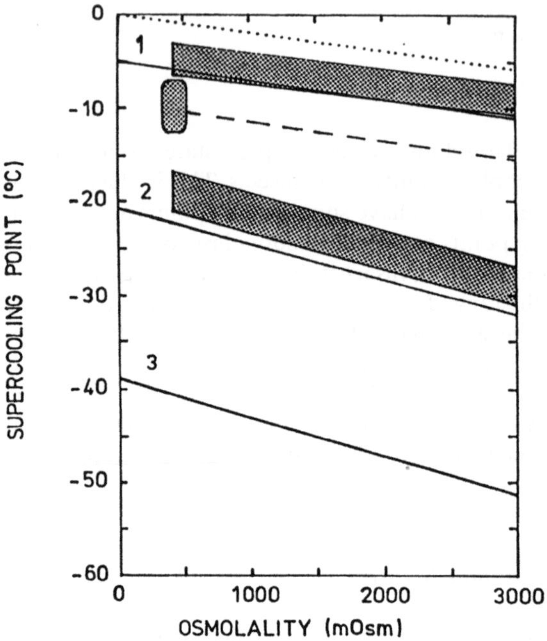


**Fig. 7.** Supercooling points of freeze-tolerant (*upper shaded area*) and freeze-avoiding (*lower shaded area*) beetles, plotted as a function of hemolymph osmolality. Values of freeze-sensitive summer beetles are represented by the *shaded area in the middle*. *Solid line 1* supercooling points of 5-µl samples of solutions of NaCl containing hemolymph INAs from a freeze-tolerant beetle. *Solid line 2* supercooling points of 5-µl samples of INA-free glycerol solution. *Solid line 3* supercooling points of highly purified solutions (homogeneous nucleation) of glycerol, sample size unknown. (MacKenzie 1977). *Broken line* hypothetical supercooling points of beetles accumulating polyols without removing INAs from body fluids. *Dotted line* melting points. (After Zachariassen 1980)

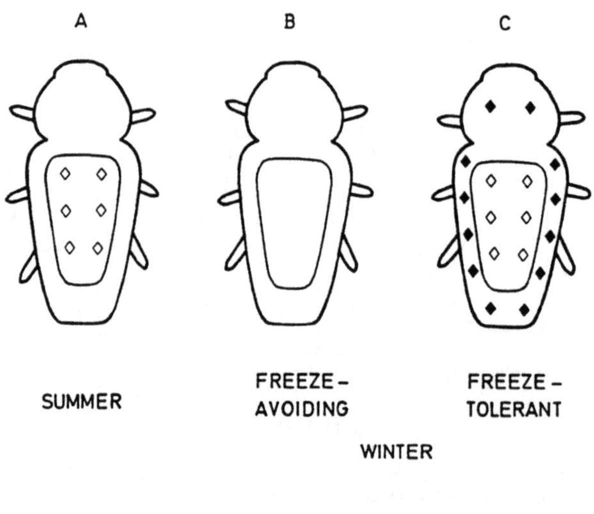


**Fig. 8A-C.** Distribution of ice nucleating agents in body fluid compartments of summer beetles and freeze-tolerant beetles and freeze-avoiding winter beetles. (Zachariassen 1980)

Salt (1953, 1961) and by Sømme (1982), who found that the supercooling capacity of insects is correlated with the presence of particles in the gut.

Evidence suggesting the existence of ice nucleating components outside the intestine of summer-adapted insects was obtained by Zachariassen (1982), who observed that the supercooling points of starving cold-hardy *Bolitophagus reticulatus* tenebrionid beetles acclimated to 20 °C in the late winter increased from about –20 to about –8 °C. There was no concomitant change in the body fluid osmolality, implying that the change is not due to colligative effects.

Apparently, components with a nucleating activity are involved in functions which are essential to active life, and insects seem only exceptionally to be able to perform normal activity in the absence of such components (one example was reported by Sømme and Zachariassen (1981). Neven et al. (1986) assumed that one such component is probably involved in lipid transport in the hemolymph, but other nucleating molecules may be enzymes of great importance to cellular metabolism. Ice nucleation is not the physiological function of these components, they just happen to carry structures which lead to freezing if the organism is sufficiently cooled.

### *Ice Nucleating Agents in Freeze-Avoiding Insects*

In order to survive cold exposure without freezing, insects must remove or inactivate all components which have the capacity to initiate ice formation. Several authors have demonstrated that the supercooling points of freeze-sensitive insects drop markedly prior to hibernation. Hansen (1973) reported that in the fall the supercooling points of larvae of the moth *Petrova resinella* dropped from about –10 °C to about –20 °C. Similar evidence has been presented by Zachariassen (1985) and Neven et al. (1986).

By removing or deactivating the ice nucleating components, freeze-avoiding insects will benefit not only from the increased supercooling capacity associated with loss of ice nucleating activity as such, but also from an increased capacity of polyols to depress supercooling points. As pointed out above, polyols depress the supercooling points induced by INAs only equivalent with the melting point depression, whereas the supercooling points of INA-free solutions are depressed by more than twice the melting point depression.

The biological expression of this phenomenon is seen in Fig. 7. Accumulation of polyols while ice nucleating components are still present in the body fluids would only lead to a moderate reduction in supercooling points, similar to that displayed by freeze-tolerant insects. Figure 7 reveals that the removal of ice nucleators will in itself reduce the supercooling point by about 10 °C, and from this level the accumulation of polyols depresses the supercooling points by the same high rate seen in INA-free physical solutions. Due to this high capacity of polyols to depress supercooling of INA free systems, freeze-avoiding hibernating insects may have supercooling points well below –30 °C, and, as reported by Ring (1982), in some cases below –50 °C.

As shown in Fig. 7, accumulation of polyols in insects with active INAs in their body fluid would cause only a moderate increase in cold hardiness. In accordance with this polyol accumulation is known to occur only in insects which lack INA activity in their body fluids.

*Removal or Masking of Ice Nucleating Agents*

The loss of nucleating activity may be due to several types of mechanisms. The ice nucleating components may be removed from the body fluids or deactivated by masking mechanisms, thus eliminating the nucleating activity while the components are still present. The nature of these mechanisms is not understood, and is still to a great extent a matter of speculation.

Sømme and Conradi-Larsen (1977) and Sømme and Block (1982) presented experimental evidence indicating that the removal of gut content of collembolans gives rise to a marked increase in organismal supercooling capacity. There seems to be some controversy regarding this phenomenon (Baust and Rojas 1985), which discussed in detail by Block (1990).

The nature of the reduction in nucleating activity outside the gut is not completely understood. One possibility is that it is caused by a chemical destruction of ice nucleating components. Experimental evidence for this has been presented by Neven et al. (1986), who reported that as the supercooling points of freeze-sensitive *Ceruchus piceus* beetles dropped in the fall, there was a concomitant removal of an ice nucleating lipoprotein from the hemolymph. Gehrken and Thorsrud (1988) reported that hibernating *Ips acuminatus* beetles with a high supercooling capacity contained substantially fewer types of protein than warm acclimated specimens. This may reflect the removal of ice nucleating proteins from the body fluids. There is also evidence that ice nucleating components may be inactivated without being removed from the organisms.

Evidence of a nucleator-masking mechanism has been presented by Xu et al. (1990), who reported that hormonal treatment (see section on freeze-tolerant insects) of larvae of the stag beetle *Ceruchus piceus* caused an increase in the levels of an ice nucleating lipoprotein, and a concomitant reduction in the nucleating activity of the same protein. Apparently, a structure involved in the organization of water molecules to form ice nuclei is inhibited without affecting the levels of the lipoprotein carrying the structure.

Baust and Zachariassen (1983) found that homogenization of cold-hardy *Rhagium inquisitor* beetles with supercooling points below –20 °C gave homogenates with supercooling points above –10 °C. Apparently, the beetles contain quite potent ice nucleating components which are inactive in the intact hibernating beetles, but which are activated by homogenization. The authors suggested that in the intact insects the ice nucleating components may be inactivated by sequestration in the lipid phase of cell membranes.

The idea of ice nucleator masking by sequestration in membrane lipids raises some problems. Duman et al. (1984) found that hemolymph INA molecules are characterized by a high content of hydrophilic amino acids, and they concluded that these hydrophilic groups are likely to be essential for the ice nucleating activity. It is difficult to imagine hydrophilic structures sequestered in a lipid phase of a membrane. However, hydrophobic structures at strategic locations on the INA molecule may be just as important as hydrophilic groups to orient and organize water molecules in the nucleation process. It is possible that the INA molecule is inactivated by sequestering essential hydrophobic structures in the cellular lipids.

It has also been suggested that thermal hysteresis antifreeze agents may be able to prevent ice nucleation induced by INAs (Duman et al. 1982). Thermal hysteresis factors (THFs) are proteinaceous antifreeze agents which counteract the growth of seeding ice crystals that usually occurs at decreasing temperature. They act in a noncolligative manner, in that they do not affect the melting point to any noticeable extent, but may reduce the temperature of crystal growth (the hysteresis freezing point) by up to 10 °C (Duman 1977). The term thermal hysteresis is derived from the separation they cause between freezing point and melting point.

Mixing of THF-containing INA-free hemolymph from the tenebrionid beetle *Iphthimus laevissimus* with THF-free INA-containing hemolymph from the tenebrionid *Eleodes blanchardi* gives the mixture supercooling points corresponding to those of pure *Eleodes* hemolymph, i.e., the THFs did not affect the nucleating activity of these INAs (Zachariassen 1985). Baust and Zachariassen (1983) tested the effect of THFs from the cerambycid beetles *Rhagium inquisitor* on the activity of INAs from the same species, and found no effect. However, with reference to Tomchaney (1981), Duman and Horwath (1983) reported that purified THFs from *Tenebrio* are able to depress at least certain types of heterogeneous nucleation. Thus, the evidence at this point seems to be somewhat contradictory.

Thermal hysteresis antifreeze agents may also act to stabilize insects in the supercooled state and prevent inoculative freezing of external ice (Zachariassen 1985).

## *Ice Nucleating Agents in Freeze-Tolerant Insects*

Many freeze-tolerant insects have a hemolymph with a high ice nucleating capacity, caused by the presence of potent INAs. The fact that the INAs are present only in the cold season may be taken as an indication that their role in cold-hardening is their main physiological function, and that it is important to these insects to initiate freezing in the hemolymph at a high subzero temperature.

There are several hypotheses as to why it is physiologically important to establish an early freezing of the extracellular fluid in these insects.

### *Inactivation of Intracellular Ice Nucleators*

The freeze-avoidance strategy may be considered as a strategy to inactivate potentially injurious INAs in cells or intestine by removing or masking them. Both removal and masking of INAs may involve complex physiological mechanisms, and the cost of removing all nucleating components may be an inoperative metabolic system, which does not allow normal biological functions.

The strategy of freezing tolerance represents an alternative means of inactivating the problematic nucleating components. By producing highly potent INAs which are distributed in the hemolymph, freeze-tolerant insects establish a protective extracellular freezing before the nucleation temperature of the other INAs is reached. Since the extracellular freezing causes an osmotic dehydration of the cells and the intestine, the fluid of these compartments will come in vapor pressure equilibrium with the extracellular ice, i.e., the fluids are no more super-cooled. This will eliminate the possibility of ice nucleation in these compartments, even when they contain ice nucleating components.

This understanding of the general principles of freezing tolerance is supported by several types of evidence. (1) The supercooling points of the freeze-tolerant stage of insects are generally higher than those of freeze-intolerant stages (Zachariassen 1980). (2) The high organismal supercooling points of the freeze-tolerant insects are reflected in the high supercooling points of isolated samples of their hemolymph (Zachariassen 1980). (3) Cells of freeze-tolerant mussels (*Mytilus edulis*) have been observed to shrink during freezing, and the ice is restricted to the extracellular tissue compartment (Kanwisher 1959).

Insects which follow the strategy of freezing tolerance do not need to remove or inactivate the other ice nucleating components. In this manner they avoid the probably complicated processes of removing or masking the ice nucleating components in cells and intestine.

Experimental evidence supporting this contention has been obtained by Gehrken and Thorsrud (1988), who found that freeze-tolerant *Chrysomela (Melasoma) collaris* beetles contain the same number of protein types as warm acclimated specimens, whereas cold-acclimated freeze-avoiding *Ips acuminatus* beetles contain substantially fewer types of protein than warm acclimated ones with high supercooling points.

The ability of freeze-tolerant insects to maintain the ice nucleating components in their body fluids when they are in a cold-hardy state may be the reason why freezing tolerance throughout the year is a common strategy of cold-hardiness among alpine insects in temperate and tropical regions (Sømme and Zachariassen 1981; van der Laak 1982). Alpine insects in these regions are not exposed to cold on seasonal basis, but every night of the year. Since these insects have to combine feeding and the complex biochemical system required for normal metabolism with exposure to cold every night, they have to meet the cold with ice nucleating components in their intestine and probably also in the cells. Thus, freeze avoidance based on removal of INAs is not a feasible cold-hardiness strategy for these insects. By having potent INAs in the hemolymph they establish a tolerable extracellular freezing at a high subzero temperature, thus preventing injurious freezing initiated by the nucleating components in intracellular compartments and the gut.

*Reduction of Osmotic Stress During Freezing*

It has also been suggested that the potent extracellular INAs protect freeze-tolerant insects by reducing the osmotic stress that develops across cell membranes during freezing (Duman et al. 1984; Storey and Storey 1988).

As pointed out above, the solutes of a freezing solution become trapped in the unfrozen fluid fraction, which subsequently becomes concentrated as the freezing proceeds. The more water which freezes, the more concentrated becomes the unfrozen fluid fraction. Assuming that the initial freezing takes place exclusively in the hemolymph, the initial concentrating effect will be limited to this fluid compartment. The change in temperature and osmolality that occurs during freezing of a supercooled insect is probably one of the fastest changes in these parameters experienced by animals. The rapid extracellular freezing creates an osmotic gradient between the concentrated extracellular fluid and the intracellular compartments, which is probably unable to reach osmotic equilibrium with the extracellular fluid during the short initial freezing phase. By inducing freezing at a more moderate degree of supercooling, the amount of initially formed ice will be reduced, and so will the transmembrane osmotic stress developed during the initial freezing phase.

The importance of this effect may be evaluated by calculating the osmotic stress formed with and without extracellular INAs present in the hemolymph. The osmotic stress can be expressed as the initial temperature increase during the freezing. The relationship between the initial temperature increase and the amount of water that freezes during the initial phase is given by the equation

$$C_{sb} \times P_b \times \Delta t = S_w \times X, \qquad (1)$$

where $C_{sb}$ is the specific heat of the insect, $P_b$ the body weight of the insect, $\Delta t$ the initial temperature increase, $S_w$ the specific heat of fusion of water, and X the amount of water that freezes out of the hemolymph during the initial freezing phase.

By assuming that the initial freezing is limited to the hemolymph, the relationship between $\Delta t$ and X is also given by the equation

$$\frac{\Delta t + t_{scp}}{t_m} = \frac{P_h}{P_h - X}, \qquad (2)$$

where $t_{scp}$ is the supercooling point, $t_m$ the melting point of the unfrozen body fluid, and $P_h$ the weight of the hemolymph.

By solving Eq. (1) with regard to X and substituting for X in Eq. (2), $\Delta t$ can be found by solving the second degree equation

$$\frac{C_{sb} \cdot P_b}{S_w} (\Delta t)^2 + \left( \frac{C_{sb} \cdot P_b \cdot t_{scp}}{S_w} - P_h \right) \cdot \Delta t + (t_m - t_{scp}) \cdot P_h = 0. \qquad (3)$$

By assuming that $C_{sb}$ is 0.9 cal/g °C, $P_b$ is 1 g, $S_w$ is 79 cal/g, $t_m$ is –1 °C, $P_h$ is 0.15 g (relative water content = 60%, hemolymph volume = 25% of total water content), we can calculate the values of $\Delta t$ corresponding to different values of organismal supercooling point, $t_{scp}$ (Fig. 9A).

By giving $t_{scp}$ the value –8 °C, representing the absence of hemolymph INAs, $\Delta t$ becomes 6.13 °C, i.e., the hemolymph melting point after the initial freezing phase will be –1.87 °C. The transmembrane osmotic stress ($\Delta t_m$), as expressed in terms of difference between the melting points of the intracellular fluid and the frozen hemolymph, would in this case be 0.87 °C. By giving $t_{scp}$ the value –6 °C, which represents the situation with INAs present in the hemolymph, $\Delta t$ will have the value 4.48 °C. This implies that the hemolymph melting point immediately after the initial freezing would be 1.52 °C, and that the osmotic stress ($\Delta t_m$) would correspond to a melting point difference of 0.52 °C. Assuming that the primary physiological function of hemolymph INAs is to reduce osmotic stress, it appears from these estimates that a transmembrane osmotic stress of 0.52 °C is tolerated, whereas 0.87 °C is not.

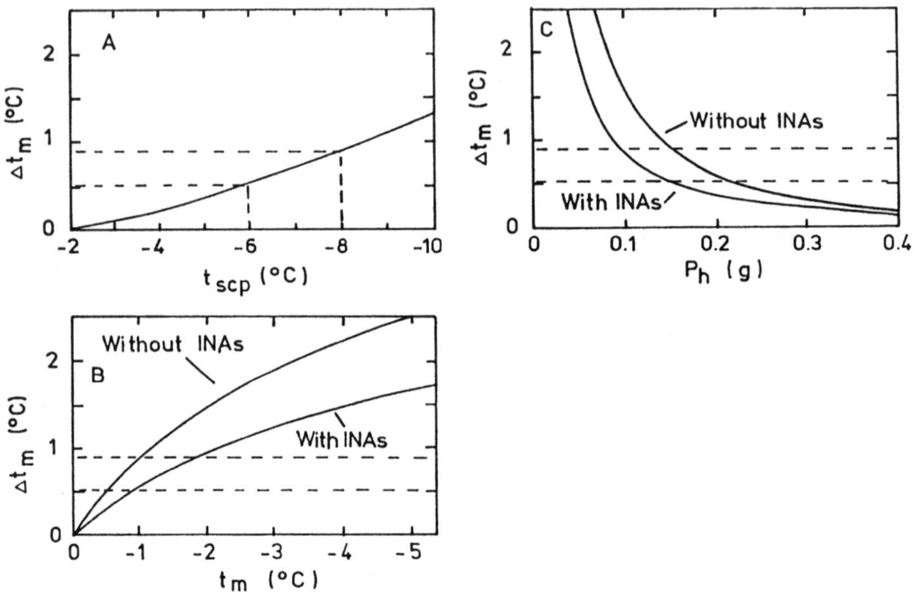


**Fig. 9A-C.** Osmotic stress across cell membranes after initial phase of extracellular freezing, plotted as a function of three different physiological parameters. Osmotic stress is expressed in terms of difference between melting points of intracellular and extracellular fluid ($\Delta t_m$). **A** Osmotic stress as a function of supercooling point ($t_{scp}$). *Broken curves* represent situation with ($T_{scp}$ = 6 °C) and without ($t_{scp}$ = –8 °C) INAs in the hemolymph. **B** Osmotic stress as a function of hemolymph melting point ($t_m$), as it varies with polyol concentration. *Broken curves* represent tolerable (*lower curve*) and intolerable stress (*upper curve*), as determined from **A.** **C** Osmotic stress as a function of hemolymph content ($P_h$). *Broken curves* as under **B.** For further information, see text

On the basis of these data, it would be interesting to compute from Eq. (3) how transmembrane osmotic stress is affected by variations in other parameters. Figure 9B shows how increasing polyol concentrations, expressed as decreasing melting point ($t_m$), influence transmembrane osmotic stress ($\Delta t_m$) when hemolymph INAs are absent ($t_{scp}$ = –8 °C) and present ($t_{scp}$ = –6 °C). The figure indicates that osmotic stress increases markedly as polyol concentrations increase, and that intolerable values develop already at moderately high polyols concentrations ($t_m$ < –2 °C). The contribution from hemolymph INAs is not sufficient to bring the $\Delta t_m$ values down to tolerated levels. However, accumulation of polyols is likely to lead to an increase in body fluid viscosity, which serves to reduce freezing rates (Baust 1973) and thus osmotic stress developed during freezing.

Another way of affecting the osmotic stress that develops across cell membranes during the initial freezing phase is to change the amount of hemolymph (Fig. 9C). It appears from the figure that by increasing the amount of hemolymph to 0.2 g (total water content = 0.6 g), the osmotic stress is brought down to the tolerated level without the presence of INAs in the hemolymph. However, this strategy would imply a shrinking of the cells before freezing, and thus make them more vulnerable to the osmotic shrinking that occurs as the freezing proceeds to lower temperatures. This may be the reason why increasing the fraction of extracellular water is not a feasible alternative to the production of extracellular INAs. The importance of transmembrane osmotic stress in relation to freezing injury and cryoprotection should be subject to further investigations.

Neven et al. (1989) discuss the phenomenon, which is also illustrated in Fig. 3, that the INAs present in the hemolymph of freeze-tolerant insects are generally less active than those produced by bacteria. Referring to Farrant (1980), who presented evidence indicating that freezing survival of a system is highest if freezing takes place at an intermediate rate, the authors point out that this is probably the case also with freeze-tolerant insects.

*Improved Water Balance During Hibernation*

The initiation of freezing at a high subzero temperature is also likely to improve the water balance of hibernating insects. Insects hibernating in a supercooled state will have body fluids with a higher vapor pressure than ice at the same temperature. Thus, they will have an evaporative water loss to ice that might be present in the hibernaculum. If the hibernation period is long, they may suffer a substantial water loss, as observed by Ring (1982). For insects hibernating in a frozen state the body water will be in vapor pressure equilibrium with ice, and consequently, they will neither lose nor gain water to a frozen hibernaculum. By entering the frozen state at the highest possible subzero temperature, the neutral water balance will extend over the longest possible period. It has been suggested that this may be an important evolutionary drive behind the development of freezing tolerance and the production of extracellular INAs (Zachariassen 1985). Recent unpublished results obtained by Lundheim reveal that frozen *Upis ceramboides* beetles incubated in a chamber with ice on the walls do display a negligible water loss, and rates of water loss which are substantially lower than those of supercooled specimens incubated under the same conditions. However, he also found that insects hibernating in a supercooled state in nature display only a very moderate drop in relative water content. Thus, the importance of water balance for the development of freezing tolerance is not settled.

## Levels of Ice Nucleating Agents and Their Regulation

Zachariassen et al. (1982) found that 5-μl samples of hemolymph from *Eleodes blanchardi* beetles had to be diluted by a factor of more than 1000 to display a significant reduction in the nucleating temperature, while hemolymph from *Upis ceramboides* displayed high activity following dilution by factors of up to $10^6$. Thus, INAs appear to be present in substantially higher amounts than those necessary to secure freezing at a high temperature. The basis of this great surplus of nucleating capacity is not obvious. Neven et al. (1989) pointed out that the ice nucleating lipoprotein in the hemolymph of freeze-tolerant *Tipula trivittata* is also likely to be involved in lipid transport. Other INAs may be involved in other physiological functions. The functions of INA molecules beyond ice nucleation should be the object of further studies.

Not much is known about the environmental cues that control the production of ice nucleating agents in freeze-tolerant insects. Zachariassen et al. (1982) found that temperature has a strong effect on the production of hemolymph INAs in freeze-tolerant *Eurosta solidaginis* larvae. The production of hemolymph ice nucleators was most efficient at 5 °C. To what extent environmental cues other than temperature (humidity and photoperiod) are involved is not known.

Hormonal processes involved in the regulation of ice nucleating activity in freeze-avoiding *Ceruchus piceus* beetles have been studied by Xu et al. (1990). These investigators found that juvenile hormone reduced nucleating activity of the hemolymph lipoprotein of this species, but increased the levels of the same lipoprotein. Apparently, juvenile hormone causes a modification of the chemical structure of the lipoprotein, making it less active as an ice nucleator. Adipokinetic hormone increased nucleating activity as well as lipoprotein levels. The endocrine regulation of insect INAs is an interesting field which should be given increased attention by investigators in the future.

*Acknowledgment.* I would like to thank Rolv Lundheim for critical comments on the manuscript.

## References

Baust JG (1973) Mechanisms of cryoprotection in freeze-tolerant animal systems. Cryobiology 10:197–205

Baust JG, Rojas RR (1985) Review – insect cold hardiness: facts and fancy. J Insect Physiol 31:755–759

Baust JG, Zachariassen KE (1983) Seasonally active cell matrix associated ice nucleators in an insect. Cryobiol Lett 4:65–71

Block W (1990) Cold tolerance of insects and other arthropods. Phil Trans R Soc Lond B 326:613–633

Duman JG (1977) The role of macromolecular antifreeze in the darkling beetle *Meracantha contracta*. J Comp Physiol 115:279–286

Duman JG, Horwath K (1983) The role of hemolymph proteins in the cold-tolerance of insects. Annu Rev Physiol 45:261–270

Duman JG, Horwath K, Tomchaney A, Patterson JL (1982) Antifreeze agents in terrestrial arthropods. Comp Biochem Physiol 73A:545–555

Duman JG, Morris JP, Castellino FJ (1984) Purification and composition of an ice nucleating protein from queens of the hornet, *Vespula maculata*. J Comp Physiol 154:79–83

Farrant J (1969) Is there a common mechanism of protection of living cells by polyvinylpyrrolidone and glycerol during freezing? Nature (Lond) 222:1175–1176

Farrant J (1980) General observations on cell preservation. In: Ashwood-Smith MJ, Farrant J (eds) Low temperature preservation in biology and medicine. Univ Park Press, Baltimore

Gehrken U, Thorsrud AK (1988) Temperature modification of protein appearance in insects in relation to cold hardiness. In: Gehrken U (ed) Mechanisms involved in insect cold tolerance. Doct Thesis, Univ Oslo

Green RL, Warren GJ (1985) Physical and functional repetition in a bacterial ice nucleation gene. Nature (Lond) 317:645–648

Hansen T (1973) Variations in glycerol content in relation to cold-hardiness in larvae of *Petrova rosinella* L. (Lepidoptera, Tortricidae). Eesti NSV Tead Akad Toim Biol 22:105–112

Husby JA, Zachariassen KE (1980) Antifreeze agents in the body fluid of winter active insects and spiders. Experientia 36:963–964

Kanwisher JW (1959) Histology and metabolism of frozen intertidal animals. Biol Bull 116:258–264

Knight CA (1967) The freezing of supercooled liquids. Van Nostrand, Princeton, New York

Kosloff LM, Lute M, Arellano F (1987) Role of phosphatidylinositol in ice nucleation. Pap 3rd Int Conf Ice Nucleation, October 1987, Newport, Oregon

Lovelock JE (1953) The mechanism of the cryoprotective effect of glycerol against freezing and thawing. Biochim Biophys Acta 11:28–36

MacKenzie AP (1977) Non-equilibrium freezing behaviour of aqueous systems. Phil Trans R Soc London Biol Sci 278:167–189

Mazur P (1963) Kinetics of water loss from cells at subzero temperatures and the likelihood of intracellular freezing. J Gen Physiol 47:347–369

Meryman HT (1971) Osmotic stress as a mechanism of freezing injury. Cryobiology 8:489–500

Moiseyev VA, Nardid OA, Belous AM (1982) On a possible mechanism of the protective action of cryoprotectants. Cryobiol Lett 3:17–26

Mueller GM, Wolber PK, Warren GJ (1990) Clustering of ice nucleation protein correlates with ice nucleation activity. Cryobiology 27:416–422

Neven LG, Duman JG, Beals JM, Castellino FJ (1986) Overwintering adaptations in the stag beetle, *Ceruchus piceus*: removal of ice nucleators in the winter to promote supercooling. J Comp Physiol 156:707–716

Neven LG, Duman JG, Low MG, Sehl CL, Castellino FJ (1989) Purification and characterization of an insect hemolymph lipoprotein ice nucleator: evidence for the importance of phosphatidylinositol and apolipoprotein in the ice nucleator activity. J Comp Physiol 159B:71–82

Ring R (1982) Freezing-tolerant insects with low supercooling points. Comp Biochem Physiol 73A:605–612

Schmidt-Nielsen K (1983) Animal physiology: adaptation and environment, 3rd edn. Cambridge Univ Press, Cambridge Lond

Salt RV (1953) The influence of food on cold hardiness of insects. Can Entomol 85:261–269

Salt RV (1961) Principles of insect cold hardiness. Annu Rev Entomol 6:55–74

Sømme L (1982) Supercooling and winter survival in terrestrial arthropods. Comp Biochem Physiol 73A:519–543

Sømme L, Block W (1982) Cold hardiness of *Collembola* at Signy Island, maritime Antarctic. Oikos 38:168–176

Sømme L, Conradi-Larsen EM (1977) Cold-hardiness of collembolans and oribatid mites from windswept mountain ridges. Oikos 29:118–126

Sømme L, Zachariassen KE (1981) Adaptations to low temperature in high altitude insects from Mount Kenya. Ecol Entomol 6:199–204

Storey KB, Storey JM (1988) Freeze tolerance in animals. Physiol Rev 68:27–83

Tomchaney AP (1981) The purification and characterization of a thermal hysteresis protein from the larvae of *Tenebrio molitor*. Master's Thesis, Univ Notre Dame

van der Laak S (1982) Physiological adaptations to low temperature in freezing-tolerant *Phyllodecta laticollis* beetles. Comp Biochem Physiol 73A:613–620

Xu L, Neven LG, Duman JG (1990) Hormonal control of hemolymph lipoprotein ice nucleators in overwintering freeze-susceptible larvae of the stag beetle *Ceruchus piceus*: adipokinetic hormone and juvenile hormone. J Comp Physiol 160:51–59

Zachariassen KE (1980) The role of polyols and nucleating agents in cold-hardy insects. J Comp Physiol 140:227–234

Zachariassen KE (1982) Nucleating agents in cold-hardy insects. Comp Biochem Physiol 73A:557–562

Zachariassen KE (1985) Physiology of cold-tolerance in insects. Physiol Rev 65:799–832

Zachariassen KE (1989) Thermal adaptations to polar environments. In: Mercer JB (ed) Thermal physiology 1989. Proc Int Symp Thermal Physiol Tromsø, Norway, 16–17 July 1989. Exerpta Med, Elsevier, Amsterdam New York

Zachariassen KE (1990) The water relations of overwintering insects. In: Lee RE, Denlinger D (eds) Insects at low temperature. Chapman and Hall, New York Lond, pp 47–63

Zachariassen KE, Hammel HT (1976) Nucleating agents in the hemolymph of insects tolerant to freezing. Nature (Lond) 262:285–287

Zachariassen KE, Hammel HT (1988) The effect of ice-nucleating agents on ice nucleating activity. Cryobiology 25:143–147

Zachariassen KE, Baust JB, Lee RE (1982) A method for quantitative determination of ice nucleating agents in insect hemolymph. Cryobiology 19:180–184

Chapter 17

# Hemolymph Proteins Involved in the Cold Tolerance of Terrestrial Arthropods: Antifreeze and Ice Nucleator Proteins

J.G. DUMAN[1], D.W. WU[1], K.L. YEUNG[2], E.E. WOLF[2]

## Introduction

Keeping an appropriate amount of body water in the liquid state is a major problem for terrestrial arthropods exposed to subzero temperatures. Freeze-susceptible species cannot survive even minimal ice formation in their body fluids and so must avoid freezing by producing antifreezes and/or removing ice nucleators. Even freeze-tolerant species survive freezing only if the ice is confined to the extracellular fluid. This extracellular freezing results in dehydration of the cells and thus major adaptations contributing to freeze tolerance, along with elimination of intracellular freezing (i.e., extracellular ice nucleators), are those concerned with combating low water availability. (For recent general reviews on this topic see Bale 1987; Baust and Rojas 1985; Cannon and Block 1988; Block 1990; Sömme 1989; Storey and Storey 1988; Zachariassen 1985; Lee and Denlinger 1991).

Ice nucleator proteins minimize supercooling by organizing water into an ice-like "embryo crystal" which, when it reaches a critical size, seeds and freezes the metastable supercooled system (Knight 1967). The details of how this occurs are not certain, but ice nucleators probably function by acting as a template to orient water monomers into the embryo crystal. In contrast, insect antifreeze proteins (AFPs) inhibit freezing by lowering the nonequilibrium freezing point of water via a noncolligative mechanism (DeVries 1988) and by promoting supercooling (Zachariassen and Husby 1982; Duman et al. 1991). While it may appear incongruous to discuss proteins with opposite functions in the same chapter, in fact there are several good reasons for doing so, in addition to the obvious point that both are involved in subzero temperature adaptations. Actually, the simultaneous study of these proteins can be quite informative, and perhaps in some situations it may be required. The ability of insect antifreeze proteins to promote supercooling, in at least some cases, requires the inhibition of ice nucleators. Also, as will be discussed later, addition of certain ice nucleator proteins to a solution of AFPs actually increases the freezing point depressing abilities of the AFPs. At the molecular level, there may be similarities and even sequence homologies between these proteins. AFPs function by adsorbing, via hydrogen bonding, to the surface of a potential seed crystal and

[1]Department of Biological Sciences, University of Notre Dame, Notre Dame, IN 46556, USA
[2]Department of Chemical Engineering, University of Notre Dame, Notre Dame, IN 46556, USA

Somero et al. (Eds.)
Water and Life

thus the structure of the AFP probably provides a lattice match to ice. In a somewhat similar fashion, ice nucleator proteins must organize water molecules into an "embryo" ice crystal. Finally, although this may seem paradoxical, the hemolymph ice nucleator proteins of freeze-tolerant insects, by inducing nucleation in the extracellular fluid, actually function to prevent lethal intracellular freezing. Thus, in a roundabout but very real sense, extracellular ice nucleators are intracellular antifreezes.

## Insect Ice Nucleator Proteins

The function of ice nucleator proteins is discussed elsewhere in this Volume (Zachariassen) and therefore this treatment will concentrate on structure/function relationships of these interesting proteins. However, a brief discussion of the function and evolution of insect ice nucleators may help to place these structure/function relationships into proper biological context.

Zachariassen and Hammel (1976) identified the presence of ice nucleating factors in the hemolymph of freeze-tolerant insects and discussed their probable function, namely to inhibit lethal intracellular ice formation by inducing freezing in the extracellular fluid at comparatively high subzero temperatures (usually –6 to –10 °C). The exclusion of solutes from the ice raises the osmotic pressure in the unfrozen fraction of the extracellular fluid, generating an osmotic outflux of water from the cells which thus lowers the freezing and supercooling points of the intracellular water. In contrast, tissue freezing following more extensive supercooling is likely to result in the formation of lethal intracellular ice (Mazur 1977).

It is interesting that the ice nucleators (INs) of freeze-tolerant insects initiate nucleation at approximately –6 to –10 °C while the ice nucleator proteins on the surface of many bacteria are more potent and induce freezing at –2 to –5 °C (Lindow 1983; Wolber and Warren 1989). Thus it appears that in freeze-tolerant insects there has been selection pressure for INs with intermediate activities rather than for the bacterial type with such high activities. A hint to understanding why this may have occurred comes from the freeze preservation literature, where it has been well established that optimal survivorship of a particular cell or tissue type usually is achieved only over a fairly narrow range of freezing rates, presumably because at faster rates lethal intracellular freezing occurs and at slower rates osmotic and other problems are maximized (Farrant 1980). The optimal rate varies with cell type. Of course, overwintering insects have no control over freezing rates, except that by appropriate choice of a thermally buffered overwintering site some species may assure slow rates of temperature change. However, the initial rate of ice formation may be quite crucial, with high rates being lethal (Claussen et al. 1990). By controlling the nucleation temperature with ice nucleators of appropriate activity the initial surge of ice formation could be controlled (i.e., low levels of undercooling result in lower initial rates of crystal growth). Hence, the common "two-step" cryopreservation protocols (Farrant 1980), which initiate freezing of the system at a fairly high temperature followed by rapid cooling to a lower storage temperature, mimic the freeze-tolerant insects.

It is important not to suggest here that all freeze-tolerant insects follow this general theme whereby freezing is initiated by hemolymph ice nucleator proteins at –6 to –10 °C, although this is the case for many (most?) freeze-tolerant insects as well as other freeze-tolerant invertebrates and lower vertebrates (Zachariassen 1985; Storey and Storey 1988). The tremendous adaptive radiation of insects assures that there are exceptions to most rules. For example, some insects, such as the arctic caterpillar *Gynaephora groenlandica*, nucleate in this temperature range (–6 to –10 °C), but the nucleator is not present in the hemolymph (Kukal et al. 1988). Also, overwintering larvae of the moth *Cisseps fulvicollis* (Fields and McNeil 1986) and *Lithobius bifurcatus* centipedes (Tursman and Duman, unpublished) are freeze-tolerant only when internal freezing is initiated across the cuticle by external ice, with little or no supercooling. Thus in some species the ultimate nucleator, ice itself, seems to be required for freeze tolerance. In contrast, others, such as several species of giant silkworm which overwinter as freeze-tolerant pupae, lack active ice nucleators and nucleate at –17 to –23 °C (Duman et al. 1991). There are even a few insect species from Alaska and the Canadian Rockies which actually supercool to –50 to –60 °C, well below the temperature of spontaneous nucleation of water (–40 °C), and yet survive freezing if and when it eventually occurs (Miller 1982; Ring 1982). In spite of these exceptions, it does seem that most freeze-tolerant insects have selected for hemolymph proteins with ice nucleator activity.

In contrast, freeze-susceptible species which inhabit regions where they experience subzero temperatures would be expected to have experienced selection pressure against proteins or other surfaces (i.e., membranes, etc) with ice nucleator activity. It is, of course, impossible to know whether a species which today lacks ice nucleators in fact never had them in their evolutionary past or whether they had ice nucleators, but lost this activity over evolutionary time because of negative selection pressure. However, certain freeze-avoiding insects, such as all the aphid species tested (O'Doherty and Bale 1985; Knight and Bale 1986; O'Doherty and Ring 1987) seem not to have ice nucleator activity. In other situations it may be that the ice nucleator activity associated with a particular protein results from the structure of a region of the protein which is intimately connected with the essential function of the protein, and thus cannot be, or was not, selected out of the population. If this occurs, and it apparently has, there are other alternatives by which this ice nucleator activity can be overcome, namely seasonal removal of the ice nucleator and/or antifreeze production. Although the protein may be essential during active periods, the insect may be able to survive over the winter without this protein (or other factor), and thus the insect may seasonally remove it. This is known to occur in a few freeze-avoiding species (Zachariassen 1982; Bakken 1985), but the best understood of these are the overwintering larvae of the stag beetle *Ceruchus piceus* (Neven et al. 1986). *Ceruchus* larvae lower their SCPs from –7 °C in summer to approximately –25 °C in winter, yet they do not produce antifreezes. Instead, the larvae cease feeding and clear the gut of potential nucleators in the autumn (as do most freeze-avoiding insects), but in addition they remove a lipoprotein with ice nucleating activity from the hemolymph and thereby gain considerable supercooling ability without the cost of antifreeze production. Because of the diapause state (reduced metabolism) of the larvae, the lipid transport function of the lipoprotein ice nucleator is apparently dispensable during winter. This process is under hormonal control (Xu et al. 1990). In contrast, overwintering adults of the beetle *Uloma impressa* avoid freezing in a very different fashion (Duman 1979b). *Uloma* retains a rather potent ice nucleator during the winter and this necessitates the production of high levels of antifreezes. The beetles produce very high concentrations of colligative antifreezes (mainly glycerol) which depress the hemolymph melting point to –10 °C and, in addition, enough antifreeze protein to lower the hysteretic freezing point of the hemolymph to –15 °C. Yet they supercool only to –22 °C, just 7 °C below the hysteretic freezing point. Recall that *Ceruchus* larvae supercool to –20 °C or less without antifreeze. These few examples should suffice to underscore the extreme diversity of adaptations to low temperature exhibited by insects, and to make the point that endogenous ice nucleators are important factors in the subzero temperature tolerance not only of freeze-tolerant species (which have presumably experienced selection pressure for hemolymph ice nucleators), but also of freeze-avoiding species which must overcome them.

### *Structure-Function Relationships of Insect Ice Nucleator Proteins*

The diversity of insect adaptations is also apparent at the molecular level when one investigates the insect ice nucleator proteins. While some convergent evolution is apparent (i.e., between the lipoprotein ice nucleator of *Tipula* larvae and the bacterial ice nucleator proteins), it is obvious that different types of proteins may initiate nucleation. Our intuition suggests that proteins with ice nucleator activity should have large highly conserved domains, much like the active sites of enzymes but encompassing a larger surface of the protein. However, this is not apparent and, although insufficient information is available to know this for certain, a closer analogy may be the antifreeze proteins of fish, where several diverse structures result in quite similar antifreeze activities (DeVries 1986, 1988; Davies et al. 1988).

Before discussing the structure of insect ice nucleator proteins, it might be useful to quickly review the structure of bacterial ice nucleators, as these are much better understood. Several species of epiphytic bacteria have proteins on their surfaces which are very potent ice nucleators and initiate freezing at –2 to –5 °C (Wolber and Warren 1989). Consequently, these bacteria often cause extensive freeze damage to crops when they nucleate condensed water on the surfaces of leaves (Lindow 1983). The *Pseudomonas syringae* ice nucleator protein gene was cloned in *E. coli* and sequenced (Green and Warren 1985). The 150 kDa predicted translation product consists largely (~81% of total sequence) of a central repeating domain which can be subdivided into three regions. Two of these have high fidelity 48-residue repeats, each of which can be subdivided into 16-residue repeats of medium fidelity, and each of these contains two octapeptide low fidelity repeats. The third region of the repeating central domain consists only of octapeptide repeats. The protein ice nucleators of other bacterial species are quite similar (Wolber and Warren 1989). It is now apparent from different lines of investigation that activity of the bacterial ice nucleators requires the aggregation of several proteins (Govindarajan and Lindow 1988a; Southworth et al. 1988; Mueller et al. 1990). Models, based on the repeat structure of the proteins and the likelihood of cooperativity between adjacent proteins in ordering water molecules into an embryo crystal, have been offered (Wolber and Warren 1989). It should be noted that the bacterial protein INs are anchored in the outer membrane and in fact are not active unless they are attached to the membrane, or combined with phospholipids into proteoliposomes (Govindarajan and Lindow 1988b).

The first insect protein ice nucleator to be purified was a 74-kDa hemolymph protein from the freeze-tolerant overwintering queens of the hornet *Vespula maculata* (Duman et al. 1984). This interesting protein is very hydrophilic, consisting of ~ 20 mol% glutamate/glutamine, suggesting that hydrogen bonding of the water to the hydrophilic side chains may be involved in its ice nucleating activity.

The best-studied insect ice nucleator proteins are those from the freeze-tolerant larvae of the cranefly *Tipula trivittata*. *Tipula* larvae have a lipoprotein and at least two other proteins with ice nucleator activity in their hemolymph (Duman et al. 1985). The lipoprotein ice nucleator (LPIN) has been characterized and the components required for activity identified (Neven et al. 1989) (Fig. 1). The globular 800-kDa molecule is approximately 45% protein, 51% lipid, and 4% carbohydrate. The apoproteins consist of 265-kDa (Apo-I) and 80-kDa (Apo-II) proteins. Phosphatidylinositol (PI), illustrated in Fig. 2, is an interesting lipid component of the LPIN for a number of reasons. (1) It had not previously been identified in hemolymph lipoproteins from other insects. (2) Phospholipids are known to form a monolayer on the surface of insect hemolymph lipoproteins (Katagiri 1985) and are thus in position to interact with surface water. (3) It has been suggested that inositol, via hydrogen bonding of its hydroxyls, may order water in an ice-like fashion (Warner 1962). (4) Original studies by Kozloff et al. (1984) identified PI as an essential component of the bacterial ice nucleator sites, although their more recent work indicates that PI functions to anchor the ice nucleator proteins in the membrane rather than to order water in the active site (Kozloff et al. 1987; Turner et al. 1990). Treatments which affect the head group of PI inactivate the *Tipula* LPIN while those that affect other phospholipids do not. Thus treatment with a PI-specific phospholipase C, which hydrolyzes the phosphodiester bond to cleave the inositol

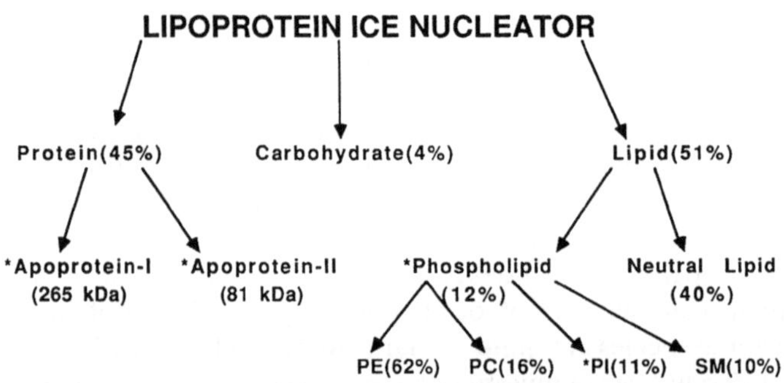


**Fig. 1.** Composition of the 800-kDa lipoprotein ice nucleator from *Tipula trivittata*. Percentages are % by weight. *PE* phosphatidylethanolamine; *PC* phosphotodylcholine; *PI* phosphatidylinositol; *SM* sphingomyelin. *Asterisks* indicate those components which are essential for ice nucleator activity (i.e., apoproteins-I and II and phosphatidylinositol)

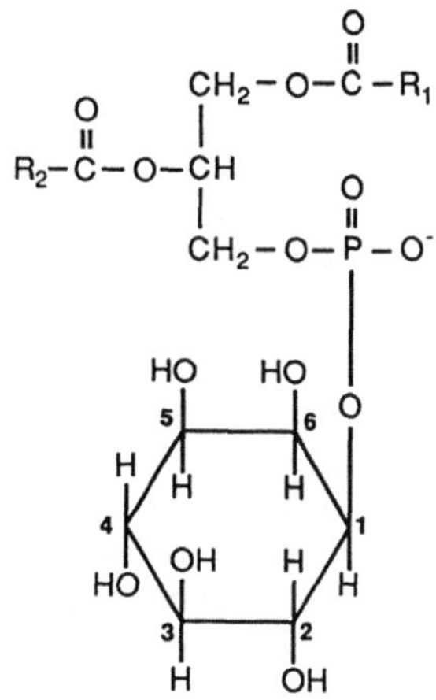


**Fig. 2.** General structure of phosphatidylinositol. The integrity of the hydroxyls of the inositol head group are essential for ice nucleator activity of the LPIN. Interestingly, the nature of the fatty acids (both $R_1$ and $R_2$) seem not to be crucial. In fact, the $R_2$ fatty acid may be removed without affecting activity of the LPIN

head group, inactivates the LPIN (Neven et al. 1989). The integrity of the hydroxyl groups of the inositol ring is apparently important, since treatment with sodium periodate, which cleaves the bond between carbons 4 and 5 and oxidizes their hydroxyls to aldehydes, inactivates the LPIN. As expected, the apoproteins are also needed for activity. While the delipidated apoproteins lack activity, addition of the LPIN lipids back to the proteins and sonication to produce proteoliposomes restores activity. In fact, proteoliposomes composed only of the LPIN apoproteins and PI have activity. Both Apo-I and II are required. Substitution of apoproteins from the hemolymph lipoprotein of *Manduca sexta*, which does not have ice nucleator activity, for the LPIN apoproteins yields inactive proteoliposomes. Substitution of other phospholipid components of the LPIN (phosphatidylcholine, phosphatidylethanolamine) for PI does not return activity and substitution of PI-4-monophosphate or PI-4,5-diphosphate greatly reduces activity. Thus, the two apoproteins and PI (probably the hydroxyl groups of inositol) are required, and sufficient, for activity of the LPIN.

From the above discussion it may seem that the apoproteins function to orient the PI so that the inositol rings of several PIs form the active site. While this may be part of the answer, other data suggest that the apoproteins may also be directly involved in ordering water into an embryo crystal. The reason for this statement is that immunological studies show common epitopes between the LPIN and bacterial ice nucleator proteins (Duman, Wolber, Mueller and Neven, unpubl.). Polyclonal antibodies to the LPIN cross-react on Western blots with the *Pseudomonas fluorescens* protein IN, and polyclonal antibodies to the bacterial protein cross-react with the LPIN. As mentioned previously, ~ 80% of the bacterial protein consists of repeat structure, and therefore it is likely that at least some of the common epitopes between the proteins are also arranged as these repeat structures in the LPIN. Indeed, this is confirmed by cross-reactivity with the LPIN of antibodies prepared to a synthetic consensus octapeptide (Leu-Thr-Ala-Gly-Tyr-Gly-Ser-Thr) of the bacterial protein. Decreased cross-reactivity of the antibodies, as compared to reactivity with the natural antigen, indicates a lower level of presence of the repeat structure in the LPIN. This is to be expected, since in addition to its ice nucleator function, the LPIN almost certainly functions to shuttle lipid in the normal fashion of hemolymph lipoproteins, and this latter function is likely to place some constraints upon its structure. However, since considerable evidence indicates that the repeat sequences in the bacterial protein INs function as a template to order water into the embryo crystal, it is likely that these sequences serve the same function in the LPIN. While this does not necessarily preclude the direct involvement of PI in water structuring, it does raise some doubts. An alternative and/or additional function for PI could be to increase the solubility of the apoproteins and/or to assure their proper higher order structure. Recall that the native LPIN is 51% lipid. Therefore, it is likely that association with lipids is required for the native structure of the apoproteins. Yet there is a specific requirement for PI, indeed for the integrity of the inositol head group, for ice nucleator activity and this may argue for a more direct role.

Another potential function for PI and/or the apoproteins concerns the apparent requirement for aggregation of LPIN molecules for activity. While we might suspect that one molecule of LPIN in a volume of water should be able to induce nucleation, Fig. 3 illustrates that this is not the case. The LPIN at the lowest concentration plotted, $10^{-11}$ M, has no effect on the SCP of the buffer. As the concentration is progressively increased, the SCP increases until at a concentration of $1.7 \times 10^{-7}$ M the SCP plateaus at –6 to –7 °C. Although these low molar concentrations produce ice nucleator activity, a simple calculation shows that at a concentration of $10^{-9}$ M, the lowest LPIN concentration which is necessary to induce a significant increase in the SCP over that of the buffer, the 1-μl droplets on which these measurements are made contain ~ $10^8$ molecules of LPIN. Obviously, one LPIN molecule does not induce nucleation, and increasing concentration above the minimum required for activity results in an increase in nucleation temperature, indicating cooperation between the LPIN molecules. Recall that similar requirements exist for the bacterial protein ice nucleators. Recent scanning tunneling microscopy studies of the *Tipula* LPIN showed an interesting aggregation behavior of the LPINs into unusual chain structures (Yeung et al. 1991). Individual spherical LPINs are arranged, like pearls on a

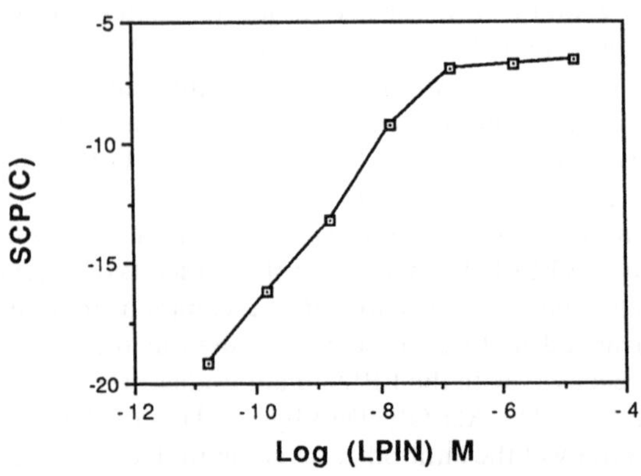


**Fig. 3.** The effect of addition of various concentrations (log molar concentration) of the *Tipula* LPIN on the supercooling point of a buffer solution (50 mM Tris, pH = 7.5, 100 mM NaCl). The SCP of the buffer alone was –18 °C

necklace, into long chains (Fig. 4). Usually two, sometimes three, chains are closely associated. Hemolymph lipoprotein from *Manduca sexta*, which lacks ice nucleator activity, does not show this aggregation behavior. Likewise, purified delipidated LPIN Apo-I (which lacks activity) does not aggregate, and neither do PI liposomes (which also lack activity). This suggests that this unusual organization behavior of the LPINs may be involved in their ice nucleation abilities. Thus, in addition to the requirement that an ice nucleator provide a template on its surface (active site) which organizes water molecules into an embryo crystal, it seems that there is a second requirement, namely that individual LPINs be structured into a cooperative unit, perhaps such that their individual active sites are closely aligned. Although it may seem counter-intuitive, the above-mentioned requirement for large numbers of LPIN molecules to initiate nucleation suggests that this alignment of active sites may be stochastic. In any event, it seems that for an LPIN to nucleate it must function as a template to organize water into an embryo crystal and aggregate with other LPINs in an appropriate fashion. Three components of the LPIN (Apo-I, Apo-II and PI) have been identified as being essential for activity. The task now is to determine what functions each of these components perform.

The apparent convergent evolution between the bacterial protein INs and the *Tipula* LPIN is quite interesting and might suggest that the bacterial-type repeat sequences are required for ice nucleating activity of a protein. However, this is not the case. Polyclonal antibodies to the *Tipula* LPIN fail to cross-react with the *Ceruchus piceus* LPIN (Neven et al. 1989), and neither do polyclonals to the bacterial proteins. Also, in addition to the LPIN, two other hemolymph proteins with ice nucleator activity have been purified from *Tipula* larvae (Duman, unpubl.). Both of these 77-kDa proteins lack lipid and both have unusual amino acid compositions. One contains 39 mol% glycine and the other is over 40 mol% tyrosine. This high tyrosine content of the latter protein indicates that it may be a "storage protein" (Levenbook 1985). Neither the anti-LPIN antibodies nor those to the bacterial PINs

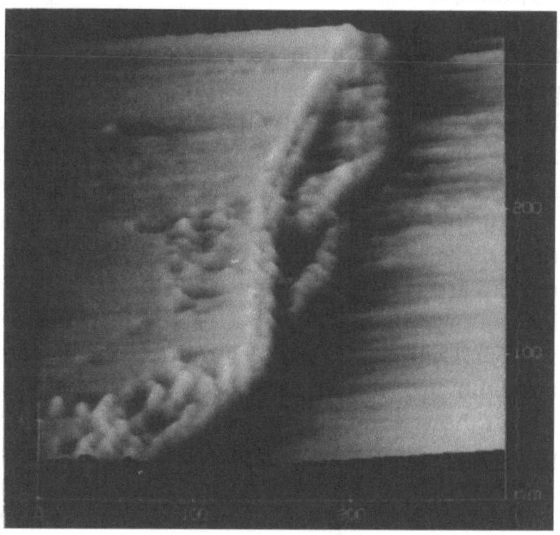

**Fig. 4.** Scanning tunneling microscope (Digital Instruments, Nanoscope II) image of the *Tipula* LPIN deposited on Highly Oriented Pyrolytic Graphite (HOPG) (for details see Yeung et al. 1991). Note the strong orientation of individual LPINs into chains, usually two or three chains wide with the LPINs in aligned conformation. The diameter of individual LPINs is 90–120 Å. An individual LPIN can be seen in the *upper left corner* of the field

cross-react with either of these *Tipula* PINs. Thus it is apparent that protein domains other than the bacterial PIN repeat structures can function to organize an embryo crystal and indeed insects, even a single species, have evolved different proteins to perform this function. It appears that *Tipula* has evolved hemolymph proteins with initial functions other than ice nucleation (i.e., lipid transport in the case of the LPIN, and storage protein function in the case of the high tyrosine PIN) into multi-functional proteins (i.e., ice nucleation plus lipid transport in the LPIN). The question of why, or if, *Tipula* larvae might require three different types of ice nucleator protein is interesting. Perhaps the function of the ice nucleator is more subtle than simply inducing nucleation at some specific temperature. Perhaps the pattern and/or type of ice crystals resulting are also important.

## Antifreeze Proteins

Antifreeze proteins (AFPs) produce a thermal hysteresis whereby they lower, by a noncolligative mechanism, the temperature at which an ice crystal will grow (hysteretic freezing point) but do not lower the melting point (see DeVries 1986). The magnitude of the thermal hysteresis (hysteretic freezing point minus melting point) is dependent on AFP concentration although, as will be discussed later, other factors may be involved. AFPs have been best studied in fish, but the thermal hysteresis phenomenon was first described by Ramsay (1964) in his classic study on the extremely efficient water reabsorbing cryptonephridial rectal complex of larvae of the beetle *Tenebrio molitor*. Ramsay found thermal hysteresis in the hemolymph, midgut, and especially high levels in the perirectal space. Thermal hysteresis was identified as being caused by proteins, but an attempt to purify the responsible proteins was unsuccessful (Grimstone et al. 1968). Because of the concentration of thermal hysteresis activity in the perirectal space, Ramsay (pers. commun.) thought that these proteins might be involved in the water reabsorption mechanism of the cryptonephridial rectal complex. Although the thermal hysteresis producing proteins are now known to function as antifreezes in a variety of animals, it is possible that Ramsay's intuition was correct. Acclimation of *Tenebrio* larvae to a high vapor pressure deficit (low relative humidity) induces AFP production. More importantly, larvae with high hemolymph AFP concentrations (induced by acclimation to low temperature, short photoperiod, or low relative humidity) had a greatly improved ability to survive very low relative humidities (Patterson and Duman 1978).

In spite of this potential water balance function, which deserves further attention, thermal hysteresis producing proteins do serve an antifreeze function in terrestrial arthropods. Studies published to date have identified AFPs in 26 species of insects (see Duman et al. 1991 for a listing), the centipede *Lithobius forficatus*, spiders (Duman 1979a; Husby and Zachariassen 1980), and an Antarctic mite (Block and Duman 1989). In most of these, thermal hysteresis activity is present only in the winter; however, in several other species low levels (~ 0.5 °C) of hysteresis are also present in summer with a several-fold increase taking place in winter (Duman 1977a, 1979b, 1980). It is unknown whether these low summer AFP levels represent residual proteins with a long half-life which are left over from the previous winter or whether

AFPs are being produced at low levels over the summer, and if the latter, whether these AFPs have a function in the summer.

### *Control of AFP Production*

The seasonal cycle of production and loss of AFPs can be cued by changes in temperature, photoperiod, or thermoperiod (Duman 1977b, 1980; Patterson and Duman 1978; Horwath and Duman 1983a,b, 1986), and the circadian system is involved in the photoperiodic control of AFP production in larvae of the beetle *Dendroides canadensis* (Horwath and Duman 1982, 1984). The endocrine system provides a link between the sensory systems which perceive these environmental cues and the fat body which produces the AFPs. Treatment of *Dendroides* (Horwath and Duman 1983c) or *Tenebrio* (Xu, Duman, Goodman and Wu, unpubl.) larvae with juvenile hormone induces AFP production. Juvenile hormone (JH) titers were elevated in *Tenebrio* acclimated to AFP inducing conditions (short photoperiod or low temperature). In vitro studies with cultured fat bodies showed that the fat body is a site of AFP production and further emphasized a role for JH in AFP induction both in *Dendroides* and in *Tenebrio* (Xu and Duman 1991; Xu, Duman, Goodman and Wu, unpubl.). However, stimulation of AFP production in cultured fat bodies occurred only when the fat bodies had been removed from larvae that had been pre-treated with JH. The presence of JH in the culture media did not stimulate fat bodies taken from larvae that had not been pre-treated. This suggests that an additional hormone(s), or other factor, is required to induce fat body production of AFP.

### *Antifreeze Functions of AFPs*

Typical values of hemolymph hysteresis in overwintering insects range from 3 to 6 °C. These are high levels when compared to the 1 to 2 °C values typical of cold water marine fish. However, overwintering terrestrial arthropods from temperate or colder regions usually experience much colder temperatures than those encountered by marine fish (–1.9 °C minimum), and thus even the comparatively high levels of thermal hysteresis seen in some overwintering insects would seem to be insufficient to afford adequate protection for a freeze avoiding species, even if colligative antifreezes such as glycerol are also produced (which is often the case). Two points are potentially important in this regard. (1) The hysteretic freezing point measured by the freezing-melting point difference technique (DeVries 1986) may not accurately reflect the temperature at which an insect might be seeded by contact with external ice, and (2) in insects, the temperature of spontaneous nucleation (supercooling point), not the freezing point, has generally been believed to be the temperature at which an individual will freeze, even if it is in contact with external ice (Sömme 1982; Block 1990). The wax-coated cuticle of insects is thought to prevent inoculative freezing by outside ice, but inoculative freezing does occur and may be more important than initially thought (Sömme 1982). However, in most cases, the supercooling point of the insect should be the relevant parameter which indicates the temperature at which mortality will occur from freezing in a freeze avoiding insect. (Note that mortality may result from chill coma at temperatures above the nucleation temperature; see Bale 1987.) Thus the effect of AFPs on SCP is an important question.

In insects which produce AFPs, an increase in supercooling ability is usually associated with the seasonal increase in AFPs. However, such correlations do not identify a cause and effect. Zachariassen and Husby (1982) demonstrated an inverse relationship between the measured thermal hysteresis of a sample containing antifreeze proteins and the size of the seed crystal used in the measurement (i.e., the smaller the crystal the larger the hysteresis). These experimental data were extrapolated beyond the lower limit of crystal size actually used in the determinations to suggest an inhibition by AFPs of potential embryo crystals that otherwise should have reached critical size. Based on this, it was suggested that AFPs should stabilize the metastable supercooled state, but the authors hypothesized that AFPs would not actually lower the SCP. Parody-Morreale et al. (1988) showed that glycoprotein antifreezes from Antarctic fish have the ability to inhibit bacterial ice nucleators. Likewise, AFPs purified from larvae of the beetle *Dendroides canadensis* inhibit certain insect hemolymph proteins with ice nucleator activity, but other hemolymph ice nucleators are not affected (Duman et al. 1991). Thus it appears that the type and concentration of ice nucleators present in an insect in relation to the concentration, and perhaps type, of AFP will determine whether the SCP is stabilized and/or lowered. Figure 5 illustrates the inhibition by *Dendroides* AFPs of ice nucleator activity present in a protein fraction, (containing six proteins) from *Dendroides* hemolymph (collected from larvae in the spring). Although complete inhibition of ice nucleation was not achieved, higher molar ratios of AFP to ice nucleator proteins may have provided more complete inhibition, or it may be that more than one type of ice nucleator protein was present, and while the AFP inhibited one type it did not affect the other. Interactions of AFPs and ice nucleator proteins will be discussed further later.

As noted above, inoculative freezing in insects and other terrestrial invertebrates may be more common than was previously thought. Certain species, perhaps those with more permeable cuticles, may be more susceptible than others. The antifreeze effect of fish AFPs in preventing the seeding of fish by contact with

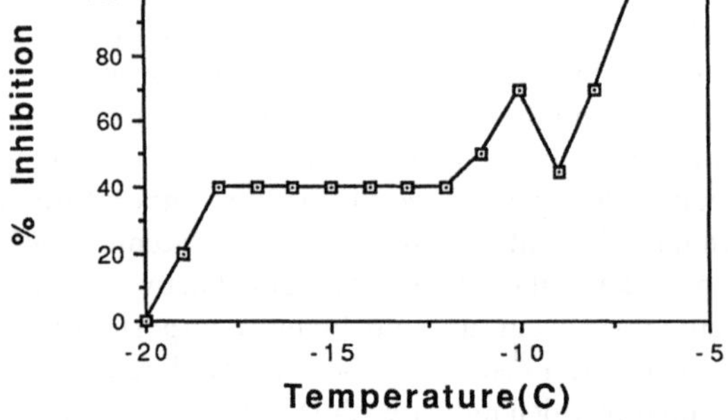


**Fig. 5.** Inhibition of ice nucleating activity of a heterogeneous solution of hemolymph proteins from *Dendroides canadensis* by *Dendroides* AFPs. The inhibition was determined by the ability of the AFPs to shift the ice nucleation spectrum of the ice nucleation proteins to lower temperatures

external ice at temperatures down to the hysteretic freezing point has long been recognized. Although the actual level of inhibition of inoculative freezing across an insect cuticle provided by AFPs has not been similarly determined, it might be expected to be greater than the amount of hysteresis measured by the standard freezing-melting point technique. As noted above, the measured hysteresis activity is inversely proportional to the size of the seed crystal used in the measurement (Zachariassen and Husby 1982). Since the size of the seed crystal (~0.25 mm) used in the standard measurements is much larger than the cuticular pores through which an insect might be seeded, an insect with 5 °C of measured hemolymph thermal hysteresis may actually be protected from inoculation well in excess of 5 °C. Using a differential scanning calorimetric technique Hansen and Baust (1988) confirmed the effect of crystal size of AFP activity, and determined greater activity in *Tenebrio* with this method than with the standard technique. Also, using fluorescent antibodies, we (Wu, Tweedell, and Duman, unpubl.) recently demonstrated that epidermal cells from under the cuticle of *Dendroides* larvae have AFPs located on the exterior surface of the cell membrane. Although it is not obvious how these AFPs might function, their location suggests that they might act to inhibit inoculative freezing. The potential effect of AFPs on inoculative freezing and the relationship between the measured hysteretic freezing point and the temperature at which an insect is affected by inoculative freezing (if at all) deserve further attention.

Antifreeze proteins can also provide protection of the gut fluid (Duman 1984) and in this regard they are an advantage over low molecular weight colligative antifreezes such as polyols and sugars. The latter have never been reported to concentrate in insect gut fluid, perhaps because the normal absorptive processes of the gut preclude their concentration. In contrast, *Dendroides* larvae in winter have levels of thermal hysteresis in the gut comparable to those in the hemolymph.

A few species of freeze-tolerant insects and the freeze-tolerant centipede *Lithobius forficatus* produce AFPs. The function of the AFPs in these species is not obvious, especially as hemolymph ice nucleators also are typically present to inhibit supercooling. The AFPs could be important in early autumn and late spring when the animals are not freeze-tolerant; however, they probably also function to inhibit recrystallization. During thawing or long-term holding of aqueous solutions at high subzero temperatures, it is common to observe a redistribution of crystal size. This process, recrystallization, may cause tissue damage from mechanical disruption by the growing crystals (Mazur 1984). This could be a serious problem in freeze-tolerant animals which often spend long periods at high subfreezing temperatures. Antifreeze proteins, both fish (Knight et al. 1984, 1988) and insect (Knight and Duman 1986), are able to prevent recrystallization, even at concentrations well below those required to detect thermal hysteresis activity. Most individuals of the centipede *Lithobius*, even in midwinter, do not have detectable levels of thermal hysteresis. In fact, only occasionally in some individuals collected in the autumn can hysteresis be measured. Yet midwinter hemolymph has strong recrystallization inhibition activity, while summer hemolymph does not (Tursman, Duman, and Knight, unpubl.). Thus, AFPs may be involved in the freeze tolerance of animals, even when thermal hysteresis is not detected by the usual techniques. However, recrystallization inhibition activity is not required for freeze tolerance in all insect species (Knight and Duman 1986).

### *AFP Activity*

An ongoing problem concerning the level of thermal hysteresis activity of the *Dendroides* AFPs has been that the activity of purified AFPs, even at very high concentrations, is considerably less (maximum of ~2.5 °C) than that generally seen in winter hemolymph (usually 3–6 °C with some individuals having 8–9 °C). This indicates the presence of a hemolymph-activating factor(s) that is missing in the test tube. Although optimal pH, inorganic ions, and polyol concentrations exerted some positive effects, the high levels of activity seen in the hemolymph could not be reached by manipulation of these factors. An unexpected result provided insight into the problem when addition of specific antibodies to the *Dendroides* AFP, rather than resulting in decreased activity, significantly enhanced (two- to threefold) thermal hysteresis (Wu et al. 1991). It is likely that some of the polyclonal antibodies bind to the AFP in such a manner that the AFP is still able to adsorb to the surface of potential seed ice crystals, but the AFP-antibody complex, being much larger than the AFP alone, blocks a larger surface of the crystal and extends further above the surface, making it more difficult to overgrow the complex and thus requiring a further decrease in temperature before crystal growth occurs.

Insects do not produce antibodies, so this enhancement cannot occur in vivo; however, it was subsequently determined that other proteins, including a 70-kDa protein purified from *Dendroides*, were also capable of significant enhancement of *Dendroides* AFPs (Wu and Duman 1991). This activation is demonstrated in Fig. 6. At this time the mechanism of the activation is unknown, but we have demonstrated that all the proteins which activate the *Dendroides* AFPs bind to the AFP, or vice versa. Thus the mechanism of activation may be similar to that proposed above for the activation by specific antibodies. It is also interesting that all of the activator proteins, except the gelatin proteins, have ice nucleator activity. If AFPs inhibit ice

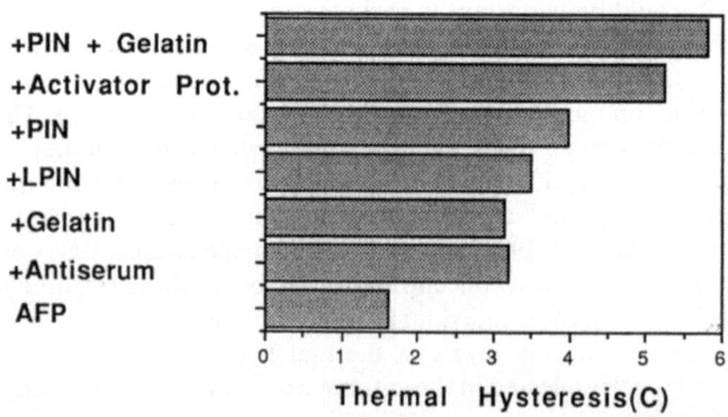


**Fig. 6.** The ability of various proteins to increase the thermal hysteresis activity of a 4 mg/ml solution of *Dendroides canadensis* AFPs. The various activators were anti-*Dendroides* AFP antiserum (rabbit), gelatin (5 mg/ml. The gelatin is soluble at this concentration. At higher concentrations where the gel state occurs, there is no activation), *Tipula* lipoprotein ice nucleator (1 mg/ml), *Tipula* protein ice nucleator (1 mg/ml PIN), and an endogenous 70-kDa protein purified from *Dendroides* (1 mg/ml)

nucleators we might expect that the AFP would bind to the ice nucleator, perhaps at the ice nucleation active site. It may be that in this process the thermal hysteresis activity of the AFP is increased because of the larger size of the AFP-IN protein complex. Of course, for this to occur the AFP must still be able to bind to the seed crystal while simultaneously binding to the activator protein.

### *Compositions and Structure-Function Relationships of Insect AFPs*

Although the structures of insect AFPs have not been as extensively investigated as have those of fishes, AFPs have been purified and partially characterized from four species. The sizes of these insect AFPs range between 14–25 kDa and they tend to have a somewhat higher component of hydrophilic amino acids than do fish AFPs. None has been identified as having the high content of alanine found in many fish AFPs (i.e., winter flounder), and none yet analyzed has had a carbohydrate component. Three of the four species produce AFPs with an unusually large amount of cysteine. The first insect AFPs to be purified were those of *Tenebrio molitor* larvae (Patterson and Duman 1979, 1982; Schneppenheim and Theede 1980; Tomchaney et al. 1982). *Tenebrio* produce several AFPs of two basic types, those with and those without cysteine. *Tenebrio* AFPs with cys contents of 15 and 28 mol% have been characterized (Patterson and Duman 1982), and that of the spruce budworm has 6 mol% cys (Hew et al. 1983). The AFP from the milkweed bug *Oncopeltus fasciatus* is interesting in that it contains 30 mol% serine and 20 mol% glycine (Patterson et al. 1981).

The best characterized of the insect AFPs are those from the larvae of the beetle *Dendroides canadensis* (Wu, Duman, Cheng, and Castellino, unpubl.). All four of these 15–25-kDa AFPs have similar amino acid compositions with high contents of hydrophilic amino acids (45–55 mol%) and cysteine (~16 mol%). Approximately half of the Cys residues are involved in disulfide bridges, and both the disulfide bridges and free sulfhydryls are essential for activity. The amino terminals of the *Dendroides* AFPs are blocked, making sequencing difficult; however, we have sequenced ~30% of the most abundant of the *Dendroides* AFPs (Wu, Duman, and Parmalee, unpubl.). Although the data are incomplete, there appear to be two distinct regions in the protein. The larger, very hydrophilic region(s) lacks cysteine, while the other is cys-rich. Although the hydrophilic region(s) lacks the obvious repeat sequences seen in most of the fish AFP's, there is some repeat structure. Most interestingly, a computer search of the NIH protein sequence data base showed that the only protein in the data base with sequence homology to a hydrophilic 19 amino acid sequence from the *Dendroides* AFP was an octapeptide repeat sequence from the ice nucleator protein of the bacterium *Pseudomonas fluorescens* (75% identity of overlap). We suspect that the side chains of the hydrophilic amino acids are involved in hydrogen bonding to the ice crystal lattice, as they seem to be in fishes, and consequently this region of the protein may have primary and higher-order structures which provide a lattice match to ice. As described earlier, the octapeptide repeat sequence of the ice nucleator protein is thought to provide a template which orders water molecules into the embryo crystal. Thus it is perhaps not surprising that similar sequences might appear in these two proteins. However, remember that the two proteins have opposite functions. *Dendroides* larvae typically supercool to –30 °C during cold winters. Obviously, the AFPs are not acting as nucleators, and presumably they function to promote supercooling. Perhaps that is why the repeat sequence is much less common in the AFP than in the nucleator, and also why the AFP is much smaller than the ice nucleator protein. It should also be noted that none of the insect ice nucleator proteins with which we have worked have thermal hysteresis activity.

If our speculation is correct that the hydrophilic region(s) binds to ice, then what is the function of the cys-rich region of the *Dendroides* AFP? One of several possibilities is that it could be involved in binding to other macromolecules. Our current hypothesis is that the activator protein(s) mentioned above combines with the AFP and thereby activates it. Also, recall that at least some AFP's are seen on the outside of the membrane of certain cells. We suggest that a certain region(s) of the AFP has evolved to allow this behavior while not interfering with the essential ice-binding process.

## Future Studies

It should now be obvious to the reader that much study is required to accomplish a proper understanding of the biology and biochemistry of ice nucleator and antifreeze proteins in insects and other terrestrial arthropods. The function of extracellular ice nucleators is reasonably well established in freeze-tolerant insects. However, while hemolymph ice nucleator proteins induce nucleation at –6 to –10 °C in most freeze-tolerant species, unknown nucleation sites perform essentially the same function in other species such as *Gynaephora groenlandica* (Kukal et al. 1988). Also, the interesting question remains as to why other freeze-tolerant species nucleate at quite different temperatures (i.e., very high temperatures, –1 to –2 °C, in the centipede *Lithobious forficatus*; moderately low temperatures, –17 to –23 °C, in silkmoth pupae; and very low temperatures, –45 to –60 °C, in some Alaskan and Canadian Rockies species.) Also, much more needs to be done concerning the seasonal removal of ice nucleators in freeze-avoiding species.

While some progress has been made in understanding the structure/ice nucleator mechanism relationships in the *Tipula* LPIN, much still remains unknown. In particular, the sequences of the apoproteins are needed, especially the locations and arrangements of the bacterial-type octapeptide repeats. Also, the structural relationships and specific functions of the apoproteins and of PI are of special concern. In addition, the role and mechanism of the interesting aggregation behavior of the LPINs must be further investigated. Other types of protein ice nucleators, such as the high tyrosine PIN in *Tipula*, should be characterized as they should provide useful comparisons to the *Tipula* LPIN.

Likewise, sequence and other structural information is essential to begin to understand the mechanism of insect AFPs. Comparisons between the insect and fish AFPs will be most interesting and may provide clues as to why certain insect AFPs have such high specific activities. Further studies on the nature of the activation of the *Dendroides* AFPs by certain proteins should be high priority. Clarification of the role of AFPs in promoting supercooling and inhibiting ice nucleators should be addressed and will require both in vitro and in vivo studies. An example of the latter is to determine in a freeze-avoiding species how the relative concentrations of ice nucleators, AFPs, and AFP-activating proteins vary with season. Information on environmental cues and hormones which control these seasonal changes will be required.

Applications of modern molecular techniques for the analysis of gene structure and function should be applied to many of the above mentioned problems. Sequencing of the genes for certain of these proteins may be essential, because of the difficulty in obtaining sufficient amounts of the proteins. Transgenic production of these proteins would provide protein for study and perhaps even for applied work involving the proteins in cryopreservation studies.

*Acknowledgments.* Many of the studies from the author's laboratory were funded by the National Science Foundation, currently by DCB-87098972.

## References

Bakken H (1985) Cold hardiness in the alpine beetles *Patrobus septentrionis* and *Calathus melanocephalus*. J Insect Physiol 31:447–453

Bale JS (1987) Insect cold hardiness: freezing and supercooling – an ecological perspective. J Insect Physiol 33:899–908

Baust JG, Rojas RR (1985) Review – insect cold hardiness: facts and fancy. J Insect Physiol 31:755–759

Block W (1990) Cold tolerance of insects and other arthropods. Phil Trans R Soc Lond B 326:613–633

Block W, Duman JG (1989) Presence of thermal hysteresis producing antifreeze proteins in the Antarctic mite, *Alaskozetes antarcticus*. J Exp Zool 250:229–231

Cannon RJC, Block W (1988) Cold tolerance of microarthropods. Biol Rev 63:23–77

Claussen DL, Townsley MD, Bausch RG (1990) Supercooling and freeze tolerance in the European wall lizard, *Podacris muralis*, with a revision of the history of the discovery of freeze-tolerance in vertebrates. J Comp Physiol B 160:137–143

Davies PI, Hew CL, Fletcher GL (1988) Fish antifreeze proteins: physiology and evolutionary biology. Can J Zool 66:2611–2617

DeVries AL (1986) Antifreeze glycopeptides and peptides: interactions with ice and water. Methods Enzymol 127:293–303

DeVries AL (1988) The role of glycopeptides and peptides in the freezing avoidance of antarctic fishes. Comp Biochem Physiol 90B:611–621

Duman JG (1977a) Variations in macromolecular antifreeze levels in larvae of the darkling beetle *Meracantha contracta*. J Exp Zool 201:85–93

Duman JG (1977b) The effects of temperature, photoperiod and relative humidity on antifreeze production in larvae of the darkling beetle *Meracantha contracta*. J Exp Zool 201:333–337

Duman JG (1979a) Subzero temperature tolerance in spiders: the role of thermal hysteresis factors. J Comp Physiol 131B:347–352

Duman JG (1979b) Thermal hysteresis factors in overwintering insects. J Insect Physiol 25:805–810

Duman JG (1980) Factors involved in the overwintering survival of the freeze-tolerant beetle *Dendroides canadensis*. J Comp Physiol 136B:53–59

Duma JG (1984) Thermal hysteresis antifreeze proteins in the midgut fluid of overwintering larvae of the beetle *Dendroides canadensis*. J Exp Zool 230:355–361

Duman JG, Morris, Castellino FJ (1984) Purification and composition of an ice nucleating protein from queens of the hornet, *Vespula maculata*. J Comp Physiol 15B4:79–83

Duman JG, Neven LG, Beals JM, Olson KO, Castellino FJ (1985) Freeze tolerance adaptations, including haemolymph protein and lipoprotein ice nucleators, in larvae of the cranefly *Tipula trivittata*. J Insect Physiol 31:1–9

Duman JG, Xu L, Neven LG, Tursman D, Wu DW (1991) Hemolymph proteins involved in insect subzero temperature tolerance: ice nucleators and antifreeze proteins. In: Lee RE, Denlinger DL (eds) Insects at low temperatures. Chapman Hall New York London, pp 94–127

Farrant J (1980) General observations on cell preservation. In: Ashwood-Smith MJ, Farrant J (eds) Low temperature preservation in biology and medicine. Univ Park Press, Baltimore, pp 1–8

Fields PG, McNeil JN (1986) Possible dual cold-hardiness strategies in *Cisseps fulvicollis* (Lepidoptera: Arctiidae). Can Entomol 118:1309–1311

Govindarajan AG, Lindow SE (1988a) Phospholipid requirement for expression of ice nuclei in *Pseudomonas syringae* and in vitro. J Biol Chem 263:9333–9338

Govindarajan AG, Lindow SE (1988b) Size of bacterial ice-nucleation sites measured in situ by radiation inactivation analysis. Proc Natl Acad Sci USA 85:1334–1338

Green RL, Warren GJ (1985) Physical and functional repetition in a bacterial ice nucleation gene. Nature (Lond) 317:645–648

Grimstone AV, Mullinger AM, Ramsay JA (1968) Further studies on the rectal complex of the mealworm, *Tenebrio molitor*. Phil Trans R Soc London B 253:343–382

Hansen TN, Baust JG (1988) Differential scanning calorimetric analysis of antifreeze protein activity in the common mealworm, *Tenebrio molitor*. Biochim Biophys Acta 957:217–221

Hew CL, Kao MH, So YP (1983) Presence of cystine-containing antifreeze proteins in the spruce budworm, *Choristoneura fumiferana*. Can J Zool 61:2324–2328

Horwath KL, Duman JG (1982) Involvement of the circadian system in photoperiodic regulation of insect antifreeze proteins. J Exp Zool 219:267–270

Horwath KL, Duman JG (1983a) Preparatory adaptations for winter survival in the cold-hardy beetles *Dendroides canadensis* and *Dendroides concolor*. J Comp Physiol 151B:225–232

Horwath KL, Duman JG (1983b) Photoperiodic and thermal regulation of antifreeze protein levels in the beetle *Dendroides canadensis*. J Insect Physiol 29:907–917

Horwath KL, Duman JG (1983c) Induction of antifreeze protein production by juvenile hormone in larvae of the beetle, *Dendroides canadensis*. J Comp Physiol 151:233–240

Horwath KL, Duman JG (1984) Further studies on the involvement of the circadian system in photoperiodic control of antifreeze production in the beetle *Dendroides canadensis*. J Insect Physiol 30:947–955

Horwath KL, Duman JG (1986) Thermoperiodic involvement in antifreeze protein production in the cold hardy beetle *Dendroides canadensis*: Implications for photoperiodic time measurement. J Insect Physiol 32:799–806

Husby JA, Zachariassen KE (1980) Antifreeze agents in the body fluid of winter active insects and spiders. Experientia 36:963–964

Katagiri C (1985) Structure of lipophorin in insect blood: location of phospholipid. Biochim Biophys Acta 834:139–143

Knight CA (1967) The freezing of supercooled liquids. Van Nostrand, New York

Knight CA, Duman JG (1986) Inhibition of recrystallization of ice by insect thermal hysteresis proteins: a possible cryoprotective role. Cryobiology 23:256:262

Knight CA, DeVries AL, Oolman LD (1984) Fish antifreeze protein and the freezing and recrystallization of ice. Nature (Lond) 308:295–296

Knight CA, Hallet J, DeVries AL (1988) Solute effect on ice recrystallization: an assessment technique. Cryobiology 25:55–60

Knight JD, Bale JS (1986) Cold hardiness and overwintering of the grain aphid *Sitobion avenae*. Ecol Entomol 11:189–197

Kozloff LM, Lute M, Westaway D (1984) Phosphatidylinositol as a component of the ice nucleating site of *Pseudomonas syringae* and *Erwinia herbicola*. Science 226:845–846

Kozloff LM, Lute M, Arellano F (1987) Role of phosphatidylinositol in ice nucleation. Pap 3rd Int Conf Biol Ice Nucleation, October 1987, Newport, Oregon

Kukal O, Serianni AS, Duman JG (1988) Glycerol production in a freeze-tolerant arctic insect, *Gynaephora groenlandica*: An in vivo $^{13}C$ NMR study. J Comp Physiol 158:175–183

Lee RE, Denlinger DL (1991) Insects at low temperature. Chapman and Hall New York London, 513 pp
Levenbook L (1985) Insect storage proteins. In: Kerkut GA, Gilbert LI (eds) Comprehensive insect physiology, vol 10. Pergamon Press, New York, pp 307–346
Lindow SE (1983) The role of bacterial ice nucleation in frost injury to plants. Annu Rev Phytopathol 21:363–384
Mazur P (1977) The role of intracellular freezing in the death of cells cooled at supraoptimal rates. Cryobiology 14:251–274
Mazur P (1984) Freezing of living cells: mechanisms and implications. Am J Physiol 247:C125–C142
Miller LK (1982) Cold hardiness strategies of some adult and immature insects overwintering in interior Alaska. Comp Biochem Physiol 73A:595–604
Mueller GM, Wolber PK, Warren GJ (1990) Clustering of ice nucleation protein correlates with ice nucleation activity. Cryobiology 27:416–422
Neven LG, Duman JG, Beals JM, Castellino FJ (1986) Overwintering adaptations of the stag beetle, *Ceruchus piceus*: removal of ice nucleators in winter to promote supercooling. J Comp Physiol 156:707–716
Neven LG, Duman JG, Low MG, Sehl LC, Castellino FJ (1989) Purification and characterization of an insect hemolymph lipoprotein ice nucleator: evidence for the importance of phosphatidylinositol and apolipoprotein in the ice nucleator activity. J Comp Physiol 159:71–82
O'Doherty R, Bale JS (1985) Factors affecting cold hardiness of the peach-potato aphid *Myzus persicae*. Ann Appl Biol 106:219–228
O'Doherty R, Ring JS (1987) Supercooling ability of aphid populations from British Columbia and the Canadian Arctic. Can J Zool 65:763–765
Parody-Morreale A, Murphy KP, DiCerca E, Fall R, DeVries AL, Gill SJ (1988) Inhibition of bacterial ice nucleators by fish antifreeze glycoproteins. Nature (Lond) 333:782–783
Patterson JL, Duman JG (1978) The role of thermal hysteresis producing proteins in the low temperature tolerance and water balance of larvae of the mealworm, *Tenebrio molitor*. J Exp Biol 74:37–45
Patterson JL, Duman JG (1979) Composition of protein antifreeze from larvae of the beetle *Tenebrio molitor*. J Exp Zool 210:361–367
Patterson JL, Duman JG (1982) Purification and composition of protein antifreezes with high cysteine contents from larvae of the beetle *Tenebrio molitor*. J Exp Zool 219:381–384
Patterson JL, Kelly TJ, Duman JG (1981) Purification and composition of a thermal hysteresis producing protein from the milkweed bug *Oncopeltus fasciatus*. J Comp Physiol 142B:539–542
Ramsay RA (1964) The rectal complex of the meal worm, *Tenebrio molitor* L., (Coleoptera, Tenebrionidae). Phil Trans R Soc London B 248:279–314
Ring JA (1982) Freezing-tolerant insects with low supercooling points. Comp Biochem Physiol 73A:605–612
Schneppenheim R, Theede H (1980) Isolation and characterization of freezing point depressing peptides from larvae of *Tenebrio molitor*. Comp Biochem Physiol B 67:561–568
Sömme L (1982) Supercooling and winter survival in terrestrial arthropods. Comp Biochem Physiol 73A:519–543
Sömme L (1989) Adaptations of terrestrial arthropods to the alpine environment. Biol Rev 64:367–407
Southworth MW, Wolber PK, Warren GJ (1988) Nonlinear relationship between concentration and activity of a bacterial ice nucleation protein. J Biol Chem 263:15211–15216
Storey KB, Storey JM (1988) Freeze tolerance in animals. Physiol Rev 68:27–84
Tomchaney AP, Morris JP, Kang SH, Duman JG (1982) Purification, composition and physical properties of a thermal hysteresis antifreeze protein from larvae of the beetle *Tenebrio molitor*. Biochemistry 21:716–721
Turner MA, Arellano F, Kozloff LM (1990) Three separate classes of bacterial ice nucleation structures. J Bacteriol 172:2521–2526
Warner DT (1962) Some possible relationships of carbohydrates and other biological components with the water structure at 37 °. Nature (Lond) 196:1055–1058
Wolber PK, Warren GJ (1989) Bacterial ice nucleation proteins. Trends Biochem Sci 14:179–182

Wu DW, Duman JG (1991) Activation of antifreeze proteins from the beetle *Dendroides candensis*. J Comp Physiol (in press)

Wu DW, Duman JG, Xu L (1991) Enhancement of insect antifreeze protein activity by antibodies. Biochim Biophys Acta 1076:416–420

Xu L, Duman JG (1991) Involvement of juvenile hormone in the induction of antifreeze protein production by the fat body of larvae of the beetle *Dendroides canadensis*. J Exp Zool (in press)

Xu L, Neven LG, Duman JG (1990) Hormonal control of hemolymph lipoprotein ice nucleators in overwintering freeze-susceptible larvae of the stag beetle, *Ceruchus piceus*: adipokinetic hormone and juvenile hormone. J Comp Physiol 160B:51–59

Yeung KL, Wolf EE, Duman JG (1991) A scanning tunneling microscopy study of an insect lipoprotein ice nucleator. J Vac Sci tech (in press)

Zachariassen KE (1982) Nucleating agents in cold-hardy insects. Comp Biochem Physiol 73A:557–562

Zachariassen KE (1985) Physiology of cold tolerance in insects. Physiol Rev 65:799–832

Zachariassen KE, Hammel HT (1976) Nucleating agents in the haemolymph of insects tolerant to freezing. Nature (Lond) 262:285–287

Zachariassen KE, Husby JA (1982) Antifreeze effect of thermal hysteresis agents protects highly supercooled insects. Nature (Lond) 298:865–867

Chapter 18

# The Role of Antifreeze Glycopeptides and Peptides in the Survival of Cold-water Fishes

A.L. DeVries and C.-H.C. Cheng

## Introduction

The necessity of maintaining a liquid state at the cellular level for life is obvious. For poikilotherms, the limits of cellular function and hence life are set by the temperatures at which phase changes of water occur. At the lower end it is 0 °C, the freezing point of water, or the freezing point of the body fluids of the organism, which is slightly lower than 0 °C, as determined by the salt content. Some organisms have evolved mechanisms which allow them to exist in a dormant state, or in an active state at temperatures well below their freezing points. This chapter is a brief account of how certain marine fishes maintain their hypo-osmotic body fluids in a liquid state while living actively in hyper-osmotic, freezing seawater laden with ice crystals, and avoid death from freezing.

## Freezing Marine Environments

The polar oceans and the near-shore water of north temperate oceans during winter are at the freezing temperature of seawater (–1.9 °C) (Littlepage 1965), and the top 30 m of the water column is often laden with microscopic ice crystals (Dayton et al. 1969). The freezing point of a typical marine teleost is about –0.8 °C (Black 1951), and therefore fish would freeze in these waters because supercooling in the presence of ice is impossible (Scholander et al. 1957). However, these waters support an extensive fish fauna, some members of which are paradoxically closely associated with ice formations (Andriashev 1970; DeVries 1982).

Some fish from these coldwater regions avoid freezing by behavioral adaptations, either migrating to warmer (+1 °C) off-shore waters during winter (Leim and Scott 1966), or remaining in the cold but ice-free deep water (>100 m) in a constantly supercooled state (Scholander et al. 1957). The permanent inhabitants of the ice-laden shallow waters, however, are able to avoid freezing because of the presence of endogenous "antifreeze" substances. These antifreeze agents were first identified in the antarctic notothenioid fishes as a special family of glycopeptides (DeVries 1968)

Department of Physiology and Biophysics, University of Illinois, 524 Burrill Hall, 407 S. Goodwin, Urbana, IL 61801, USA

Somero et al. (Eds.)
Water and Life

and as peptides in the North American winter flounder (Duman and DeVries 1974, 1976).

Fishes without biological antifreeze agents freeze in the presence of ice at about the equilibrium freezing point of their blood (DeVries 1982). The equilibrium freezing point of the blood of these fishes is related to the concentration of salts and other small organic molecules dissolved in it (DeVries 1982), governed by the same colligative property that determines the equilibrium freezing point of ordinary salt or biological solutions.

The equilibrium freezing point of a solution is the same as its equilibrium melting point. In practice, the equilibrium m.p. is taken as the temperature at which a small crystal (<50 μm) in the solution of interest melts as the solution is being slowly warmed (0.02 °C/min), and the equilibrium f.p. is taken as the temperature at which the ice crystal begins to grow again when the solution is slowly cooled before the ice completely melts (DeVries 1986). For ordinary solutions, the ice growth and melting are observed to occur within 0.01 °C of each other, i.e., the equilibrium f.p. and m.p. are essentially the same. However, for blood from fishes with antifreezes, the "freezing point" (temperature of ice growth) and the equilibrium m.p. are not the same (DeVries 1971; DeVries and Lin 1977a). The temperature of ice growth in the presence of antifreeze is therefore a "nonequilibrium freezing point", and for some high latitude antarctic fishes, it can be as low as –2.7 °C (Table 1). The equilibrium m.p. of their blood is about –1.1 °C and is accounted for largely by the salt and small organic molecule content. This separation of f.p. and m.p. is called thermal hysteresis. The large depression (1.6 °C) of f.p. from m.p. by the antifreezes indicated that a noncolligative mechanism is involved, which has now been shown to be a process of adsorption-inhibition (Raymond and DeVries 1977).

Ice growth in the blood serum of temperate fishes cooled to its equilibrium f.p. is slow and in the form of dendritic plates, which is similar to that in ordinary

**Table 1.** Comparison of freezing temperatures and blood freezing-melting points of antarctic fishes inhabiting different thermal environments and depths

| Antarctic species | Water temperature (°C) (Depth) | Organismal freezing temperature (°C) | Blood serum Freezing point[a] (°C) | Melting point[a] (°C) |
|---|---|---|---|---|
| *Pagothenia borchgrevinki* | –1.9 (Shallow, ice-laden) | –2.5 | –2.7 | –1.1 |
| *Rhigophila dearborni* | –1.9 (Deep, ice-free) | –1.9 | –2.0 | –0.9 |
| *Chaenocephalus aceratus* | –1.0 (Shallow, limited ice) | –1.5 | –1.5 | –0.9 |

[a]Freezing points were the temperatures of initial ice propagation from a 50-μm diameter polycrystalline seed crystal, and melting points were the temperatures at which the same size seed crystal melted, in a 10-μl capillary tube containing about 4 μl of serum.

solutions. By contrast, ice growth in the presence of antifreeze compounds is very rapid and in the form of fine spicules when the nonequilibrium f.p. is reached (DeVries 1971). The rapid propagation of ice during freezing of fish possessing antifreezes leads to massive disruption of the tissues resulting in death. Even partial freezing of fishes results in death, regardless of the presence or absence of antifreeze, although the time between thawing and death may be several days (Scholander et al. 1953).

## Glycopeptide and Peptide Antifreeze Agents

The antifreeze agents in fishes are either glycopeptides or peptides. The antifreeze glycopeptides have been most thoroughly characterized in the antarctic notothenioid fishes. They are composed of repeats of the glycotripeptide unit, alanyl-alanyl-threonine, with the disaccharide, N-acetylgalactosamine and galactose, linked to the threonines (Fig. 1) (DeVries 1971; Komatsu et al. 1970; DeVries et al. 1971; Shier et al. 1972, 1975). Early studies showed that there are eight discrete sizes in the small antarctic cryopelagic *Pagothenia borchgrevinki*, depending on the number of the glycotripeptide repeats. They are named AFGP 1–8 with molecular weights ranging from 2600 to 34000 Da, AFGP 8 being the smallest. As many as 16 sizes, but within the same molecular weight range, have now been identified in the blood serum of other antarctic notothenioid fishes (Ahlgren and DeVries 1984). Identical AFGPs have been isolated from the gadoid fishes from the arctic waters. Both the Greenland cod, *Gadus ogac* and the arctic cod, *Boreogadus saida*, have AFGPs which are identical to those found in the unrelated antarctic notothenioids (Van Voorhies et al. 1978). Similar AFGPs have also been found in two subarctic cods, *Eliginus gracilis* and *Microgadus tomcod*, which differ somewhat from those described above in that there are fewer sizes, and that a few threonines are replaced by arginine (Raymond et al. 1975; Osuga and Feeney 1978; O'Grady et al. 1982c; Fletcher et al. 1982). The small AFGPs 7 and 8 are the most abundant and they have less antifreeze activity (weaker inhibitor of ice crystal growth) than the larger AFGPs. They also differ in that a few of the alanines are replaced by proline (Lin et al. 1972; Morris et al. 1978).

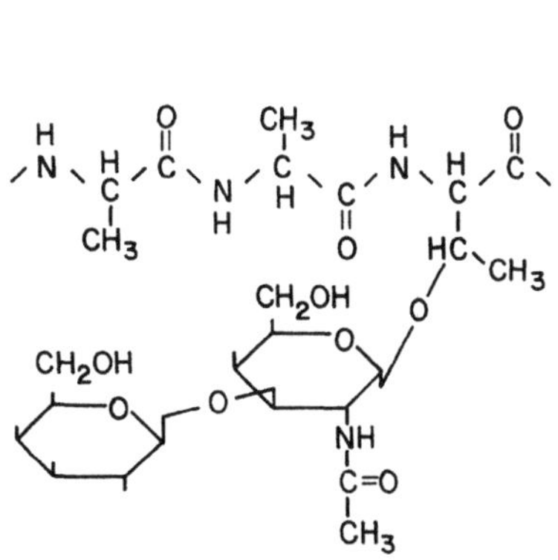


**Fig. 1.** Basic repeating structural unit of glycopeptide antifreezes (AFGPs) isolated from the blood of antarctic notothenioid fishes. The glycotripeptide unit is alanyl-alanyl-threonine with threonine joined by a glycosidic linkage to the disaccharide N-acetylgalactosamine and galactose. The smaller AFGPs have proline replacing a few of the alanine

The antifreeze peptides (AFPs), in contrast to the AFGPs, vary in composition and structure, but there is less variation in size within each structural group, of which there are three. The first group of AFPs are alanine-rich, alpha-helical peptides with molecular weights of 3300 to 5000 Da (DeVries and Lin 1977b; Raymond et al. 1977; Davies and Hew 1990). These are found in the northern flatfishes and sculpins (Duman and DeVries 1976; Gourlie et al. 1984; Hew et al. 1985; Scott et al. 1987). The helical AFPs have been most thoroughly characterized in the winter flounder, *Pseudopleuronectes americanus*. Two-thirds of its amino acids are alanine, and the peptide contains a repeating structure of 11 amino acids, Thr-$X_2$-polar aa-$X_7$, where X is mostly alanine (Gourlie et al. 1984; Hew et al. 1987). The 11-residue repeat (Fig. 2) is conserved in other species of flatfish such as the Alaskan plaice. The composition of the sculpin AFP is very similar to flounder and plaice AFPs, but the 11-residue repeat is not present (Fig. 2).

The second group of AFPs are found in the eel pouts or zoarcid fishes, which are about twice as large as the helical peptides. They are made up of all the common amino acids except histidine and tryptophan. They lack any significant helical and beta structure and there is no apparent sequence repeat (Fig. 2). All the eel pout AFPs studied have molecular weights of about 7000 daltons and there is about 60 to 80% sequence homology between species. Each fish produces multiple AFPs with some compositional differences, which are mostly from conservative replacements of amino acids of both the polar and nonpolar types (Li et al. 1985; Ananthanarayanan et al. 1986; Schrag et al. 1987; Cheng and DeVries 1989).

The AFPs in the sea raven, a cottid, have molecular weights of about 14 000 Da, are cysteine-rich, and appear to have significant beta-structure (Slaughter et al. 1981;

A Winter Flounder:

DTASDAAAAAALTAANAAAAAKLTADNAAAAAAATAA

Alaskan Plaice:

DTASDAAAAAAATAAAAKAAAEKTARDAAAAAAATAAAAR

Short-horn Sculpin:

MDGETPAQKAARLAAAAAALAAKTAADAAAKAAAIAAAAASA

B Antarctic Eel pout:

TKSVVASQLIPINTALTPAMMKAKEVSPKGIPAEEMSKIVGMQVN
RAVNLDETLMPDMVKTYQ

**Fig. 2.** **A** Amino acid sequences of the alanine-rich helical AFPs from winter flounder, Alaskan plaice, and short-horn sculpin. The 11-residue repeats in the peptide are indicated by *brackets*. **B** Amino acid sequence of the major AFP from antarctic eel pout *Austrolycichthys brachycephalus*. The amino acid composition is unbiased and there is no sequence repeat in the peptide

Ng et al. 1986). No other definitive structural information of this group of AFP is known.

Although the antifreezes are structurally very diverse, they appear to perform the same functioin. That is, they all adsorb to ice and significantly lower the temperature of ice growth below the equilibrium freezing point, although there is a size dependency in the AFGP antifreeze activity (Schrag et al. 1982). At the same concentrations, AFGPs 1–5 and the antarctic eel pout AFPs produce similar freezing point depression, while the small AFGPs 7 and 8 are less potent and produce only about two-thirds of the freezing point depression (Fig. 3).

## Genomic Basis of Antifreeze Heterogeneity

The extent of molecular heterogeneity exhibited by both the AFGPs and AFPs is truly striking. The AFGPs are found in as many as 16 different sizes within a given antarctic notothenioid fish. For the AFPs, the greatest heterogeneity is seen in the zoarcid fishes. The antarctic eel pout *Rhigophila dearborni* has three major and at least four minor AFPs (in terms of abundance in serum), the Atlantic ocean pout *Macrozoarces americanus* has 12 AFPs, and the arctic eel pout *Lycodes polaris* has four major and at least nine minor AFPs. The heterogeneous population of AFP molecules in a number of northern fishes is found to be encoded by a multigene family (Davies et al. 1984; Gourlie et al. 1984; Scott et al. 1985; Scott et al. 1988; Hew

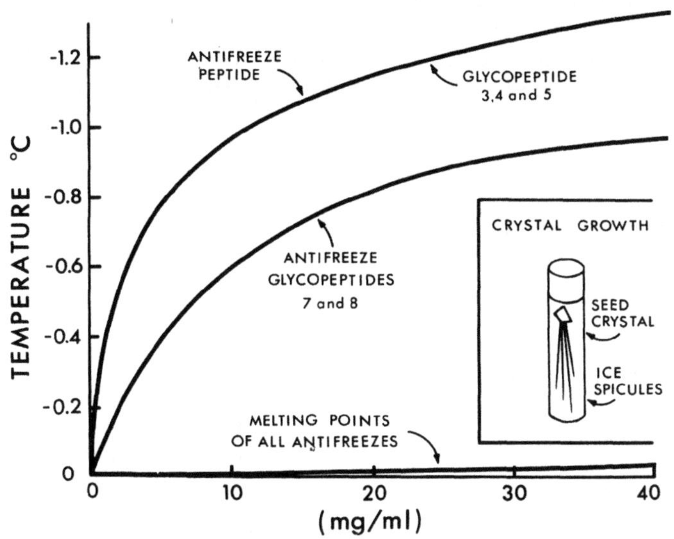


**Fig. 3.** Freezing and melting points of aqueous solutions of the peptide and glycopeptide antifreezes at several concentrations. The high molecular weight AFGPs and the AFPs (from antarctic eel pouts) have essentially the same activity, while the low molecular weight AFGPs have much less. The melting points are the same for all antifreezes. *Insert* ice growth is in the form of spicules parallel to the c-axis

et al. 1988). The multigene family of winter flounder AFPs is estimated to contain 30–40 member genes (Scott et al. 1985) and that of Atlantic ocean pout, 150 (Hew et al. 1988). Each gene encodes a single AFP, and for the majority of AFP gene families, the member genes are closely linked and arrayed in tandem repeats.

The AFGPs of the antarctic notothenioid fish are encoded by multigene families as well. However, unlike the AFP genes of the northern fishes, each AFGP gene encodes multiple copies of the AFGP peptide. In *Notothenia coriiceps neglecta* (Hsaio et al. 1990), one of the AFGP genes encodes 46 copies of the mature peptide, 44 of which are AFGP8 and the remaining two AFGP7. The peptides are separated by 3-amino acid spacers of the sequence Leu/Phe-Xaa-Phe and presumably are the sites for cleavage by chymotrypsin-like proteases (Fig. 4). In other words, the AFGP gene is a polyprotein gene of very high peptide copy number, and the gene product is a large polyprotein that is processed post-translationally to yield the individual peptides (Hsaio et al. 1990). Such a multigene family, multiple peptide per gene organization, conceivably can provide a very high gene dosage for the production of the constant high levels of AFGPs (3.5% w/v in blood) in the antarctic notothenioid fishes to permit survival in their perennially freezing environment.

## Antifreeze Synthesis and Distribution

The antifreezes are synthesized by the liver (Hudson et al. 1979; O'Grady et al. 1982a) and secreted into the blood, where they reach peak winter levels of 1.0 to 2.0% (w/v) in the northern fishes (Petzel et al. 1980; Fletcher et al. 1985; Reisman et al. 1987), to as high as 3.5% perennially in the antarctic notothenioid fishes and eel pouts. In the case of the notothenioid fishes, the AFGPs then become distributed from the blood to the other fluid compartments and the entire interstitial space of

| Positions | Sequence | Flanking | No. of Copies |
|---|---|---|---|
| 10 | AATAATAATPATAA | *LNF*<br>*LHF*<br>*FNF*<br>*CNF* | 20 |
| 7, 13 | AATAATPATAATPA | *LIF* | 17 |
| 10, 13 | AATAATAATPATPA | *FHF*<br>*FNF* | 4 |
| 7, 10, 13 | AATAATPATPATPA | *LIF* | 1 |
| 10 | AAAAATAATPATAA | *LNF* | 1 |
| 10, 12, 13, 14 | AATAATAATTAARG | | 1 |
| 10, 16 | AATAATAATPATAATPA | *LIF* | 1 |
| 10, 13, 16 | AATAATAATPATPATPA | *LIF* | 1 |

**Fig. 4.** The different AFGPs and their copy number in the AFGP8-polyprotein gene. There are six variants of AFGP8 and two of AFGP7 in this gene, depending on the placement of the proline residues. A single copy of AFGP8 at the 3′ end of the structural gene contains unusual amino acids (R, and G). The *numbers above* various amino acids indicate their positions in the AFGP. The three-amino acid segments flanking the AFGP are in italics

the fish; no antifreeze is found intracellularly (Ahlgren et al. 1988). The pericardial, peritoneal, and extradural fluid all contain AFGPs 1–8, at comparable or somewhat lower concentrations as those found in blood. The distribution of AFGPs into these fluid compartments is passive, possibly by diffusion through the capillary pores down their concentration gradients as do many of the other secreted blood proteins (Ahlgren et al. 1988). The intestinal fluids are fortified with AFGPs 7 and 8 and some AFGP 6 (O'Grady et al. 1982b, 1983) via a different route. They are translocated from the blood to the bile via a paracellular route, and then enter along with the bile into the anterior end of the intestine when the gall bladder evacuates. The AFGPs are neither digested nor reabsorbed in the intestinal tract. As the intestinal fluid transits the tract, water and salts are reabsorbed, and AFGPs become concentrated to 40 to 50 mg/ml at the distal end and contribute substantially to the intestinal fluid's freezing point of –2.4 °C (O'Grady et al. 1982b, 1983). Prevention of the freezing of the intestinal fluid is necessary and important because it is hypo-osmotic to seawater (f.p. = –1 °C), and it acquires ice crystals as the fish ingests ice-laden seawater.

The secreted fluids which include the endolymph and urine are protein-free, and show no detectable amounts of antifreeze (Dobbs et al. 1974; Dobbs and DeVries 1975a; Ahlgren et al. 1988). Endolymph is ordinarily in no danger of freezing because ice propagation into its deep-seated location in the inner ear is unlikely, and in addition, the surrounding tissues are fortified with antifreeze.

Urine is antifreeze-free in the antarctic notothenioid fish, because no filtration is involved in urine formation in their aglomerular kidneys (Dobbs et al. 1974; Dobbs and DeVries 1975b; Eastman and DeVries 1986). The antarctic eel pouts have glomerular kidneys, and their AFPs are small enough to filter through. However, no filtration occurs because the filtration barrier is very thick, and the Bowman's space does not appear to be connected with tubule lumen (Eastman et al. 1979). These adaptations effectively conserve their antifreezes which are needed to survive the perennially freezing environment.

In the winter flounder, glomerular filtration rates are 1 ml/kg/h, a rate sufficient to clear most of the AFP from the blood in 24 h. However, very little is cleared from the blood because the negatively charged AFP is repulsed by the negative charges associated with the pores in the basement membrane of the glomerular capillary walls (Boyd and DeVries 1983).

The ocular fluids in the notothenioid fish contain small amounts of the small AFGPs (Turner et al. 1985; Ahlgren et al. 1988), but insufficient to depress the freezing point by any significant amount. The freezing points of both urine and ocular fluids are approximately –1 °C, mostly due to the ion content (Dobbs and DeVries 1975a; Turner et al. 1985; Ahlgren et al. 1988). These two fluids are therefore supercooled by 0.9 °C and would freeze in the presence of ice. In the antarctic fish, urine does not freeze because the urethra, which could be the route of inward ice propagation, is normally closed by a strong muscular sphincter with a substantial amount of mucous. The ocular fluid cavity, on the other hand, does not communicate with the environment and therefore the only way for ice entry is across the integument. The head skin surrounding the ocular orbit extends over the cornea as a transparent tissue and has been shown to act as a physical barrier to ice entry (Turner et al. 1985).

The cerebro-spinal fluid (CSF) is a secreted fluid, but interestingly contains all eight AFGPs in the case of the notothenioid fish, at levels slightly lower than the blood (Cheng and DeVries, unpubl.). $H^3$-labeled AFGP injected into the systemic circulation does not appear in the CSF, indicating intact blood-brain, blood-CSF barriers (Ahlgren et al. 1988). A logical source for CSF AFGPs would be local synthesis in the brain. The functional role of antifreeze in the CSF is unclear, since like urine and endolymph, it can exist in a supercooled state, as the brain appears to be well protected by the antifreeze-fortified extradural fluid and cartilaginous skull. It is likely that, since the antarctic fishes are oviparous (Moyle and Cech 1988), antifreeze would be needed in the rudimentary brain during embryogenesis and larval development before the antifreeze-fortified skull is developed to provide freezing protection. Continued synthesis of AFGPs in the adult brain may be less costly than regulation.

Overall, the passive distribution following secretion, and the blood-to-bile translocation, endow the major fluid compartments, and all the interstitial spaces with AFGPs. AFGPs are not found intracellularly, except, of course, in the liver cells, the synthetic sites. Thus the antifreeze essentially functions to prevent freezing of the fish's body fluids, which in turn confers organismal freezing avoidance. The secreted fluids, except the CSF, avoid freezing by a combination of physical barriers and protection by surrounding antifreeze-fortified tissues.

## Mechanism of Antifreeze Action – Adsorption-Inhibition

### *Adsorption*

When solutions of AFGPs and AFPs are being frozen, the antifreeze molecules are preferentially retained in the ice, unlike other molecules of similar size and shape, which are excluded into the unfrozen solution (Duman and DeVries 1972; Lin et al. 1976; Tomimatsu et al. 1976; Raymond and DeVries 1977). This led to our proposed mechanism of adsorption-inhibition, that is, antifreezes bind to ice crystals and inhibit ice crystal growth (Raymond and DeVries 1977). The adsorption of AF molecules was proposed to be by hydrogen bonding, and probably involves a lattice match between the hydrogen-bonding groups in the AF molecules and the water molecules in the ice lattice (DeVries and Lin 1977b; DeVries 1984). The involvement of hydrogen bonding in binding was shown by the fact that chemical modifications of almost any of the potential hydrogen-bonding groups, i.e., the hydroxyls of the disaccharides of AFGPs (Duman and DeVries 1972; Shier et al. 1972; Lin et al. 1972), or the carboxyl groups of the aspartic and glutamic acid residues in some AFPs (Duman and DeVries 1976), all lead to loss of the antifreeze property. In space-filling models of AFGPs, many of the hydroxyls of the disaccharides are separated from each other by 4.5 Å, which is the same as the repeat spacing of water molecules in the ice lattice parallel to the a-axes, forming a lattice match. In addition, assuming a completely extended conformation of the AFGP molecule, the alternate carbonyl groups on the peptide backbone are 7.3 Å apart, which matches the repeat spacing (7.36 Å) along the c-axis in the ice lattice (Fletcher 1970).

The lattice match appeared most obvious in the helical AFPs from the winter flounder, Alaskan plaice, and short-horn sculpin. These AFPs contain clusters of polar amino acids separated by long sequences of nonpolar alanine residues. The polar clusters usually contain threonine and aspartate separated by two alanines (DeVries and Lin 1977b). These helical AFPs are amphiphilic, with the nonpolar and polar residues positioned on opposite side of the helix. The distance between the polar side chains of the aspartate and threonine are separated by 4.5 Å, matching the repeat spacing in the ice lattice parallel to the a-axes, suggesting that these AFPs may bind to faces parallel to the a-axes, i.e., the primary prism planes (DeVries 1984).

Adsorption of antifreezes to ice dramatically alters the crystal habit of ice such that faces which normally are not seen in ice grown from the melt become expressed (Knight et al. 1984; Raymond et al. 1989). These faces are probably where antifreeze molecules bind. Binding of AF retards the growth of these faces, and they become expressed when growth layers are deposited on adjacent faster-growing faces. We have examined in detail the effect of AF on ice crystal habits using large single ice crystals (Raymond et al. 1989). At low supercooling (temperatures within the hysteresis gap), antifreezes inhibit growth of the prism faces but allow limited growth on the basal plane. During growth, pyramidal faces $\{10\bar{1}X\}$ develop on the exterior of the ice crystal, and hexagonal pits with pit faces of $\{11\bar{2}X\}$ develop within the basal plane. Growth becomes completely halted when the basal plane becomes fully pitted in the case of AFGP1–5. For AFGP6 and some AFPs, growth on the pit faces continue, until the ice crystal assumes the shape of a hexagonal bipyramid. With the less potent antifreeze, AFGP7 and 8, pits do not form, but limited growth occurs on both the primary and secondary prism faces until the crystal becomes a hexagonal bipyramid. The expression of $\{10\bar{1}X\}$ and $\{11\bar{2}X\}$ faces suggests that these faces contain sites of adsorption.

We recently developed a method which directly determines the ice crystal plane where adsorption occurs (Knight and DeVries 1988; Knight et al. 1991). Different antifreezes are found to have different preferential planes of adsorption. The helical AFPs were found not to adsorb to the prism faces as previously thought (DeVries 1984; Yang et al. 1988). The winter flounder and Alaskan plaice AFPs adsorb to the pyramidal planes $\{20\bar{2}1\}$, and the short-horn sculpin AFP adsorbs to the secondary prism planes $\{2\bar{1}\bar{1}0\}$. All three AFPs are deduced to align along the $<01\bar{1}2>$ direction in the ice lattice, which has a 16.7A repeat spacing. The helical AFPs in the flatfishes contain three to four similar sequence repeats of eleven amino acids ending with a polar residue, which constitutes a distance of 16.5A per repeat (Fig. 2). Although lacking repeats, the helical sculpin AFP does contain three 11-amino acid sequences where the residue at position 11 is a polar residue which projects from the same side of the helical coil. The rotational freedom of the side chain of the polar residues, and/or the flexibility in the helix itself, can readily compensate for the 0.2A difference to allow for a lattice match (Knight et al. 1991).

The fact that antifreezes have specific planes of adsorption and molecular alignment strongly supports the involvement of lattice or structural matching between the AF molecule and ice in the adsorption process. However, although a structural match appears evident with antifreezes that have repetitive structures such as the helical AFPs, it is not so with the zoarcid AFPs, which have no repetitive sequences or obvious secondary structure (Ananthanarayanan et al. 1986; Schrag et al. 1987; Cheng and DeVries 1989). Nevertheless, the AFPs from the antarctic eel pouts are found to specifically adsorb on the primary prism planes $\{10\bar{1}0\}$ (Knight and DeVries 1988). Whether a structural match is involved in their adsorption cannot be ascertained until their secondary structures and/or tertiary structures are definitively determined.

Antifreeze also alters the habit of ice crystal growth during freezing, such that ice propagates rapidly parallel to the c-axis, the thermodynamically nonpreferred direction of growth, forming fine spicules (DeVries 1982; Knight et al. 1984). Growth parallel to the a-axes, the preferred direction, i.e., prism plane growth, which would result in the dendritic growth commonly seen in freezing of pure water or ordinary solutions, is inhibited (DeVries 1982). This suggests that AF adsorb to the prism planes.

The three lines of evidence for the plane of adsorption by antifreezes, i.e., that from expression of certain crystal faces using large single ice crystals (Raymond et al. 1989), the direct determination of specific planes of adsorption (Knight et al. 1991), and that from the spicular growth implying adsorption to prism planes (DeVries 1982), appear to contradict each other. Further work is required to reconcile these differences.

### *Inhibition*

The noncolligative lowering of the freezing point by adsorbed antifreezes probably involves the Kelvin effect. Our initial model (Raymond and DeVries 1977) is depicted in Fig. 5a, which shows in perspective a growth step on an ice surface, pinned in place by antifreeze molecules. It proposes that antifreeze adsorbs to the faces in the path of growing steps forcing growth to occur in the regions between them, and resulting in many small curved fronts. The increased interfacial curvature raises the surface free energy which inhibits further growth, and greater supercooling is required to remove the energy from the system to allow the small fronts to propagate, i.e., for freezing to continue. This is the Kelvin effect, which acts to lower the local freezing point, but not the bulk freezing point, producing the freezing hysteresis. The spacing between the adsorbed antifreeze molecules appears to be a function of their concentration, size and shape. If certain assumptions are made about the density and randomness of the antifreeze molecules on the crystal face, the amount of freezing hysteresis can be taken as proportional to the square root of the concentration. Using this relationship, it has been shown that there is generally good agreement between the predicted freezing hysteresis and that from freezing curves obtained experimentally (Raymond and DeVries 1977).

Our recently refined model (Knight et al. 1991) is shown in Fig. 5b, which shows a cross-section of an ice surface with antifreeze molecules pinning growth normal to the surface. In perspective, this would look like the surface of a mattress with the buttons representing the antifreeze molecules. The Kelvin effect acts similarly here to produce the freezing hysteresis. The fact that some of the adsorption planes determined are fairly high-indexed (secondary prism planes, pyramidal planes)

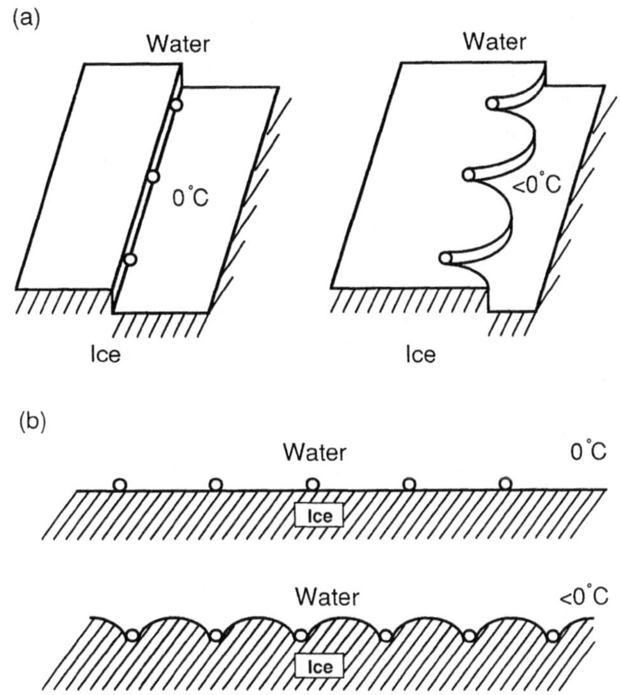


**Fig. 5a,b.** The two models of growth inhibition and their geometric assumptions. Antifreeze molecules inhibit in **a** step growth across the ice surface, and in **b** a surface cross-section, growth normal to surface. In both cases at 0 °C there is no curvature, but below 0 °C, growth requires interface curvature and this brings into effect a change in the local freezing point (Knight et al. 1991)

rather strongly favors the three-dimensional picture in Fig. 5b, because if the crystal surface on which step growth is inhibited (as in Fig. 5a) is the adsorption plane, there is no evident reason why growth normal to the adsorption plane would not occur.

## Freezing Avoidance in the Presence of Endogenous Ice

The mechanism of adsorption-inhibition for the freezing avoidance of fishes implies that ice is present somewhere in the live fish. The surface waters of the high latitude polar oceans are laden with minute ice crystals which make intimate contact with the gills and skin as the fish swim about. Ingestion of ice also occurs during drinking of seawater for osmotic and ion regulation. Using the antarctic species from ice-laden shallow waters, we observed that these fishes freeze immediately at –2.7 °C in ice-free seawater, which confirms the presence of ice in these specimens. During freezing, ice visibly propagates through their superficial tissues as well as through the clear ocular fluids. Freezing still occurs in these fishes even if they have been held for several hours at –1.2 °C, a temperature that would be expected to have melted any external ice crystals associated with their gills or integument, indicating presence of endogenous ice crystals somewhere within their bodies. At higher temperatures the endogenous ice appears to melt within a few minutes at 0 °C. Such ice-free fish can be supercooled to at least –6 °C in ice-free seawater without spontaneous freezing and with no apparent ill effect due to low temperature per se. Subsequent exposure to ice-laden seawater at –1.9 °C for 1 h "reinoculates" the fish and they again freeze at –2.7 °C in ice-free seawater. Systematic studies of the location of ice have revealed that blood, urine, bile, and the ocular fluids are ice-free and can exist in a supercooled state indefinitely. Liver, heart, white and red muscle tissue are ice-free as well. The integument (skin and scales), gills, and intestinal tract, including the intestinal fluid contain ice (Table 2) (Tien et al., submitted). Despite the presence of ice on the gills and skin, it does not propagate into the blood because the epidermal layer of cells, and possibly the overlying mucous, act as physical barriers. Although it is not known for certain, it appears that the ice is associated with the mucous overlying the cells of the gills and that coating the scales and epidermis. The transparent head skin over the eye (part of the integument) has been shown to act as a physical barrier to ice propagation down to temperatures as low as –3 °C (Turner et al. 1985). The rest of the integument may be an equally effective barrier to ice propagation. Assuming that the integument is an effective barrier to ice, the primary function of the antifreezes would appear to be one of preventing inward ice propagation when the integument is not as effective, such as during embryonic stages, or when it sustains injury, which is often observed in adult fish. The ice associated with the intestinal tract has as its origin the ingestion of ice-laden seawater. The large amounts of AFGPs 7 and 8 in the intestinal fluid inhibit ice growth and thereby prevent ice propagation through the intestinal wall, as well as preventing freezing of the intestinal fluid. The intestinal

**Table 2.** Location of ice in high latitude shallow water antarctic fish

| Tissues or fluids | Ice present (+) or absent (–) |
|---|---|
| External tissue | |
| Scaled integument | + |
| Scale-less integument | + |
| Gills | + |
| Internal tissue | |
| Intestine | + |
| Liver | – |
| Kidney | – |
| White muscle | – |
| Red muscle | – |
| Heart | – |
| Fluid | |
| Intestinal fluid | + |
| Blood | – |
| Bile | – |
| Urine | – |
| Vitreous humor | – |

wall can be considered as an external surface since it can be directly exposed to the external seawater during drinking. It appears then that the elaborate distribution of antifreeze from the blood to the entire interstitial space and intestinal fluid is to prevent ice propagation through the external surfaces.

*Acknowledgment.* This work was supported in part by National Science Foundation grant DPP87-16296 to ALD.

## References


Ahlgren JA, DeVries AL (1984) Comparison of antifreeze glycopeptides from several antarctic fishes. Polar Biol 3:3–97

Ahlgren JA, Cheng CC, Schrag JD, DeVries AL (1988) Freezing avoidance and the distribution of antifreeze glycopeptides in blody fluids and tissues of antarctic fish. J Exp Biol 137:549–563

Ananthanarayanan VS, Slaughter D, Hew CL (1986) Antifreeze proteins from the ocean pout, *Macrozoarces americanus*: circular diochroism spectral studies on the native and denatured states. Biochim Biophys Acta 870:154–159

Andriashev AP (1970) Cryopelagic fishes in the arctic and antarctic and their significance in polar ecosystems. In: Holdgate MW (ed) Antarctic ecology, vol 1. Academic Press, Lond New York, p 297

Black VS (1951) Some aspects of the physiology of fish. II. Osmotic regulation in teleost fishes. Univ Toronto Stud Biol Ser 59 71:53–89

Boyd RB, DeVries AL (1983) The seasonal distribution of anionic binding sites in the basement membrane of the kidney glomerulus of the winter flounder *Pseudopleuronectes americanus*. Cell Tissue Res 234:271–277

Cheng CC, DeVries AL (1989) Structures of antifreeze peptides from the antarctic eel pout, *Austrolycichthys brachycephalus*. Biochim Biophys Acta 997:55–64

Davies PL, Hew CL (1990) Biochemistry of fish antifreeze proteins. FASEB J 4:2460–2468

Davies PL, Hough C, Scott GK, Ng N, White BN, Hew CL (1984) Antifreeze protein genes of the winter flounder. J Biol Chem 259:9241–9247

Dayton PK, Robbilliard GA, DeVries AL (1969) Anchor ice formation in McMurdo Sound, Antarctica, and its biological effects. Science 163:273–274

DeVries AL (1968) Freezing resistance in some antarctic fishes. PhD Thesis, Stanford Univ, Standford, California

DeVries AL (1971) Glycoproteins as biological antifreeze agents in antarctic fishes. Science 172:152–1155

DeVries AL (1982) Biological antifreeze agents in coldwater fishes. Comp Biochem Physiol A73:627–640

DeVries AL (1984) Role of glycopeptides and peptides in inhibition of crystallization of water in polar fishes. Phil Trans R Soc Lond B304:575–588

DeVries AL (1986) Antifreeze glycopeptides and peptides: interactions with ice and water. In: Packer L (ed) Methods of enzymology, vol 127. Academic Press, New York, 293 pp

DeVries AL, Lin Y (1977a) The role of glycoprotein antifreezes in the survival of antarctic fishes. In: Llano GA (ed) Adaptations within antarctic ecosystems. Gulf, Houston, Texas, p 439

DeVries AL, Lin Y (1977b) Structure of a peptide antifreeze and mechanism of adsorption to ice. Biochim Biophys Acta 495:88–392

DeVries AL, Vandenheede J, Feeney RE (1971) Primary structure of freezing point-depressing glycoproteins. J Biol Chem 246:305–308

Dobbs GH, DeVries AL (1975a) Renal function in antarctic teleost fishes: serum and urine composition. Mar Biol 29:59–70

Dobbs GH, DeVries AL (1975b) Aglomerular nephron of antarctic teleosts: a light electron microscopic study. Tissue Cell 7:159–170

Dobbs GH, Lin Y, DeVries AL (1974) Aglomerularism in antarctic fish. Science 185:793–794

Duman JG, DeVries AL (1972) Freezing behavior of aqueous solutions of glycoproteins from the blood of antarctic fish. Cryobiology 9:469–472

Duman JG, DeVries AL (1974) Freezing resistance in winter flounder, *Pseudopleuronectes americanus*. Nature (Lond) 247:237–238

Duman JG, DeVries AL (1976) Isolation, characterization and physical properties of protein antifreezes from the winter flounder, *Pseudopleuronectes americanus*. Comp Biochem Physiol 53b:375–380

Eastman JT, DeVries AL (1986) Renal glomerular evolution in antarctic notothenioid fishes. J Fish Biol 29:649–662

Eastman JT, DeVries AL, Coalson RE, Nordquist RE, Boyd RB (1979) Renal conservation of antifreeze peptide in antarctic eel pout, *Rhigophila dearborni*. Nature (Lond) 282:217–218

Fletcher GL, Hew CL, Joshi SB (1982) Isolation and characterization of antifreeze glycopeptides from the frost fish, *Microgadus tomcod*. Can J Zool 60:348–355

Fletcher GL, Hew CL, Li X, Haya K, Kao MH (1985) Year-round presence of high levels of plasma antifreeze peptides in a temperate fish, ocean pout (*Macrozoarces americanus*). Can J Zool 63:488–493

Fletcher NH (1970) The chemical physics of ice. Cambridge Univ Press, Cambridge, p 11

Gourlie B, Lin Y, Powers D, DeVries AL, Huang RC (1984) Winter flounder antifreeze protein: evidence for a multigene family. J Biol Chem 259:14960–14965

Hew CL, Joshi S, Wang NC, Kao MH, Ananthanaryanan VS (1985) Structures of shorthorn sculpin antifreeze polypeptides. Eur J Biochem 151:167–172

Hew CL, Chakrabartty A, Yang D (1987) Biochemical adaptation to the freezing environment structure-function relationship of antifreeze polypeptides. In: Kon OL et al. (eds) Integration and control of metabolic processes. ICSU Press Symp Ser 7, Cambridge Univ Press, Cambridge, pp 299–309

Hew CL, Wang NC, Joshi S, Fletcher GL, Scott GK, Hayes PH, Buettner B, Davies PL (1988) Multiple genes provide the basis for antifreeze protein diversity and dosage in the ocean pout, *Macrozoarces americanus*. J Biol Chem 263:12409–12055

Hudson AP, DeVries AL, Haschemeyer AEV (1979) Antifreeze glycoprotein biosynthesis in antarctic fishes. Comp Biochem Physiol 62B:179–183

Hsaio K, Cheng CC, Fernandes IE, Detrich HW, DeVries AL (1990) An antifreeze glycopeptide gene from the antarctic cod *Notothenia coriiceps neglecta* encodes a polyprotein of high peptide copy number. Proc Natl Acad Sci USA 87:9265–9269

Knight CA, DeVries AL (1988) The prevention of ice crystal growth from water by "antifreeze proteins". In: Wagner PE, Valli G (eds) Atmospheric aerosol and nucleation. Springer, Berlin Heidelberg New York, p 717

Knight CA, DeVries AL, Oolman LD (1984) Fish antifreeze protein and the freezing and recrystallization of ice. Nature (Lond) 308:295–296

Knight CA, Cheng CC, DeVries AL (1991) Adsorption of α-helical antifreeze peptides on specific ice crystal surface planes. Biophys J 59:409–418

Komatsu SK, DeVries AL, Feeney RE(1970) Studies of the structure of the freezing point-depressing glycoproteins from an antarctic fish. J Biol Chem 245:2901–2908

Leim AH, Scott WB (1966) Fishes of the Atlantic coast of Canada. Fish Res Board Can, Ottawa, 357 pp

Li XM, Trinh KY, Hew CL, Buettner B, Baenziger J, Davies PL (1985) Structure of an antifreeze polypeptide and its precursor from the ocean pout, *Macrozoarces americanus*. J Biol Chem 260:12904–12909

Lin Y, Duman JG, DeVries AL (1972) Studies on the structure and activity of low molecular weight glycoproteins from an antarctic fish. Biochim Biophys Res Commun 46:87–92

Lin Y, Raymond JA, Duman JG, DeVries AL (1976) Compartmentalization of NaCl in frozen solutions of antifreeze glycoproteins. Cryobiology 13:334–340

Littlepage JL (1965) Oceanographic investigations in McMurdo Sound, Antarctica. In: Llano GA (ed) Antarctic research series, vol 5, Biology of antarctic seas II. Am Geophys Union, Washington DC, p 1

Morris HR, Thompson MR, Osuga DT, Ahmed AI, Chan SM, Vandenheede JR, Feeney RE (1978) Antifreeze glycoproteins from the blood of an Antarctic fish. J Biol Chem 253:5155–5162

Moyle PB, Cech JJ (1988) Fishes – an introduction to ichthyology. Prentice Hall, Englewood Cliffs New York, p 479

Ng N, Trinh YK, Hew CL (1986) Structure of an antifreeze polypeptide precursor from the sea raven, *Hemitripterus americanus*. J Biol Chem 261:15690–15696

O'Grady SM, Clarke A, DeVries AL (1982a) Characterization of glycoprotein antifreeze biosynthesis in isolated hepatocytes from *Pagothenia borchgrevinki*. J Exp Zool 220:179–189

O'Grady SM, Ellory JC, DeVries AL (1982b) Protein and glycoprotein antifreezes in the intestinal fluid of polar fishes. J Exp Biol 98:429–438

O'Grady SM, Schrag JD, Raymond JA, DeVries AL (1982c) Comparison of antifreeze glycopeptides from arctic and antarctic fishes. J Exp Zool 224:177–185

O'Grady SM, Ellory JC, DeVries AL (1983) The role of low molecular weight antifreeze glycopeptides in the bile and intestinal fluid of antarctic fishes. J Exp Biol 104:149–162

Osuga DT, Feeney RE (1978) Antifreeze glycoproteins from arctic fish. J Biol Chem 253:5338–5343

Petzel D, Reisman H, DeVries AL (1980) Seasonal variation of antifreeze peptide in the winter flounder, *Pseudopleuronectes americanus*. J Exp Zool 211:63–69

Raymond JA, DeVries AL (1977) Adsorption inhibition as a mechanism of freezing resistance in polar fishes. Proc Natl Acad Sci USA 74:2589–2593

Raymond JA, Lin Y, DeVries AL (1975) Glycoproteins and protein antifreeze in two Alaskan fishes. J Exp Zool 193:25–130

Raymond JA, Radding W, DeVries AL (1977) Circular dichroism of protein and glycoprotein fish antifreeze. Biopolymers 16:2575–2578

Raymond JA, Wilson P, DeVries AL (1989) Inhibition of growth of non-basal planes in ice by fish antifreezes. Proc Natl Acad Sci 86:881–885

Reisman HM, Fletcher GL, Kao MH, Shears MA (1987) Antifreeze proteins in the grubby sculpin, *Myoxocephalus aenaeus*, and the tomcod, *Microgadus tomcod*. Environ Biol Fish 18:295–301

Scholander PF, Flaff W, Hock RJ, Irving L (1953) Studies on the physiology of frozen plants and animals in the arctic. J Cell Comp Physiol 42:1–56

Scholander PF, Vandam L, Kanwisher JW, Hammel HT, Gordon MS (1957) Supercooling and osmoregulation in arctic fish. J Cell Comp Physiol 49:5–24

Schrag JD, O'Grady SM, DeVries AL (1982) Relationship of amino acid composition and molecular weight of antifreeze glycopeptides to noncolligative freezing point depression. Biochim Biophys Acta 717:322–326

Schrag JD, Cheng CC, Panico M, Morris HR, DeVries AL (1987) Primary and secondary structure of antifreeze peptides from arctic and antarctic zoarcid fishes. Biochim Biophys Acta 915:357–370

Scott GK, Hew CL, Davies PL (1985) Antifreeze protein genes are tandemly linked and clustered in the genome of the winter flounder. Proc Natl Acad Sci USA 82:2613–2617

Scott GK, Davies PL, Shears MA, Fletcher GL (1987) Structural variations in the alanine-rich antifreeze proteins of the pleuronectinae. Eur J Biochem 168:629–633

Scott GK, Hayes PH, Fletcher GL, Davies PL (1988) Wolffish antifreeze protein genes are primarily organized as tandem repeats that each contain two genes in inverted orientation. Mol Cell Biol 8:3670–3675

Shier WT, Lin Y, DeVries AL (1972) Structure and mode of action of glycoproteins from an antarctic fish. Biochim Biophys Acta 263:406–413

Shier WT, Lin Y, DeVries AL (1975) Structure of the carbohydrate of antifreeze glycoproteins from an antarctic fish. FEBS Lett 54:135–138

Slaughter D, Fletcher GL, Ananthanarayanan VS, Hew CL (1981) Antifreeze proteins from the sea raven, *Hemitripterus americanus*. J Biol Chem 256:2022–2026

Tien R, Wilson PW, DeVries AL (1991) Ice in antarctic fishes (submitted)

Tomimatsu Y, Scherer J, Yeh Y, Feeney RE (1976) Raman spectra of a solid antifreeze glycoprotein and its liquid and frozen aqueous solutions. J Biol Chem 251:2290–2298

Turner JD, Schrag JD, DeVries AL (1985) Ocular freezing avoidance in antarctic fish. J Exp Biol 118:121–131

Van Voorhies WV, Raymond JA, DeVries AL (1978) Glycoproteins as biological antifreeze agents in the cod *Gadus ogac* (Richardson). Physiol Zool 51:347–353

Yang DSC, Sax M, Chakrabartty A, Hew CL (1988) Crystal structure of an antifreeze polypeptide and its mechanistic implications. Nature (Lond) 333:232–237

Chapter 19

# Freeze-Thaw Injury and Cryoprotection of Thylakoid Membranes

D.K. Hincha[1,2] and J.M. Schmitt[1]

## Introduction

### *Dehydration of Leaf Cells During Freezing*

Freeze-thaw injury to plants is a highly complex process. While there is a good understanding of the physics and chemistry associated with the freezing of aqueous solutions (Franks 1981) the physiology and biochemistry of the freezing of whole plants or organs are poorly understood. Cells in plant tissues are, unlike animal cells, encased by a rigid cell wall. Leaf tissue contains large air-filled intercellular spaces (Fig. 1a,c). When a leaf is slowly frozen, ice crystallizes first in the dilute apoplastic (extracellular) solution (Beck et al. 1984; Pearce and Willison 1985). Since the water potential of ice is lower at the same temperature than that of liquid water, cellular water diffuses from the cells to the extracellular ice crystals (Olien and Smith 1981). The cells are thereby dehydrated, until an equilibrium is reached. During thawing, the water potential gradients are reversed and water diffuses back to the cells, provided that the plasmamembrane has not been injured.

The drastic effects of such a dehydration-rehydration cycle can be seen in Fig. 1. By using ultra-rapid freezing in liquid propane and chemical fixing additives to the organic solvent during freeze substitution, we were able to use transmission electron microscopy to analyze the effects of a lethal freeze-thaw cycle on spinach leaf ultrastructure. Previous investigations were either limited to conventionally fixed tissues before and after a freeze-thaw cycle (Martin and Öquist 1979; Soikkeli 1980; Krause et al. 1984; Wisniewski and Ashworth 1986), or $OsO_4$ was directly added to the partially frozen samples (Singh 1979). These methods, however, give ambiguous results, since conventional chemical fixation methods lead to alterations of the permeability properties of the membranes before fixation is complete. This can influence the relative volume of cellular compartments in the fixed samples (Soikkeli 1980; Holopainen and Holopainen 1988). By fixation during freeze substitution such artifacts can be largely avoided (Humbel and Müller 1986; Steinbrecht and Müller 1987).

[1]Institut für Pflanzenphysiologie und Mikrobiologie, Freie Universität, Königin Luise-Str. 12–16, 1000 Berlin 33, FRG
[2]Institut für Botanik und Pharmazeutische Biologie, Universität Würzburg, Mittlerer Dallenbergweg 64, 8700 Würzburg, FRG

Somero et al. (Eds.)
Water and Life

The unfrozen cells (Fig. 1a) show the characteristic smooth cell walls and a large central vacuole. After freezing to –9 °C approximately 90% of the cellular water has been removed. The cell walls have collapsed with the cells, and vacuoles are no longer visible (Fig. 1b). The only internal membrane structures which still seem intact, although also strongly dehydrated, are the chloroplasts (Hincha et al. 1989b). Their characteristic organization into stacks of thylakoid membranes embedded in a soluble matrix, the chloroplast stroma, is clearly visible (Fig. 1b). After thawing (Fig. 1c), damage is apparent. The cells are rehydrated and the cell walls have regained their original shape. However, no central vacuole is discernable. Likewise, all other internal membrane systems seem disrupted. Thylakoids can still be distinguished but are no longer organized in their characteristic stacks. No chloroplasts with intact envelope membranes can be found (Fig. 1c).

Such electron micrographs lend credit to the general assumption that the sites of cellular freezing damage are the biomembranes (Steponkus 1984; Steponkus, this Vol.). These structural data, however, are only of limited use to investigate the molecular mechanisms of membrane damage. Since much is known about the composition, structure, and biochemical activities of photosynthetic membranes, we have focused our attention on the responses of thylakoids to freezing in vitro and in vivo.

In the following we shall analyze possible mechanisms of freeze-thaw damage to isolated thylakoids and compare these results with the manifestations of injury suffered by chloroplasts which were frozen and thawed in intact leaves. We will then discuss how cold-hardening of plants can confer greater freezing resistance to photosynthetic membranes in situ.

## *The Freezing Process in Vitro*

Thylakoid membranes can be readily isolated in a functionally intact form. Photosynthetic functions like electron transport, generation of an electrochemical gradient, or photophosphorylation can be easily measured in vitro. The body of knowledge which has been accumulated by many groups working on photosynthesis facilitates the interpretation of changes brought about by freezing or cold acclimation.

When an aqueous solution freezes, the major part of the water crystallizes to pure ice. Above the eutectic temperature of the system, solutes are concentrated in an unfrozen residual volume. This results in a severe osmotic dehydration of any cell or organelle which will be entrapped in the unfrozen solution. Under equilibrium conditions, the osmolality of a solution coexisting with ice is a function of the temperature. For any given temperature, the volume of the unfrozen phase is therefore determined by the initial solute concentration of the unfrozen solution. If the initial solution contains more than one solute, the final concentration of any given solute is determined by its molar ratio to all other solutes for a given freezing temperature.

In a simple in vitro experiment where membranes, organelles, or cells are suspended in a solution and frozen and thawed, there are therefore three variables

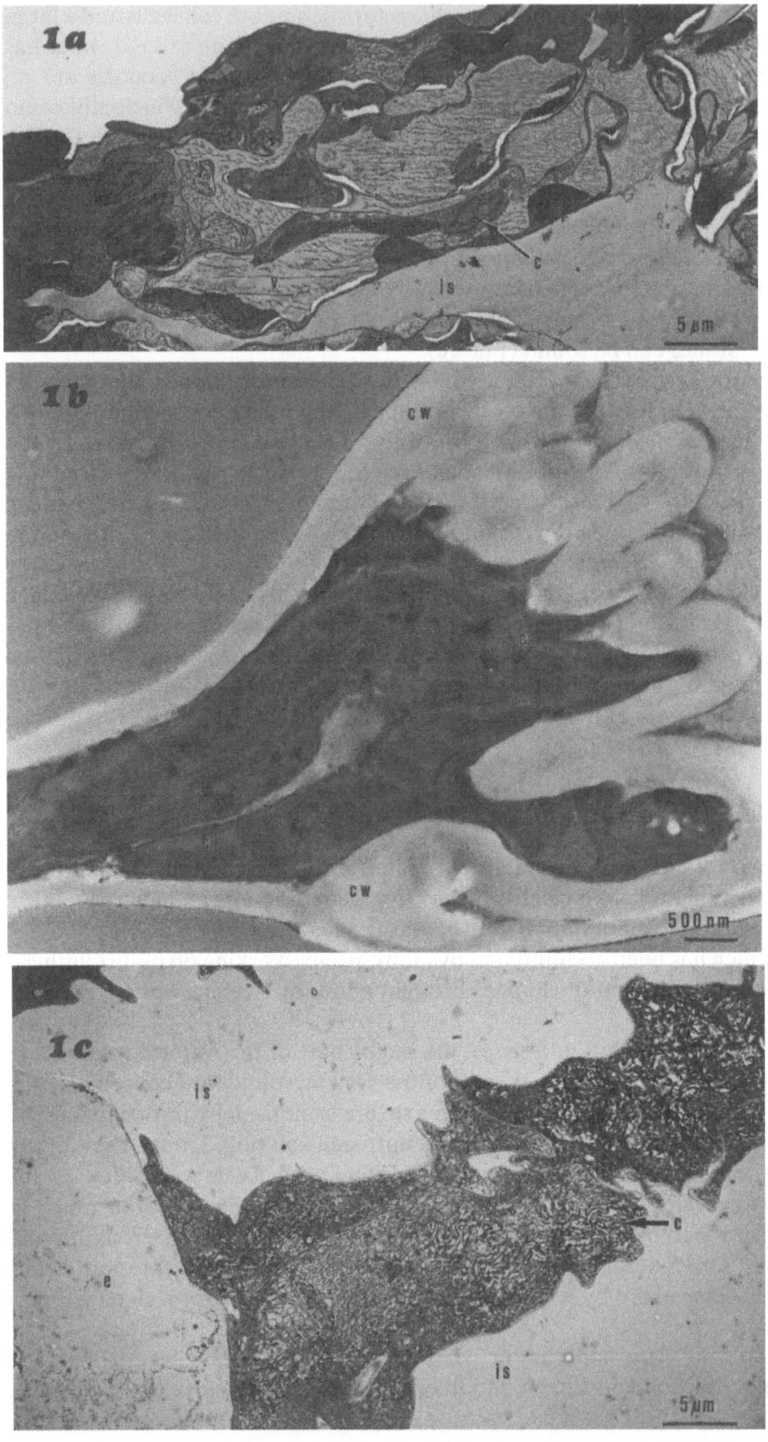
1a
v
v
c
is
5 µm
1b
cw
cw
500 nm
1c
is
c
e
is
5 µm

which can be controlled by the investigator: temperature, time, and solute composition of the medium.

## Simulating Damage in Vitro

The physicochemical changes described above lead to several potential stresses which can be broadly classified into physical and chemical (Hincha and Schmitt 1985). The distinction is not always clearcut, but the two types of stress may be distinguished operationally. Injury inflicted by chemical stresses must be modifiable by the chemical composition of the system. Physical damage, in contrast, should be largely independent of the chemical nature of the solutes. The presence of potentially membrane-toxic chemicals is no prerequisite for physical injury. Thylakoid membranes can be inactivated when frozen-thawed in sugar solutions in the absence of salt (Hincha et al. 1984).

### *Chemical Damage by Freezing*

The severity of freezing damage to thylakoids frozen in vitro can be controlled to a large extent by the solute composition of the medium. Biochemical activity can be fully preserved after a freeze-thaw cycle to –20 °C in, for instance, 0.5 M sucrose with 50 mM sodium chloride also present. Addition of increasing amounts of sodium chloride will lead to injury when a molar ratio of salt over sugar of about 1.5 is exceeded. A further increase in salt concentration will lead to severe injury, with total breakdown of some biochemical activities at ratios of about 5:1 (Mollenhauer et al. 1983). Chemical damage controlled by the ratio of cryotoxic over cryoprotective solute is shown schematically in Fig. 2.

The chemical properties of the salt also play an important role. Different halogenide anions can be used to control the severity of injury. The anions can be ranked in the order of increasing cryotoxicity from fluoride < chloride < bromide < iodide (Heber et al. 1981; Mollenhauer et al. 1983). This is exactly the order in which the capability of these ions to dissolve proteins is expressed (Hofmeister 1888). In

**Fig. 1a-c.** Electron micrographs of thin sections of spinach leaves. The leaves were frozen at 4 °C/h to –9 °C and were thawed at the same rate. At –2 °C nucleation of extracellular ice was initiated. Samples from the leaves (approx. 2 × 2 mm) were taken at the beginning of the freeze-thaw cycle (**a** 0 °C), after 1 h at the lowest temperature (**b** –9 °C) and after thawing (**c** 0 °C). The samples were rapidly frozen in liquid propane (–180 °C). They were fixed by freeze substitution in methanol containing 1% $OsO_4$, 0.5% uranyl acetate and 3% glutaraldehyde following the procedure of Humbel and Müller (1986). The fixed samples were infiltrated with Epon at 0 °C. The resin was cured at 60 °C for 24 h. Thin sections were stained with uranyl acetate and lead citrate, and were examined with a Philips EM 301 at 100 kV. *v* vacuole; *c* chloroplasts; *is* intercellular space; *cw* cell wall; *e* epidermal cell

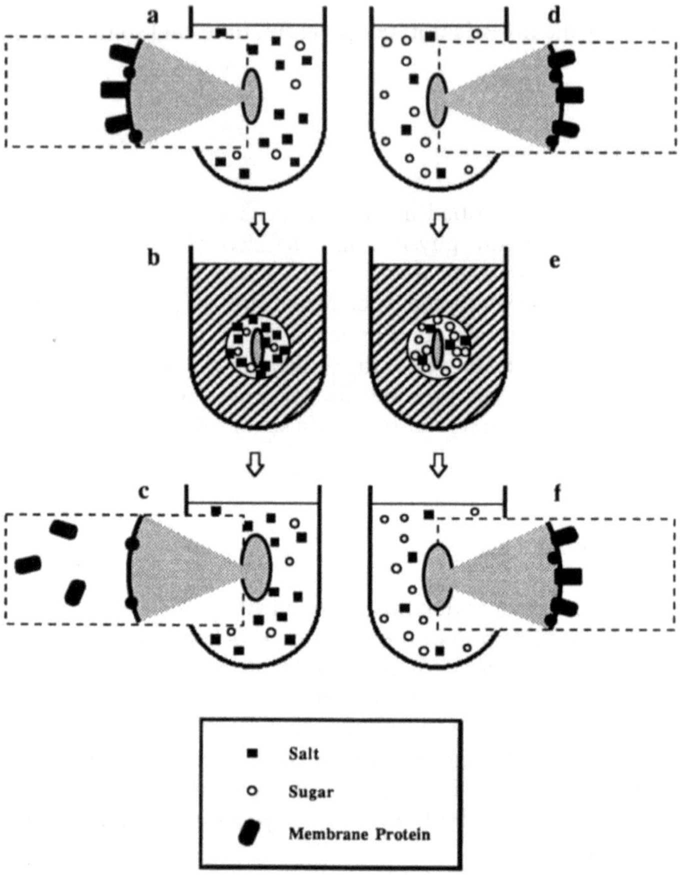


**Fig. 2a-f.** Schematic representation of an in vitro freeze-thaw cycle in the presence of different ratios of cryoprotective and cryotoxic solutes. **a** and **d** Thylakoid membrane vesicles (*shaded ovals*) are suspended in equiosmolar solutions of salt (*black squares*, potentially membrane toxic) and sugar (*open circles*, membrane-compatible). The surface of the membrane with the peripheral proteins is shown in an expanded view (**a,c,d,f**). After freezing (panels **b** and **e**), solutes are excluded from the ice (*hatched area*) and concentrated in the vicinity of the membranes. When the ratio of salt over sugar is high (**a**), the *final* concentration of salt (**b**) is injurious to the membranes (**c**) releasing peripheral membrane proteins. When the ratio of salt over sugar is low, salts are "diluted" by sugar at subzero temperatures (**e**) and peripheral membrane proteins stay attached (**f**)

other words, fluoride is least effective, and iodide is most effective in keeping proteins in solution.

This coincidence is more than fortuitous. A freeze-thaw cycle can solubilize membrane proteins (Garber and Steponkus 1976a; Volger et al. 1978; Mollenhauer et al. 1983) and iodide is more cryotoxic to photophosphorylation and releases more protein than the less cryotoxic fluoride during freezing (Mollenhauer et al. 1983). Preservation of cyclic photophosphorylation is linearly correlated with the retention of protein on the membranes (Hincha et al. 1984; Hincha and Schmitt 1985).

Up to about 35 different polypeptides are released during freeze-thawing under injurious conditions, as could be shown by immunoelectrophoresis and polyacrylamide gel electrophoresis in the presence of SDS (Hincha et al. 1984; Hincha and Schmitt 1985; Mollenhauer et al. 1983; Volger et al. 1978). The pattern of proteins released during freeze-thawing closely resembles the pattern of polypeptides released by a mild EDTA treatment at ice-bath temperature (Wolter et al. 1984). This shows that loosely bound, peripheral proteins rather than hydrophobic intrinsic proteins are detached during freezing. Among the proteins which have been identified is the coupling factor (synonyms: $CF_1$, peripheral part of the ATPase) (Garber and Steponkus 1976a; Volger et al. 1978; Hincha et al. 1985; Santarius 1987a).

The chloroplast ATPase is an $F_o/F_1$-type ATPase. The peripheral $CF_1$ part is localized on the stroma side of the thylakoids and protrudes into the medium after isolation (Strotmann and Bickel-Sandkötter 1984). It is removed from the membranes at high salt concentrations under nonfreezing (Santarius 1984a) as well as freezing conditions (Hincha et al. 1985; Santarius 1984b). Its release leads to an uncoupling of electron transport from ATP synthesis and a breakdown of the light-induced pH gradient across the membrane due to an opening of $CF_o$ proton channels (Schmitt et al. 1985; Coughlan and Pfanz 1986). $CF_1$ is specifically stabilized on the membranes by ATP (Santarius 1984b) and sulfate (Santarius 1987a).

During freezing in concentrated solutions, inactivation of photophosphorylation due to the release of $CF_1$ is dependent not only on the ratio of cryotoxic over cryoprotective solutes, but also on the total initial solute concentration at a given molar ratio (Hincha et al. 1984; Hincha and Schmitt 1985). This becomes clear when the binding of a peripheral protein to a membrane surface is viewed as an association/dissociation equilibrium. Since the final solute concentration during freezing is dependent only on the freezing temperature, an increase in the initial concentration will lead to a larger final volume at a given temperature (Fig. 3). Therefore, an increase in the initial solute concentration at a given salt-to-sugar ratio increases the release of $CF_1$ from thylakoid membranes (Hincha et al. 1985). The conclusion that $CF_1$ release is governed by the unfrozen volume available for dissociation is corroborated by the fact that an increase in membrane concentration results in reduced damage to photophosphorylation (Hincha et al. 1984) and membrane bound ATPase activity (Steponkus et al. 1977) by shifting the binding equilibrium of $CF_1$ towards association.

Chemical damage as determined by the ratio of cryotoxic to cryoprotective solutes was first invoked to explain freeze-thaw injury in red blood cells (Lovelock 1953a,b, 1954). Taken to the extreme, Lovelock's theory predicts that the severity of freezing damage at a given temperature is determined only by the composition but not by the total solute concentration of the initial solution. A schematic series of straight lines illustrating this is given in Fig. 3a.

### *Mechanical Damage Caused by Osmotic Contraction/Expansion*

Thylakoid membranes do not respond as predicted by Lovelock's theory, as shown schematically in Fig. 3b (Hincha and Schmitt 1985). At low initial solute concentra-

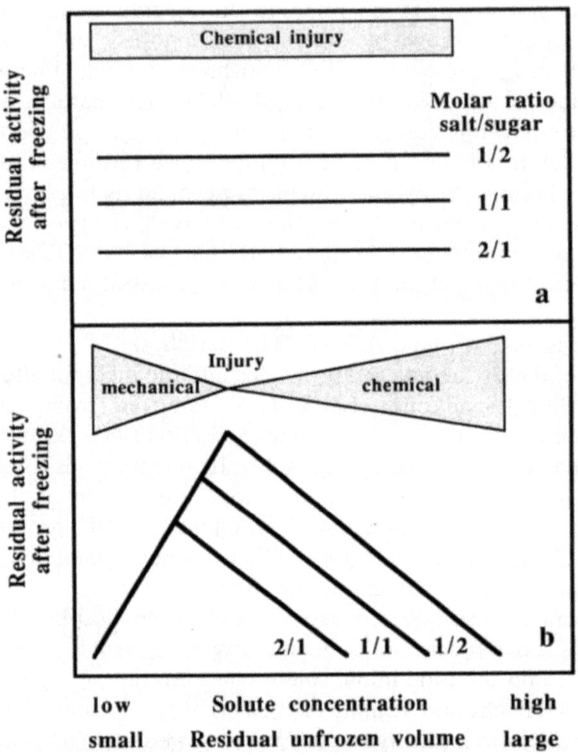


**Fig. 3a,b.** Schematic view of freeze-thaw preservation of biochemical activity as a function of total solute concentration at different molar ratios of cryotoxic over cryoprotective solute. **a** Membrane vesicles are suspended in solutions of salt (potentially membrane-toxic) and sugar (membrane-neutral) at the molar ratios indicated. The colligative theory predicts that injury is dependent on the solute ratio but not on initial solute concentration. **b** Thylakoid membranes show optima of cryopreservation at intermediate solute concentrations (Hincha et al. 1984; Hincha and Schmitt 1985), indicating the interplay of mechanical and chemical injury. The slopes of the curves under conditions of chemical injury are thought to be a consequence of the higher residual volume of the unfrozen solution at high initial solute concentration. A schematic representation of the injurious processes at low initial solute concentration is given in Fig. 4, right column. Injury at high initial solute concentration and high ratio of salt over sugar is shown in Fig. 2a-c

tions (under about 0.1 Osm/kg) freeze-thaw survival is essentially independent of the ratio of salt over sugar of the suspending medium. It is, in contrast, linearly correlated with solute concentration (Hincha and Schmitt 1985). Membranes frozen in low concentrations (<20 mM) of the cryoprotectant sucrose alone can be severely injured (Hincha et al. 1984). According to the operational definition given above, this indicates a physical mode of cryoinjury or cryoprotection, respectively.

Two possible causes for physical injury have been proposed: piercing of the membranes by the ice crystals (Santarius and Giersch 1983, 1984) and osmotic contraction/expansion caused by the drastic solute concentration differences between the frozen and the unfrozen state (Hincha et al. 1985). Although direct mechanical effects of the ice crystals cannot be totally excluded, the contraction/expansion hypothesis seems more likely.

When thylakoids are frozen and thawed in media of low osmolality (10 to 100 mOsm), their ability for cyclic photophosphorylation is reduced. This reduction is linearly correlated with the release of the electron transport protein plastocyanin from the membranes (Hincha et al. 1985). Plastocyanin is a soluble protein (MW 10 500) which is located in the intrathylakoid space (Haehnel 1984, 1986). The release of a protein from the membrane vesicle lumen can obviously only occur when the permeability barrier has been broken down, at least temporarily.

Rupture, however, is transient and the membranes regain their semi-permeability properties after thawing (Hincha 1986). Under these conditions, damage to photophosphorylation can be mitigated by freezing the membranes in solutions containing plastocyanin (Garber and Steponkus 1976a; Hincha et al. 1985). Added plastocyanin reduces the net efflux of this protein from the thylakoids. This indicates that even after a damaging freeze-thaw cycle the membranes are capable of generating the electrochemical gradient which is necessary for ATP formation. Therefore, they must have resealed after rupture.

From the analysis of the volumetric behavior of thylakoids after a freeze-thaw cycle, it was found that two cooperative effects contribute to the mechanical rupture of the membrane vesicles (Hincha 1986). Firstly, the ability of the vesicles to expand during thawing is reduced. While unfrozen controls responded as ideal osmometers when challenged with a wide range of sucrose concentrations, the frozen-thawed samples ruptured when suspended in dilute solutions before freezing. Their capacity to re-expand was less than that of unfrozen controls (Hincha 1986; Hincha et al. 1989b).

A similar volumetric behavior during a freeze-thaw cycle has been found in isolated rye protoplasts, where it was termed expansion-induced lysis (Steponkus 1984). It has been explained by the appearance of membrane vesicles extruded from the plasmamembrane during hypertonic shrinkage (Wiest and Steponkus 1978). Upon thawing, the vesicles are not reincorporated at a sufficiently high rate to compensate for the osmotic swelling (Wolfe et al. 1985). Vesiculation under hypertonic conditions has also been reported for phospholipid vesicles (Callow and McGrath 1985), for vacuoles in isolated plant cells (Johnson-Flanagan and Singh 1986; Singh et al. 1987), and for unilamellar vesicles of rye plasmamembrane lipids (Steponkus and Lynch 1989). Direct evidence for vesiculation as the mechanism underlying the reduced extensibility of isolated thylakoids after freezing is, however, still lacking.

In addition, thylakoids take up external solutes during freezing. This has been reported for sorbitol (Williams and Meryman 1970), NaCl (Jensen and Oettmeier 1984) and sucrose (Hincha 1986). This influx has been attributed to membrane breakage during freeze-induced dehydration (Jensen and Oettmeier 1984). Williams and Meryman (1970) have proposed that the vesicles have a minimum critical volume below which they resist further shrinkage. This would result in a hydrostatic pressure across the membrane which would facilitate transient membrane breakage and solute influx.

It has been shown, however, that thylakoids are permeable at a low rate even to molecules as hydrophilic and bulky as the disaccharide sucrose (Hincha 1986). Influx is linearly dependent on the concentration gradient between both sides of the membranes. Freezing-induced dehydration increases solute concentration and therefore makes any preexisting concentration gradient much steeper, as shown schematically in Fig. 4. Solutes permeate the membranes driven by diffusion, (Fig. 4b). Using the permeability coefficients for sucrose and the temperature dependence of this process, it was shown that solute influx into thylakoids during freezing can be described in terms of passive diffusion (Hincha 1986; Hincha et al. 1989b). No permeability changes of the membranes during freezing (Williams and Meryman 1970) need to be invoked.

During thawing, water follows the solutes and the vesicles swell osmotically (Fig. 4c). Thylakoids which have initially been suspended at a high solute concentration will swell but have enough osmotic support from the medium so that rupture is

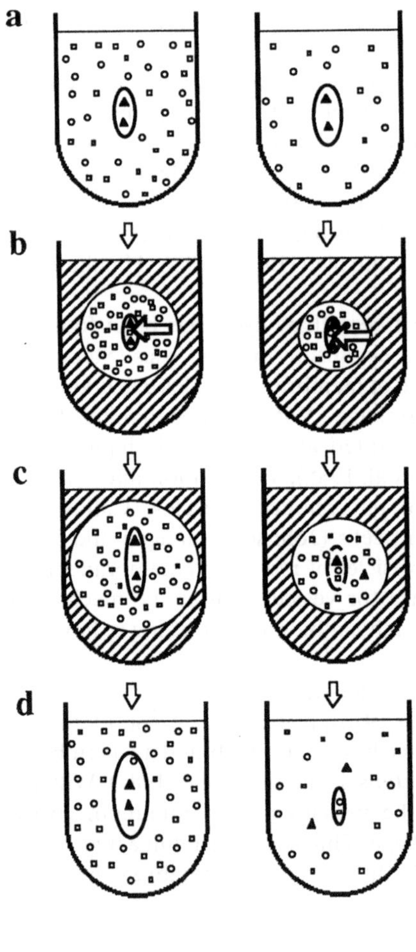


**Fig. 4a-d.** Schematic representation of an in vitro freeze-thaw process. **a** Thylakoid membrane vesicles (*open ovals*) are suspended in solutions of high (*left column*) and low (*right column*) solute concentrations. **b** Freezing removes water to give pure ice (*hatched area*). Membranes and solutes are concentrated in a small volume. Driven by the steep gradient, solutes permeate the membranes. During thawing **c** solute concentration is decreased by the melting ice. The thylakoids swell osmotically and may rupture (*second column*). **d** After thawing, thylakoids have either increased in volume due to solute influx (*first column*) or, after rupturing, resealed and decreased in volume (*second column*)

prevented (Fig. 4, left column). Thylakoids which have initially been suspended in dilute solutions will swell excessively, and finally rupture and collapse (Fig. 4, right column).

### *An in Vitro Model for the Intact System*

It is obvious that simple salt-sugar solutions do not provide an adequate model system to simulate all aspects of an in situ freeze-thaw cycle for thylakoid membranes. More complex, stroma-like media, have been introduced in order to simulate freeze-thaw injury under more realistic conditions (Grafflage and Krause 1986; Santarius 1986a; Hincha and Schmitt 1988a). Changes in the solute composition of complex media modulate freeze-thaw damage in complex and not readily predictable ways (Santarius 1986a,b,c, 1987b, 1990). The sequence of the inactivation of several partial reactions of photosynthesis was shown to be similar after freezing in an artificial stroma medium when compared to the sequence that is found after a lethal freeze-thaw cycle in vivo (Grafflage and Krause 1986).

Mechanical membrane rupture can be generated in vitro in a medium comprising salts, sugar, and amino acid at solute concentrations which are comparable to the ones of plant cells (Hincha and Schmitt 1988a). When the NaCl concentration is raised above approximately 250 mM, chemical freezing damage, as measured by the release of $CF_1$ from the membranes, can be observed (Hincha and Schmitt 1988a).

In the presence of such a "stroma" medium, plastocyanin release can be detected even under nonfreezing conditions (Hincha and Schmitt 1988a). It is linearly time-dependent. A time-dependent plastocyanin release is also found in samples frozen to various subzero temperatures. Here, however, an additional rapid, temperature-dependent component contributes to the total plastocyanin release of thylakoids (Hincha and Schmitt 1988a).

The rate of plastocyanin release at ice bath temperature depends on the permeability of the solute present (Hincha et al. 1989b). Higher permeability leads to a faster influx of solutes into the membrane vesicles. It seems likely that in addition colloid osmotic effects are contributing to the rupture syndrome.

Colloid osmotic effects have been shown to play an important role in the lysis of ion-permeabilized red blood cells under various nonfreezing conditions (Pooler 1985a,b). Hemolysis has also been found to depend on the solute gradient over the cell membrane and on the duration of exposure to hypertonic conditions (Zade-Oppen 1968), similar to the situation in thylakoids. Solute loading has also been observed in erythrocytes during freezing, but since solutes such as sucrose and sodium chloride are deemed "impermeable", membrane leaks were invoked to explain their influx (Pegg and Diaper 1988). As in the case of thylakoids (Hincha 1986), it seems possible that careful measurements over longer incubation times (several hours) will reveal whether low permeability coefficients would be sufficient to explain solute loading in red blood cells under strongly hypertonic conditions.

## Determining the Type of Injury in Vivo

When leaves are frozen to lethal temperatures, photosynthesis is irreversibly damaged (Krause et al. 1988). Thylakoids isolated from such leaves show an inactivation of photophosphorylation (Heber and Santarius 1964). Whether thylakoid membranes have been injured by mechanical or chemical stresses cannot be distinguished by measuring photophosphorylation after thawing. Different causes lead to the same effect (Fig. 2b). The in vitro borderline concentration between the modes of mechanical and chemical damage in simple salt/sugar mixtures is approximately 0.1 Osm/kg (Fig. 3). This is far below the solute concentration commonly found in glycophytic plant cells. Consequently, we concluded earlier that chemical, but not physical stresses lead to injury during freezing in situ (Schmitt et al. 1985). Several lines of recent evidence indicate that this is not the case.

One of the consequences of chemical freezing injury is the loss of the $CF_1$ part of the ATP synthase (Hincha et al. 1985; Schmitt et al. 1985; Santarius 1987a). This leads to an uncoupling of electron transport from ATP synthesis (Coughlan and Pfanz 1986). In thylakoids isolated from frost-damaged leaves, however, no uncoupling could be found. Loss of photophosphorylation was attributed to an inhibition of electron transport (Klosson and Krause 1981; Krause et al. 1988; Rumich-Bayer and Krause 1986). Also, the concentration of NaCl necessary to lead to a freeze-induced loss of $CF_1$ in an artificial stroma medium, approximately 250 mM (Hincha and Schmitt 1988a), is far higher than any concentration that has ever been found in chloroplasts even in halotolerant (Kaiser et al. 1983; Robinson et al. 1983) or halophytic plants (Robinson and Downton 1985; Demmig and Winter 1986) under salt stress.

When spinach leaves are homogenized after a freeze-thaw cycle and the distribution of the marker proteins for chemical and mechanical damage ($CF_1$ and plastocyanin) between the membranes and the soluble supernatant are determined, very little loss of $CF_1$ can be found. Plastocyanin loss, in contrast, occurs parallel to the inactivation of leaf photosynthesis (Hincha et al. 1987; Schmidt et al. 1986). This indicates that mechanical membrane rupture is the predominant form of injury suffered by thylakoids during freezing and thawing in vivo.

However, it could not be totally excluded that the thylakoids are made mechanically more labile by freezing, and that the observed release of plastocyanin is an artifact due to homogenization of the leaves. We have therefore used immunogold labeling on embedded sections of spinach leaves after low temperature fixation to verify the localization of plastocyanin during an in vivo freeze-thaw cycle (Fig. 5). Plastocyanin is clearly associated with the thylakoid membranes in unfrozen controls (Fig. 5a) and in leaves frozen slowly to –9 °C (Fig. 5b). This corroborates the apparent structural integrity of chloroplasts in frozen leaves (Hincha et al. 1989b, Fig. 1b). During thawing, membrane disintegration occurs (Fig. 1c) and plastocyanin is lost from the membranes (Fig. 5c) and is found free in the cells (Schmitt et al. 1987). This is unequivocal evidence for mechanical membrane rupture in thylakoids during an in vivo freeze-thaw cycle.

There is evidence from volumetric (Hincha 1986; Hincha and Schmitt 1988a) and electron microscopic studies (Hincha et al. 1989b; Holopainen and Holopainen 1988; Krause et al. 1984; Martin and Öquist 1979) that the stresses leading to

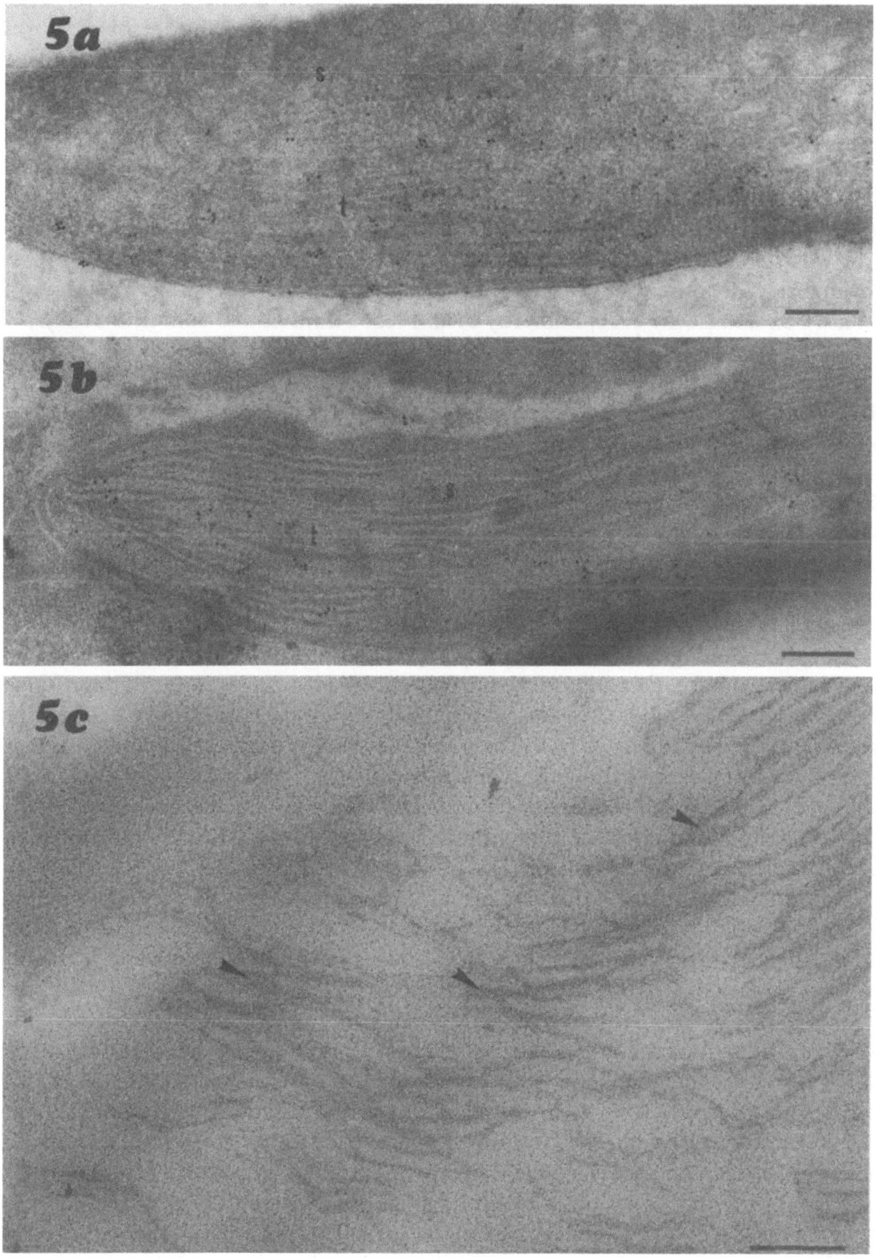


**Fig. 5a-c.** Electron micrographs of thin sections of spinach leaves. Freeze-thaw conditions were as in Fig. 1. Leaf samples were taken prior to freezing (**a** 0 °C), after 1 h at the minimum temperature (**b** -9 °C) and during thawing (**c** -3 °C). After rapid freezing in liquid propane (cf. Fig. 1), freeze substitution was performed in methanol with 2% formaldehyde and 0.2% glutaraldehyde following the protocol of Humbel and Müller (1986). The fixed samples were infiltrated at -30 °C with Lowicryl K4M. Polymerization by UV irradiation was for 24 h at the same temperature. Thin sections were labeled with affinity purified antibodies against plastocyanin. Bound antibodies were visualized with goat-anti-rabbit antibodies carrying 10 nm gold particles. *Bars* = 200 nm. *Arrowheads* point to thylakoid membranes. *t* thylakoids; *s* stroma membrane rupture in vivo and in vitro are osmotic in nature. The linearly time-dependent, diffusion-induced release of plastocyanin from thylakoids frozen in the presence of an artificial stroma medium (Hincha and Schmitt 1988a) has also been found when leaves were frozen to sublethal temperatures and kept frozen for extended durations (Hincha et al. 1989a; Hincha and Schmitt 1988b). This is in agreement with the finding that the survival of leaves at sublethal freezing temperatures was strongly dependent on the duration of freezing. As in the case of plastocyanin release, other parameters of cellular injury increased over several days to weeks (Cox and Levitt 1976; Pomeroy et al. 1985; Schulteis and Santarius 1989).

## Effects of Frost Hardening on Freeze-Thaw Damage to Photosynthetic Membranes

The following discussion is limited to physiological adaptations which enable plants to withstand freezing of their leaf tissues. Mechanisms of chilling resistance and freeze avoidance will not be considered. Physiognomic and structural adaptations of plants from cold climates to the specific stresses of mountain and tundra habitats have been reviewed in detail recently (Körner and Larcher 1988).

Frost-hardening can be achieved by exposing non-hardy plants to an environment of low, nonfreezing temperatures (typically between 10 and 0 °C) and short photoperiods (Levitt 1980). In this way spinach and many other herbaceous plants can be hardened by 5 to 25 °C (Klosson and Krause 1981; Rumich-Bayer and Krause 1986; Bauer and Kofler 1987; Steponkus et al. 1983; Yelenosky and Guy 1989; Fennell et al. 1990). Some species of shrubs and trees can be hardened substantially more, some to liquid nitrogen temperature (Hirsh et al. 1985; Rütten and Santarius 1988; Strand and Öquist 1988).

In some species frost-hardening can also be achieved by salt stress (Schmidt et al. 1986) or mild desiccation (Cox and Levitt 1976; Cloutier and Siminovitch 1982; Siminovitch and Cloutier 1983; Vigh et al. 1986). The hardening process is readily reversible when plants are transferred back to higher temperatures (Greer and Stanley 1985; Guy and Haskell 1987; Strand and Öquist 1988).

Even though frost-hardening may occur at normal growth temperature when plants are stressed by applying NaCl to their hydroponic culture medium (Schmidt et al. 1986), no signals from the roots seem necessary to induce frost hardiness in the leaves. In spinach, only an exposure of the shoot to low temperature leads to increased hardening, not a cooling of the roots alone (Fennell et al. 1990). In cabbage, hardening is induced by cutting the stem from the roots and thereby applying rapid water stress (Cox and Levitt 1976). This is similar to the induction of CAM-specific enzymes in an inducible CAM plant by wilting of detached leaves (Schmitt 1990).

### *The Contribution of Osmotic Adaptation to the Frost Hardiness of Chloroplasts in Situ*

In many plant species, frost hardening is accompanied by an increase in the cellular osmolality. This could be advantageous since an increased osmotic potential would reduce the freeze-induced contraction of the cells as less water is removed from the cells to the extracellular ice at any given temperature (Levitt 1980). In some cases it could be shown that the cells were killed at a constant residual cell volume, while the killing temperature varied with hardening (Meryman et al. 1977; Schmidt et al. 1986). Other investigators, however, found that this correlation is not general (Yelenosky and Guy 1989; Fennell et al. 1990).

Also, several reports suggested that besides osmotic effects more specific factors must play a role in hardening (O'Neill 1983; van Swaaij et al. 1985; Rütten and Santarius 1988; Yelenosky and Guy 1989; Fennell et al. 1990).

Increased solute levels during cold acclimation are mainly achieved by increases in soluble sugars, proline, and glycine betaine. In most cases the intracellular localization of the accumulated solutes was not determined (Kandler and Hopf 1982). Localization in the chloroplasts was shown only in the case of sucrose and raffinose accumulation in cabbage (Santarius and Milde 1977) during cold acclimation and glycine betaine in spinach (Robinson and Jones 1986) under salt stress. Glycine betaine (Coughlan and Heber 1982), as well as sugars, is cryoprotective for isolated thylakoid membranes (Schmitt et al. 1985; Steponkus 1984). The cryoprotective effectiveness of different sugars has been compared in vitro under different experimental conditions (Santarius 1973; Steponkus et al. 1977; Lineberger and Steponkus 1980). The protective effect of sugars commonly found in higher plants during frost hardening is not significantly different (Hincha 1990).

Apart from a possible role in cellular dehydration avoidance during freezing, as discussed above, accumulated solutes may also osmotically stabilize internal membranes such as thylakoids during thawing and thereby reduce rupture and loss of plastocyanin (cf. Fig. 4). Since solute influx during freezing can lead to membrane rupture in vivo (Hincha and Schmitt 1988a, Fig. 1) the permeability and intracellular distribution of all solutes will influence the extent of damage.

An additional function of cellular cryoprotectants could be the colligative reduction of metabolite concentration gradients across internal cellular membranes (Hincha and Schmitt 1988b). Newly synthesized, membrane-neutral solutes such as proline or sugars will reduce the effective concentration of any pre-existing cellular solute in the same compartment at a given subzero temperature (Franks 1981). Therefore, metabolite gradients across the numerous cellular membranes will be reduced and osmotically driven redistribution of metabolites to other cellular compartments or the apoplast will be slowed down, reducing compartmentation stress (Hincha and Schmitt 1988b) after thawing. A detailed knowledge of intracellular solute gradients and of the cellular compartmentation of low molecular weight cryoprotectants during hardening will be necessary to elucidate the role that the accumulated solutes play in the frost hardiness of specific cell organelles.

### *Frost-Hardening at the Thylakoid Membrane Level*

As discussed above, the frost hardiness of plant species cannot be accounted for by osmotic adaptation alone. An involvement of functional membrane alterations has been shown conclusively for the plasma membrane (see Steponkus, this Vol.).

Thylakoids from spinach leaves acclimated by low temperature treatment showed a better preservation of proton uptake in comparison to thylakoids from nonacclimated leaves (Garber and Steponkus 1976b), when frozen in media of low osmolality. Likewise, frost hardening either by salt treatment or by cold acclimation is accompanied by an increased resistance of isolated thylakoids against loss of plastocyanin induced by a freeze-thaw cycle at low initial solute concentrations (Schmidt et al. 1986).

No evidence for an increased resistance of isolated thylakoids against chemical freezing damage could be found (Krause et al. 1984; Schmidt et al. 1986). After freezing in an artificial stroma medium, Grafflage and Krause (1986) reported that cold-hardening had no effect on the preservation of linear electron transport, while time dependent plastocyanin release at 0 °C in a similar medium was clearly reduced in membranes isolated from hardy leaves (Hincha and Schmitt 1988b). This discrepancy remains to be resolved.

We have shown above that two mechanisms contribute to mechanical freeze-thaw damage: a slow component attributed to solute permeation into the thylakoids, and a rapid component attributed to a decreased membrane extensibility after freezing in dilute solutions. Frost hardening of spinach leaves involves an increased resistance of isolated thylakoids against both types of injury (Hincha and Schmitt 1988b).

The maximum volume that thylakoids from hardy leaves were able to reach after freezing was approximately 25% higher than the maximum volume of thylakoids from non-hardy leaves. The permeability of thylakoids for sucrose, sorbitol and an artificial stroma medium was decreased by an average of 45% both at 0 and –20 °C (Hincha and Schmitt 1988b). The decrease in permeability measured with isolated thylakoids is paralleled by a lower rate of plastocyanin release in leaves frozen to different temperatures and kept at these temperatures for up to 11 days (Hincha et al. 1989a; Hincha and Schmitt 1988b).

The structural or compositional changes in the thylakoid membranes that lead to an increased freeze-thaw resistance are still unclear. Although changes in the lipid and fatty acid composition of thylakoids from different plants under hardening conditions have been reported (Vigh et al. 1985), no functional role in hardening could be assigned to any of these observed changes. At the present state of knowledge, we agree with Huner's contention (Huner 1988) that subtle changes in membrane composition and organization rather than bulk changes in the lipid or protein components of the membranes are likely to be responsible for differences in freezing resistance. In rye plasma membranes, no lipid component was found, which was unique to either the cold acclimated or the non-acclimated state of the plants (Lynch and Steponkus 1987).

### *The Role of Soluble Proteins in Frost Hardiness*

Over the last few years the regulation of frost hardiness by changes in gene expression has attracted widespread interest. There is a general consensus now that frost hardening leads to a variety of subtle qualitative and quantitative changes in the electrophoretic pattern of cellular proteins. These changes are due both to the disappearance of proteins which are present in non-hardy tissue and to the de novo synthesis of proteins during low temperature acclimation (Guy et al. 1985, 1988; Guy and Haskell 1987, 1988; Mohapatra et al. 1987a,b; Gilmour et al. 1988; Hughes and Pearce 1988; Kurkela et al. 1988). The analysis of the in vitro translation products of isolated mRNA suggests that changes in protein synthesis patterns are paralleled by changes in mRNA levels (Guy et al. 1985, 1988; Gilmour et al. 1988; Hughes and Pearce 1988; Kurkela et al. 1988; Mohapatra et al. 1987a,b).

Some low temperature-induced genes have now been cloned and partially sequenced (Mohapatra et al. 1989). The gene for one cold-inducible protein from *Arabidopsis thaliana* has recently been completely sequenced. It codes for a 6.5-kDa protein which shows a low (28%) homology with a class of low molecular weight anti-freeze proteins from the arctic flounder (Kurkela and Franck 1990). This homology was evident only at the amino acid but not at the DNA level. Whether the *Arabidopsis* protein has any antifreeze activity has not yet been determined.

Cell suspension cultures have also been used to study protein synthesis under hardening conditions. Here, as in some plant species, low temperatures can in many instances be substituted by the application of the plant hormone abscisic acid (Johnson-Flanagan and Singh 1987; Reaney and Gusta 1987; Robertson et al. 1988; Askman et al. 1990). The results obtained with cultured cells are qualitatively similar to those from whole-plant studies.

Although such studies may shed some light on the question of the regulation of protein synthesis by temperature, it is so far totally unclear whether any of the identified gene products have a functional role in frost hardiness (Guy 1990). We therefore decided to use isolated thylakoids as a test system to screen for proteins which are able to stabilize membranes during a freeze-thaw cycle. The isolation of such proteins would enable us to study frost hardening at the molecular level using probes (antibodies, cDNA) specific for a functional component of plant freezing resistance.

The identification of cryoprotective proteins requires an in vitro assay which models the in vivo situation. In earlier attempts to isolate such proteins (Heber and Kempfle 1970; Volger and Heber 1975), cryoprotection was measured by photophosphorylation activity. This test, however, is ambiguous, since cyclic photophosphorylation is damaged under conditions of chemical as well as mechanical freeze-thaw injury. Release of plastocyanin, on the other hand, is an indicator of mechanical membrane rupture, and we have used the retention of plastocyanin in the membranes after a freeze-thaw cycle in the presence of different protein fractions as an assay for cryoprotective proteins (Hincha et al. 1989a, 1990).

For the protein isolation we have used the procedure of Heber and Kempfle (1970) with some modifications (Hincha et al. 1990). We were able to isolate cryoprotective proteins from cold acclimated spinach and cabbage leaves. When leaves from non-acclimated plants were used for the same extraction, no cryoprotective activity was found (Hincha et al. 1990). This is evidence that the proteins are indeed involved in leaf frost hardening.

A preliminary characterization has revealed that the proteins are stable to boiling (Heber and Kempfle 1970; Hincha et al. 1990). They are inactivated by tryptic digestion (Hincha et al. 1990). Their molecular mass, as estimated by gel filtration chromatography, is close to 28 kDa (Hincha et al. 1989a). This value, however, is tentative, since the proteins are glycosylated (D.K. Hincha, unpubl.), and this could significantly influence their chromatographic behavior.

On a molar basis, the protective proteins are at least 20 000 to 40 000 times more effective than sucrose in preventing mechanical freeze-thaw rupture of isolated thylakoids (Hincha et al. 1989a). The mechanism by which this extremely high efficiency is achieved is completely obscure. It has been shown, however, that the proteins protect thylakoids against both components of mechanical freeze-thaw damage. Solute loading during freezing and time dependent plastocyanin release at –20 °C were completely suppressed by the addition of protective protein (Hincha et al. 1990). In an artificial stroma medium, the rapid component of plastocyanin release was also strongly reduced (Hincha et al. 1990). There is evidence from ion exchange chromatography that at least two different proteins are responsible for the two different protective effects (D.K. Hincha, unpubl.).

How these proteins interact with the membranes to achieve this dramatic decrease in permeability and increase in extensibility remains to be elucidated. Experiments with different sugars showed that their cryoprotective efficiency only varied by a factor of 10 and that the sugars only acted on solute loading (Hincha 1989, 1990).

It should be stressed that plant cryoprotective proteins are active only in frozen samples, but do not prevent plastocyanin leakage at 0 °C (Hincha et al. 1990). This clearly distinguishes the plant proteins from the protective proteins in arctic and antarctic fish which inhibit the growth of ice crystals (DeVries 1971). Their protection is therefore based on freezing avoidance, while the plant proteins are protective only in a frozen system.

*Acknowledgments.* We are grateful to Dr. M. Müller and T. Hillmann (ETH Zürich) for their help with the ultrastructural investigations and to Prof. U. Heber (Universität Würzburg) for many helpful discussions and his continuous support and encouragement. Financial support was provided by the Deutsche Forschungsgemeinschaft through Forschergruppe Ökophysiologie and SFB 251 at the Universität Würzburg.

## References

Askman A, Abromeit M, Sarnighausen E, Dörffling K (1990) Formation of polypeptides related to frost tolerance in response to cold hardening and abscisic acid treatment in winter wheat. Physiol Plant 79:A105

Bauer H, Kofler R (1987) Photosynthesis in frost-hardened and frost-stressed leaves of *Hedera helix* L. Plant Cell Environ 10:339–346

Beck E, Schulze ED, Senser M, Scheibe R (1984) Equilibrium freezing of leaf water and extracellular ice formation in afroalpine "giant rosette" plants. Planta 162:276–282

Callow RA, McGrath JJ (1985) Thermodynamic modeling and cryomicroscopy of cell size, unilamellar, and paucilamellar liposomes. Cryobiology 22:251–267
Cloutier Y, Siminovitch D (1982) Correlation between cold- and drought-induced frost hardiness in winter wheat and rye varieties. Plant Physiol 69:256–258
Coughlan SJ, Heber U (1982) The role of glycinebetaine in the protection of spinach thylakoids against freezing stress. Planta 156:62–69
Coughlan SJ, Pfanz H (1986) The reversibility of freeze/thaw injury to spinach thylakoids, restoration of light-induced proton pumping, membrane-conformational changes and proton gradient formation. Biochim Biophys Acta 849:32–40
Cox W, Levitt J (1976) Interrelations between environmental factors and freezing resistance of cabbage leaves. Plant Physiol 57:553–555
Demmig B, Winter K (1986) Sodium, potassium, chloride and proline concentrations of chloroplasts isolated from a halophyte, *Mesembryanthemum crystallinum* L. Planta 168:421–426
DeVries AL (1971) Glycoproteins as biological antifreeze agents in antarctic fishes. Science 172:1152–1155
Fennell A, Li PH, Markhart III AH (1990) Influence of air and soil temperature on water relations and freezing tolerance of spinach (*Spinacia oleracea*). Physiol Plant 78:51–56
Franks F (1981) Biophysics and biochemistry of low temperatures and freezing. In: Morris GJ, Clarke A (eds) Effects of low temperatures on biological membranes. Academic Press, Lond New York, pp 3–19
Garber MP, Steponkus PL (1976a) Alterations in chloroplast thylakoids during an in vitro freeze-thaw cycle. Plant Physiol 57:673–680
Garber MP, Steponkus PL (1976b) Alterations in chloroplast thylakoids during cold acclimation. Plant Physiol 57:681–686
Gilmour S, Hajela RK, Thomashow MF (1988) Cold acclimation in *Arabidopsis thaliana*. Plant Physiol 87:745–750
Grafflage S, Krause GH (1986) Simulation of in situ freezing damage of the photosynthetic apparatus by freezing in vitro of thylakoids suspended in complex media. Planta 168:67–76
Greer DH, Stanley CJ (1985) Regulation of the loss of frost hardiness in *Pinus radiata* by photoperiod and temperature. Plant Cell Environ 8:111–116
Guy CL (1990) Cold acclimation and freezing stress tolerance: role of protein metabolism. Annu Rev Plant Physiol Plant Mol Biol 41:187–223
Guy CL, Haskell D (1987) Induction of freezing tolerance in spinach is associated with the synthesis of cold acclimation induced proteins. Plant Physiol 84:872–878
Guy CL, Haskell D (1988) Detection of polypeptides associated with the cold acclimation process in spinach. Electrophoresis 9:787–796
Guy CL, Niemi KJ, Brambl R (1985) Altered gene expression during cold acclimation of spinach. Proc Natl Acad Sci USA 82:3673–3677
Guy CL, Haskell D, Yelenosky G (1988) Changes in freezing tolerance and polypeptide content of spinach and citrus at 5 °C. Cryobiology 25:264–271
Haehnel W (1984) Photosynthetic electron transport in higher plants. Annu Rev Plant Physiol 35:659–693
Haehnel W (1986) Plastocyanin. In: Staehelin LA, Arntzen CJ (eds) Encyclopedia of plant physiology, New Series. Springer, Berlin Heidelberg New York, 19:547–559
Heber U, Kempfle M (1970) Proteine als Schutzstoffe gegenüber dem Gefriertod der Zelle. Z Naturforsch 25b:834–842
Heber U, Santarius KA (1964) Loss of adenosine triphosphate synthesis and its relationship to frost hardiness problems. Plant Physiol 39:712–719
Heber U, Schmitt JM, Krause GH, Klosson RJ, Santarius KA (1981) Freezing damage to thylakoid membranes in vitro and in vivo. In: Morris GJ, Clarke A (eds) Effects of low temperatures on biological membranes. Academic Press, Lond New York, pp 264–287
Hincha DK (1986) Sucrose influx and mechanical damage by osmotic stress to thylakoid membranes during an in vitro freeze-thaw cycle. Biochim Biophys Acta 861:152–158
Hincha DK (1989) Low concentrations of trehalose protect isolated thylakoids against mechanical freeze-thaw damage. Biochim Biophys Acta 987:231–234

Hincha DK (1990) Differential effects of galactose containing saccharides on mechanical freeze-thaw damage to isolated thylakoid membranes. Cryo-Lett 11:437–444
Hincha DK, Schmitt JM (1985) Mechanical and chemical injury to thylakoid membranes during freezing in vitro. Biochim Biophys Acta 812:173–180
Hincha DK, Schmitt JM (1988a) Mechanical freeze-thaw damage and frost hardening in leaves and isolated thylakoids from spinach. I. Mechanical freeze-thaw damage in an artificial stroma medium. Plant Cell Environ 11:41–46
Hincha DK, Schmitt JM (1988b) Mechanical freeze-thaw damage and frost hardening in leaves and isolated thylakoids from spinach. II. Frost hardening reduces solute permeability and increases extensibility of thylakoid membranes. Plant Cell Environ 11:47–50
Hincha DK, Schmidt JE, Heber U, Schmitt JM (1984) Colligative and non-colligative freezing damage to thylakoid membranes. Biochim Biophys Acta 769:8–14
Hincha DK, Heber U, Schmitt JM (1985) Antibodies against individual thylakoid membrane proteins as molecular probes to study chemical and mechanical freezing damage in vitro. Biochim Biophys Acta 809:337–344
Hincha DK, Höfner R, Schwab KB, Heber U, Schmitt JM (1987) Membrane rupture is the common cause of damage to chloroplast membranes in leaves injured by freezing or excessive wilting. Plant Physiol 83:251–253
Hincha DK, Heber U, Schmitt JM (1989a) Freezing ruptures thylakoid membranes in leaves, and rupture can be prevented in vitro by cryoprotective proteins. Plant Physiol Biochem 27:795–801
Hincha DK, Müller M, Hillmann T, Schmitt JM (1989b) Osmotic stress causes mechanical freeze-thaw damage to thylakoids in vitro and in vivo. In: Cherry JH (ed) Environmental stress in plants. Springer, Berlin Heidelberg New York, pp 303–315
Hincha DK, Heber U, Schmitt JM (1990) Proteins from frost-hardy leaves protect thylakoids against mechanical freeze-thaw damage in vitro. Planta 180:416–419
Hirsh AG, Williams RJ, Meryman HT (1985) A novel method of natural cryoprotection. Plant Physiol 79:41–56
Hofmeister F (1888) Zur Lehre von der Wirkung der Salze. Arch Exp Pathol 24:247–260
Holopainen JK, Holopainen T (1988) Cellular responses of Scots pine (*Pinus sylvestris* L.) seedlings to simulated summer frost. Eur J For Pathol 18:207–216
Hughes MA, Pearce RS (1988) Low temperature treatment of barley plants causes altered gene expression in shoot meristems. J Exp Bot 39:1461–1467
Humbel B, Müller M (1986) Freeze substitution and low temperature embedding. In: Science of biological specimen preparation. SEM Inc AMF O'Hare, Chicago, pp 171–183
Huner NPA (1988) Low-temperature-induced alterations in photosynthetic membranes. CRC Crit Rev Plant Sci 7:257–278
Jensen M, Oettmeier W (1984) Effects of freezing on the structure of chloroplast membranes. Cryobiology 21:465–473
Johnson-Flanagan AM, Singh J (1986) Membrane deletion during plasmolysis in hardened and non-hardened plant cells. Plant Cell Environ 9:299–305
Johnson-Flanagan AM, Singh J (1987) Alteration of gene expression during the induction of freezing tolerance in *Brassica napus* suspension cultures. Plant Physiol 85:699–705
Kaiser WM, Weber H, Sauer M (1983) Photosynthetic capacity, osmotic response and solute content of leaves and chloroplasts from *Spinacia oleraceae* under salt stress. Z Pflanzenphysiol 113:15–27
Kandler O, Hopf H (1982) Oligosaccharides based on sucrose (sucrosyl oligosaccharides). In: Loewus FA, Tanner W (eds) Encyclopedia of plant physiol New Series 13A. Springer, Berlin Heidelberg New York, pp 348–382
Klosson RJ, Krause GH (1981) Freezing injury in cold-acclimated and unhardened spinach leaves I. Photosynthetic reactions of thylakoids isolated from frost-damaged leaves. Planta 151:339–346
Körner C, Larcher W (1988) Plant life in cold climates. In: Long SF, Woodward FI (eds) Plants and temperature. Company Biol Ltd, Cambridge, pp 25–57
Krause GH, Klosson RJ, Justenhoven A, Ahrer-Steller V (1984) Effects of low temperatures on the photosynthetic system in vivo. In: Sybesma C (ed) Advances in photosynthesis research. Nijhoff, Brussels, pp 349–358

Krause GH, Grafflage S, Rumich-Bayer S, Somersalo S (1988) Effects of freezing on plant mesophyll cells. Symp Soc Exp Biol 42:311–327

Kurkela S, Franck M (1990) Cloning and characterization of a cold- and ABA-inducible *Arabidopsis* gene. Plant Mol Biol 15:137–144

Kurkela S, Franck M, Heino P, Lång V, Palva ET (1988) Cold-induced gene expression in *Arabidopsis thaliana* L. Plant Cell Rep 7:495–498

Levitt J (1980) Responses of plants to environmental stresses. Vol I: Chilling, freezing, and high temperature stresses. Academic Press, Orlando

Lineberger RD, Steponkus PL (1980) Cryoprotection by glucose, sucrose, and raffinose to chloroplast thylakoids. Plant Physiol 65:298–304

Lovelock JE (1953a) The haemolysis of human red blood-cells by freezing and thawing. Biochim Biophys Acta 10:414–426

Lovelock JE (1953b) Heat mechanism of the protective action of glycerol against haemolysis by freezing and thawing. Biochim Biophys Acta 11:28–36

Lovelock JE (1954) The protective action of neutral solutes against haemolysis by freezing and thawing. Biochem J 56:265–270

Lynch DV, Steponkus PL (1987) Plasma membrane lipid alterations associated with cold acclimation of winter rye seedlings (*Secale cereale* L. cv Puma). Plant Physiol 83:761–767

Martin B, Öquist G (1979) Seasonal and experimentally induced changes in the ultrastructure of chloroplasts of *Pinus silvestris*. Physiol Plant 46:42–49

Meryman HT, Williams RJ, Douglas MSJ (1977) Freezing injury from "solution effects" and its prevention by natural or artificial cryoprotection. Cryobiology 14:287–302

Mohapatra SS, Poole RJ, Dhindsa RS (1987a) Cold acclimation, freezing resistance and protein synthesis in alfalfa (*Medicago sativa* L. cv. Saranac). J Exp Bot 38:1697–1703

Mohapatra SS, Poole RJ, Dhindsa RS (1987b) Changes in protein patterns and translatable messenger RNA populations during cold acclimation of alfalfa. Plant Physiol 84:1172–1176

Mohapatra SS, Wolfraim L, Poole RJ, Dhindsa RS (1989) Molecular cloning and relationship to freezing tolerance of cold-acclimation specific genes of alfalfa. Plant Physiol 89:375–380

Mollenhauer A, Schmitt JM, Coughlan S, Heber U (1983) Loss of membrane proteins from thylakoids during freezing. Biochim Biophys Acta 728:331–338

Olien CR, Smith MN (1981) Protective systems that have evolved in plants. In: Olien C R, Smith MN (eds) Analysis and improvement of plant cold hardiness. CRC Press, Cleveland, pp 61–87

O'Neill SD (1983) Osmotic adjustment and the development of freezing resistance in *Fragaria virginia*. Plant Physiol 72:938–944

Pearce RS, Willison JHM (1985) Wheat tissues freeze-etched during exposure to extracellular freezing: distribution of ice. Planta 163:295–303

Pegg DE, Diaper MP (1988) On the mechanism of injury to slowly frozen erythrocytes. Biophys J 54:471–488

Pomeroy MK, Andrews CJ, Stanley KP, Ji-Yin-Gao (1985) Physiological and metabolic responses of winter wheat to prolonged freezing stress. Plant Physiol 78:207–210

Pooler JP (1985a) The kinetics of colloid osmotic hemolysis. I. Nystatin-induced lysis. Biochim Biophys Acta 812:193–198

Pooler JP (1985b) The kinetics of colloid osmotic hemoloysis II. Photohemolysis. Biochim Biophys Acta 812:199–205

Reaney MJT, Gusta LV (1987) Factors influencing the induction of freezing tolerance by abscisic acid in cell suspension cultures of *Bromus inermis* Leyss and *Medicago sativa* L. Plant Physiol 83:423–427

Robertson AJ, Gusta LV, Reaney MJT, Ishikawa M (1988) Identification of proteins correlated with increased freezing tolerance in Bromegrass (*Bromus inermis* Leyss. cv. Manchar) cell cultures. Plant Physiol 86:433–447

Robinson SP, Downtown WJS (1985) Potassium, sodium and chloride ion concentrations in leaves and isolated chloroplasts of the halophyte *Suaeda australis* R. Br. Aust J Plant Physiol 12:471–479

Robinson SP, Jones GP (1986) Accumulation of glycinebetaine in chloroplasts provides osmotic adjustment during salt stress. Aust J Plant Physiol 13:659–668

Robinson SP, Downtown WJS, Millhouse JA (1983) Photosynthesis and ion content of leaves and isolated chloroplasts of salt-stressed spinach. Plant Physiol 73:238–242
Rumich-Bayer S, Krause GH (1986) Freezing damage and frost tolerance of the photosynthetic apparatus studied with isolated mesophyll protoplasts of *Valerianella locusta* L. Photosynth Res 8:161–174
Rütten D, Santarius KA (1988) Cold acclimation of *Ilex aquifolium* under natural conditions with special regard to the photosynthetic apparatus. Physiol Plant 72:807–815
Santarius KA (1973) The protective effect of sugars on chloroplast membranes during temperature and water stress and its relationship to frost desiccation and heat resistance. Planta 113:105–114
Santarius KA (1984a) The role of the chloroplast coupling factor in the inactivation of thylakoid membranes at low temperatures. Physiol Plant 61:591–598
Santarius KA (1984b) Effective cryoprotection of thylakoid membranes by ATP. Planta 161:555–561
Santarius KA (1986a) Freezing of isolated thylakoid membranes in complex media I. The effect of potassium and sodium chloride, nitrate, and sulfate. Cryobiology 23:168–176
Santarius KA (1986b) Freezing of isolated thylakoid membranes in complex media II. Simulation of the conditions in the chloroplast stroma. Cryo-Lett 7:31–40
Santarius KA (1986c) Freezing of isolated thylakoid membranes in complex media III. Differences in the pattern of inactivation of photosynthetic reactions. Planta 168:281–286
Santarius KA (1987a) Freezing of isolated thylakoid membranes in complex media IV. Stabilization of $CF_1$ by ATP and sulfate. J Plant Physiol 126:409–420
Santarius KA (1987b) Relative contribution of inorganic electrolytes to damage and protection of thylakoid membranes during freezing in complex media. In: Li PH (ed) Liss, New York, pp 229–242
Santarius KA (1990) Freezing of isolated thylakoid membranes in complex media V. Inactivation and protection of electron transport reactions. Photosynth Res 23:49–58
Santarius KA, Giersch C (1983) Cryopreservation of spinach chloroplast membranes by low-molecular-weight carbohydrates II. Discrimination between colligative and noncolligative protection. Cryobiology 20:90–99
Santarius KA, Giersch C (1984) Factors contributing to inactivation of isolated thylakoid membranes during freezing in the presence of variable amounts of glucose and NaCl. Biophys J 46:129–139
Santarius KA, Milde H (1977) Sugar compartmentation in frost-hardy and partially dehardened cabbage leaf cells. Planta 136:163–166
Schmidt JE, Schmitt JM, Kaiser WM, Hincha DK (1986) Salt treatment induces frost hardiness in leaves and isolated thylakoids from spinach. Planta 168:50–55
Schmitt JM (1990) Rapid concentration changes of phosphoenolpyuvate carboxylase mRNA in detached leaves of *Mesembryanthemum crystallinum* in response to wilting and rehydration. Plant Cell Environ 13:845–850
Schmitt JM, Schramm MJ, Pfanz H, Coughlan S, Heber U (1985) Damage to chloroplast membranes during dehydration and freezing. Cryobiology 22:99–104
Schmitt JM, Müller M, Hincha DK (1987) Mechanischer Schaden an der Thylakoidmembran beim Tauen gefrorener Spinatblätter. Biomed Tech 32:53–54
Schulteis C, Santarius KA (1989) Effects of prolonged freezing stress on the photosynthetic apparatus of moderately hardy leaves as assayed by chlorophyll fluorescence kinetics. Plant Cell Environ 12:819–823
Siminovitch D, Cloutier Y (1983) Drought and freezing tolerance and adaptation in plants: some evidence of near equivalences. Cryobiology 20:487–503
Singh J (1979) Ultrastructural alterations in cells of hardened and non-hardened winter rye during hyperosmotic and extracellular freezing stresses. Protoplasma 98:329–341
Singh J, Iu B, Johnson-Flanagan AM (1987) Membrane alterations in winter rye and *Brassica napus* cells during lethal freezing and plasmolysis. Plant Cell Environ 10:163–168
Soikkeli S (1980) Ultrastructure of the mesophyll in Scots pine and Norway spruce: seasonal variation and molarity of the fixative buffer. Protoplasma 103:241–252
Steinbrecht RA, Müller M (1987) Freeze-substitution and freeze-drying. In: Steinbrecht RA, Zierold K (eds) Cryotechniques in biological electron microscopy. Springer, Berlin Heidelberg New York, pp 149–172

Steponkus PL (1984) Role of the plasma membrane in freezing injury and cold acclimation. Annu Rev Plant Physiol 35:543–584

Steponkus PL, Lynch DV (1989) The behaviour of large unilamellar vesicles of rye plasma membrane lipids during freeze/thaw-induced osmotic excursions. Cryo-Lett 10:43–50

Steponkus PL, Dowgert MF, Gordon-Kamm WJ (1983) Destabilization of the plasma membrane of isolated plant protoplasts during a freeze-thaw cycle: the influence of cold acclimation. Cryobiology 20:448–465

Steponkus PL, Garber MP, Myers SP, Lineberger DR (1977) Effects of cold acclimation and freezing on structure and function of chloroplast thylakoids. Cryobiology 14:303–321

Strand M, Öquist G (1988) Effects of frost hardening, dehardening and freezing stress on in vivo chlorophyll fluorescence of seedlings of Scots pine (*Pinus sylvestris* L.). Plant Cell Environ 11:231–238

Strotmann H, Bickel-Sandkötter S (1984) Structure, function, and regulation of chloroplast ATPase. Annu Rev Plant Physiol 35:97–120

van Swaaij AC, Jacobsen E, Feenstra WJ (1985) Effect of cold hardening, wilting and exogenously applied proline on leaf proline content and frost tolerance of several genotypes of *Solanum*. Physiol Plant 64:230–236

Vigh L, Horvàth I, van Hasselt PR, Kuiper PJC (1985) Effect of frost hardening on lipid and fatty acid composition of chloroplast thylakoid membranes in two wheat varieties of contrasting hardiness. Plant Physiol 79:756–759

Vigh L, Huitema H, Woltjes J, van Hasselt PR (1986) Drought stress-induced changes in the composition and physical state of phospholipids in wheat. Physiol Plant 67:92–96

Volger HG, Heber U (1975) Cryoprotective leaf proteins. Biochim Biophys Acta 412:335–349

Volger H, Heber U, Berzborn RJ (1978) Loss of function of biomembranes and solubilization of membrane proteins during freezing. Biochim Biophys Acta 511:455–469

Wiest SC, Steponkus PL (1978) Freeze-thaw injury to isolated spinach protoplasts and its simulation at above freezing temperatures. Plant Physiol 62:699–705

Williams RJ, Meryman HT (1970) Freezing injury and resistance in spinach chloroplast grana. Plant Physiol 45:752–755

Wisniewski M, Ashworth EN (1986) A comparison of seasonal ultrastructural changes in stem tissues of peach (*Prunus persica*) that exhibit contrasting mechanisms of cold hardiness. Bot Gaz 147:407–417

Wolfe J, Dowgert MF, Steponkus PL (1985) Dynamics of membrane exchange of the plasma membrane and the lysis of isolated protoplasts during rapid expansions in area. J Membr Biol 86:127–138

Wolter FP, Schmitt JM, Bohnert HJ, Tsugita A (1984) Simultaneous isolation of three peripheral proteins – a 32 kDa protein, ferredoxin NADP+ reductase and coupling factor – from spinach thylakoids and partial characterization of a 32 kDa protein. Plant Sci Lett 34:323–334

Yelenosky G, Guy CL (1989) Freezing tolerance of citrus, spinach, and petunia leaf tissue. Plant Physiol 89:444–451

Zade-Oppen AMM (1968) Posthypertonic hemolysis in sodium chloride systems. Acta Physiol Scand 73:341–364

Chapter 20

# Freeze-Induced Dehydration and Membrane Destabilization in Plants

P.L. Steponkus and M.S. Webb

## Introduction

A multitude of potentially lethal stresses occurs during a freeze/thaw cycle, including thermal, mechanical, chemical, osmotic, and possibly even electrical perturbations. Nevertheless, there is a general consensus that, in the absence of intracellular ice formation, freeze-induced cell dehydration is a primary cause of freezing injury. Although this was first proposed by Muller-Thurgau in 1886, the cellular and molecular mechanisms responsible for dehydration-induced injury have begun to be elucidated only recently (Steponkus 1984; Steponkus and Lynch 1989b).

Injury resulting from freeze-induced dehydration is a consequence of the destabilization of cellular membranes, with the plasma membrane a primary site of injury (Steponkus 1984). However, alterations in the ultrastructure of the plasma membrane resulting from dehydration are manifested in several different forms, depending on the freeze-thaw protocol and the extent of cell dehydration. In some instances, injury can result from dehydration at the cellular level, which involves the flux of bulk water and large osmotic excursions; whereas in other instances, membrane destabilization is a consequence of the removal of water that is closely associated with membranes. Most important, the cryostability of cellular membranes, in general, and the plasma membrane, in particular, is increased during cold acclimation.

Cold acclimation is a complex developmental phenomenon involving hormonal responses, altered gene activity, and numerous alterations in metabolism (Steponkus 1978). Ultimately, these changes result in the minimization of the freeze-induced destabilization of cellular membranes and include factors such as osmotic adjustment, which ameliorates the extent of freeze-induced cell dehydration (Steponkus 1979), and the accumulation of cryoprotective solutes such as sucrose and proline, which stabilize membranes at low hydrations (Crowe LM et al. 1984; Crowe JH et al. 1987; Rudolph and Crowe 1985; Rudolph et al. 1986). However, cold acclimation also results in changes in the lipid composition of the plasma membrane, which alter the cryobehavior of the plasma membrane and increase its tolerance to dehydration at both the cellular and molecular level (Steponkus and Lynch 1989b; Steponkus et al. 1990).

Department of Soil, Crop and Atmospheric Sciences, Cornell University, Ithaca, NY 14853, USA

Somero et al. (Eds.)
Water and Life

The focus of this review is on the cellular and molecular repercussions of freeze-induced dehydration from a perspective of the plasma membrane, largely deduced from studies of protoplasts isolated from leaves of winter rye (*Secale cereale* L. cv. Puma) at various stages of cold acclimation. The rationale for using isolated protoplasts has been elaborated in previous publications (Steponkus et al. 1982; Steponkus 1984). Elucidation of the cellular and molecular aspects of freezing injury has provided a foundation for the direct assessment of the role of plasma membrane lipid alterations in the cold acclimation process – a longstanding and controversial issue.

It is important to note that although cell dehydration is the primary cause of freezing injury, cell dehydration is also essential for survival in that the cell must be dehydrated to preclude excessive supercooling of the cytosol, which predisposes the cell to intracellular ice formation. Thus, the degree of cell dehydration determines the fate of the cell: insufficient dehydration results in injury because of intracellular ice formation, whereas excessive dehydration results in injury because of membrane destabilization. Therefore, increasing survival of tissues subjected to a freeze/thaw cycle requires an increased tolerance rather than avoidance of cell dehydration. Elucidation of the factors that ameliorate the deleterious effects of cell dehydration is critical in considering both cold hardiness of plant species and cryopreservation procedures used for the conservation of plant germplasm – either conventional procedures that involve freeze-induced dehydration or those that are referred to as "vitrification" procedures that involve osmotic dehydration of the specimens prior to quenching in liquid nitrogen.

## The Freezing Process and Cell Dehydration

### *Freezing of Aqueous Solutions*

During the cooling of an aqueous solution, ice formation will occur as a result of either heterogeneous nucleation or seeding of the solution by an ice crystal. Growth of the ice crystals occurs because the chemical potential of ice is lower than that of water at the same temperature. During the growth of ice crystals, solutes and gases are excluded from the ice matrix and accumulate in an unfrozen portion of the partially frozen mixture. Crystallization continues until the chemical potential of the unfrozen portion of the ice-solution mixture is in equilibrium with that of the ice, which is a function of the subzero temperature. Following thermal equilibration of the partially frozen mixture, the osmolality of the unfrozen solution is equal to (273 -T)/1.86 and is independent of the initial solute concentration (see Pitt 1990 for a comparison of different methods for calculating osmolality of partially frozen solutions.) The proportion of the solution that remains unfrozen depends on the initial solute concentration and can be estimated as the ratio of the initial to final osmolality. However, this varies considerably for different solutes because of differences in the osmotic coefficient and can only be accurately determined from the liquidus curve of the phase diagram for the solution.

Freeze-induced concentration of the solutes is not a continuous function of the subzero temperature. Ultimately, the concentration will become so great that the remaining unfrozen solution will form a glass (MacFarlane 1987). Because the mixture also contains ice crystals, it is referred to as a partially crystallized glass. The solute concentration at which the glass transformation occurs depends on the solute and the cooling rate. Because the occurrence of a glass transition in an aqueous solution will establish a limit to the increase in osmotic pressure that can be effected by freezing, its occurrence – either in the suspending medium or the cytoplasm – will influence the extent of cell dehydration.

### *Freezing of Protoplast Suspensions in Aqueous Media*

During cooling of a protoplast suspension, ice formation will occur first in the aqueous suspending medium because protoplasts lack heterogeneous nucleating agents that are effective at temperatures above –10 to –15 °C. Ice formation in the suspending medium creates a gradient in the chemical potential of the intracellular and extracellular solutions. The protoplasts will then respond osmotically to the gradient in chemical potential and begin to dehydrate. During cooling, the intracellular solution will be transiently supercooled, the extent being dependent on the rate of water efflux relative to the rate of cooling. The rate of water efflux is a function of the magnitude of the chemical potential gradient, the water permeability of the plasma membrane, and the area/volume ratio of the protoplast. When cooled at a rate of 3 °C $min^{-1}$ or less, the extent of supercooling will be sufficiently low that intracellular ice formation is precluded and the protoplasts contract to a volume predicted from the Boyle van't Hoff relationship (Dowgert and Steponkus 1983).

Isolated protoplasts behave as ideal osmometers and exhibit characteristic Boyle van't Hoff behavior over a wide range of osmolalities, i.e., their volume varies linearly as a function of $osmolality^{-1}$. For protoplasts isolated from nonacclimated rye leaves, the fractional volume FV = 0.084 + 0.485 ($Osm^{-1}$) as measured over the range of 0.3 to 3.0 osmolal (Dowgert and Steponkus 1983). For protoplasts isolated from cold-acclimated rye leaves, FV = 0.146 + 0.879 ($Osm^{-1}$) over the range of 0.4 to 6.0 osmolal. The increased slope of the Boyle van't Hoff relationship following cold acclimation is a consequence of the increase in the internal solute concentration, i.e., osmotic adjustment. An 0.53 osmolal solution is isotonic for protoplasts isolated from nonacclimated rye leaves, and a 1.03 osmolal solution is isotonic for those isolated from leaves following four weeks of cold acclimation, with the solute concentration increasing in a linear manner throughout the acclimation period (Uemura and Steponkus 1989). The fractional osmotic volume in an isotonic solution, as determined from the Boyle van't Hoff relationship ($1 - V_b$), is 0.915 for nonacclimated protoplasts and 0.853 for acclimated protoplasts.

Because the osmolality of the partially frozen solution increases linearly as a function of the subfreezing temperature and volume decreases linearly as a function of the reciprocal of the osmolality, large decreases in protoplast volume occur at relatively high subfreezing temperatures. For example, for protoplasts isolated from nonacclimated rye leaves, approximately 82% of the osmotically active water is removed following freezing to –5 °C. With protoplasts isolated from cold-acclimated leaves, the extent of dehydration is decreased in direct proportion to the increase in the internal solute concentration.

### *Freezing in Intact Leaves*

Although it is universally acknowledged that ice formation occurs in the extracellular spaces of plant tissues, there are remarkably few published studies documenting the location and morphology of the ice crystals and the morphology of the cells in the frozen state. Many of the studies were conducted in the late 19th century and involved light microscopy studies of specimens sectioned from material frozen in situ with poorly defined thermal histories (Steponkus et al. 1984). Quite often, storage tissues of plants such as beet or potato were used for the studies. More recent studies have used improved cold stages for better control of the thermal history of the samples, but the excised specimens are usually suspended in an aqueous medium, which infiltrates the tissue and also freezes. Alternatively, specimens can be mounted in oil. Nevertheless, the preparation of specimens for light microscopy can alter the pattern of ice formation such that it does not reflect the pattern of ice formation during freezing in situ.

Low-temperature scanning electron microscopy (LTSEM) is a powerful procedure that can be used to analyze, at a reasonably high spatial and temporal resolution (Jeffree et al. 1987), the location and morphology of ice crystals formed in situ and their effects on cell morphology. Recently, LTSEM has been used to determine the location of ice crystal formation in both woody twigs (Ashworth et al. 1988) and leaves of winter cereals (Allan et al. 1990).

In unfrozen leaves of winter rye, the mesophyll cells form a complex, porous network of interconnected, thin-walled cells (Allan et al. 1990). In leaves frozen at –2.5 °C, extracellular ice does not encapsulate the cells. Rather, large, discrete, and often angular extracellular ice crystals are formed within the intercellular air spaces, and only a small proportion of the surface of the individual cells is in direct contact with the ice crystals. Nevertheless, the cells are substantially dehydrated. Because only a small proportion of the surface area of the cells is in contact with the ice crystals, cell dehydration occurs by vapor phase equilibration with the ice crystals. The diffusion of water from the surface of the cells to the surface of the remote ice crystals is a function of the gradient in the concentration of the water phase at each interface. Although diffusion in the gas phase is more rapid than in the liquid phase, the distance between the cell surface and the surface of the ice crystals is considerably greater than when the cells are encapsulated in ice. As a result, the diffusion of water from the cell surface to the ice crystals can be two to three orders of magnitude slower than that which would occur if the cells were in a partially frozen aqueous medium at the same temperature. This large difference in water diffusion would, in large part, account for the large differences in the critical cooling rate necessary to preclude intracellular ice formation in isolated protoplasts suspended in an aqueous medium (3 °C $min^{-1}$) in comparison to that which is required to prevent intracellular ice formation in intact leaves (~3 °C $hr^{-1}$). Thus, the porosity of the leaf tissue and the characteristic pattern of ice crystal formation will greatly affect the rate of cell dehydration and hence the critical cooling rate required to preclude intracellular ice formation.

## The Plasma Membrane as the Primary Site of Freezing Injury

Freezing injury is a consequence of the freeze-induced destabilization of cellular membranes. The flaccid, water-soaked appearance of tissues immediately after thawing leaves little doubt that disruption of the plasma membrane has occurred. Most of the commonly used techniques for assessing tissue viability following a freeze-thaw cycle (plasmolytic techniques; vital staining with neutral red, Evans blue, or fluorescein diacetate; leakage of electrolytes; and reduction of triphenyl tetrazolium chloride) are based on the retention of the semi-permeable characteristics of the plasma membrane. Manifestations of destabilization of other cellular membranes are also apparent. For example, discoloration and browning of tissues reflects the loss of cellular compartmentalization. Measurements of photosynthetic and respiratory activities reveal that disruption of chloroplasts and mitochondria also occurs.

Although all cellular membranes are vulnerable to dehydration-induced destabilization, the plasma membrane is of primary importance because of the central role that it plays during a freeze/thaw cycle (Steponkus 1984). As the principal semi-permeable barrier between the cytoplasm and the extracellular milieu, the plasma membrane allows for the efflux/influx of water during a freeze/thaw cycle while serving as a barrier to preclude seeding of the intracellular solution by extracellular ice. Thus, maintenance of the structural and functional integrity of the plasma membrane during a freeze/thaw cycle is the first prerequisite to survival. Hence, alterations in the semi-permeable characteristics or lysis of the plasma membrane may be considered as a primary cause of freezing injury. However, destabilization of the plasma membrane will be manifested as different lesions depending on the stage of acclimation and the extent of freeze-induced cell dehydration. Injury resulting from cell dehydration occurs on two different scales: that which occurs as a result of the osmotic excursions associated with the removal of bulk water and that which occurs because of the removal of water closely associated with cellular membranes.

### *Destabilization of the Plasma Membrane During Osmotic Excursions*

During a freeze/thaw cycle, protoplasts are subjected to extremely large osmotic excursions, which involve osmotic contraction during cooling and freezing of the suspending medium and subsequent osmotic expansion following warming and melting of the suspending medium. For example, during cooling to –5 °C, approximately 80% of the osmotically active water is removed from the protoplasts. However, this extent of dehydration per se is not injurious because the protoplasts are osmotically responsive following melting of the suspending medium. Nevertheless, in protoplasts isolated from nonacclimated rye leaves, lysis occurs during subsequent osmotic expansion before the protoplasts return to their initial volume. Hence, this form of injury is referred to as expansion-induced lysis (Steponkus and Wiest 1979). Although expansion-induced lysis is the primary cause of freezing injury in protoplasts isolated from nonacclimated rye leaves, it is not observed in protoplasts isolated from cold-acclimated leaves.

Characterization of the mechanical properties and stress-strain relationship of the plasma membrane in conjunction with cryomicroscopic and electron microscopic studies of the behavior of the plasma membrane during osmotic excursions have resulted in a mechanistic understanding of expansion-induced lysis (see Steponkus 1990 for a comprehensive review) and elucidation of the changes in the plasma membrane lipid composition that are responsible for the increased tolerance to osmotic excursions following cold acclimation (Steponkus and Uemura 1989; Uemura and Steponkus 1989; Steponkus et al. 1991).

### *Behavior of Protoplasts Isolated From Non-Acclimated Leaves*

Protoplasts are spherical and have a turgid appearance, characteristics that indicate that the plasma membrane is under tension and supporting a hydrostatic pressure. In isotonic solutions, the resting tension, $\gamma_r$, of the plasma membrane is approximately 0.1 mN $m^{-1}$ (Wolfe and Steponkus 1981, 1983). Commencing with osmotic contraction, the tension in the plasma membrane is rapidly decreased to zero and the protoplasts appear flaccid and irregular in shape. Because of the large difference in the resting tension and the area elastic modulus, $k_A$ = 230 mN $m^{-1}$, only a small reduction in the protoplast volume is required for the rapid relaxation of tension in the plasma membrane. At zero tension, the plasma membrane undergoes endocytotic vesiculation (Dowgert and Steponkus 1984; Gordon-Kamm and Steponkus 1984a). Invagination and vesiculation of the plasma membrane occurs within seconds following the initial efflux of water and is readily observed using high resolution video microscopy techniques. The vesicles range in diameter from 0.3 to 1.0 µm and tend to occur in clusters subtending the plasma membrane. The process of endocytotic vesiculation continues after osmotic equilibration has occurred, with the duration ranging from tens of seconds to tens of minutes, depending on the extent of osmotic contraction. As the surface area of the plasma membrane is decreased, the protoplast returns to its spherical shape and again appears turgid – albeit at a reduced volume. At this time, the tension in the plasma membrane is again established to approximately 0.1 mN $m^{-1}$.

Thus, the surface area of the plasma membrane is not conserved during osmotic contraction because of the tension-mediated process of endocytotic vesiculation. A reduction in the area of the plasma membrane per se is not injurious, as the protoplasts are osmotically responsive following osmotic contraction. However, endocytotic vesiculation predisposes the plasma membrane to lysis during subsequent osmotic expansion if the area reductions are sufficiently large.

The irreversibility of sufficiently large area reductions derives from the stress strain relationship of the plasma membrane during osmotic expansion (Wolfe et al. 1985, 1986a,b). During the initial influx of water into the protoplast, the plasma membrane behaves elastically and changes in the surface tension are related to changes in area by the equation $\gamma = k_A(\Delta A/A)$, where $\Delta A/A$ is the fractional change in area. However, intrinsic elastic expansion of the plasma membrane is limited to 2 to 3% because the probability of lysis increases greatly at tensions of 4 to 6 mN $m^{-1}$. Therefore, to accommodate the much larger increases in surface area that can occur during osmotic expansion (approximately 34%), additional membrane material must be incorporated into the plasma membrane. The incorporation of additional membrane material into the plasma membrane is a function of the tension in the plasma membrane: the rate of the proportional increase in area increases exponentially with tension (Wolfe et al. 1985). At tensions over the range of 0.1 to 3.0 mN $m^{-1}$, the proportional increase in area is less than 0.05% $s^{-1}$; over the range of 3.0 to 6.0 mN $m^{-1}$, it increases from 0.05 to 2.0% $s^{-1}$. However, the probability of lysis also increases exponentially with tension: the predicted lifetime is 43 s at $\gamma = 5.0$ mN $m^{-1}$, which is halved by each increment of 0.5 mN $m^{-1}$. Interestingly, the co-dependence of both incorporation of membrane material and lysis on tension results in the apparent lack of any influence of the rate of osmotic expansion on the surface area at which lysis occurs (Wolfe et al. 1986a). At more rapid rates of expansion, the increase in tension is greater, but so is the rate of incorporation of membrane material. At slower rates of expansion, the increase in tension is less, but the rate at which material is incorporated into the membrane is also less; as a result, the duration of time over which lysis can occur is longer (Wolfe et al. 1986a). Thus, because the area expansion potential is a constant increment, regardless of the extent of osmotic contraction (Steponkus and Wiest 1978; Wiest and Steponkus 1978), it appears that lysis occurs because of the finite amount of material that is available for incorporation. For protoplasts isolated from nonacclimated rye leaves this is equivalent to 34% of the surface area of the plasma membrane (Dowgert and Steponkus 1984).

### *Behavior of Protoplasts Isolated from Cold-Acclimated Leaves*

Although expansion-induced lysis is the predominant form of injury in protoplasts isolated from nonacclimated rye leaves, it does not occur in protoplasts isolated from cold-acclimated leaves (Dowgert and Steponkus 1984). This is because osmotic contraction of protoplasts isolated from cold-acclimated leaves results in the formation of exocytotic extrusions rather than endocytotic vesicles. This transformation in the behavior of the plasma membrane occurs within the first 7 to 10 days of the acclimation period, after which the incidence of expansion-induced lysis is less than 10% at any subzero temperature (Uemura and Steponkus 1989).

In scanning electron micrographs, the extrusions appear as tethered spheres or polyps on the surface of the protoplasts (Gordon-Kamm and Steponkus 1984b). The diameter of the spheres is approximately 0.5 µm, with considerable variation in the length of the tethers. In some instances the spheres appear sessile; in others, the tethers may be as long as 5 µm. In thin sections, the interior of the extrusions is densely osmiophilic and bounded by the plasma membrane. In freeze-fracture replicas of the protoplasmic fracture face of the plasma membrane there is an increase in the intramembrane particle density in both the regions that extend over the extrusions and in regions between the extrusions. In contrast, there is no apparent change in the intramembrane particle frequency in the cytoplasmic fracture face. The densely osmiophilic core of the extrusions and the increase in the intramembrane particle density may be explained by the preferential subduction of lipids from the plasma membrane.

Both the formation of exocytotic extrusions and osmotic contraction are readily reversible in protoplasts isolated from cold-acclimated leaves, and a causal relationship between the two events has been established (Dowgert et al. 1987). Following equilibration in hypertonic solutions, the tension in the plasma membrane is reestablished to a low, but nonzero value. During subsequent osmotic expansion, the tension remains at this value until the isotonic surface area is exceeded. Thus, the formation of exocytotic extrusions results in the conservation of the area of the plasma membrane and osmotic contraction is readily reversible.

### *Role of Plasma Membrane Lipid Alterations*

The transformation in the behavior of the plasma membrane during osmotic contraction is the result of alterations in the lipid composition of the plasma membrane (see Steponkus et al. 1990). This conclusion is based on studies of the cryobehavior of liposomes prepared from plasma membrane lipid extracts (Steponkus and Lynch 1989a) and membrane engineering studies in which the lipid composition of the plasma membrane was selectively altered (Steponkus et al. 1988).

Initial studies to establish that the differential behavior of the plasma membrane during osmotic contraction is the result of alterations in the lipid composition of the plasma membrane as opposed to alterations in the cytoskeleton involved studies of the cryobehavior of large unilamellar vesicles (LUVs) prepared from the total lipid extract of the plasma membrane fraction isolated from rye leaves (Steponkus and Lynch 1989a). With LUVs prepared from the plasma membrane lipids isolated from nonacclimated leaves, osmotic contraction results in the formation of numerous daughter vesicles that are sequestered in the interior of the mother LUV. Following osmotic equilibration, the total surface area of the daughter vesicles is equal to the reduction in the surface area of the mother LUV. In contrast, osmotic contraction of LUVs prepared from the plasma membrane lipids of cold-acclimated leaves results in the formation of tubular extrusions or vesicles that are attached to the exterior of the mother LUV. Thus, liposomes prepared from plasma membrane lipid extracts exhibit the same differential behavior as the plasma membrane in protoplasts isolated from nonacclimated versus cold-acclimated leaves, which suggests that changes in the lipid composition following cold acclimation are responsible for the differential behavior. Further, because there are no lipid species that are unique to plasma membrane fractions isolated from either nonacclimated or cold-acclimated rye leaves (Lynch and Steponkus 1987), the differential cryobehavior is a consequence of altered lipid-lipid interactions resulting from differences in the proportions of the various lipid species.

To establish specific structure-function relationships, the lipid composition of the plasma membrane was selectively modified using a protoplast-liposome fusion technique (Arvinte and Steponkus 1988). Enrichment of the plasma membrane of protoplasts isolated from nonacclimated rye leaves with phospholipids isolated from the plasma membrane of cold-acclimated leaves transforms the behavior of the plasma membrane such that exocytotic extrusions, rather than endocytotic vesicles, are formed during osmotic contraction and preclude expansion-induced lysis (Steponkus et al. 1988). As a result, survival after a freeze/thaw cycle is increased to approximately 100% over the range of 0 to –5 °C, which is the temperature range over which expansion-induced lysis is the predominant form of injury. The same transformation is also effected by enrichment of the plasma membrane with either mono- or di-unsaturated species of phosphatidylcholine (1-palmitoyl-2-oleoylphosphatidylcholine, 1-palmitoyl-2-linoleoylphosphatidylcholine, dioleoylphosphatidylcholine, dilinoleoylphosphatidylcholine, or dilinolenoylphosphatidylcholine). However, enrichment with disaturated species of phosphatidylcholine (either dipalmitoylphosphatidylcholine or dimyristoylphosphatidylcholine) has no effect – neither positive nor negative. Similarly, enrichment with phosphatidylethanolamine species has no effect on either the freezing tolerance or the behavior of the plasma membrane during osmotic contraction.

Collectively, these studies establish that alterations in the phospholipid composition of the plasma membrane during cold acclimation are responsible for the transformation in the cryobehavior of the plasma membrane during osmotic contraction. Both the changes in the phospholipid composition (Cahoon et al. 1989) and the transformation in the behavior of the plasma membrane (Uemura and Steponkus 1989) occur during the first 7 to 10 days of the cold acclimation period.

### Freeze-Induced Membrane Dehydration

When frozen to temperatures over the range of –10 to –40 °C, isolated protoplasts are subjected to osmotic potentials of –12 to –48 MPa. Under these conditions, cell dehydration becomes very severe, and two forms of injury are apparent: (1) in protoplasts isolated from nonacclimated rye leaves severe dehydration results in lamellar-to-hexagonal$_{II}$ phase transitions in the plasma membrane and subtending lamellae and (2) in protoplasts isolated from cold-acclimated rye leaves, the dehydration-induced close approach of the plasma membrane with the subtending lamellae results in a form of injury referred to as the "fracture-jump phenomenon", which is described in below.

#### *Lamellar-to-Hexagonal*$_{II}$ Phase Transitions

During a freeze thaw cycle to temperatures of –10 °C or lower, protoplasts isolated from nonacclimated leaves of rye respond osmotically during cooling and reach a minimum volume predicted from the Boyle van't Hoff relationship (Dowgert and Steponkus 1984). However, the protoplasts are osmotically unresponsive following melting of the suspending medium. This loss of osmotic responsiveness is associated with several alterations in the ultrastructure of the plasma membrane including the formation of aparticulate domains in the plasma membrane, aparticulate lamellae subtending the plasma membrane, and lamellar-to-hexagonal$_{II}$ phase transitions in the plasma membrane and subtending lamellae (Gordon-Kamm and Steponkus 1984c). Lamellar-to-hexagonal$_{II}$ phase transitions are also observed in the chloroplast envelope and endoplasmic reticulum in regions that are in close apposition to the plasma membrane as a result of freeze-induced dehydration (Sugawara and Steponkus 1990).

These changes in membrane ultrastructure are a consequence of freeze-induced dehydration rather than exposure to subzero temperatures per se and can also be effected by osmotic dehydration (Gordon-Kamm and Steponkus 1984c; Steponkus and Gordon-Kamm 1985). Nevertheless, there is a temperature dependence for the phenomenon, with the incidence of lamellar-to-hexagonal$_{II}$ phase transitions greater if the dehydration is effected at 0 °C than at the freezing point of the solution (e.g., –10 °C for a 5.37 osmolal solution).

Studies of the co-variance of the loss of osmotic responsiveness and the incidence of lamellar-to-hexagonal$_{II}$ phase transitions as a function of the freezing temperature under a wide variety of conditions suggest a causal relationship between formation of the hexagonal$_{II}$ phase and the loss of osmotic responsiveness. Normally with protoplasts isolated from nonacclimated rye leaves, loss of osmotic responsiveness occurs in 50% of the protoplasts frozen to –10 °C, with the hexagonal$_{II}$ phase observed in a similar percentage of the fracture faces. However, if the protoplasts are dehydrated in a 5.37 osmolal solution at 0 °C, both the incidence of the hexagonal$_{II}$ phase and the loss of osmotic responsiveness are greater (Gordon-Kamm and Steponkus 1985; Steponkus and Gordon-Kamm 1985). In contrast, if the survival at –10 °C is increased by the addition of a cryoprotectant such as dimethylsulfoxide (DMSO), the incidence of the hexagonal$_{II}$ phase is also decreased (Pihakaski and Steponkus 1987).

Most important, freeze-induced lamellar-to-hexagonal$_{II}$ transitions are not observed in protoplasts isolated from cold-acclimated rye leaves, with as little as one week of cold acclimation sufficient to preclude their occurrence (Sugawara and Steponkus 1990). However, it is important to note that lamellar-to-hexagonal$_{II}$ phase transitions can be effected in protoplasts isolated from cold-acclimated leaves if dehydration occurs 0 °C rather than at the freezing point of the solution. This observation has important ramifications in considering the molecular bases responsible for the differential propensity for freeze-induced lamellar-to-hexagonal$_{II}$ phase transitions in protoplasts isolated from nonacclimated leaves versus those isolated from cold-acclimated leaves.

To provide a mechanistic understanding of the dehydration-induced lamellar-to-hexagonal$_{II}$ phase transitions in protoplasts isolated from nonacclimated leaves and the amelioration of this form of injury following cold acclimation, it is necessary to understand the behavior of lipids and the interactions of bilayers at low hydrations.

*Hydration Characteristics of Lipids and Hydration Forces Between Bilayers*

Biological membranes are subject to a wide variety of long- and short-range forces that determine the extent to which bilayers or membranes may approach and interact with each other (Rand 1981; Israelachvili 1985). The primary attractive force is the van der Waals force. This force decays approximately with the inverse sixth power of the distance and, hence, has an effective range of about 15 nm. Bilayers and membranes possessing fixed or adsorbed charges also demonstrate a repulsive electrostatic potential, the magnitude and decay characteristics of which exhibit a complex dependence on pH, ionic strength, and both concentration and valency of aqueous phase counter-ions.

The van der Waals forces of attraction and repulsive electrostatic forces are the principal components at medium- and long-range interbilayer separations. However, at very short separations of ≤ 2–3 nm, bilayer interactions are dominated by a hydration repulsion force that arises from the work required to dehydrate hydrophilic surface groups (Rand 1981; Marra and Israelachvili 1985). This repulsive force increases exponentially with decreasing interbilayer separation distance ($D_w$) and typically has an exponential decay length ($\lambda$) on the order of 0.2 to 0.3 nm:

$$P = P_o \bullet \exp(-D_w/\lambda),$$

where P is the force required to bring bilayers into close approach and $P_o$ is an intrinsic hydration characteristic equal to the force required to bring the bilayers to zero separation (Rand 1981; Rand and Parsegian 1989). It should be clear from these considerations that bilayers with a low $P_o$ and/or low $\lambda$ value have an equilibrium separation distance, $D_w$, that is smaller than that of a bilayer with larger values of $P_o$ and $\lambda$. Conversely, bilayers with larger $P_o$ and $\lambda$ values will require higher applied pressures to achieve similar interbilayer separation distances as those with low $P_o$ and $\lambda$ values. The lipids known to comprise the plasma membrane from rye leaf cells differ significantly in both $P_o$ and $\lambda$ values.

There is a distinct specificity to lipid hydration. In studies of the hydration characteristics of rye plasma membrane lipids by the construction of desorption isotherms (Lynch and Steponkus 1989a,b), it was determined that the hydration of the plasma membrane components decreases in the order: phosphatidylcholine > phosphatidylethanolamine > free sterols ≈ steryl glycosides > cerebrosides.

At an osmotic potential of −10 MPa, the wt% water retained by these lipids was 19.5, 14, 6, 5, and 1, respectively (Lynch and Steponkus 1989b). These values correspond to molar ratios of water/lipid of 10.6, 6.5, 1.5, 1.7, and 1, respectively. The greater hydration of phosphatidylcholine compared to phosphatidylethanolamine and the low hydration of cholesterol is well known from a variety of previous studies (Jendrasiak and Hasty 1974). The differential hydration of phosphatidylcholine and phosphatidylethanolamine is reflected in the higher number of water molecules associated with the head groups of these lipids, 10–12 mol of water $mol^{-1}$ of phosphatidylcholine (Crowe and Crowe 1984) and 7 mol of water $mol^{-1}$ of phosphatidylethanolamine (Cevc and Marsh 1985). These differences in hydration can be expressed as the pressure required to completely remove water from aqueous dispersions of these lipids. Values of this pressure, $P_o$, are reported to be ≈ 360–540 MPa for phosphatidylcholine and 140 MPa for phosphatidylethanolamine (Marsh 1989).

In general, the magnitude of the hydration decay length, $\lambda$, is expected to be primarily a function of the properties of the solvent and not of the lipid component.

There are, however, significant differences in the measured decay lengths of various lipids (Rand and Parsegian 1989). Phosphatidylethanolamine species typically show decay lengths much shorter than those for phosphatidylcholine, approximately 0.1 nm and 0.25–0.30 nm, respectively. This observation is the basis of the recently proposed hydration-attraction force (Rand et al. 1988) in which solvent-mediated hydrogen bonding facilitates close interbilayer approach. This mechanism notwithstanding, the combination of low $P_o$ and small $\lambda$ values for phosphatidylethanolamine species indicates that phosphatidylethanolamine bilayers should have smaller separation distances than those composed of phosphatidylcholine at the same dehydration pressure. Marra and Israelachvili (1985) measured a separation distance of approximately 0.9 nm between dipalmitoylphosphatidylethanolamine (DPPE) bilayers compared to about 1.7 nm for dipalmitoylphosphatidylcholine (DPPC) bilayers at an applied pressure of 1 mN $m^{-1}$.

A variety of structural consequences resulting from the interactions of bilayers at close approach have been predicted (Rand 1981; Bryant and Wolfe 1989). These include: (1) changes in association of ions with the bilayer surface; (2) increased lateral pressure within the bilayer; (3) consequent decreases in lipid molecular area and possibly the $L_\alpha$ to $L_\beta$ phase transition of the acyl chains; (4) demixing of membrane components such as proteins (Bryant and Wolfe 1989) or highly hydrated lipids such as phosphatidylcholine (Rand 1981) from regions of close approach; and (5) the formation of nonbilayer structures, including the hexagonal$_{II}$ phase.

*Lyotropic Phase Behavior of Lipids*

The most common phase adopted by phospholipids dispersed in water is the lamellar liquid-crystalline ($L_\alpha$) phase, which is characteristic of biological membranes. The $L_\alpha$ phase is characterized by a disordered hydrocarbon region with highly isotropic motion of the hydrocarbon chains, low order parameters, and fast lateral diffusion of lipids within the plane of the bilayer. A variant of the lamellar phases is the highly ordered, lamellar gel ($L_\beta$) phase. The $L_\beta$ phase is characterized by highly anisotropic acyl chain motion, high order parameters, and low lipid diffusion rates. The hexagonal$_{II}$ ($H_{II}$) phase is an inverted cylindrical micellar phase in which the lipid head groups occupy the inner side of the cylinder and surround a central water core. The cylinders are packed hexagonally. More complete descriptions of these phases may be found in Chapman et al. (1967) and Seddon (1990).

The transition from the $L_\alpha$ to the $L_\beta$ phase is promoted by factors that decrease trans-gauche acyl chain isomerizations and also by those that increase attractive head group interactions. Such factors include decreased unsaturation of the acyl chains, lower temperatures, elevated ion levels and/or increased ionic strength, decreased pH, and decreased hydration.

The transition from the lamellar to the $H_{II}$ phase may be qualitatively understood by considering the relative shape of the lipid (Cullis and de Kruijff 1979). The tendency for a lipid to undergo the $L_\alpha$-to-$H_{II}$ transition is increased by factors that increase the ratio of the swept hydrophobic area/swept hydrophilic area. Thus, a lipid with a time-averaged wedge shape will have a stronger preference to adopt the $H_{II}$ phase than one with an overall cylindrical shape. Clearly, factors that would promote the lamellar-to-$H_{II}$ phase transition will be those that increase the volume of the hydrophobic domain and/or decrease that of the head group domain. These include cis isomers of unsaturated acyl chains, elevated temperatures, small head group size and/or low head group hydration, and electrostatic screening of charged lipid head groups. Nevertheless, the lamellar-to-$H_{II}$ phase transition is an interbilayer event and requires close apposition of adjacent bilayers for the transition to occur.

A more recent approach to the understanding of the $L_\alpha$-to-$H_{II}$ phase transition has been advanced by Gruner and co-workers (Gruner 1989). In this analysis, the stability of the lamellar or $H_{II}$ phase is determined by a balance between the free energy of packing of the acyl chains in the hydrophobic domain and the free energy of packing in the head group domain. It is argued that depth-dependent variation in the magnitude of these energies yields an intrinsic curvature to a lipid monolayer. At critical thresholds, the overall free energy of the monolayer is lowered by the $L_\alpha$-to-$H_{II}$ phase transition or vice versa. This concept is supported, in part, by the observation that the addition of small quantities of either di-22:1 phosphatidylcholine or di-24:1 phosphatidylcholine to dioleoylphosphatidyl ethanolamine significantly stabilize the $H_{II}$ phase adopted by the mixture (Tate and Gruner 1987). Gruner's intrinsic curvature concept suggests that the long-chain acyl groups preferentially partition into the interstitial regions of the $H_{II}$ lattice and relieve the packing stress associated with the need for short-chain fatty acids to stretch and fill these areas. The relief of this packing stress and stabilization of the $H_{II}$ phase is manifested as a decrease in the free energy of packing the hydrophobic domain in the $H_{II}$ phase. The possible relevance of these observations to the freeze-induced lamellar-to-$H_{II}$ phase transition in rye plasma membranes is discussed below.

Overall, it should be noted that several factors relevant to freeze-induced dehydration change the $L_\alpha$-to-$L_\beta$ and $L_\alpha$-to-$H_{II}$ phase transition temperatures (i.e., the $T_m$ and the $T_{bh}$, respectively) such that the $L_\alpha$ phase is more likely to enter either the $L_\beta$ or $H_{II}$ phase. Specifically, lowered hydration increases the $L_\alpha$-to-$L_\beta$ phase transition temperature ($T_m$) (Chapman et al. 1967) and lowers the $L_\alpha$-to-hexagonal$_{II}$ phase transition temperature ($T_{bh}$) (Luzzati 1968). Similar effects occur when elevated salt levels and/or increased ionic strength screen the electrostatic repulsion of charged lipid head groups. Such conditions may occur during the dehydration-induced increase in intracellular solute levels. Depending on the specifics of temperature, hydration, lipid composition and aqueous phase composition, the $L_\alpha$ phase lipids in the plasma membrane may be expected to enter either the $H_{II}$ or $L_\beta$ phase, or possibly go directly from the $H_{II}$ to the $L_\beta$ phase.

*Lipid-Lipid Interactions*

Many of the hydration and phase characteristics presented above have been determined from studies of pure lipids or binary lipid mixtures. The behavior of these lipids in complex mixtures typical of biological membranes in which there are dozens or hundreds of molecular species of lipids is unknown. Our current goal is to determine the effect of lipid-lipid interactions on both bilayer hydration and lipid phase properties, beginning with studies of simple mixtures of phospholipids before considering the influence of other lipid classes such as sterols and cerebrosides, which are present in the plasma membrane of rye.

Desorption isotherms of mixtures of 1-palmitoyl-2-oleoylphosphatidylethanolamine (POPE) and 1-palmitoyl-2-oleoylphosphatidylcholine (POPC) reveal that the wt% water of the mixture varies linearly with the mol% of phosphatidylcholine following equilibration at vapor pressures over the range of –2.7 to –286 MPa. Similarly, mixtures of dioleoylphosphatidylethanolamine and dioleoylphosphatidylcholine (DOPE:DOPC) that are in the lamellar phase, as determined by freeze-fracture electron microscopy, show a linear relationship between wt% water and mol% DOPC (Webb and Steponkus 1990b). Therefore, it would appear that the hydration characteristics of phosphatidylethanolamine and phosphatidylcholine are additive when both are in the lamellar phase. Although pure DOPE is in the $H_{II}$ phase at all hydrations at 20 °C, the addition of as little as 25 mol% DOPC to DOPE significantly stabilizes the lamellar phase as seen by freeze-fracture electron microscopy (Webb and Steponkus 1990a). A 3:1 mixture of DOPE:DOPC at 20 °C remains in the $L_\alpha$ phase at all hydrations above 12 wt% water. The further increase of DOPC to equimolar levels with DOPE yields a mixture that is lamellar at all water contents > 7.5 wt% water at 20 °C. The sensitivity of DOPE to small additions of DOPC is similar to the dramatic effect that the single methylation of DOPE has on stabilizing the lamellar phase with respect to the $H_{II}$ phase (Gruner et al. 1988).

Using an equimolar DOPE:DOPC dispersion as a starting point for further work with complex mixtures, we have examined the ability of other plasma membrane lipid components to alter the hydration and phase behavior of lipid bilayers. It would be expected that membranes possessing higher proportions of the poorly hydrated lipids would have lowered water contents and decreased interbilayer separations – characteristics that may facilitate the $L_\alpha$-to-$H_{II}$ phase transition. The addition of 50 mol% of free sterols, primarily β-sitosterol and campesterol, to a 1:1 mixture of DOPE:DOPC significantly reduces the wt% water in the desorbed mixtures. At a vapor pressure of –13 MPa and 20 °C, the DOPE:DOPC 1:1 dispersion retains 14.9 wt% water; under similar conditions, a DOPE:DOPC:sterol (1:1:2) dispersion retains 10.3 wt% water (Webb and Steponkus 1990b). That is, the addition of 50 mol% of free sterols reduced the wt% water by approximately 30%. The addition of rye cerebrosides (5 mol%) and reduction of total phospholipids to 45 mol% (DOPE:DOPC:sterols:cerebrosides 22.5:22.5:50:5) has little effect on the hydration of the mixture (9.5 wt% at –13 MPa). However, increasing the proportion of cerebrosides to 20 mol% with a concomitant decrease in total phospholipids to 30 mol% (DOPE:DOPC:sterols:cerebrosides 15:15:50:20) significantly decreases the hydration of the mixture, i.e., 7.4 wt% water at –13 MPa (Webb and Steponkus 1990b). In both cases, the water content of these complex mixtures decreases linearly with increasing proportions of either sterols or cerebrosides in the mixtures. Further, the variation in water content as a function of either the sterol/phospholipid or cerebroside/phospholipid ratio is identical, indicating that the two components are equally effective in reducing bilayer hydration. Once again, these data suggest that the hydration characteristics of individual lipids are additive in complex mixtures.

Using the same approach, we have examined the effects of plant sterols and cerebrosides on the lyotropic phase behavior of complex lipid mixtures. In dehydrated mixtures of DOPE:DOPC containing 50 mol% free sterols and variable levels of cerebrosides as described above, the presence of the sterols reduces the $T_m$ for the mixture to an extent that is linearly dependent on the phospholipid/sterol ratio. Perhaps more importantly, the presence of free sterols at 50 mol% precludes the dehydration-induced increase of $T_m$ observed in pure phospholipid dispersions. The $T_m$ of a 1:1 mixture of DOPE:DOPC increased from –11 °C at 16 wt% to –2 °C at 3 wt% water. With the addition of 50 mol% sterols, the $T_m$ was stable at –12 to –14 °C over the hydration range of 14 wt% to 3 wt% water. Similar results were obtained when cerebrosides were added at 5 or 20 mol%. Therefore, the presence of sterols intercalated between the acyl lipids in the hydrophobic domain of the bilayer may prevent the decrease in lipid molecular area and increase in acyl chain order associated with the lyotropic $L_\alpha$-to-$L_\beta$ phase transition. In addition, the presence of rye cerebrosides at 5 or 20 mol% (accompanied by decreased phospholipid levels to 45 and 30 mol%) had no effect on the $T_m$ of the mixture, nor did it show evidence of phase separation of the cerebrosides. It would appear, then, that the rye cerebrosides are miscible with the other lipid components of the rye plasma membrane at 0–20 mol% and all hydrations from 1–16 wt% water at 20 °C, despite their high $T_m$ value when dispersed in pure form (see Lynch et al. 1991).

Very little is known about the effect of cerebrosides on the propensity of a mixture to adopt the $H_{II}$ phase during dehydration. However, some predictions may be made. First, the low hydration of the cerebroside head group and consequent lowering of total bilayer hydration is expected to promote the close approach of bilayers that is necessary for the lamellar-to-$H_{II}$ phase transition. Secondly, the low hydration of cerebrosides and the extensive ability of the head groups to hydrogen bond with other head groups would be expected to lower lateral pressure in the head group region and, hence, lower the free energy of packing in the hydrophilic domain and promote the $H_{II}$ phase transition. Finally, the acyl chains of rye cerebrosides are significantly longer than those of the phospholipid and sterol components of the rye plasma membranes (Cahoon and Lynch 1988). Consequently, the acyl chains of the cerebrosides might be expected to preferentially occupy the interstitial areas of the $H_{II}$ lattice, thus relieving packing stress in this domain and stabilizing the $H_{II}$ phase (Tate and Gruner 1987; Gruner 1989).

In contrast with the situation for cerebrosides, the effect of sterols on the stability of the lamellar phase has been extensively studied (see Tilcock 1986). Cholesterol is known to destabilize the lamellar phase and consequently favor formation of the $H_{II}$ phase in mixtures consisting of phosphatidylcholine and phosphatidylethanolamine and in a wide variety of other mixed phospholipid systems. In addition, cholesterol appears to increase the sensitivity of mixed phospholipid systems to cation-induced $H_{II}$ phase formation. Because a large variety of other sterols also facilitates the formation of the $H_{II}$ phase (Tilcock 1986), it can be predicted that the free sterols in the plasma membrane of rye would promote the formation of the $H_{II}$ phase. However, it is paradoxical that the free sterol content of the plasma membrane increases after cold acclimation (Lynch and Steponkus 1987), which results in a decreased propensity for freeze-induced lamellar-to-$H_{II}$ phase transitions. Therefore, the effects of sterols and cerebrosides on the tendency of DOPE:DOPC mixtures to adopt the $H_{II}$ phase are currently being investigated.

*Mechanism of Lamellar-to-$H_{II}$ Phase Transitions in the Plasma Membrane*

The formation of aparticulate domains in the plasma membrane, aparticulate lamellae subtending the plasma membrane, and lamellar-to-hexagonal$_{II}$ phase transitions in the plasma membrane and the subtending lamellae are manifestations of hydration-dependent bilayer-bilayer interactions. During cell dehydration, the plasma membrane is brought into close approach with several different endomembranes, including the vesicles that originate from the plasma membrane during osmotic contraction, the outer envelope of chloroplasts, and the endoplasmic reticulum. Normally, the close approach of these membranes is precluded by various repulsive forces, including electrostatic forces at long ranges and hydration/steric forces at short ranges. However, cell dehydration results in substantial increases in the concentration of intracellular ions, and interbilayer electrostatic forces are decreased because of charge screening. Moreover, freezing results in osmotic potentials that are sufficiently large to overcome the hydration forces associated with bilayers, resulting in the removal of water from between the bilayers. As previously discussed, Rand (1981) has suggested that the close approach of bilayers will result in several events, including $L_\alpha$-to-$L_\beta$ phase transitions, demixing of membrane components, and lamellar-to-$H_{II}$ phase transitions.

In suspensions of protoplasts frozen to –10 °C, the formation of aparticulate domains in the plasma membrane and aparticulate lamellae subtending the plasma membrane precede the lamellar-to-$H_{II}$ phase transitions (Gordon-Kamm and Steponkus 1984c). Usually, the formation of aparticulate domains is attributed to the exclusion of membrane proteins from regions of the membrane that have undergone an $L_\alpha$-to-$L_\beta$ phase transition, in which case the aparticulate domains would be referred to as lateral phase separations. However, we consider it unlikely that the aparticulate domains observed during freezing to –10 °C are a consequence of lyotropic $L_\alpha$-to-$L_\beta$ phase transitions. In dispersions of the total lipid extract of the plasma membrane at the hydration levels that occur during freezing at –10 °C, there is no phase transition detectable by either differential scanning calorimetry (Lynch and Steponkus 1989a) or fluorescence polarization studies using trans-parinaric acid (M. Uemura and P. Steponkus, unpubl. results). In both cases, it is likely that the high proportion of free sterols precludes any detectable $L_\alpha$-to-$L_\beta$ phase transition. Similarly, in fluorescence polarization studies using TMA-DPH, there is no evidence for an $L_\alpha$-to-$L_\beta$ phase transition in the plasma membrane of isolated protoplasts. Instead, Bryant and Wolfe (1989) have proposed that the formation of aparticulate domains is a consequence of the exclusion of membrane proteins from regions of close approach because of large differences in the hydration/steric characteristics of membrane proteins and lipids.

The mechanism responsible for the lamellar-to-$H_{II}$ phase transition is more enigmatic. Of the plasma membrane lipids, unsaturated species of phosphatidylethanolamine are the predominant "nonbilayer-forming lipids". However, phosphatidylethanolamine species constitute only 11 mol% of the total lipids in the plasma membrane (Lynch and Steponkus 1987). Under conditions of full hydration, these lipids are maintained in a bilayer configuration because of the influence of other lipid species such as phosphatidylcholine. Therefore, a priori, one would expect that lamellar-to-$H_{II}$ phase transitions require demixing of the lipid species and enrichment of phosphatidylethanolamine in localized domains of the plasma membrane. For example, J.H. Crowe and L.M. Crowe (1984, 1986) and L.M. Crowe and J.H. Crowe (1986) have proposed that, under conditions of severe dehydration, demixing of membrane lipids (e.g., phosphatidylcholine and phosphatidylethanolamine) is the result of a lyotropic $L_\alpha$-to-$L_\beta$ phase transition in phosphatidylcholine species, which results in the localized enrichment and lamellar-to-$H_{II}$ phase transitions in phosphatidylethanolamine (also see Quinn 1985). However, the extent of dehydration that occurs at –10 °C does not result in a sufficiently large increase in the $T_m$ of the mono-and di-unsaturated species of phosphatidylcholine that are found in the plasma membrane of rye, and a lyotropic $L_\alpha$-to-$L_\beta$ phase transition in the plasma membrane is unlikely at –10 °C (Lynch and Steponkus 1989a). Instead, we have suggested that demixing of the lipid components occurs because of differences in the hydration characteristics of the various lipid species (Steponkus et al. 1990). Thus, dehydration and the close approach of bilayers composed of lipid species that vary in their hydration characteristics will result in lateral diffusion of lipids into or out of regions of close approach.

Experimental evidence for dehydration-induced demixing is limited to studies of simple mixtures of phospholipids and is inconsistent. In X-ray diffraction studies of DOPE:DOPC mixtures, Tamura-Lis et al. (1986) reported the coexistence of two separate lamellar phases with different repeat spacings at low hydrations, whereas only a single lamellar phase was observed at higher water contents. In contrast, Eriksson et al. (1985) reported the existence of a single lamellar phase, as detected by $^{31}$P-NMR in various DOPE:DOPC mixtures equilibrated in 10 wt% water. Also, in DSC studies of DOPE:DOPC, we have not been able to detect any bulk demixing prior to the $L_\alpha$-to-$H_{II}$ phase transition, which was determined by freeze-fracture electron microscopy studies (Webb and Steponkus 1990b). Thus, the evidence for hydration-dependent demixing in simple mixtures of DOPE:DOPC is equivocal. It is possible that in simple phospholipid mixtures that are sufficiently dehydrated, demixing or phase separations are not required for the lamellar-to-hexagonal$_{II}$ phase transition. This is consistent with the observation that unsaturated species of phosphatidylcholine, traditionally considered to be "bilayer-forming" lipids, can adopt the $H_{II}$ phase at low levels of hydration (Luzzati 1968; Bradshaw et al. 1989). This behavior is consistent with the stability of a bilayer being determined by the intrinsic radius of curvature of the monolayer, which is influenced by hydration (Gruner 1989). Nevertheless, the behavior of a simple mixture of phospholipids may not be indicative of complex mixtures such as the plasma membrane, in which the difference in hydration characteristics of the various lipid classes is substantially greater.

*Effect of Cold Acclimation on Dehydration-Induced Lamellar-to-$H_{II}$ Phase Transitions in the Plasma Membrane*

Whereas freeze-induced dehydration at –10 °C results in lamellar-to-$H_{II}$ phase transitions in the plasma membrane and subtending lamellae of protoplasts isolated from nonacclimated rye leaves, $H_{II}$ phase is not observed in protoplasts isolated from cold-acclimated leaves regardless of the extent of freeze-induced dehydration (Gordon-Kamm and Steponkus 1984c). Recent studies have established that the decreased propensity for dehydration-induced lamellar-to-$H_{II}$ phase transitions is a consequence of alterations in the lipid composition of the plasma membrane (Steponkus et al. 1990). First, studies of the phase behavior of multilamellar lipid vesicles prepared from the total lipid extracts of plasma membrane fractions isolated from either nonacclimated or cold-acclimated rye leaves reveal that osmotic dehydration of the vesicles prepared from the plasma membrane lipids of nonacclimated leaves results in the formation of $H_{II}$ phase, whereas there is no evidence of the $H_{II}$ phase in vesicles prepared from cold-acclimated leaves (Cudd and Steponkus 1988). Second, the decreased propensity for dehydration-induced lamellar-to-$H_{II}$ phase transitions occurs during the first 7 to 10 days of the cold acclimation period (Sugawara and Steponkus 1990), the period during which the major changes in the phospholipids occur (including the quantitative increase in the total phospholipids and increase in the proportion of di-unsaturated species of phosphatidylcholine and phosphatidylethanolamine) (Cahoon et al. 1989). Third, freeze-induced lamellar-to-$H_{II}$ phase transitions in the plasma membrane of protoplasts isolated from nonacclimated rye leaves can be precluded by artificial enrichment of the plasma membrane with dilinoleoylphosphatidylcholine (Sugawara and Steponkus 1990).

We believe that the decreased propensity for dehydration-induced lamellar-to-$H_{II}$ phase transitions resulting from increased proportions of phosphatidylcholine in the plasma membrane is, in part, the result of alterations in the hydration characteristics of the plasma membrane. Phosphatidylcholine is the most hydrated lipid present in the plasma membrane, and its desorption characteristics are such that it retains the greatest amount of water at any given osmotic potential (Lynch and Steponkus 1989a). Therefore, an increase in the proportion of phosphatidylcholine in the plasma membrane will increase the hydration of the plasma membrane. As a result, a greater spatial separation between the plasma membrane and endomembranes will be maintained during freeze-induced dehydration.

Although an increase in the hydration characteristics of the plasma membrane would be expected to defer the lamellar-to-$H_{II}$ phase transition to lower osmotic potentials, cold acclimation does not merely defer the formation of $H_{II}$ phase to lower temperatures. Rather, cold acclimation precludes freeze-induced formation of the $H_{II}$ phase. To account for this behavior it is necessary to consider that (1) during freezing, the increase in osmolality of the partially frozen suspending medium is a function of the subzero temperature, i.e., the osmotic potential and the extent of dehydration are coupled to the subfreezing temperature, (2) the lamellar-to-$H_{II}$ phase transition is dependent on both hydration and temperature, and (3) dehydration decreases the lamellar-to-$H_{II}$ phase transition temperature ($T_{bh}$) and increases the $L_\alpha$-to-$L_\beta$ phase transition temperature ($T_m$).

Thus, during freezing to –10 °C, the osmotic potential decreases to –12 MPa and results in the critical degree of dehydration of the plasma membrane required for formation of the $H_{II}$ phase in protoplasts isolated from nonacclimated leaves. In contrast, when protoplasts isolated from cold-acclimated leaves are frozen to –10 °C, less water is removed from the plasma membrane because of the increased hydration characteristics and the bilayer remains in the lamellar phase. To remove additional water, the temperature must be decreased. However, at a lower temperature, the sample is below the $T_{bh}$. Further decreases in temperature to effect greater extents of dehydration will result in an $L_\alpha$-to-$L_\beta$ phase transition rather than a lamellar-to-

HII phase transition. A corollary to this explanation is that $H_{II}$ phase will occur in protoplasts isolated from cold-acclimated leaves if dehydration is uncoupled from temperature, i.e., if dehydration is effected osmotically at higher temperatures. This does, in fact, occur: lamellar-to-$H_{II}$ phase transitions can be effected in protoplasts isolated from cold-acclimated leaves if they are subjected to a 5.37 osmolal (–12 MPa) solution of 0 °C (Gordon-Kamm and Steponkus 1984c). However, further experimental support of this proposal requires characterization of the hydration characteristics and the temperature/hydration-dependent phase behavior of the plasma membrane lipids isolated from nonacclimated and cold-acclimated rye leaves, which is a nontrivial task given the quantities of the plasma membrane lipids that can be isolated.

### *Ultrastructural Manifestations of Injury in Protoplasts Isolated from Cold-Acclimated Leaves: the "Fracture-Jump Phenomenon"*

Expansion-induced lysis and loss of osmotic responsiveness resulting from lamellar-to-hexagonal$_{II}$ phase transitions are the two forms of injury that limit the freezing tolerance of protoplasts isolated from nonacclimated rye leaves. As such, they represent the first stress barriers that must be overcome during the process of cold acclimation. For both, this occurs during the first week of the cold acclimation period (see Uemura and Steponkus 1989 for studies of expansion-induced lysis and the transformation in the behavior of the plasma membrane from endocytotic vesiculation to the formation of exocytotic extrusions during osmotic contraction; see Sugawara and Steponkus 1990 for studies of the decreased propensity for lamellar-to-hexagonal$_{II}$ phase transitions). As a result, the freezing tolerance ($LT_{50}$) of the protoplasts decreases from –5 to –12 °C; however, an additional three weeks of acclimation are required to attain the maximum freezing tolerance ($LT_{50}$ = –27 °C). Recently, we have initiated studies to characterize the manifestations and cause(s) of injury in protoplasts isolated from fully acclimated leaves in order to determine what factors limit the maximum freezing tolerance of winter rye.

Phenomenologically, injury in protoplasts isolated from fully acclimated leaves is manifested as a loss of osmotic responsiveness and lethally injured protoplasts are not stained by fluorescein diacetate. However, freeze-fracture electron microscopy studies reveal remarkably few changes in the ultrastructure of the plasma membrane when the protoplasts are frozen over the range of –20 to –40 °C (Fujikawa and Steponkus 1990). The only noticeable change was a deviation in the fracture plane in localized regions of the plasma membrane, i.e., the fracture plane "jumped" from the plasma membrane to aparticulate lamellae and endomembranes (endoplasmic reticulum and chloroplast envelopes) that were in close apposition to the plasma membrane. This phenomenon is referred to as a "fracture jump". The size of the region involved in the fracture jump ranged from 0.1 to 2.5 μm in diameter. The number of these regions increased over the range of –20 to –40 °C, the range over which survival decreased from >80% to <20%.

In protoplasmic fracture faces, the fracture jump resulted in a void in the plasma membrane, with the subtending lamellae visible in this region. In exoplasmic fracture faces, the fracture jump resulted in the appearance of stacks of aparticulate lamellae on the plasma membrane. In several instances, the intramembrane particle (IMP) distribution in the fracture face of the plasma membrane was altered in the region around the fracture jump. Around the perimeter of the fracture jump, the IMP frequency was greatly reduced; in some instances, this region was aparticulate. In more distal regions of the periphery, the IMP frequency was increased; in some instances, the IMPs appeared to be closely packed. At greater distances from the periphery of the fracture jump, the IMP distribution was normal. The localized alterations in the IMP frequencies are interpreted to reflect the exclusion of the IMPs from the regions of close approach. Within the region of the fracture jump, several fracture planes were apparent in the subtending lamellae. However, it could not be discerned whether these fracture planes were through the middle of the lamellae or between closely appressed lamellae. In most instances, the distance between the plasma membrane and the subtending lamellae was very small, with some regions of the plasma membrane appearing to merge into the plane of the subtending lamellae.

In preliminary studies to characterize the incidence of the fracture-jump phenomenon as a function of cold acclimation, the incidence of the phenomenon is greater in protoplasts isolated from plants acclimated for one week than in protoplasts isolated from fully acclimated seedlings (Fujikawa and Steponkus 1990). Also, the fracture jump occurs at higher subzero temperatures (–10 to –20 °C) in protoplasts isolated from leaves after one week of cold acclimation than in protoplasts isolated from fully acclimated leaves (–20 to –40 °C), which coincides with the range of temperatures that result in a decrease in survival from >80% to <20% (see Uemura and Steponkus 1989). Although a more systematic and quantitative analysis is required to establish the occurrence of the fracture-jump phenomenon as a function of the stage of acclimation and the subfreezing temperature, these preliminary studies suggest that the fracture-jump phenomenon is closely correlated with freezing injury in protoplasts isolated from leaves of rye seedlings that are in both intermediate stages of acclimation and at the maximum freezing tolerance.

As a working hypothesis, we propose that the fracture-jump phenomenon is a consequence of the enrichment and interdigitation of asymmetric lipid species in localized domains of the plasma membrane. This hypothesis assumes the following: (1) Freezing to temperatures over the range of –10 to –40 °C results in large osmotic potentials (–12 to –48 MPa) that are sufficient to overcome the hydration forces associated with the hydrophilic regions of the membrane proteins and the polar head groups of the lipids (see Rand and Parsegian 1989). As a result, the plasma membrane is in close approach with subtending lamellae and endomembranes such as the chloroplast envelope and endoplasmic reticulum. (2) The close approach of the bilayers results in demixing of the membrane components because of differences in the repulsive forces (both hydration and steric forces) among the membrane components. Because membrane proteins have the largest repulsive forces, they will be excluded from regions of close approach, which will result in the formation of aparticulate domains. Within these domains, further dehydration will result in demixing of lipid components that have large differences in hydration characteristics. For example, at –10 MPa, phosphatidylcholine species, which are strongly hydrated, retain 10 to 11 mol of water/mol of lipid, whereas glucocerebrosides, which are weakly hydrated, retain only 1 to 2 mol of water/mol of lipid (Steponkus and Lynch 1989b). (3) Demixing of the lipid components under conditions of low hydration and low temperature will result in liquid crystalline-to-gel ($L_\alpha$-to-$L_\beta$) phase transitions in domains that are enriched in lipids with a high $T_m$ and/or low hydration. (4) In domains that are enriched in highly asymmetric, mixed-chain lipid species in the $L_\beta$ phase, the lipids will become interdigitated (either partially interdigitated, mixed interdigitated, or fully interdigitated; see Hui et al. 1984; Hui and Huang 1986). (5) The fracture plane, which normally traverses through the midplane of the plasma membrane, i.e., at the plane delineated by the terminal methyl groups of apposing acyl chains, is "disrupted" when an interdigitated region is encountered and it (the fracture plane) jumps to either the mid-plane of a subtending lamellae or between closely appressed lamellae (see Hui et al. 1984; Slater and Huang 1988). (6) The fracture-jump phenomenon only occurs during the freeze-fracture procedure; it is not the lesion per se. Injury is attributed to the formation of gel-phase domains within the plasma membrane. "Leakiness" of the plasma membrane resulting in a loss of osmotic responsiveness is a consequence of packing defects at the boundaries of the coexisting lamellar ($L_\alpha$ and $L_\beta$) phases.

There are several reasons to suggest that cerebrosides, a class of lipids unique to the plasma membrane and the tonoplast (see Lynch and Steponkus 1987), play a key role in this phenomenon: (1) cerebrosides are the least hydrated lipid species in the plasma membrane and would be expected to be the most prone to undergo dehydration-induced demixing; (2) because of strong hydrogen-bonding characteristics, cerebrosides tend to self-associate, which would increase the tendency to form localized domains; (3) cerebrosides have the highest $L_\alpha$-to-$L_\beta$ phase transition temperature of the plasma membrane lipids (55 to 65 °C) (Lynch et al. 1991); and (4) cerebrosides are the most asymmetrical lipid species present in the plasma membrane. The predominant molecular species, which comprises 40% of the total cerebroside fraction, is composed of 4-hydroxysphingenine (an 18-carbon base) and 2-hydroxynervonic acid (a 24-carbon hydroxy acyl chain) (Cahoon and Lynch 1988). X-ray crystallographic studies of sphingosine, a closely related 18-carbon base, indicate that this base penetrates to a depth of only 13 to 14 carbon atoms (see Boggs et al. 1988). Therefore, in the gel phase, there is a large degree of asymmetry between the long chain base and the acyl chain in the cerebroside.

An extension of the working hypothesis to the process of cold acclimation would suggest that the decreased incidence of the fracture-jump phenomenon during weeks one through four of the cold acclimation period is a consequence of a decrease in the cerebroside content of the plasma membrane. This is consistent with preliminary studies of the temporal changes in the plasma membrane lipid composition during cold acclimation (Cahoon et al. 1989): whereas the majority of the phospholipid changes (both the increase in total phospholipids and the changes in the molecular species) occurs during the first week of cold acclimation, the cerebroside content does not begin to change until *after* the first week of cold acclimation – after which there is a progressive decline during the subsequent three weeks of cold acclimation.

## Conclusions

Freeze-induced dehydration is a major cause of freezing injury in plants, with the plasma membrane the primary site of injury. At relatively high subzero temperatures, destabilization of the plasma membrane results from cellular dehydration and the resulting osmotic excursions incurred during the freeze/thaw cycle. At lower temperatures, which result in more severe dehydration, injury is a consequence of lamellar-to-hexagonal$_{II}$ phase transitions in the plasma membrane. Both of these forms of injury are precluded in protoplasts isolated from cold-acclimated leaves; this is a consequence of alterations in the lipid composition of the plasma membrane and specifically the result of an increase in the phosphatidylcholine content of the plasma membrane. Preliminary studies of injury in protoplasts isolated from cold-acclimated leaves suggest that freeze-induced dehydration results in $L_\alpha$-to-$L_\beta$ phase transitions and possibly interdigitation of asymmetrical lipid species such as cerebrosides in the $L_\beta$ phase. A molecular understanding of freezing injury and cold acclimation of plants – a long-standing enigma – is emerging.

*Acknowledgments.* Portions of this work were supported by the United States Department of Energy (Grant No. DE-FG02-84ER13214) and the United States Department of Agriculture Competitive Research Grant Program (Grant No. 88-37264-3988).

## References

Allan WTG, Read ND, Jeffree CE, Steponkus PL (1990) Cryo-scanning electron microscopy of ice formation in rye leaves. Cryobiology 27:664

Arvinte T, Steponkus PL (1988) Characterization of the pH-induced fusion of liposomes with the plasma membrane of rye protoplasts. Biochemistry 27:5671–5677

Ashworth EN, Echlin P, Pearce RS, Hayes TL (1988) Ice formation and tissue response in apple twigs. Plant Cell Environ 11:703–710

Boggs JM, Koshy KM, Rangaraj G (1988) Interdigitated lipid bilayers of long acyl chain species of cerebroside sulfate. A fatty acid spin label study. Biochim Biophys Acta 938:373–385

Bryant G, Wolfe J (1989) Can hydration forces induce lateral phase separations in lamellar phases? Eur J Biophys 16:369–372

Bradshaw JP, Edenborough MS, Sizer PJH, Watts A (1989) A description of the phospholipid arrangement intermediate to the humidity produced $L_\alpha$ and $H_{II}$ phases in dioleoylphosphatidylcholine and its modification by dioleoylphosphatidylethanolamine as studied by X-ray diffraction. Biochim Biophys Acta 987:104–110

Cahoon EB, Lynch DV (1988) Molecular species analysis of plasma membrane glucocerebrosides. Plant Physiol 86:S-53

Cahoon EB, Steponkus PL, Lynch DV (1989) Temporal changes in plasma membrane lipid composition during cold acclimation of rye (*Secale cereale* L. cv Puma). Plant Physiol 89:S-28

Cevc G, Marsh D (1985) Hydration of noncharged lipid bilayer membranes. Theory and experiments with phosphatidylethanolamines. Biophys J 47:21–31

Chapman D, Williams RM, Ladbrooke BD (1967) Physical studies of phospholipids. VI. Thermotropic and lyotropic mesomorphism of some 1,2-diacyl-phosphatidylcholines (lecithins). Chem Phys Lipids 1:445–475

Crowe JH, Crowe LM (1984) Effects of dehydration on membranes and membrane stabilization at low water activities. In: Chapman D (ed) Biological membranes, vol 5. Academic Press, Lond New York, pp 57–103

Crowe JH, Crowe LM (1986) Stabilization of membranes in anhydrobiotic organisms. In: Leopold AC (ed) Membranes, metabolism and dry organisms. Comstock, Ithaca NY, pp 188–209
Crowe JH, Crowe LM, Carpenter JF, Aurell Wistrom C (1987) Stabilization of dry phospholipid bilayers and proteins by sugars. Biochem J 242:1–10
Crowe LM, Crowe JH (1986) Hydration-dependent phase transitions and permeability properties of biological membranes. In: Leopold AC (ed) Membranes, metabolism and dry organisms. Comstock, Ithaca NY, pp 210–230
Crowe LM, Mouradian R, Crowe JH, Jackson SA, Womersley C (1984) Effects of carbohydrates on membrane stability at low water activity. Biochim Biophys Acta 769:141–150
Cudd A, Steponkus PL (1988) Lamellar-to-hexagonal$_{II}$ phase transitions in liposomes of rye plasma membrane lipids after osmotic dehydration. Biochim Biophys Acta 941:278–286
Cullis PR, de Kruijff B (1979) Lipid polymorphism and the functional roles of lipids in biological membranes. Biochim Biophys Acta 559:399–420
Dowgert MF, Steponkus PL (1983) Effect of cold acclimation on intracellular ice formation in isolated protoplasts. Plant Physiol 72:978–988
Dowgert MF, Steponkus PL (1984) The behavior of the plasma membrane of isolated plant protoplasts during a freeze-thaw cycle. Plant Physiol 75:1139–1151
Dowgert MF, Wolfe J, Steponkus PL (1987) The mechanics of injury to isolated protoplasts following osmotic contraction. Plant Physiol 83:1001–1007
Eriksson PO, Rilfors L, Lindblom G, Arvidson G (1985) Multicomponent spectra from $^{31}$P-NMR studies of the phase equilibria in the system dioleoylphosphatidylcholine-dioleoylphosphatidylethanolamine-water. Chem Phys Lipids 37:357–371
Fujikawa S, Steponkus PL (1990) Freeze-induced alterations in the ultrastructure of the plasma membrane of rye protoplasts isolated from cold-acclimated leaves. Cryobiology 27:665–666
Gordon-Kamm WJ, Steponkus PL (1984a) The behavior of the plasma membrane following osmotic contraction of isolated protoplasts: implications in freezing injury. Protoplasma 123:83–94
Gordon-Kamm WJ, Steponkus PL (1984b) The influence of cold acclimation on the behavior of the plasma membrane following osmotic contraction of isolated protoplasts. Protoplasma 123:161–173
Gordon-Kamm WJ, Steponkus PL (1984c) Lamellar-to-hexagonal$_{II}$ phase transitions in the plasma membrane of isolated protoplasts following freeze-induced dehydration. Proc Natl Acad Sci USA 81:6373–6377
Gordon-Kamm WJ, Steponkus PL (1985) Freeze-induced bilayer-to-hexagonal$_{II}$ phase transitions of the plasma membrane of isolated protoplasts: influence of the composition of the suspending medium. Plant Physiol 77:S-155
Gruner SM (1989) Stability of lyotropic phases with curved interfaces. J Phys Chem 93:7562–7570
Gruner SM, Tate MW, Kirk GL, So PTC, Turner DC, Keane DT, Tilcock CPS, Cullis PR (1988) X-ray diffraction study of the polymorphic behavior of N-methylated dioleoylphosphatidylethanolamine. Biochemistry 27:2853–2866
Hui SW, Huang CH (1986) X-ray diffraction evidence for fully interdigitated bilayers of 1-stearoyllysophosphatidylcholine. Biochemistry 25:1330–1335
Hui SW, Mason JT, Huang CH (1984) Acyl chain interdigitation in saturated mixed-chain phosphatidylcholine bilayer dispersions. Biochemistry 23:5570–5577
Israelachvili JN (1985) Thermodynamic and geometric aspects of amphiphile aggregation into micelles, vesicles and bilayers, and the interactions between them. In: Degiorgio V (ed) Physics of amphiphiles: micelles, vesicles and microemulsions. North Holland, Amsterdam, pp 24–58
Jeffree CE, Read ND, Smith JAC, Dale JE (1987) Water droplets and ice deposits in leaf intercellular spaces: redistribution of water during cryofixation for scanning electron microscopy. Planta 172:20–37
Jendrasiak GL, Hasty JH (1974) The hydration of phospholipids. Biochim Biophys Acta 337:79–91
Luzzati V (1968) X-ray diffraction studies of lipid-water systems. In: Chapman D (ed) Biological membranes, vol 1. Academic Press, Lond New York, pp 71–123
Lynch DV, Steponkus PL (1987) Plasma membrane lipid alterations associated with cold acclimation of winter rye seedlings (*Secale cereale* L. cv. Puma). Plant Physiol 83:761–767

Lynch DV, Steponkus PL (1989a) Lyotropic phase behavior of unsaturated phosphatidylcholine species: relevance to the mechanism of plasma membrane destabilization and freezing injury. Biochim Biophys Acta 984:267–272

Lynch DV, Steponkus PL (1989b) The behavior of rye plasma membrane lipids at low hydrations. Cryobiology (Abstr) 26:556

Lynch DV, Caffrey M, Hogan J, Steponkus PL (1991) Calorimetric and X-ray diffraction studies of rye cerebroside mesomorphism. Biophys J (in press)

MacFarlane DR (1987) Physical aspects of vitrification in aqueous solutions. Cryobiology 24:181–195

Marra J, Israelachvili J (1985) Direct measurements of forces between phosphatidylcholine and phosphatidylethanolamine bilayers in aqueous electrolyte solutions. Biochemistry 24:4608–4618

Marsh D (1989) Water adsorption isotherms and hydration forces for lysolipids and diacyl phospholipids. Biophys J 55:1093–1100

Muller-Thurgau H (1886) Über das Gefrieren und Erfrieren der Pflanzen. Landwirtsch Jahrb 15: 453–610

Pihakaski K, Steponkus PL (1987) Freeze-induced phase transitions in the plasma membrane of isolated protoplasts. Physiol Plant 69:666–674

Pitt RE (1990) Cryobiological implications of different methods of calculating the chemical potential of water in partially frozen suspending media. Cryo-Lett 11:227–240

Quinn PJ (1985) A lipid-phase separation model of low-temperature damage to biological membranes. Cryobiology 22:128–146

Rand RP (1981) Interacting phospholipid bilayers: measured forces and induced structural changes. Annu Rev Biophys Bioeng 10:277–314

Rand RP, Parsegian VA (1989) Hydration forces between phospholipid bilayers. Biochim Biophys Acta 988:351–376

Rand RP, Fuller N, Parsegian VA, Rau DC (1988) Variation in hydration forces between neutral phospholipid bilayers: evidence for hydration attraction. Biochemistry 27:7711–7722

Rudolph AS, Crowe JH (1985) Membrane stabilization during freezing: the role of two natural cryoprotectants, trehalose and proline. Cryobiology 22:367–377

Rudolph AS, Crowe JH, Crowe LM (1986) Effects of three stabilizing agents – proline, betaine, and trehalose – on membrane phospholipids. Arch Biochem Biophys 245:134–143

Seddon JM (1990) Structure of the inverted hexagonal ($H_{II}$) phase, and non-lamellar phase transitions of lipids. Biochim Biophys Acta 1031:1–69

Slater JL, Huang CH (1988) Interdigitated bilayer membranes. Prog Lipid Res 27:325–359

Steponkus PL (1978) Cold hardiness and freezing injury of agronomic crops. Adv Agron 30:51–98

Steponkus PL (1979) A unified concept of stress in plants? In: Rains DW, Valentine RC, Hollander A (eds) Genetic engineering of osmoregulation. Plenum Press, New York, pp 235–255

Steponkus PL (1984) Role of the plasma membrane in freezing injury and cold acclimation. Annu Rev Plant Physiol 35:543–584

Steponkus PL (1991) Behavior of the plasma membrane during osmotic excursions. In: Hawes C, Coleman J, Evans D (eds) Endocytosis, exocytosis and vesicle traffic. Soc Exp Biol Sem Ser, Cambridge Univ Press, Cambridge (in press)

Steponkus PL, Gordon-Kamm WJ (1985) Cryoinjury of isolated protoplasts: a consequence of dehydration or the fraction of the suspending medium that is frozen? Cryo-Lett 6:217–226

Steponkus PL, Lynch DV (1989a) The behavior of large unilamellar vesicles of rye plasma membranes during freeze-induced osmotic excursions. Cryo-Lett 10:43–50

Steponkus PL, Lynch DV (1989b) Freeze/thaw-induced destabilization of the plasma membrane and the effects of cold acclimation. J Bioenerg Biomembr 21:21–41

Steponkus PL, Uemura M (1989) Behavior of the plasma membrane during osmotic excursions: the effect of alterations in the plasma membrane lipid composition. In: Tazawa M, Katsumi M, Masuda Y, Okamoto H (eds) Plant water relations and growth under stress. Proc XXII Yamada Conf, Osaka and Myu KK, Tokyo, pp 75–82

Steponkus PL, Wiest SC (1978) Plasma membrane alterations following cold acclimation and freezing. In: Li PH, Sakai A (eds) Plant cold hardiness and freezing stress. Academic Press, Lond New York, pp 75–91

Steponkus PL, Wiest SC (1979) Freeze-thaw-induced lesions in the plasma membrane. In: Lyons JM, Graham D, Raison JK (eds) Low temperature stress in crop plants. Academic Press, Lond New York, pp 231–254

Steponkus PL, Dowgert MF, Evans RY, Gordon-Kamm WJ (1982) Cryobiology of isolated protoplasts. In: Li PH, Sakai A (eds) Plant cold hardiness and freezing stress. Academic Press, Lond New York, pp 459–474

Steponkus PL, Dowgert MF, Ferguson JR, Levin RL (1984) Cryomicroscopy of isolated plant protoplasts. Cryobiology 21:209–233

Steponkus PL, Uemura M, Balsamo RA, Arvinte T, Lynch DV (1988) Transformation of the cryobehavior of rye protoplasts by modification of the plasma membrane lipid composition. Proc Natl Acad Sci USA 85:9026–9030

Steponkus PL, Lynch DV, Uemura M (1990) The influence of cold acclimation on the lipid composition and cryobehavior of the plasma membrane of isolated rye protoplasts. Phil Trans R Soc Lond B 326:571–583

Sugawara Y, Steponkus PL (1990) Effect of cold acclimation and modification of the plasma membrane lipid composition on lamellar-to-hexagonal$_{II}$ phase transitions in rye protoplasts. Cryobiology 27:667

Tamura-Lis W, Reber EJ, Cunningham BA, Collins JM, Lis LJ (1986) $Ca^{2+}$ induced phase separation in phospholipid mixtures. Chem Phys Lipids 39:119–124

Tate MW, Gruner SM (1987) Lipid polymorphism of mixtures of dioleoylphosphatidylethanolamine and saturated and monounsaturated phosphatidylcholines of various chain lengths. Biochemistry 26:231–236

Tilcock CPS (1986) Lipid polymorphism. Chem Phys Lipids 40:109–125

Uemura M, Steponkus PL (1989) Effect of cold acclimation on the incidence of two forms of freezing injury in protoplasts isolated from rye leaves. Plant Physiol 91:1131–1137

Webb MS, Steponkus PL (1990a) Hydration characteristics of phospholipid mixtures. Cryobiology 27:665

Webb MS, Steponkus PL (1990b) Dehydration-induced hexagonal$_{II}$ phase formation in phospholipid bilayers. Cryobiology 27:666–667

Wiest SC, Steponkus PL (1978) Freeze-thaw injury to isolated spinach protoplasts and its simulation at above freezing temperatures. Plant Physiol 62:699–705

Wolfe J, Steponkus PL (1981) The stress-strain relationship of the plasma membrane of isolated plant protoplasts. Biochim Biophys Acta 643:663–668

Wolfe J, Steponkus PL (1983) Mechanical properties of the plasma membrane of isolated protoplasts mechanism of hyperosmotic and extracellular freezing injury. Plant Physiol 71:276–285

Wolfe J, Dowgert MF, Steponkus PL (1985) Dynamics of membrane exchange of the plasma membrane and the lysis of isolated protoplasts during rapid expansions in area. J Membr Biol 86:127–138

Wolfe J, Dowgert MF, Steponkus PL (1986a) Mechanical study of the deformation and rupture of the plasma membranes of protoplasts during osmotic expansions. J Membr Biol 93:63–74

Wolfe J, Dowgert MF, Maier B, Steponkus PL (1986b) Hydration, dehydration and the stresses and strains in membranes. In: Leopold AC (ed) Membranes, metabolism and anhydrous organisms. Comstock, Ithaca, pp 286–305

# Subject Index

MIX
Papier aus verantwortungsvollen Quellen
Paper from responsible sources
FSC® C105338

If you have any concerns about our products,
you can contact us on
**ProductSafety@springernature.com**

In case Publisher is established outside the EU,
the EU authorized representative is:
**Springer Nature Customer Service Center GmbH**
**Europaplatz 3, 69115 Heidelberg, Germany**

Printed by Libri Plureos GmbH
in Hamburg, Germany